T0324047

Neural Mechanisms of Addiction

Neural Mechanisms
of Addiction

Edited by

Mary Torregrossa

Academic Press is an imprint of Elsevier
125 London Wall, London EC2Y 5AS, United Kingdom
525 B Street, Suite 1650, San Diego, CA 92101, United States
50 Hampshire Street, 5th Floor, Cambridge, MA 02139, United States
The Boulevard, Langford Lane, Kidlington, Oxford OX5 1GB, United Kingdom

Copyright © 2019 Elsevier Inc. All rights reserved.

No part of this publication may be reproduced or transmitted in any form or by any means, electronic or mechanical, including photocopying, recording, or any information storage and retrieval system, without permission in writing from the publisher. Details on how to seek permission, further information about the Publisher's permissions policies and our arrangements with organizations such as the Copyright Clearance Center and the Copyright Licensing Agency, can be found at our website: www.elsevier.com/permissions.

This book and the individual contributions contained in it are protected under copyright by the Publisher (other than as may be noted herein).

Notices
Knowledge and best practice in this field are constantly changing. As new research and experience broaden our understanding, changes in research methods, professional practices, or medical treatment may become necessary.

Practitioners and researchers must always rely on their own experience and knowledge in evaluating and using any information, methods, compounds, or experiments described herein. In using such information or methods they should be mindful of their own safety and the safety of others, including parties for whom they have a professional responsibility.

To the fullest extent of the law, neither the Publisher nor the authors, contributors, or editors, assume any liability for any injury and/or damage to persons or property as a matter of products liability, negligence or otherwise, or from any use or operation of any methods, products, instructions, or ideas contained in the material herein.

Library of Congress Cataloging-in-Publication Data
A catalog record for this book is available from the Library of Congress

British Library Cataloguing-in-Publication Data
A catalogue record for this book is available from the British Library

ISBN 978-0-12-812202-0

For information on all Academic Press publications visit our website at https://www.elsevier.com/books-and-journals

  Working together
to grow libraries in
developing countries

www.elsevier.com • www.bookaid.org

Publisher: Nikki Levy
Acquisition Editor: Joslyn Chaiprasert-Paguio
Editorial Project Manager: Samuel Young
Production Project Manager: Anusha Sambamoorthy
Cover Designer: Christian Bilbow

Typeset by SPi Global, India

Contents

6. Fos-Expressing Neuronal Ensembles in Addiction Research

Bruce T. Hope

7. Interoceptive Stimulus Effects of Drugs of Abuse

Brady M. Thompson, Rick A. Bevins, and
Jennifer E. Murray

8. Maladaptive Memory Mechanisms in Addiction and Relapse

Matthew T. Rich and Mary M. Torregrossa

12. Negative Reinforcement Mechanisms in Addiction

Olivier George

13. Circadian Rhythms and Addiction

*Kelly Barko, Micah A. Shelton,
Joseph A. Seggio, and Ryan W. Logan*

14. Role of Oxytocin in Countering Addiction-Associated Behaviors Exacerbated by Stress

*Jacqueline F. McGinty, Courtney E. King,
Casey E. O'Neill, and Howard C. Becker*

15. The Role of Norepinephrine in Drug Addiction: Past, Present, and Future

Stephanie L. Foster and David Weinshenker

Contributors

Numbers in parentheses indicate the pages on which the authors' contributions begin.

Basma Al Masraf (23), Department of Physiology, Michigan State University, East Lansing, MI, United States

Eden Anderson (35), Department of Biomedical Sciences, Marquette University, Milwaukee, WI, United States

David A. Baker (237), Department of Biomedical Sciences, Marquette University, Milwaukee, WI, United States

Kelly Barko (189), Translational Neuroscience Program, Department of Psychiatry, University of Pittsburgh School of Medicine, Pittsburgh, PA, United States

Christelle Baunez (271), Institut de Neurosciences de la Timone, UMR7289, CNRS and Aix Marseille Université, Marseille, France

Allison R. Bechard (61), Psychology Department, University of Florida, Gainesville, FL, United States

Howard C. Becker (213), Department of Psychiatry and Behavioral Sciences, Medical University of South Carolina, Charleston, SC, United States

Lauren N. Beloate (247), Department of Neuroscience, Medical University of South Carolina, Charleston, SC, United States

Rick A. Bevins (89), Department of Psychology, University of Nebraska-Lincoln, Lincoln, NE, United States

David M. Dietz (137), Department of Pharmacology and Toxicology, Research Institute on Addictions, Program in Neuroscience, The State University of New York at Buffalo, Buffalo, NY, United States

Elizabeth M. Doncheck (157), Department of Biomedical Sciences, Marquette University, Milwaukee, WI, United States

Andrew Eagle (23), Department of Physiology, Michigan State University, East Lansing, MI, United States

Michel Engeln (259), Department of Anatomy and Neurobiology, University of Maryland School of Medicine, Baltimore, MD, United States

Stephanie L. Foster (221), Department of Human Genetics, Emory University School of Medicine, Atlanta, GA, United States

T. Chase Francis (259), Intramural Research Program, National Institute on Drug Abuse; Biomedical Research Center, US National Institutes of Health, Baltimore, MD, United States

Rita A. Fuchs (3), Integrative Physiology and Neuroscience; Alcohol and Drug Abuse Research Program; Translational Addiction Research Center, Washington State University, Pullman, WA, United States

Amy M. Gancarz (137), Department of Psychology, California State University Bakersfield, Bakersfield, CA, United States

Olivier George (179), Department of Neuroscience, The Scripps Research Institute, La Jolla, CA, United States

Ethan J. Hansen (3), Integrative Physiology and Neuroscience, Washington State University, Pullman, WA, United States

Matthew Hearing (35, 237), Department of Biomedical Sciences, Marquette University, Milwaukee, WI, United States

Evan Hess (237), Department of Biomedical Sciences, Marquette University, Milwaukee, WI, United States

Jessica A. Higginbotham (3), Integrative Physiology and Neuroscience, Washington State University, Pullman, WA, United States

Bruce T. Hope (75), Behavioral Neuroscience Research Branch, Intramural Research Program, National Institute on Drug Abuse-National Institutes of Health, Baltimore, MD, United States

Peter W. Kalivas (247), Department of Neuroscience, Medical University of South Carolina, Charleston, SC, United States

Courtney E. King (213), Department of Psychiatry and Behavioral Sciences, Medical University of South Carolina, Charleston, SC, United States

Lori A. Knackstedt (61), Psychology Department, University of Florida, Gainesville, FL, United States

Mary Kay Lobo (259), Department of Anatomy and Neurobiology, University of Maryland School of Medicine, Baltimore, MD, United States

Ryan W. Logan (189), Translational Neuroscience Program, Department of Psychiatry, University of Pittsburgh School of Medicine, Pittsburgh, PA; The Jackson Laboratory, Bar Harbor, ME, United States

Aric Madayag (237), Department of Biomedical Sciences, Marquette University, Milwaukee, WI, United States

John R. Mantsch (157), Department of Biomedical Sciences, Marquette University, Milwaukee, WI, United States

Jacqueline F. McGinty (213), Department of Neuroscience, Medical University of South Carolina, Charleston, SC, United States

Natalie S. McGuier (123), Department of Biological Sciences, Carnegie Mellon University, Pittsburgh, PA, United States

Patrick J. Mulholland (123), Departments of Neuroscience and Psychiatry & Behavioral Science, Charleston Alcohol Research Center, Medical University of South Carolina, Charleston, SC, United States

Jennifer E. Murray (89), Department of Psychology, University of Guelph, Guelph, ON, Canada

Casey E. O'Neill (213), Department of Neuroscience, Medical University of South Carolina, Charleston, SC, United States

Yann Pelloux (271), Institut de Neurosciences de la Timone, UMR7289, CNRS and Aix Marseille Université, Marseille, France

Matthew T. Rich (103), Department of Psychiatry; Center for Neuroscience; Center for the Neural Basis of Cognition, University of Pittsburgh, Pittsburgh, PA, United States

A.J. Robison (23), Department of Physiology, Michigan State University, East Lansing, MI, United States

Joseph A. Seggio (189), Department of Biological Sciences, Bridgewater State University, Bridgewater, MA, United States

Micah A. Shelton (189), Translational Neuroscience Program, Department of Psychiatry, University of Pittsburgh School of Medicine, Pittsburgh, PA, United States

Brady M. Thompson (89), Department of Psychology, University of Nebraska-Lincoln, Lincoln, NE, United States

Mary M. Torregrossa (1, 103, 287), Department of Psychiatry; Center for Neuroscience, University of Pittsburgh, Pittsburgh, PA, United States

Joachim D. Uys (123), Department of Cell and Molecular Pharmacology, Medical University of South Carolina, Charleston, SC, United States

David Weinshenker (221), Department of Human Genetics, Emory University School of Medicine, Atlanta, GA, United States

Craig T. Werner (137), Department of Pharmacology and Toxicology, Research Institute on Addictions, Program in Neuroscience, The State University of New York at Buffalo, Buffalo, NY, United States

Chapter 1

Introduction to the Neural Basis of Addiction

Mary M. Torregrossa

Department of Psychiatry, University of Pittsburgh, Pittsburgh, PA, United States

Drug addiction is a devastating illness characterized by repeated cycles of uncontrolled consumption, periods of abstinence or withdrawal, and relapse to use. According to the latest survey by the Substance Abuse and Mental Health Services Administration, in the United States, 7.8% of the adult population met the criteria for diagnosis of a substance use disorder for alcohol, prescription drugs, or illicit substances in 2016. Lifetime prevalence rates are even higher, and many children and adolescents under the age of 18 also suffer from substance use disorders/addiction, making the actual number of people affected by addiction much higher. Moreover, substance abuse produces a huge cost to society, with a reported $740 billion dollars lost annually due to crime, health care, and lost productivity. Thus, drug addiction is a serious and growing public health problem in our society, but unfortunately very few effective treatments exist.

In order to develop effective treatment strategies, a substantial amount of clinical and preclinical research has been conducted over the past several decades. However, for many years, addiction research suffered from the popularly held belief that the "addict" was to blame for their problem, and that individuals could quit using if they simply had enough will power. With the advent of animal models of addiction in nonhuman primates and in rodents, where the animal could choose to perform a behavior to receive the drug, scientists began to understand that drugs of abuse exert powerful control over behavior in a manner that cannot simply be attributed to a character flaw. Animal models allowed researchers to recognize addiction as a progressive disease that has a biological basis in the brain. Early studies used pharmacological and biochemical approaches to determine which biological molecules were targeted by different classes of drugs of abuse, which allowed researchers to identify the receptors and transporters that were responsible for the euphoric effects of drugs like heroin, cocaine, and amphetamine. Research was further advanced by the development of in vivo microdialysis, which allowed scientists to measure changes in extracellular neurotransmitter release in response to drug exposure. Dopamine was found to increase dramatically in several brain regions, most notably the nucleus accumbens, in response to the administration of most classes of abused substances. The rodent studies were supported by neuroimaging studies in human drug users, providing evidence for cross-species mechanisms of action and the translational value of the animal models. Henceforth, research into the neuroscience of addiction rapidly expanded.

Initial neuroscience research in addiction focused on the dopamine system due to its obvious importance in mediating the reinforcing or rewarding effects of abused drugs. Studies using electrical stimulation or lesions also began to isolate the specific brain regions that mediated the ability of drugs to motivate behavior. Over the years, numerous scientific advances, particularly in genomics and molecular biology, further increased our understanding of the long-term effects of drugs of abuse on the brain and the body. In particular, the identification of long-lasting changes in gene expression and synaptic function across multiple brain regions further confirmed that addiction is a progressive and chronic brain disorder. Unfortunately, despite this increase in our understanding of the biology of addiction, there still has been little advance in developing effective treatments.

In more recent years, however, there has been an astoundingly rapid increase in our understanding of the neural mechanisms of addiction, which is providing hope that better addiction treatments may be on the horizon. First, the sophistication of our animal models has increased exponentially in the past 10 years. Many recent studies have used animal models that better reproduce the core features of addiction, including progressive increases in intake, negative affect, and use despite adverse consequences. In other words, instead of simply studying the neural mechanisms underlying the initial reinforcing effects of drugs that are observed in almost all animals (and humans), animal models now study the development of uncontrolled and compulsive drug use. Second, many studies now recognize that there are large individual differences in the propensity for animals to exhibit an addiction phenotype. These studies have improved our understanding of the genetic vulnerabilities and brain mechanisms that lead some people to develop addiction, while others can use socially without problem. Furthermore, there have been substantial technological advances that have allowed scientists to investigate the neural mechanisms of addiction with finer detail and precision than was possible even 5 years ago. These technological advances in research tools are beginning to lead to completely novel methods for treating neurological diseases that may not require the development of a "magic pill," but rather will allow us to rewire disrupted brain circuits.

Thus, the purpose of this book is to provide the reader an updated view on our understanding of the neural mechanisms underlying addiction, with a focus on how emerging technologies, applied to improved animal models, is leading to a new understanding of this complex disorder. Each chapter was written by an expert in addiction neuroscience, who provides a unique perspective on emerging themes in the field based on their own ongoing research. The first chapter provides a comprehensive overview of the most common animal models used in addiction neuroscience, including a description of the most recently developed models, which are leading to an increased understanding of the neural mechanisms underlying the core features required for diagnosis of a substance use disorder. Subsequent chapters focus on the latest developments in our understanding of how the brain is altered by drug exposure with descriptions of changes in molecules, synaptic function, neuronal morphology, circuits, and systems. In addition, several chapters discuss how drugs of abuse influence specific systems such as learning and memory, decision making, and even circadian rhythms to further perpetuate the disease. Finally, the book includes chapters describing novel treatment strategies, such as neuropeptide and stimulation-based therapies. Several chapters included discussion of drug class-specific neuroadaptations, such as effects that are specific to alcohol, opioids, or cocaine, which will clearly be an important consideration for determining effective treatment strategies for each individual. However, much work still needs to be done to compare across classes of drugs, and a substantial number of recent and exciting findings in the field may only apply to one type of drug. As technologies continue to emerge and their application to the neuroscience of addiction progresses, it is clear that our understanding of the disorder will improve. Thus, this is an exciting time to be investigating such a critical health problem in our society, and the developments described in this book are paving the way for true progress in tackling this devastating disease.

Chapter 2

Animal Models of Addiction

Rita A. Fuchs[*,†,‡], **Jessica A. Higginbotham**[*], **and Ethan J. Hansen**[*]

[*]*Integrative Physiology and Neuroscience, Washington State University, Pullman, WA, United States,* [†]*Alcohol and Drug Abuse Research Program, Washington State University, Pullman, WA, United States,* [‡]*Translational Addiction Research Center, Washington State University, Pullman, WA, United States*

INTRODUCTION

Drug addiction is a chronic disease characterized by compulsive drug seeking and drug taking despite adverse consequences [1]. Animal models have provided crucial insights into the associative learning, functional neuroanatomical, and cellular mechanisms of this complex disorder. Historically, procedures focused on acquisition and maintenance of consummatory and goal-directed behaviors, as well as Pavlovian-conditioned responses, that are associated with drug exposure, voluntary drug taking, or drug withdrawal. More recent efforts aimed to capture cardinal features of drug addiction, including the escalation of consummatory and goal-directed behaviors, the development of compulsive and habitual drug taking, and increased reactivity to relapse triggers, like environmental stimuli, stress, and drug itself. This chapter provides an overview of several addiction-related constructs and various behavioral procedures used to probe the neurobiological underpinnings of addiction. Readers are referred to Table 1 for a glossary of italicized terms.

VOLITIONAL CONTROL OVER DRUG EXPOSURE

Recreational drug use and substance use disorders are characterized by voluntary drug taking. It is essential for animal models of addiction to capture this construct because volitional drug taking prompts different neurochemical responses and neuroadaptations compared with passive drug exposure [2–9]. Nonetheless, passive drug regimens have been useful for studying the effects of prenatal and early adolescent drug exposure on propensity for drug taking and drug seeking in adulthood [10–12]. Furthermore, passive drug regimens have been employed to facilitate the acquisition of $\Delta(9)$-tetrahydrocannabinol (THC) and ethanol self-administration in rodent models by prompting the development of tolerance to aversive drug effects [13,14] and by establishing a state of drug dependence [15,16], respectively.

PAVLOVIAN MODELS OF REWARDING AND AVERSIVE DRUG EFFECTS

Passive drug exposure regimens are also integral to several animal models designed to assess the conditioned *rewarding* and aversive effects of drugs of abuse. One such model is the **place conditioning paradigm** [17,18]. This procedure involves repeated pairing of drug effects with confinement to a distinct environmental *context* plus repeated pairing of vehicle effects with confinement to a different context, that is, initially equally preferred (in an unbiased procedure) or differentially preferred (in a biased procedure) (Fig. 1A). Approach [i.e., *conditioned place preference* (CPP)] or avoidance [i.e., *conditioned place aversion* (CPA)] of the drug-paired context during a drug-free choice test is theorized to indicate that the drug-paired context has acquired conditioned rewarding or aversive effects through association with unconditioned rewarding or aversive drug effects, respectively (Fig. 1B). However, when a biased conditioning procedure is used, increase in the time spent in the drug-paired context may indicate rewarding drug effects or a reduction in the anxiogenic (or other aversive) effects of the initially nonpreferred context after repeated pairing with anxiolytic drug effects (Fig. 1C). Furthermore, failure to acquire CPP can indicate the absence of detectable rewarding drug effects or impairment in associative learning—the latter possibility is especially worth considering when studying genetically modified organisms.

Drugs of abuse that produce self-reports of euphoria readily elicit CPP in a variety of species and strains of rodents. Such drugs include cocaine, amphetamine, methamphetamine, morphine, and heroin [19–22,186–201]. In contrast, drugs of abuse with modest to no acute euphoric effects, such as ethanol, nicotine, or THC, can produce either CPP or CPA,

Neural Mechanisms of Addiction. https://doi.org/10.1016/B978-0-12-812202-0.00002-6
Copyright © 2019 Elsevier Inc. All rights reserved.

TABLE 1 Glossary of Terms

Anhedonia	Impaired ability to experience pleasure
Behavioral sensitization	Increase in the motor-stimulant effects of drugs of abuse following repeated drug administration
Break point	The final schedule of reinforcement completed under a progressive reinforcement schedule before responding ceases for a predefined period of time
Compulsive	Insensitive to (anticipated) adverse consequences
Conditioned place aversion (CPA)	Acquired aversion to an initially neutral environmental context after confinement to that context is repeatedly paired with aversive drug effects
Conditioned place preference (CPP)	Acquired preference for an initially neutral/nonpreferred environmental context after confinement to that context is repeatedly paired with rewarding drug effects
Conditioned reinforcement	Increase in the probability of an instrumental response that results in the presentation of a conditioned reinforcer, in the absence of the primary reinforcer
Conditioned reinforce	Initially neutral stimulus that gains the ability to increase the probability of a response that results in its own presentation through previous pairing with a primary reinforce
Conditioned rewarding effect	Conditioned positive affective state (i.e., euphoria, positive affective state) elicited by the presentation of a CS paired previously with a rewarding US
Conditioned stimulus (CS)	Initially neutral environmental stimulus (e.g., light) that comes to elicit a conditioned response through its repeated pairing with a US (e.g., drug)
Consummatory behavior	Behavior that leads to the satisfaction of an innate need state (e.g., drinking, eating)
Context	Static multimodal stimuli which constitute a setting where Pavlovian or instrumental conditioning takes place. Contextual stimuli are presented independent of the subject's behavior
Discriminative stimulus	Temporally discrete stimulus that signals when a particular reinforcement contingency is in effect. Discriminative stimuli are presented independent of the subject's behavior
Escalation	Gradual increase in drug taking and drug intake in the course of drug self-administration training that is, not related to the acquisition of instrumental responding
Extended-access procedure	In this procedure, subjects are moved from a drug self-administration regimen with limited access to drug (1-hour daily sessions) to one with extended access to drug ($\geq$6-hour daily sessions)
Extinction	Prescribed reinforcement contingency under which instrumental responses are no longer reinforced (instrumental) or the CS-US relationship is eliminated (Pavlovian)
Face validity	A desirable quality of animal models of drug addiction. It refers to the extent of overt similarity between the animal model and the clinical phenomenon under study
Forced abstinence	Experimenter-imposed drug-free period
Free-operant session	Experimental session during which a subject can freely perform an instrumental behavior
Habitual behavior	Automatic behavior or fixed sequence of behaviors acquired through repetition
Hedonic state	Affective or emotional state
Interoceptive effect	Distinct internal state produced by a stimulus or event
Negative reinforcement	Increase in the probability of a response that terminates or prevents an aversive event
Occasion setter	Discriminative or contextual stimulus that signals the instrumental contingency currently in effect
Operandum	Response apparatus (e.g., lever, nose poke aperture, chain); *pl.* operanda
Positive reinforcement	Increase in the probability of a response that results in the presentation of a rewarding event
Primary reinforce	Stimulus with innate ability to increase the probability of a response that results in its own presentation (e.g., natural reward, drug of abuse)
Rate-altering effects	Stimulant or sedative effects of a drug that alter behavioral performance
Reinforcement rate	The rate at which reinforcer presentations occur in an instrumental procedure
Reinforcement schedule	Prescribed contingency between instrumental responding and reinforcement (e.g. fixed ratio schedule)

TABLE 1 Glossary of Terms—cont'd

Reinstatement	Increase in drug seeking or Pavlovian conditioned place preference during a test session compared to responding during an extinction training session that immediately precedes the test session
Renewal	Recovery of a previously extinguished instrumental response upon re-exposure to a CS in the previously drug-paired context after the CS is extinguished in a different context
Response cost	Energy expended or opportunity lost in association with performance of an instrumental response
Reward threshold	Index of hedonic state in the intra-cranial self-stimulation paradigm. It is defined as the minimum frequency at which intra-cranial electrical stimulation reinforces instrumental responding
Rewarding effect	Positive hedonic or euphoric effect produced by a stimulus
Run time	Inverse measure of drug-induced motivation in the runway drug self-administration paradigm. It is defined as the time it takes from leaving the start box until entry into the goal box
Schedule completion	Performance of the set of responses required for a single delivery of reinforce
Schedule demand	Response cost associated with schedule completion under a particular reinforcement schedule
Self-imposed abstinence	Drug-free period initiated by the subject
Single-order reinforcement schedule	Simple prescribed contingency between the instrumental response and reinforcement (e.g., ratio and progressive ratio schedules, as well as interval and time schedules described elsewhere)
Social-defeat stress	Aversive state produced by losing in a confrontation with a conspecific
Start latency	An inverse measure of drug-induced motivational effects in the runway drug self-administration paradigm. It is the time it takes to leave the start box and enter the runway after the door separating the two is retracted by the experimenter
Sucrose/saccharin fading	Procedure used to facilitate ethanol drinking by presenting ethanol mixed with initially high then gradually reduced concentrations of sucrose or saccharin
Unconditioned stimulus (US)	Stimulus with inherent ability to elicit the response of interest (without prior conditioning)

depending on species- and strain-specific differences in sensitivity to the drug's rewarding, anxiogenic, or aversive effects, and depending on dosing and other experimental parameters [14,17,18,23–25]. Moreover, when the *interoceptive effects* of pharmacologically precipitated opioid or nicotine withdrawal are conditioned to a distinct context, the end result is CPA [26,27]. A shortcoming of the place conditioning paradigm is limited passive drug exposure. However, it is the most time-efficient and convenient procedure for assessing the hedonic effects of drugs of abuse and drug withdrawal, especially for use with mice and adolescent rodents. More importantly, the place conditioning paradigm has strong predictive validity for detecting drugs with abuse potential [17].

Sign-tracking behavior (e.g., autoshaping, conditioned orientating), another Pavlovian-conditioned response, has been studied extensively to elucidate the neurobiological mechanisms by which environmental stimuli acquire *conditioned rewarding effects* [28]. Sign tracking is based on the subject's innate tendency to approach and make contact with a *conditioned stimulus* (CS; e.g., cue light, lever stimulus), that is, paired with a rewarding *unconditioned stimulus* (US, i.e., natural reward) (Fig. 2). Importantly, stable individual differences exist in sign tracking versus goal tracking, a subject's tendency to approach and make contact with a localizable rewarding US (e.g., food) or the location where this stimulus is expected to appear (e.g., food tray). These individual differences, which are assessed using a food US, predict differences in sensitivity to cocaine-induced neuroadaptations, *conditioned reinforcement*, *behavioral sensitization*, as well as propensity for drug seeking [29–33].

MODELS OF DRUG-INDUCED INTEROCEPTIVE EFFECTS AND AFFECTIVE STATES

Drugs of abuse elicit distinct interoceptive states that are not necessarily limited to the subjective experience of rewarding or aversive drug effects. Historically, interoceptive effects of novel pharmacological compounds, including drugs of abuse, were assessed to classify these compounds, using the **drug-discrimination paradigm** [34]. In this model, subjects are trained to identify interoceptive effects of a particular drug of abuse (training drug) using a food-reinforced procedure (Fig. 3A). To this end, responses on one *operandum* (e.g., lever) are selectively reinforced with food after passive exposure to the

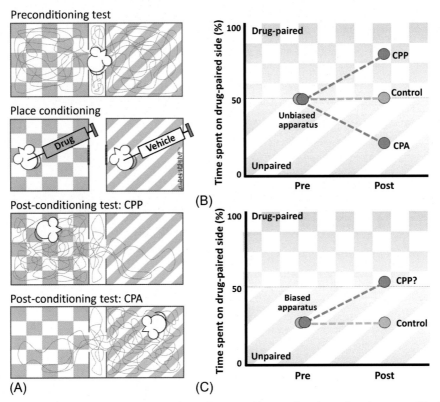

FIG. 1 Place conditioning paradigm. (A) In this procedure, experimentally naïve subjects are first given a drug-free preconditioning test with free access to all chambers of the place conditioning apparatus. Time spent in each side chamber is measured to assess the subjects' unconditioned place preferences. During conditioning, the effects of drug are paired with confinement into a side chamber, and the effects of vehicle are paired with confinement into the other side chamber, typically on alternating days. At test, the subjects receive free access to all chambers in a drug-free state. Time spent in each side chamber is measured to evaluate the extent to which the subjects have acquired conditioned place preference (CPP) or conditioned place aversion (CPA) for the drug-paired chamber. (B) Schematic illustrating CPP and CPA following conditioning (*Post*) in the unbiased place conditioning procedure, where a group of subjects show no unconditioned place preferences (*Pre*). (C) Schematic illustrating CPP or, alternatively, a drug-induced reduction in unconditioned place aversion in the biased place conditioning procedure.

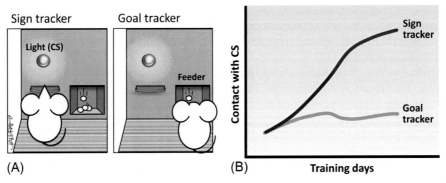

FIG. 2 Sign tracking. In this procedure, subjects are exposed to repeated Pavlovian-conditioned stimulus (CS)-food reward pairings. (A) Sign trackers maintain more contact with a reward-predictive CS than goal trackers. Goal trackers orient toward the reward (unconditioned stimulus) if it is spatially localizable. (B) Schematic illustrating difference in contact with the CS exhibited by sign trackers and goal trackers across training sessions.

training drug (i.e., in the interoceptive state elicited by the training drug). Responses on another operandum are selectively reinforced with food after exposure to vehicle (i.e., in interoceptive states other than that produced by the training drug). At test, it is assessed whether the psychoactive effects of a test drug generalize to those of the training drug, as indicated by drug-appropriate instrumental responses [35] (Fig. 3B). This procedure can serve as a useful screen for identifying novel compounds with abuse potential [35]. However, it has not been utilized extensively in this capacity because of its low-throughput nature.

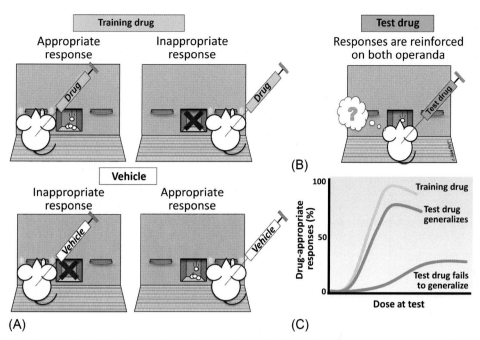

FIG. 3 Drug-discrimination paradigm. (A) In this procedure, subjects are first trained to make instrumental responses on one of two operanda depending on their interoceptive state. Responses on one operandum are reinforced with food after passive exposure to a training drug. Responses on the other operandum are reinforced with food after passive exposure to vehicle. Responses on the inappropriate operandum are not reinforced. (B) At test, subjects receive passive exposure to a test drug, and responses on both operanda are reinforced with food. (C) Schematic illustrating effects produced by the training drug and two different test drugs. Maximal responding is obtained with the training dose of the training drug. The test drug that generalizes to the training drug (i.e., elicit robust drug-appropriate responding) is assumed to elicit similar interoceptive effects as the training drug.

Chronic drug use significantly alters the affective state of organisms, which in turn profoundly impacts the subjective rewarding and motivational effects of drugs of abuse and propensity for drug taking and drug seeking [35a]. The **intracranial self-stimulation paradigm** provides an assessment of *hedonic state* [36]. In this procedure, instrumental responses are reinforced with brief electrical stimulation of brain tissue through an indwelling stimulating electrode (Fig. 4A). Stimulation frequency is titrated so that the subject's *reward threshold*, an index of hedonic state, can be determined [37] (Fig. 4B). This procedure has been used to monitor changes in hedonic state following administration of drugs of abuse (Fig. 4C, E, and F). Such studies have shown increases in reward threshold and decreases in the rewarding effects of drugs of abuse following chronic drug administration [38,39], paralleling reports of anhedonia in chronic drug users [40–42,43]. The intracranial self-stimulation procedure can also be employed to demonstrate the emergence of aversive affective states during drug withdrawal [44–46] and conditioned drug withdrawal [47] (Fig. 4D–F). Thus, a major advantage of this paradigm is its exceptionally sensitive ability to detect shifts in affective state across time, although it is subject to the *rate-altering* effects of drugs of abuse.

CONSUMMATORY AND QUASIINSTRUMENTAL MODELS OF DRUG REINFORCEMENT

Volitional drug taking involves motivated behaviors (drug seeking) and, depending on the substance, distinct *consummatory behaviors* associated with drug taking. These behaviors are considered primarily in the context of modeling ethanol and edible drug consumption. For instance, in the **two-bottle choice model** of alcohol drinking, subjects are given free-choice access to ethanol and water in separate bottles in the home cage. Ethanol, water, and total fluid consumption are monitored to assess the reinforcing effects of ethanol. Similarly, edible drug, regular chow, and total food consumption can be compared. Free-choice consummatory behaviors are relatively insensitive to the rate-altering effects of drugs (e.g., sedative effects of ethanol) and have strong translational appeal as models of oral drug taking. Considerable strain differences exist among rodents in ethanol drinking, which facilitate the study of genetic factors in alcohol-use disorders. However, it can be necessary to mix sucrose or saccharin into the ethanol solution initially (i.e., *sucrose/saccharin fading*) to stimulate drinking behavior and produce pharmacologically relevant blood ethanol levels (50–200 mg%) for low alcohol-preferring nondependent subjects [48]. Behavioral microstructure and reinforcer unit size can vary as a function

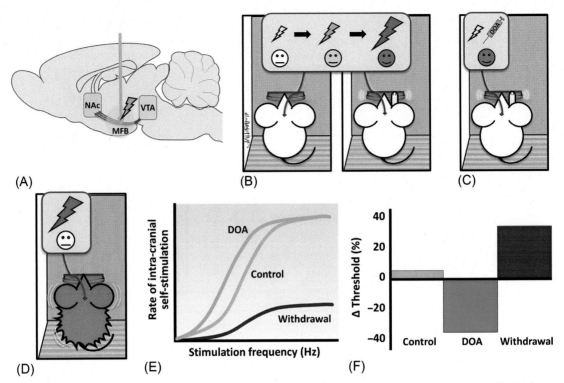

FIG. 4 Intracranial self-stimulation paradigm. (A) In this procedure, a chronically indwelling stimulating electrode is positioned in a brain reward center [e.g., medial forebrain bundle (MFB), a group of dopaminergic axons that project from the ventral tegmental area (VTA) to the nucleus accumbens (NAc)] prior to training. (B) Instrumental responses result in electrical stimulation of varying frequency during training. Stimulation frequency is titrated systematically to identify the minimum frequency that reinforces instrumental responding (reward threshold). (C) Acute exposure to drugs of abuse (DOA) reduces the reward threshold. (D) Aversive affective states associated with drug withdrawal increase the reward threshold. Schematic illustrating DOA- and drug withdrawal-associated shifts in (E) stimulation frequency effect curves and (F) reward thresholds.

of the subject's motivational state. Despite these limitations, ethanol drinking reliably correlates with instrumental ethanol self-administration, a quantitative measure of drug reinforcement [49]. Moreover, the two-bottle choice procedure is a particularly convenient high-throughput procedure for use with mice.

The **runway paradigm** offers a quasiinstrumental measure of the reinforcing effects of drugs of abuse. Runway conditioning involves a discrete trial procedure that starts with placement of the subject into the start box on one end of a runway (Fig. 5A). During each trial, a single injection of drug is available contingent upon entry into the goal box, on the other end of the runway. *Start latency* and *run time* are measured to evaluate the motivational effects of drugs of abuse

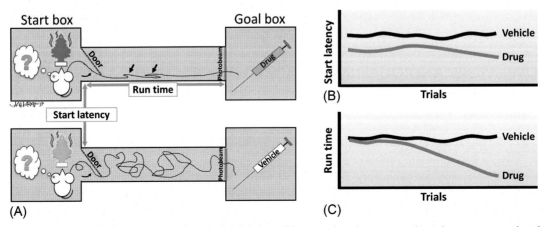

FIG. 5 Runway paradigm. (A) During the runway drug self-administration training procedure, drug exposure is contingent upon entry into the goal box. Olfactory stimuli in the start box can be conditioned to signal drug availability. *Arrows* indicate retreats (i.e., abrupt stops and reversals in running direction) that are theorized to indicate aversive drug effects. (B) Schematic illustrating the effects of DOA on start latency (i.e., time until exit from the start box) and (C) run time (i.e., time to traverse the runway) relative to vehicle.

(Fig. 5B and C). While drug exposure during runway drug self-administration is limited, it is volitional. Furthermore, start latencies predict the magnitude of *escalation* in cocaine intake during subsequent *free-operant* drug self-administration sessions [50]. Retreats (i.e., abrupt stops and brief reversals in running direction during runway drug self-administration, Fig. 5A) are also recorded and theorized to signal aversive drug effects [51]. Retreats are less frequent in rats with a history of *extended-access* cocaine self-administration [52]. Thus, the runway paradigm is unique in that it can detect the reinforcing and aversive effects of drugs of abuse as well as the development of tolerance to anxiogenic and other aversive drug effects during drug self-administration.

INSTRUMENTAL MODELS OF DRUG REINFORCEMENT

Drug addiction typically involves drug taking through intravenous or intrapulmonary routes of administration in order to maximize the euphoric effects of drugs of abuse other than ethanol. Experimental procedures for **instrumental drug self-administration** through the oral and intravenous routes have been employed since the 1960s [53,54], and recent technological advances related to the development of e-cigarettes have led to the establishment of vaporized drug self-administration procedures for rodents. In instrumental drug self-administration studies, responses on an operandum activate a liquid dipper (Fig. 6A), an infusion pump (Fig. 6B), or a drug vaporizer (Fig. 6C). Notably, self-administration of vaporized methamphetamine, MDMA, methylenedioxypyrovalerone, ethanol, sufentanil, and cannabis has been reported thus far [55,56]. However, drug vapor self-administration data are difficult to interpret because the dose and main route of drug administration (i.e., intrapulmonary, intranasal, oral) may vary depending on drug particle size in a species-specific manner [57]. Future drug pharmacokinetic studies may resolve this ambiguity. Independent of the route of drug administration, drug self-administration behaviors are sensitive to performance deficits produced by the rate-altering effects of drugs of abuse. Nonetheless, drug self-administration procedures remain the gold standard for assessing drug reinforcement and establishing drug-taking history in animal models of addiction.

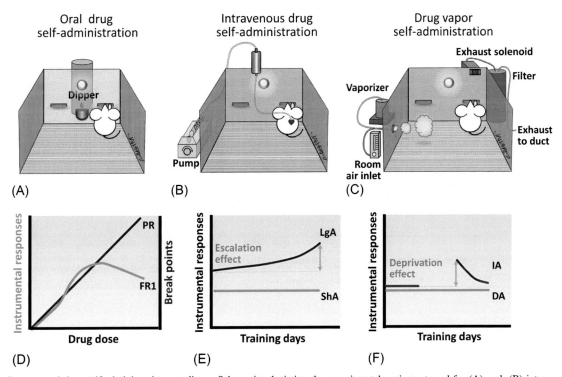

FIG. 6 Instrumental drug self-administration paradigms. Schematics depicting the experimental equipment used for (A) oral, (B) intravenous, and (C) intrapulmonary drug self-administration paradigms. Instrumental responses are reinforced with small amounts of ethanol, intravenous drug infusion, or brief drug vapor delivery. Computer-activated solenoids to control the duration of vaporized drug exposure in airtight operant conditioning chambers. (D) Schematic illustrating that progressive ratio (PR) reinforcement schedules result in linear dose-effect curves for break points, whereas fixed ratio (FR) reinforcement schedules yield inverted U-shaped dose-response curves. (E) Schematic illustrating that short-access drug regimens (e.g., 1 h/day, ShA) result in stable instrumental responding and drug intake, whereas long-access drug regimens (e.g., ≥6 h/day, LgA) result in a gradual increase in these measures (escalation effect). (F) Schematic illustrating that intermittent access (IA) ethanol drug self-administration regimens result in an increase in instrumental responding and drug intake (deprivation effect), relative to daily regimens (DA).

During drug self-administration, the *reinforcement schedule* critically shapes *response cost* and rate, *reinforcement rate*, and associative learning [58]. Thus, the reinforcement schedule should be selected according to the study goals. Most studies employ **fixed ratio (FR)** reinforcement schedules, under which each subsequent presentation of the reinforcer is contingent upon performance of a specific number of instrumental responses (e.g., FR1, one response required) (Fig. 7A).

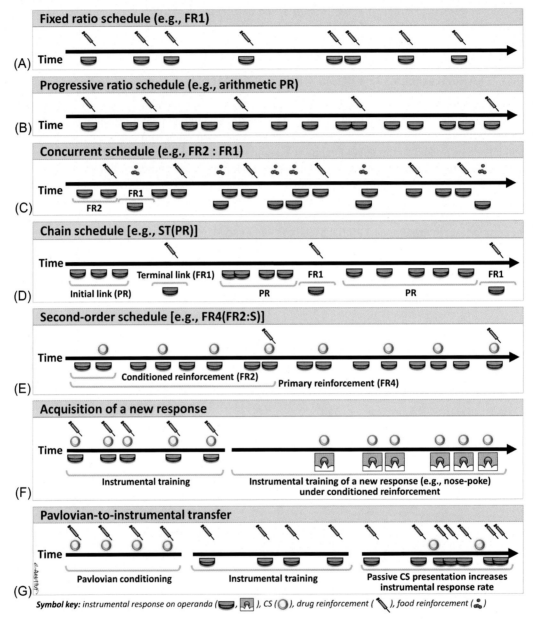

FIG. 7 Reinforcement schedules and models of conditioned reinforcement. (A) Under FR schedules, each presentation of the primary reinforcer (e.g., drug) is contingent upon a fixed number of instrumental responses. (B) Under PR schedules, subsequent presentations of the primary reinforcer are contingent upon an increasing number of instrumental responses. Under arithmetic PR schedules, the schedule demand increases by a constant (e.g., 1, as shown in B). (C) Concurrent schedules involve the simultaneous availability of the same or different reinforcers contingent on responses on two different operanda under either the same or different reinforcement schedules. (D) Chain schedules {e.g., seeking-taking PR [ST(PR)] schedule} involve sequential access to two or more schedules of reinforcement (links) on different operanda or on the same operandum (signaled by a discriminative stimulus, not shown). A single schedule completion on the initial link provides access to the next (or the final, reinforced) link. The schedule then reverts back to the initial link. (E) Second-order schedules involve the completion of responses guided by presentations of the *conditioned reinforcer* under one schedule prior to delivery of the primary reinforcer under a second schedule. (F) The acquisition of a new response procedure involves instrumental training reinforced by drug paired with the CS presentations. At test, the effects of response-contingent CS presentation on the acquisition of a different instrumental response are assessed. (G) Pavlovian-to-instrumental transfer involves separate Pavlovian CS-drug conditioning and instrumental response-drug training. At test, the effects of passive CS presentation on ongoing instrumental responding are assessed.

FR schedules are easy for subjects to acquire and engender strong response-drug associations and high drug intake. However, the *single-order reinforcement schedules* typically tend to generate inverted U-shaped dose-effect curves for drugs of abuse, with the descending limb of these dose-effect curves reflecting a decrease in instrumental responding due to aversive and/or other rate-altering effects of drugs at high doses (Fig. 6D). **Progressive ratio (PR)** schedules represent an exception to this rule. Under PR schedules, *schedule demand* is gradually increased after each *schedule completion* until responding ceases for a predetermined time period. Thus, PR schedules yield a more straightforward index of drug-induced motivation in the form of linear dose-effect curves for break points (i.e., most demanding schedule completed) (Figs. 6D and 7B). However, drug-specific performance impairments can occur at the start of the PR session, when schedule demand is low. Thus, study parameters need to be selected with care to minimize such effects, particularly when comparisons are planned across drugs [59,60]. Unlike PR schedules, **concurrent** schedules can be conveniently optimized to compare the relative reinforcing efficacy of different drugs of abuse (Fig. 7C). Under concurrent schedules, the two or more reinforcement schedules are implemented simultaneously via distinct operanda, permitting the observation of choice behavior (i.e., differential distribution of responses across operanda). Concurrent schedules can also be applied to detect changes in the motivational effects of a drug relative to natural reinforcer after chronic drug exposure [61,62]. Finally, **chain** schedules involve the two or more reinforcement schedules (i.e., links) implemented in a fixed sequence (Fig. 7D). Progression between the links is dependent on schedule completion, and completion of the terminal link results in *primary reinforcement*. Chain schedules have been successfully employed to study the distinct neurobiological mechanisms of drug seeking and drug taking. To achieve this, schedule completion on a designated "seeking lever" (initial link) provides access to the "taking lever" (terminal link), on which schedule completion results in drug reinforcement [63–66].

In addition to *positive reinforcement* mechanisms, *negative reinforcement* mechanisms can stimulate drug taking in a drug-specific manner. Aversive affective states associated with spontaneous or pharmacologically precipitated ethanol, opiate, and nicotine withdrawal are particularly effective in increasing drug self-administration and in stimulating drug choice over a natural reinforcer [67]. Similarly, persistent inflammatory pain in nondependent subjects maintains stable morphine-seeking behavior after the removal of drug reinforcement, highlighting the important role of negative reinforcement mechanisms associated with chronic pain conditions [68]. However, overall, the contribution of negative reinforcement mechanisms to self-medication has not received sufficient attention in the field of addiction research.

MODELS OF EXCESSIVE DRUG INTAKE

Loss of control over drug taking is a DSM-5 criterion for substance use disorders [69]. Multiple mechanisms likely contribute to this phenomenon, including development of pharmacological tolerance to the rewarding effects of the drug and neuroadaptations that potentiate drug-induced motivation, promote negative reinforcement, and/or expedite *habitual* or *compulsive* drug-seeking behaviors [70].

Drug self-administration procedures have been adapted to model and study the mechanisms of excessive drug use in laboratory animals. One such procedure is the **extended-access drug self-administration model**, in which access to drug is extended by moving subjects from a daily short-access drug self-administration regimen (1–2h/day) to a daily long-access regimen ($\geq$6h/day) [71]. The resulting **escalation effect**, or increase in drug intake during extended-access drug self-administration (Fig. 6E), can be observed for a number of drugs of abuse, including cocaine, methamphetamine, methylphenidate, nicotine, methylone, mephedrone, heroin, and fentanyl [71–74]. Notably, escalation of natural reward intake also has been reported [75].

The rate and magnitude of escalation appear to be sensitive to experimental parameters that vary across drugs of abuse [75–77]. For instance, extended-access drug self-administration regimens do not produce sustained escalation in ethanol intake unless ethanol dependence is established first through passive ethanol vapor exposure [16,78] or the ethanol access regimen is intermittent [79–82]. Intermittent ethanol access elicits the so-called **alcohol deprivation effect** (Fig. 6F), a transient increase in ethanol intake after a single period or repeated periods of *forced abstinence*, although a similar effect can be observed with other drugs of abuse (i.e., cocaine, heroin; Fuchs, unpublished results). The magnitude and duration of the alcohol deprivation effect is strain and species specific [83]. As an alternative to intermittent ethanol access, binge intoxication can be achieved in nondependent, alcohol-preferring strains of mice [84–86] and rats [84], using variations of the **drinking in the dark procedure** [86]. This high-throughput procedure typically consists of daily 2–4-h access to high-concentration ethanol solution starting 3h into the subjects' dark cycle. The procedure involves limited water deprivation, but robust ethanol drinking is achieved without sucrose/saccharin fading. Furthermore, the effects are strain specific in mice, facilitating the study of genetic predispositions to alcoholism [87].

Escalation in drug intake is theorized to reflect the development of drug-induced neuroadaptations and an associated increase in sensitivity to the motivational effects of the drug [70]. In strong support of this idea, both increases in brain stimulation reward threshold (an index of anhedonia in a drug-free state) [88,89] and drug-induced devaluation of natural rewards (an index of drug-value amplification) [90,91] positively correlate with escalation of drug intake.

MODELS OF COMPULSIVE AND HABITUAL DRUG TAKING AND DRUG SEEKING

Substance use disorders can be distinguished from recreational drug use based on the emergence of drug taking, that is, **compulsive**, or insensitive to adverse consequences (DSM-5) [69]. Emergence of compulsive drug seeking and drug taking is theorized to indicate experience-based and/or drug-induced neuroplasticity that shifts behavioral control to the dorsal striatum [92]. Persistent drug seeking following drug devaluation or under (threat of) punishment elegantly captures the development of compulsion [65,93–95]. Such procedures have been used to demonstrate that rats with intermittent ethanol or extended cocaine self-administration history exhibit compulsive drug seeking [65,96,97].

Drug seeking and taking behaviors can also become **habitual**, or only partially goal directed or automatic in nature, in the course of chronic drug use. In animal models, the development of habitual responding can be expedited by training subjects under the reinforcement schedules that maintain a weak response-outcome relationship, and it can be indicated by persistent drug seeking after the removal of drug reinforcement or reinforcement satiation [94,98–100]. Procedures that are employed to assess habitual responding do not involve aversive consequences, unlike procedures that measure compulsive behavior. This distinction is less clear cut when procedures used to assess habitual responding are viewed from a behavioral economical perspective (e.g., if the cost of unreinforced response is considered a loss). However, habit and compulsion are distinct constructs with nonoverlapping neurobiological mechanisms, including differential recruitment of the dorsolateral striatum and of hyperpolarization-active N-methyl-d-aspartate receptors in corticostriatal pathways [95,101].

MODELS OF CUE-INDUCED DRUG SEEKING

Motivation for drug seeking produced by environmental stimuli has been extensively studied using a variety of drug self-administration procedures. For example, drug self-administration under the **second-order reinforcement schedules** (Fig. 7E) involves extended periods of drug seeking maintained by drug-associated CSs. Each set of schedule completions is supported by cue presentations (i.e., conditioned reinforcement), and schedule completion is required for primary reinforcement delivery according to the second-order reinforcement schedule. Hence, instrumental responding during the first schedule completion provides an index of cue-induced motivation without *extinction* learning [102,103]. This permits the application of within-subject experimental designs, offsetting the limitation that training rodents to respond under second-order schedules is time consuming. **Acquisition of a new response** (Fig. 7F) under conditioned reinforcement is a stringent alternative for the assessment of cue-induced motivation [104,105]. In this procedure, subjects are trained to respond on an operandum for drug reinforcement paired with the CS presentations. At test, it is assessed whether the CS can reinforce the acquisition of instrumental responding on a different type of operandum. While this procedure can convincingly demonstrate the conditioned reinforcing effects of drug-paired stimuli, extinction learning during the test session precludes the use of within-subject designs and complicates data interpretation. Notably, second-order schedule and acquisition of a new response procedures are not designed to model cue-induced drug relapse *per se* (i.e., resumption of drug seeking or drug taking after a drug-free period).

MODELS OF CUE-INDUCED DRUG RELAPSE

Exposure to drug-associated environmental stimuli (i.e., drug-taking neighborhood, drug paraphernalia) is a key trigger for drug craving and relapse even following extended periods of abstinence [106–109]. The associative learning and neural mechanisms of this phenomenon have been investigated using *reinstatement* and *renewal* paradigms in laboratory animals [110].

In the **cue-** and **context-induced reinstatement models**, response-contingent drug delivery is paired with the presentation of a discrete CS or occurs in a distinct environmental context during drug self-administration training [110,111] (Fig. 8). Subjects then receive *extinction training* in the absence of the discrete or contextual stimuli. Extinction training weakens the influence of response-outcome associations on instrumental behavior without altering the impact of cue-related associations. Extinction training also establishes a behavioral reference point that permits demonstration of the **reinstatement effect**, an absolute increase in the magnitude of drug-seeking behavior at test (Fig. 8A). Finally, extinction

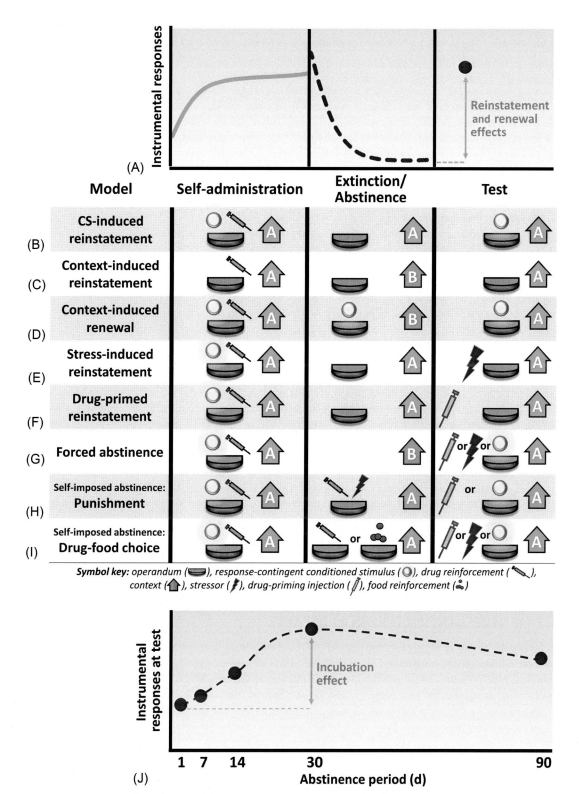

Symbol key: *operandum ( ), response-contingent conditioned stimulus ( ), drug reinforcement ( ), context ( ), stressor ( ), drug-priming injection ( ), food reinforcement ( )*

FIG. 8 Instrumental models of drug craving and relapse. (A) Schematic indicating typical patterns of instrumental responding during the self-administration, extinction/abstinence, and test phases in extinction/reinstatement, renewal, and abstinence models. (B–I) During drug self-administration training, instrumental responses are reinforced with drug infusions with or without the CS presentations. During the second phase of the procedure, in extinction paradigms, responses no longer have programed consequences (B, E, F) in the previously drug-paired context or (C,D) in an alternate context. Conversely, in abstinence paradigms, subjects (G) remain in an alternate context with no access to the operandum, (H) have access to drug reinforcement coupled with stressor presentation, or (I) make choices between drug or natural reinforcement. At test, drug-seeking behavior (i.e., non-reinforced instrumental responses) is assessed in the drug-paired context with or without response-contingent presentation of the CS, after exposure to a stressor, or following passive drug priming. (J) Schematic illustrating the incubation of drug-seeking behavior as a function of the duration of the abstinence phase prior to testing.

training permits the experimental isolation of cue-induced motivational effects. Accordingly, cue-induced drug-seeking behaviors can be assessed at test, upon response-contingent presentation of the CS [112] (Fig. 8B), passive exposure to the previously drug-paired context [113] (Fig. 8C), or a combination of drug-predictive stimuli. However, a disadvantage of reinstatement models is that extinction learning-related neuroplasticity may alter the specific neural substrates and signaling pathways recruited for drug-seeking behavior [99,114]. This may limit the translational relevance of the findings obtained using these models given that drug users do not typically experience explicit extinction training.

An idiosyncrasy of the CS-induced reinstatement model is that passive exposure to discrete drug-paired CSs is ineffective for reinstating instrumental drug-seeking behavior [115,116], even though cue-induced relapse typically occurs following inadvertent cue exposure in drug users attempting to achieve abstinence. Similarly, passive CS presentation, especially after extensive Pavlovian conditioning, fails to reliably support **Pavlovian-to-instrumental transfer** and to enhance ongoing drug-taking or drug-seeking behaviors (Fig. 7G) [117,118]. Together, these findings suggest that the motivational effects of the CS in relapse models are dependent on instrumental, as opposed to purely Pavlovian, associations and thus on conditioned reinforcement.

Importantly, *discriminative stimuli* that predict drug availability decrease the start latency and run time after passive presentation in the start box in the runway version of the extinction/reinstatement model (Fig. 5A) [119,120] and restore extinguished ethanol-seeking behavior in the reinstatement model [121]. Similarly, passive exposure to an environmental context reinstates instrumental cocaine-seeking behavior when the context is established as an *occasion setter*, a stimulus that signals drug availability contingent upon instrumental responding [113,122]. A salient demonstration of this property of the drug-paired context is the **renewal effect** [123], in which exposure to the drug-predictive context can disambiguate a CS previously extinguished in a different context and thus restore drug-seeking behavior (Fig. 8A) [124]. Accordingly, responding in the corresponding **context-induced renewal model** of drug relapse (Fig. 8D) [124–126] is likely dependent both on context exposure and response-contingent CS presentation at test.

The extinction/reinstatement procedure has been adapted to study associative learning and memory mechanisms that control cue-induced drug seeking. The contribution of associative learning to the CS-induced reinstatement has been elegantly demonstrated using a novel procedure in which subjects are trained to self-administer cocaine without an associated CS. The CS-drug associations are then established separately, during a single Pavlovian conditioning session. As a result, response-contingent CS presentation reinstates drug seeking at test [127,128]. Manipulations administered before or after the Pavlovian conditioning session can be employed to study the neurobiological mechanisms of the CS-drug associative learning or *memory consolidation* (i.e., the storage of newly formed long-term memories), respectively [129,130].

Lasting environmental stimulus control over drug craving and relapse requires the maintenance of drug-related memories overtime through *memory reconsolidation* (i.e., the restabilization of previously acquired memories after retrieval-induced destabilization) [131]. Hence, memory reconsolidation interference is of interest from an addiction treatment perspective. Instrumental and Pavlovian reinstatement procedures have been adapted to study CS-drug [132–134] and context-drug [135–137] memory reconsolidation. Subjects in these procedures are briefly re-exposed to drug-paired stimuli to elicit memory retrieval and destabilization [138,139]. Manipulations administered before or immediately after this session can be used to reveal neural mechanisms of memory destabilization and reconsolidation, respectively.

MODELS OF STRESS-INDUCED DRUG RELAPSE

Psychosocial stress is a leading trigger for drug relapse [140–142] that has been captured with somewhat limited success using **stress-induced reinstatement paradigms** [143] (Fig. 8E). Physical stressors (e.g., intermittent foot shock) [144,145] and pharmacological manipulations (e.g., yohimbine, an α2 adrenergic antagonist) [146–148] produce the most reliable reinstatement of extinguished instrumental behavior or CPP in laboratory animals. In contrast, stimuli predictive of imminent *social-defeat stress* only modestly reinstate ethanol- and cocaine-seeking behaviors in the most translationally relevant models of stress-induced drug relapse [149,150]. In addition, some stressors reinstate drug-seeking behavior selectively in instrumental paradigms (e.g., food deprivation) [151,152] or through stress-independent mechanisms (e.g., yohimbine) [153]. Thus, the development of new models and the cross-validation of findings obtained using traditional stress-induced reinstatement procedures will be particularly important in the future.

MODELS OF DRUG-PRIMED DRUG RELAPSE

Smokers and cocaine users report increases in craving upon exposure to small amounts of drug [154–156]. In laboratory animals, **drug-primed reinstatement** of extinguished instrumental responding or CPP is achieved by passive administration of a single drug-priming injection immediately before testing [157,158] (Fig. 8F). In the instrumental version of the paradigm, drug priming often occurs at doses and via routes of administration other than those previously experienced by the subject [159–162]. However, the main limitation of this model is conceptual. In the natural environment of the drug user, motivation to seek drug produced by other triggers (i.e., cues, stress) and associated corticolimbic activity precedes drug intake and thus drug-primed craving. Drug priming in laboratory animals likely circumvents this process by directly stimulating prelimbic cortical inputs to the nucleus accumbens to initiate drug-seeking behavior [163].

MODELS OF COCAINE SEEKING FOLLOWING ABSTINENCE

Drug addiction is characterized by cycles of *self-imposed* or *forced abstinence* followed by relapse. In the **forced abstinence model** of drug relapse, subjects are withdrawn from the drug self-administration regimen for a period of time by the experimenter (Fig. 8G). Drug-seeking behavior (i.e., nonreinforced instrumental responding) after forced abstinence may reflect multiple factors, including habitual behavior, novelty stress due to time away from the context, internal motivation, and drug context-induced motivation for drug [99]. Given that abstinence is not self-imposed, drug-seeking behavior cannot be considered a measure of relapse *per se*. However, the model has strong *face validity* for the resumption of drug seeking after abstinence precipitated by external factors, such as incarceration.

Modeling **self-imposed abstinence** has been challenging using laboratory animals; however, important advances have been made (Fig. 8H and I). Procedures have been developed in which subjects reduce their drug taking in response to intermittent punishment [97,164,165], conflict (e.g., electrified barrier) [166], or when they are offered discrete drug-food choice opportunities [167,168]. These methodologies yield more sophisticated models of the cognitive processes contributing to, and possibly unique neuroadaptations associated with, self-imposed abstinence.

Cue-induced craving in drug users [169–172] and cue- and stress-induced drug-seeking behaviors in laboratory animals become elevated as a function of the duration of the drug-free abstinence period, a phenomenon referred to as the **incubation effect** [173,174] (Fig. 8J). Incubation of cue-induced drug seeking can be observed not only for drugs of abuse, including cocaine, methamphetamine, nicotine, heroin, and cannabinoid receptor agonists [164,173,175–179], but also for palatable natural reinforcers [164,180].

Interestingly, the type of abstinence procedure utilized critically influences the development of incubation, such that incubation of heroin craving is observed after forced abstinence, but not self-imposed, choice-based abstinence, in rats [168]. One possibility is that the development of neuroadaptations required for incubation [181], which are distinct from those underlying the escalation of drug intake [182], are thwarted by some, although negligible, exposure to drug during the self-imposed abstinence period. Drug exposure at test may also have similar effects given that, unlike cue- and stress-induced drug seeking, drug-primed drug seeking does not exhibit incubation [183].

CONCLUSION

In the addiction research field, attention has shifted from the study of unconditioned drug effects to the study of neurobiological mechanisms involved in more complex drug addiction-related phenomena and to the study of individual differences in addiction vulnerability. This has been made possible by the development of sophisticated animal models that incorporate simpler Pavlovian and instrumental conditioning procedures. In the future, animal models will need to evolve further to meet new challenges, including the accurate evaluation of sex differences and related neuroendocrine mechanisms in drug addiction-related phenomena. In addition, more research will be needed using animal models that include relapse testing with drug reinforcement [184,185] since drug relapse involves the resumption of drug taking *per se*. The efficiency of sophisticated animal models will also need to be considered carefully as we aim to conserve resources. Finally, the efficacy of experimental therapies for drug addiction will need to be evaluated using not only valid animal models but also clinically relevant (i.e., chronic, systemic) treatment regimens in order to optimize the translatability of preclinical research findings.

ACKNOWLEDGMENTS

RAF's research is supported in part by funds provided by the National Institutes of Health grants NIDA 2 R01 DA025646-07, NIDA R01 DA033231, NIAAA R01 AA026078, NIDA F31 DA045430 (JAH)and the State of Washington Initiative Measure No. 171 and 502 administered through the Washington State University Alcohol and Drug Abuse Research Program grants WSU ADARP 127226, 125966. The authors declare no conflict of interest.

REFERENCES

[1] National Institute on Drug Abuse. Understanding drug use and addiction. Retrieved from, https://www.drugabuse.gov/publications/drugfacts/undertsnading-drug-use-addiction; 2016.

[2] Donny EC, Caggiula AR, Rose C, Jacobs KS, Mielke MM, Sved AF. Differential effects of response-contingent and response-independent nicotine in rats. Eur J Pharmacol 2000;402(3):231–40.

[3] Kantak KM, Barlow N, Tassin DH, Brisotti MF, Jordan CJ. Performance on a strategy set shifting task in rats following adult or adolescent cocaine exposure. Psychopharmacology (Berl) 2014;231(23):4489–501.

[4] Kiyatkin EA, Wise RA, Gratton A. Drug- and behavior-associated changes in dopamine-related electrochemical signals during intravenous heroin self-administration in rats. Synapse 1993;14(1):60–72.

[5] Mutschler NH, Miczek KA. Withdrawal from a self-administered or non-contingent cocaine binge: differences in ultrasonic distress vocalizations in rats. Psychopharmacology (Berl) 1998;136(4):402–8.

[6] Palamarchouk V, Smagin G, Goeders NE. Self-administered and passive cocaine infusions produce different effects on corticosterone concentrations in the medial prefrontal cortex (MPC) of rats. Pharmacol Biochem Behav 2009;94(1):163–8.

[7] Stefanski R, Ladenheim B, Lee SH, Cadet JL, Goldberg SR. Neuroadaptations in the dopaminergic system after active self-administration but not after passive administration of methamphetamine. Eur J Pharmacol 1999;371(2-3):123–35.

[8] Stefański R, Ziółkowska B, Kuśmider M, Mierzejewski P, Wyszogrodzka E, Kołomańska P, Dziedzicka-Wasylewska M, Przewłocki R, Kostowski W. Active versus passive cocaine administration: differences in the neuroadaptive changes in the brain dopaminergic system. Brain Res 2007;1157:1–10.

[9] Wilson JM, Nobrega JN, Corrigall WA, Coen KM, Shannak K, Kish SJ. Amygdala dopamine levels are markedly elevated after self- but not passive-administration of cocaine. Brain Res 1994;668(1-2):39–45.

[10] Barbier E, Houchi H, Warnault V, Pierrefiche O, Daoust M, Naassila M. Effects of prenatal and postnatal maternal ethanol on offspring response to alcohol and psychostimulants in long evans rats. Neuroscience 2009;161(2):427–40.

[11] Scherma M, Dessi C, Muntoni AL, Lecca S, Satta V, Luchicchi A, Pistis M, Panlilio LV, Fattore L, Goldberg SR, Fratta W, Fadda P. Adolescent delta(9)-tetrahydrocannabinol exposure alters WIN55,212-2 self-administration in adult rats. Neuropsychopharmacology 2016;41(5):1416–26.

[12] Shen YL, Chen ST, Chan TY, Hung TW, Tao PL, Liao RM, Chan MH, Chen HH. Delayed extinction and stronger drug-primed reinstatement of methamphetamine seeking in rats prenatally exposed to morphine. Neurobiol Learn Mem 2016;128:56–64.

[13] Melis M, Frau R, Kalivas PW, Spencer S, Chioma V, Zamberletti E, Rubino T, Parolaro D. New vistas on cannabis use disorder. *Neuropharmacology* 2017. https://doi.org/10.1016/j.neuropharm.2017.03.033.

[14] Valjent E, Maldonado R. A behavioural model to reveal place preference to delta 9-tetrahydrocannabinol in mice. Psychopharmacology (Berl) 2000;147(4):436–8.

[15] Becker HC, Lopez MF. An animal model of alcohol dependence to screen medications for treating alcoholism. Int Rev Neurobiol 2016;126:157–77.

[16] Griffin 3rd WC, Lopez MF, Becker HC. Intensity and duration of chronic ethanol exposure is critical for subsequent escalation of voluntary ethanol drinking in mice. Alcohol Clin Exp Res 2009;33(11):1893–900.

[17] Tzschentke TM. Measuring reward with the conditioned place preference paradigm: a comprehensive review of drug effects, recent progress and new issues. Prog Neurobiol 1998;56(6):613–72.

[18] Tzschentke TM. Measuring reward with the conditioned place preference (CPP) paradigm: update of the last decade. Addict Biol 2007;12(3-4):227–462.

[19] Foltin RW, Evans SM. A novel protocol for studying food or drug seeking in rhesus monkeys. Psychopharmacology (Berl) 1997;132(3):209–16.

[20] O'Dell LE, Khroyan TV, Neisewander JL. Dose-dependent characterization of the rewarding and stimulant properties of cocaine following intraperitoneal and intravenous administration in rats. Psychopharmacology (Berl) 1996;123(2):144–53.

[21] Spyraki C, Fibiger HC, Phillips AG. Cocaine-induced place preference conditioning: lack of effects of neuroleptics and 6-hydroxydopamine lesions. Brain Res 1982;253(1-2):195–203.

[22] Spyraki C, Fibiger HC, Phillips AG. Dopaminergic substrates of amphetamine-induced place preference conditioning. Brain Res 1982;253(1–2):185–93.

[23] Braida D, Iosue S, Pegorini S, Sala M. Delta9-tetrahydrocannabinol-induced conditioned place preference and intracerebroventricular self-administration in rats. Eur J Pharmacol 2004;506(1):63–9.

[24] Lepore M, Vorel SR, Lowinson J, Gardner EL. Conditioned place preference induced by delta 9-tetrahydrocannabinol: comparison with cocaine, morphine, and food reward. Life Sci 1995;56(23-24):2073–80.

[25] Murray JE, Bevins RA. Cannabinoid conditioned reward and aversion: behavioral and neural processes. ACS Chem Neurosci 2010;1(4):265–78.

[26] Hand TH, Koob GF, Stinus L, Le Moal M. Aversive properties of opiate receptor blockade: evidence for exclusively central mediation in naive and morphine-dependent rats. Brain Res 1988;474(2):364–8.

[27] Suzuki T, Ise Y, Tsuda M, Maeda J, Misawa M. Mecamylamine-precipitated nicotine-withdrawal aversion in rats. Eur J Pharmacol 1996;314(3):281–4.

[28] Everitt BJ, Parkinson JA, Olmstead MC, Arroyo M, Robledo P, Robbins TW. Associative processes in addiction and reward. The role of amygdala-ventral striatal subsystems. Ann N Y Acad Sci 1999;877:412–38.

[29] Flagel SB, Akil H, Robinson TE. Individual differences in the attribution of incentive salience to reward-related cues: implications for addiction. Neuropharmacology 2009;56(Suppl 1):139–48.

[30] Flagel SB, Watson SJ, Akil H, Robinson TE. Individual differences in the attribution of incentive salience to a reward-related cue: influence on cocaine sensitization. Behav Brain Res 2008;186(1):48–56.

[31] Robinson TE, Flagel SB. Dissociating the predictive and incentive motivational properties of reward-related cues through the study of individual differences. Biol Psychiatry 2009;65(10):869–73.

[32] Robinson TE, Yager LM, Cogan ES, Saunders BT. On the motivational properties of reward cues: individual differences. Neuropharmacology 2014;76(Pt B):450–9.

[33] Saunders BT, O'Donnell EG, Aurbach EL, Robinson TE. A cocaine context renews drug seeking preferentially in a subset of individuals. Neuropsychopharmacology 2014;39(12):2816–23.

[34] Glennon RA, Rosecrans JA, Young R. Drug-induced discrimination: a description of the paradigm and a review of its specific application to the study of hallucinogenic agents. Med Res Rev 1983;3(3):289–340.

[35] Solinas M, Panlilio LV, Justinova Z, Yasar S, Goldberg SR. Using drug-discrimination techniques to study the abuse-related effects of psychoactive drugs in rats. Nat Protoc 2006;1(3):1194–206.

[35a] Koob GF. The dark side of emotion: the addiction perspective. Eur J Pharmacol 2015;15(753):73–87.

[36] Carlezon Jr. WA, Chartoff EH. Intracranial self-stimulation (ICSS) in rodents to study the neurobiology of motivation. Nat Protoc 2007;2(11):2987–95.

[37] Gomita Y, Ichimaru Y, Moriyama M, Araki H, Futagami K. Effects of anxiolytic drugs on rewarding and aversive behaviors induced by intracranial stimulation. Acta Med Okayama 2003;57(3):95–108.

[38] Negus SS, Miller LL. Intracranial self-stimulation to evaluate abuse potential of drugs. Pharmacol Rev 2014;66(3):869–917.

[39] Wiebelhaus JM, Grim TW, Owens RA, Lazenka MF, Sim-Selley LJ, Abdullah RA, Niphakis MJ, Vann RE, Cravatt BF, Wiley JL, Negus SS, Lichtman AH. Delta9-tetrahydrocannabinol and endocannabinoid degradative enzyme inhibitors attenuate intracranial self-stimulation in mice. J Pharmacol Exp Ther 2015;352(2):195–207.

[40] Cook JW, Lanza ST, Chu W, Baker TB, Piper ME. Anhedonia: Its dynamic relations with craving, negative affect, and treatment during a quit smoking attempt. Nicotine Tob Res 2017;19(6):703–9.

[41] Hatzigiakoumis DS, Martinotti G, Giannantonio MD, Janiri L. Anhedonia and substance dependence: clinical correlates and treatment options. Front Psychiatry 2011;2. https://doi.org/10.3389/fpsyt.2011.00010.

[42] Volkow ND, Wang GJ, Telang F, Fowler JS, Alexoff D, Logan J, Jayne M, Wong C, Tomasi D. Decreased dopamine brain reactivity in marijuana abusers is associated with negative emotionality and addiction severity. Proc Natl Acad Sci U S A 2014;111(30):E3149–3156.

[43] Wang GJ, Volkow ND, Chang L, Miller E, Sedler M, Hitzemann R, Zhu W, Logan J, Ma Y, Fowler JS. Partial recovery of brain metabolism in methamphetamine abusers after protracted abstinence. Am J Psychiatry 2004;161(2):242–8. https://doi.org/10.1176/appi.ajp.161.2.242.

[44] Boutros N, Semenova S, Markou A. Adolescent intermittent ethanol exposure diminishes anhedonia during ethanol withdrawal in adulthood. Eur Neuropsychopharmacol 2014;24(6):856–64.

[45] Markou A, Koob GF. Postcocaine anhedonia. An animal model of cocaine withdrawal. Neuropsychopharmacology 1991;4(1):17–26.

[46] Russell SE, Puttick DJ, Sawyer AM, Potter DN, Mague S, Carlezon Jr. WA, Chartoff EH. Nucleus accumbens AMPA receptors are necessary for morphine-withdrawal-induced negative-affective states in rats. J Neurosci 2016;36(21):5748–62.

[47] Kenny PJ, Markou A. Conditioned nicotine withdrawal profoundly decreases the activity of brain reward systems. J Neurosci 2005;25(26):6208–12.

[48] McBride WJ, Li TK. Animal models of alcoholism: neurobiology of high alcohol-drinking behavior in rodents. Crit Rev Neurobiol 1998;12(4):339–69.

[49] Green AS, Grahame NJ. Ethanol drinking in rodents: is free-choice drinking related to the reinforcing effects of ethanol? Alcohol 2008;42(1):1–11.

[50] Ettenberg A, Fomenko V, Kaganovsky K, Shelton K, Wenzel JM. On the positive and negative affective responses to cocaine and their relation to drug self-administration in rats. Psychopharmacology (Berl) 2015;232(13):2363–75.

[51] Ettenberg A. Opponent process properties of self-administered cocaine. Neurosci Biobehav Rev 2004;27(8):721–8.

[52] Ben-Shahar O, Posthumus EJ, Waldroup SA, Ettenberg A. Heightened drug-seeking motivation following extended daily access to self-administered cocaine. Prog Neuropsychopharmacol Biol Psychiatry 2008;32(3):863–9.

[53] Weeks JR. Experimental morphine addiction: method for automatic intravenous injections in unrestrained rats. Science 1962;138(3537):143–4.

[54] Wise RA, Bozarth MA. Brain substrates for reinforcement and drug self-administration. Prog Neuropsychopharmacol 1981;5(5–6):467–74.

[55] Taffe MA, Nguyen JD, Vandewater SA, Cole M. Self-administration of psychstimulants via vapor inhalation in rats. No. 457.09. In: Neuroscience Meeting Planner. San Diego: Society for Neuroscience; 2016. p. 2016.

[56] Vendruscolo JCM, Vendruscolo LF, Tunstall BJ, Carmack SA, Cole M, Vandewater S, Tasffe MA, George O, Koob GF. Evaluation of a novel "e-vape" system as a model of opioid self-administration in rats. No. 78.19. In: Neuroscience Meeting Planner. San Diego: Society for Neuroscience; 2016. p. 2016.

[57] Snyder CA, Wood RW, Graefe JF, Bowers A, Magar K. "Crack smoke" is a respirable aerosol of cocaine base. Pharmacol Biochem Behav 1988;29(1):93–5.

[58] Panlilio LV, Goldberg SR. Self-administration of drugs in animals and humans as a model and an investigative tool. Addiction 2007;102(12):1863–70.

[59] Arnold JM, Roberts DC. A critique of fixed and progressive ratio schedules used to examine the neural substrates of drug reinforcement. Pharmacol Biochem Behav 1997;57(3):441–7.

[60] Killeen PR, Posadas-Sanchez D, Johansen EB, Thrailkill EA. Progressive ratio schedules of reinforcement. J Exp Psychol Anim Behav Process 2009;35(1):35–50.

[61] Johnson AR, Banks ML, Blough BE, Lile JA, Nicholson KL, Negus SS. Development of a translational model to screen medications for cocaine use disorder I: choice between cocaine and food in rhesus monkeys. Drug Alcohol Depend 2016;165:103–10.

[62] Kerstetter KA, Ballis MA, Duffin-Lutgen S, Carr AE, Behrens AM, Kippin TE. Sex differences in selecting between food and cocaine reinforcement are mediated by estrogen. Neuropsychopharmacology 2012;37(12):2605–14.

[63] Hellemans KG, Dickinson A, Everitt BJ. Motivational control of heroin seeking by conditioned stimuli associated with withdrawal and heroin taking by rats. Behav Neurosci 2006;120(1):103–14.

[64] Saddoris MP, Cacciapaglia F, Wightman RM, Carelli RM. Differential dopamine release dynamics in the nucleus accumbens core and shell reveal complementary signals for error prediction and incentive motivation. J Neurosci 2015;35(33):11572–82.

[65] Vanderschuren LJ, Everitt BJ. Drug seeking becomes compulsive after prolonged cocaine self-administration. Science 2004;305(5686):1017–9.

[66] Veeneman MM, van Ast M, Broekhoven MH, Limpens JH, Vanderschuren LJ. Seeking-taking chain schedules of cocaine and sucrose self-administration: effects of reward size, reward omission, and alpha-flupenthixol. Psychopharmacology (Berl) 2012;220(4):771–85.

[67] Negus SS, Banks ML. Modulation of drug choice by extended drug access and withdrawal in rhesus monkeys: Implications for negative reinforcement as a driver of addiction and target for medications development. Pharmacol Biochem Behav 2017. https://doi.org/10.1016/j.pbb.2017.04.006.

[68] Hou YY, Cai YQ, Pan ZZ. Persistent pain maintains morphine-seeking behavior after morphine withdrawal through reduced MeCP2 repression of GluA1 in rat central amygdala. J Neurosci 2015;35(8):3689–700.

[69] American Psychiatric Association. Diagnostic and statistical manual of mental disorders. 5th ed. Washigton, DC: Author; 2013.

[70] Edwards S, Koob GF. Escalation of drug self-administration as a hallmark of persistent addiction liability. Behav Pharmacol 2013;24(5-6):356–62.

[71] Ahmed SH, Koob GF. Transition from moderate to excessive drug intake: change in hedonic set point. Science 1998;282(5387):298–300.

[72] Kitamura O, Wee S, Specio SE, Koob GF, Pulvirenti L. Escalation of methamphetamine self-administration in rats: a dose-effect function. Psychopharmacology (Berl) 2006;186(1):48–53.

[73] Marusich JA, Beckmann JS, Gipson CD, Bardo MT. Methylphenidate as a reinforcer for rats: contingent delivery and intake escalation. Exp Clin Psychopharmacol 2010;18(3):257–66.

[74] Nguyen JD, Grant Y, Creehan KM, Vandewater SA, Taffe MA. Escalation of intravenous self-administration of methylone and mephedrone under extended access conditions. *Addict Biol* 2016. https://doi.org/10.1111/adb.12398.

[75] Goeders JE, Murnane KS, Banks ML, Fantegrossi WE. Escalation of food-maintained responding and sensitivity to the locomotor stimulant effects of cocaine in mice. Pharmacol Biochem Behav 2009;93(1):67–74.

[76] Kippin TE, Fuchs RA, See RE. Contributions of prolonged contingent and noncontingent cocaine exposure to enhanced reinstatement of cocaine seeking in rats. Psychopharmacology (Berl) 2006;187(1):60–7.

[77] Morgan AD, Dess NK, Carroll ME. Escalation of intravenous cocaine self-administration, progressive-ratio performance, and reinstatement in rats selectively bred for high (HiS) and low (LoS) saccharin intake. Psychopharmacology (Berl) 2005;178(1):41–51.

[78] Lopez MF, Becker HC. Operant ethanol self-administration in ethanol dependent mice. Alcohol 2014;48(3):295–9.

[79] Carnicella S, Ron D, Barak S. Intermittent ethanol access schedule in rats as a preclinical model of alcohol abuse. Alcohol 2014;48(3):243–52.

[80] Hwa LS, Chu A, Levinson SA, Kayyali TM, DeBold JF, Miczek KA. Persistent escalation of alcohol drinking in C57BL/6J mice with intermittent access to 20% ethanol. Alcohol Clin Exp Res 2011;35(11):1938–47.

[81] Melendez RI. Intermittent (every-other-day) drinking induces rapid escalation of ethanol intake and preference in adolescent and adult C57BL/6J mice. Alcohol Clin Exp Res 2011;35(4):652–8.

[82] Rosenwasser AM, Fixaris MC, Crabbe JC, Brooks PC, Ascheid S. Escalation of intake under intermittent ethanol access in diverse mouse genotypes. Addict Biol 2013;18(3):496–507.

[83] Vengeliene V, Bilbao A, Spanagel R. The alcohol deprivation effect model for studying relapse behavior: a comparison between rats and mice. Alcohol 2014;48(3):313–20.

[84] Holgate JY, Shariff M, Mu EW, Bartlett S. A rat drinking in the dark model for studying ethanol and sucrose consumption. Front Behav Neurosci 2017;11:29.

[85] Rhodes JS, Best K, Belknap JK, Finn DA, Crabbe JC. Evaluation of a simple model of ethanol drinking to intoxication in C57BL/6J mice. Physiol Behav 2005;84(1):53–63.

[86] Thiele TE, Navarro M. "Drinking in the dark" (DID) procedures: a model of binge-like ethanol drinking in non-dependent mice. Alcohol 2014;48(3):235–41.

[87] Rhodes JS, Ford MM, Yu CH, Brown LL, Finn DA, Garland Jr. T, Crabbe JC. Mouse inbred strain differences in ethanol drinking to intoxication. Genes Brain Behav 2007;6(1):1–18.

[88] Ahmed SH, Kenny PJ, Koob GF, Markou A. Neurobiological evidence for hedonic allostasis associated with escalating cocaine use. Nat Neurosci 2002;5(7):625–6.

[89] Kenny PJ, Chen SA, Kitamura O, Markou A, Koob GF. Conditioned withdrawal drives heroin consumption and decreases reward sensitivity. J Neurosci 2006;26(22):5894–900.

[90] Imperio CG, Grigson PS. Greater avoidance of a heroin-paired taste cue is associated with greater escalation of heroin self-administration in rats. Behav Neurosci 2015;129(4):380–8.

[91] McFalls AJ, Imperio CG, Bixler G, Freeman WM, Grigson PS, Vrana KE. Reward devaluation and heroin escalation is associated with differential expression of CRF signaling genes. Brain Res Bull 2016;123:81–93.

[92] Everitt BJ, Belin D, Economidou D, Pelloux Y, Dalley JW, Robbins TW. Review. Neural mechanisms underlying the vulnerability to develop compulsive drug-seeking habits and addiction. Philos Trans R Soc Lond B Biol Sci 2008;363(1507):3125–35.

[93] Dickinson A, Wood N, Smith JW. Alcohol seeking by rats: action or habit? Q J Exp Psychol B 2002;55(4):331–48.

[94] Hopf FW, Lesscher HM. Rodent models for compulsive alcohol intake. Alcohol 2014;48(3):253–64.

[95] Jonkman S, Pelloux Y, Everitt BJ. Differential roles of the dorsolateral and midlateral striatum in punished cocaine seeking. J Neurosci 2012;32(13):4645–50.

[96] Lesscher HM, van Kerkhof LW, Vanderschuren LJ. Inflexible and indifferent alcohol drinking in male mice. Alcohol Clin Exp Res 2010;34(7):1219–25.

[97] Pelloux Y, Everitt BJ, Dickinson A. Compulsive drug seeking by rats under punishment: effects of drug taking history. Psychopharmacology (Berl) 2007;194(1):127–37.

[98] Fanelli RR, Klein JT, Reese RM, Robinson DL. Dorsomedial and dorsolateral striatum exhibit distinct phasic neuronal activity during alcohol self-administration in rats. Eur J Neurosci 2013;38(4):2637–48.

[99] Fuchs RA, Branham RK, See RE. Different neural substrates mediate cocaine seeking after abstinence versus extinction training: a critical role for the dorsolateral caudate-putamen. J Neurosci 2006;26(13):3584–8.

[100] Murray JE, Belin-Rauscent A, Simon M, Giuliano C, Benoit-Marand M, Everitt BJ, Belin D. Basolateral and central amygdala differentially recruit and maintain dorsolateral striatum-dependent cocaine-seeking habits. Nat Commun 2015;6:10088.

[101] Seif T, Chang SJ, Simms JA, Gibb SL, Dadgar J, Chen BT, Harvey BK, Ron D, Messing RO, Bonci A, Hopf FW. Cortical activation of accumbens hyperpolarization-active NMDARs mediates aversion-resistant alcohol intake. Nat Neurosci 2013;16(8):1094–100.

[102] Arroyo M, Markou A, Robbins TW, Everitt BJ. Acquisition, maintenance and reinstatement of intravenous cocaine self-administration under a second-order schedule of reinforcement in rats: effects of conditioned cues and continuous access to cocaine. Psychopharmacology (Berl) 1998;140(3):331–44.

[103] John WS, Martin TJ, Nader MA. Behavioral determinants of cannabinoid self-administration in old world monkeys. Neuropsychopharmacology 2017;42(7):1522–30.

[104] Di Ciano P, Everitt BJ. Conditioned reinforcing properties of stimuli paired with self-administered cocaine, heroin or sucrose: implications for the persistence of addictive behaviour. Neuropharmacology 2004;47(Suppl 1):202–13.

[105] Taylor JR, Robbins TW. Enhanced behavioural control by conditioned reinforcers following microinjections of d-amphetamine into the nucleus accumbens. Psychopharmacology (Berl) 1984;84(3):405–12.

[106] Childress AR, McLellan AT, Ehrman R, O'Brien CP. Classically conditioned responses in opioid and cocaine dependence: a role in relapse? NIDA Res Monogr 1988;84:25–43.

[107] Sinha R. Modeling stress and drug craving in the laboratory: implications for addiction treatment development. Addict Biol 2009;14(1):84–98.

[108] Sinha R, Li CS. Imaging stress- and cue-induced drug and alcohol craving: association with relapse and clinical implications. Drug Alcohol Rev 2007;26(1):25–31.

[109] Volkow ND, Wang GJ, Telang F, Fowler JS, Logan J, Childress AR, Jayne M, Ma Y, Wong C. Cocaine cues and dopamine in dorsal striatum: mechanism of craving in cocaine addiction. J Neurosci 2006;26(24):6583–8.

[110] Fuchs RA, Lasseter HC, Ramirez DR, Xie X. Relapse to drug seeking following prolonged abstinence: the role of environmental stimuli. Drug Discov Today Dis Models 2008;5(4):251–8.

[111] Crombag HS, Bossert JM, Koya E, Shaham Y. Review. Context-induced relapse to drug seeking: a review. Philos Trans R Soc Lond B Biol Sci 2008;363(1507):3233–43.

[112] Davis WM, Smith SG. Role of conditioned reinforcers in the initiation, maintenance and extinction of drug-seeking behavior. Pavlov J Biol Sci 1976;11(4):222–36.

[113] Fuchs RA, Evans KA, Ledford CC, Parker MP, Case JM, Mehta RH, See RE. The role of the dorsomedial prefrontal cortex, basolateral amygdala, and dorsal hippocampus in contextual reinstatement of cocaine seeking in rats. Neuropsychopharmacology 2005;30(2):296–309.

[114] Self DW, Choi KH. Extinction-induced neuroplasticity attenuates stress-induced cocaine seeking: a state-dependent learning hypothesis. Stress 2004;7(3):145–55.

[115] de Wit H, Stewart J. Reinstatement of cocaine-reinforced responding in the rat. Psychopharmacology (Berl) 1981;75(2):134–43.

[116] Fuchs RA, Tran-Nguyen LT, Specio SE, Groff RS, Neisewander JL. Predictive validity of the extinction/reinstatement model of drug craving. Psychopharmacology (Berl) 1998;135(2):151–60.

[117] Holmes NM, Marchand AR, Coutureau E. Pavlovian to instrumental transfer: a neurobehavioural perspective. Neurosci Biobehav Rev 2010;34(8):1277–95.

[118] Lamb RJ, Schindler CW, Pinkston JW. Conditioned stimuli's role in relapse: preclinical research on Pavlovian-Instrumental-Transfer. Psychopharmacology (Berl) 2016;233(10):1933–44.

[119] McFarland K, Ettenberg A. Reinstatement of drug-seeking behavior produced by heroin-predictive environmental stimuli. Psychopharmacology (Berl) 1997;131(1):86–92.

[120] Su ZI, Wenzel J, Baird R, Ettenberg A. Comparison of self-administration behavior and responsiveness to drug-paired cues in rats running an alley for intravenous heroin and cocaine. Psychopharmacology (Berl) 2011;214(3):769–78.

[121] Katner SN, Magalong JG, Weiss F. Reinstatement of alcohol-seeking behavior by drug-associated discriminative stimuli after prolonged extinction in the rat. Neuropsychopharmacology 1999;20(5):471–9.

[122] Hamlin AS, Newby J, McNally GP. The neural correlates and role of D1 dopamine receptors in renewal of extinguished alcohol-seeking. Neuroscience 2007;146(2):525–36.

[123] Bouton ME. Context and ambiguity in the extinction of emotional learning: implications for exposure therapy. Behav Res Ther 1988;26(2):137–49.

[124] Crombag HS, Shaham Y. Renewal of drug seeking by contextual cues after prolonged extinction in rats. Behav Neurosci 2002;116(1):169–73.

[125] Diergaarde L, de Vries W, Raaso H, Schoffelmeer AN, De Vries TJ. Contextual renewal of nicotine seeking in rats and its suppression by the cannabinoid-1 receptor antagonist Rimonabant (SR141716A). Neuropharmacology 2008;55(5):712–6.

[126] Marinelli PW, Funk D, Juzytsch W, Li Z, Le AD. Effects of opioid receptor blockade on the renewal of alcohol seeking induced by context: relationship to c-fos mRNA expression. Eur J Neurosci 2007;26(10):2815–23.

[127] Gabriele A, See RE. Reversible inactivation of the basolateral amygdala, but not the dorsolateral caudate putamen, attenuates consolidation of cocaine-cue associative learning in a reinstatement model of drug-seeking. Eur J Neurosci 2010;32(6):1024–9.

[128] Kruzich PJ, See RE. Differential contributions of the basolateral and central amygdala in the acquisition and expression of conditioned relapse to cocaine-seeking behavior. J Neurosci 2001;21(14):RC155.

[129] Feltenstein MW, See RE. NMDA receptor blockade in the basolateral amygdala disrupts consolidation of stimulus-reward memory and extinction learning during reinstatement of cocaine-seeking in an animal model of relapse. Neurobiol Learn Mem 2007;88(4):435–44.

[130] Fuchs RA, Feltenstein MW, See RE. The role of the basolateral amygdala in stimulus-reward memory and extinction memory consolidation and in subsequent conditioned cued reinstatement of cocaine seeking. Eur J Neurosci 2006;23(10):2809–13.

[131] Haubrich J, Nader K. Memory reconsolidation. Curr Top Behav Neurosci 2016. https://doi.org/10.1007/7854_2016_463.

[132] Dunbar AB, Taylor JR. Garcinol blocks the reconsolidation of multiple cocaine-paired cues after a single cocaine-reactivation session. Neuropsychopharmacology 2017. https://doi.org/10.1038/npp.2017.27.

[133] Lee JL, Di Ciano P, Thomas KL, Everitt BJ. Disrupting reconsolidation of drug memories reduces cocaine-seeking behavior. Neuron 2005;47(6):795–801.

[134] Sanchez H, Quinn JJ, Torregrossa MM, Taylor JR. Reconsolidation of a cocaine-associated stimulus requires amygdalar protein kinase A. J Neurosci 2010;30(12):4401–7.

[135] Brown TE, Forquer MR, Cocking DL, Jansen HT, Harding JW, Sorg BA. Role of matrix metalloproteinases in the acquisition and reconsolidation of cocaine-induced conditioned place preference. Learn Mem 2007;14(3):214–23.

[136] Fuchs RA, Bell GH, Ramirez DR, Eaddy JL, Su ZI. Basolateral amygdala involvement in memory reconsolidation processes that facilitate drug context-induced cocaine seeking. Eur J Neurosci 2009;30(5):889–900.

[137] Taubenfeld SM, Muravieva EV, Garcia-Osta A, Alberini CM. Disrupting the memory of places induced by drugs of abuse weakens motivational withdrawal in a context-dependent manner. Proc Natl Acad Sci U S A 2010;107(27):12345–50.

[138] Sorg BA. Reconsolidation of drug memories. Neurosci Biobehav Rev 2012;36(5):1400–17.

[139] Torregrossa MM, Taylor JR. Learning to forget: manipulating extinction and reconsolidation processes to treat addiction. Psychopharmacology (Berl) 2013;226(4):659–72.

[140] Back SE, Hartwell K, DeSantis SM, Saladin M, McRae-Clark AL, Price KL, Moran-Santa Maria MM, Baker NL, Spratt E, Kreek MJ, Brady KT. Reactivity to laboratory stress provocation predicts relapse to cocaine. Drug Alcohol Depend 2010;106(1):21–7.

[141] Cummings KM, Jaen CR, Giovino G. Circumstances surrounding relapse in a group of recent exsmokers. Prev Med 1985;14(2):195–202.

[142] Preston KL, Epstein DH. Stress in the daily lives of cocaine and heroin users: relationship to mood, craving, relapse triggers, and cocaine use. Psychopharmacology (Berl) 2011;218(1):29–37.

[143] Mantsch JR, Baker DA, Funk D, Le AD, Shaham Y. Stress-induced reinstatement of drug seeking: 20 years of progress. Neuropsychopharmacology 2016;41(1):335–56.

[144] Shaham Y, Stewart J. Stress reinstates heroin-seeking in drug-free animals: an effect mimicking heroin, not withdrawal. Psychopharmacology (Berl) 1995;119(3):334–41.

[145] Wang B, Luo F, Zhang WT, Han JS. Stress or drug priming induces reinstatement of extinguished conditioned place preference. Neuroreport 2000;11(12):2781–4.

[146] Le AD, Harding S, Juzytsch W, Funk D, Shaham Y. Role of alpha-2 adrenoceptors in stress-induced reinstatement of alcohol seeking and alcohol self-administration in rats. Psychopharmacology (Berl) 2005;179(2):366–73.

[147] Lee B, Tiefenbacher S, Platt DM, Spealman RD. Pharmacological blockade of alpha2-adrenoceptors induces reinstatement of cocaine-seeking behavior in squirrel monkeys. Neuropsychopharmacology 2004;29(4):686–93.

[148] Mantsch JR, Weyer A, Vranjkovic O, Beyer CE, Baker DA, Caretta H. Involvement of noradrenergic neurotransmission in the stress- but not cocaine-induced reinstatement of extinguished cocaine-induced conditioned place preference in mice: role for beta-2 adrenergic receptors. Neuropsychopharmacology 2010;35(11):2165–78.

[149] Funk D, Harding S, Juzytsch W, Le AD. Effects of unconditioned and conditioned social defeat on alcohol self-administration and reinstatement of alcohol seeking in rats. Psychopharmacology (Berl) 2005;183(3):341–9.

[150] Manvich DF, Stowe TA, Godfrey JR, Weinshenker D. A method for psychosocial stress-induced reinstatement of cocaine seeking in rats. Biol Psychiatry 2016;79(11):940–6.

[151] Ma YY, Chu NN, Guo CY, Han JS, Cui CL. NR2B-containing NMDA receptor is required for morphine-but not stress-induced reinstatement. Exp Neurol 2007;203(2):309–19.

[152] Shalev U, Highfield D, Yap J, Shaham Y. Stress and relapse to drug seeking in rats: studies on the generality of the effect. Psychopharmacology (Berl) 2000;150(3):337–46.

[153] Chen YW, Fiscella KA, Bacharach SZ, Tanda G, Shaham Y, Calu DJ. Effect of yohimbine on reinstatement of operant responding in rats is dependent on cue contingency but not food reward history. Addict Biol 2015;20(4):690–700.

[154] Hodgson R, Rankin H, Stockwell T. Alcohol dependence and the priming effect. Behav Res Ther 1979;17(4):379–87.

[155] Jaffe JH, Cascella NG, Kumor KM, Sherer MA. Cocaine-induced cocaine craving. Psychopharmacology (Berl) 1989;97(1):59–64.

[156] Volkow ND, Wang GJ, Ma Y, Fowler JS, Wong C, Ding YS, Hitzemann R, Swanson JM, Kalivas P. Activation of orbital and medial prefrontal cortex by methylphenidate in cocaine-addicted subjects but not in controls: relevance to addiction. J Neurosci 2005;25(15):3932–9.

[157] Gerber GJ, Stretch R. Drug-induced reinstatement of extinguished self-administration behavior in monkeys. Pharmacol Biochem Behav 1975;3(6):1055–61.

[158] Mueller D, Stewart J. Cocaine-induced conditioned place preference: reinstatement by priming injections of cocaine after extinction. Behav Brain Res 2000;115(1):39–47.

[159] Chiamulera C, Borgo C, Falchetto S, Valerio E, Tessari M. Nicotine reinstatement of nicotine self-administration after long-term extinction. Psychopharmacology (Berl) 1996;127(2):102–7.

[160] de Wit H, Stewart J. Drug reinstatement of heroin-reinforced responding in the rat. Psychopharmacology (Berl) 1983;79(1):29–31.

[161] Le AD, Quan B, Juzytch W, Fletcher PJ, Joharchi N, Shaham Y. Reinstatement of alcohol-seeking by priming injections of alcohol and exposure to stress in rats. Psychopharmacology (Berl) 1998;135(2):169–74.

[162] Mantsch JR, Goeders NE. Ketoconazole does not block cocaine discrimination or the cocaine-induced reinstatement of cocaine-seeking behavior. Pharmacol Biochem Behav 1999;64(1):65–73.

[163] Kalivas PW, McFarland K. Brain circuitry and the reinstatement of cocaine-seeking behavior. Psychopharmacology (Berl) 2003;168(1-2):44–56.

[164] Krasnova IN, Marchant NJ, Ladenheim B, McCoy MT, Panlilio LV, Bossert JM, Shaham Y, Cadet JL. Incubation of methamphetamine and palatable food craving after punishment-induced abstinence. Neuropsychopharmacology 2014;39(8):2008–16.

[165] Panlilio LV, Thorndike EB, Schindler CW. Lorazepam reinstates punishment-suppressed remifentanil self-administration in rats. Psychopharmacology (Berl) 2005;179(2):374–82.

[166] Cooper A, Barnea-Ygael N, Levy D, Shaham Y, Zangen A. A conflict rat model of cue-induced relapse to cocaine seeking. Psychopharmacology (Berl) 2007;194(1):117–25.

[167] Caprioli D, Venniro M, Zhang M, Bossert JM, Warren BL, Hope BT, Shaham Y. Role of dorsomedial striatum neuronal ensembles in incubation of methamphetamine craving after voluntary abstinence. J Neurosci 2017;37(4):1014–27.

[168] Venniro M, Zhang M, Shaham Y, Caprioli D. Incubation of methamphetamine but not heroin craving after voluntary abstinence in male and female rats. Neuropsychopharmacology 2017;42(5):1126–35.

[169] Bedi G, Preston KL, Epstein DH, Heishman SJ, Marrone GF, Shaham Y, de Wit H. Incubation of cue-induced cigarette craving during abstinence in human smokers. Biol Psychiatry 2011;69(7):708–11.

[170] Gawin F, Kleber H. Pharmacologic treatments of cocaine abuse. Psychiatr Clin North Am 1986;9(3):573–83.

[171] Li P, Wu P, Xin X, Fan YL, Wang GB, Wang F, Ma MY, Xue MM, Luo YX, Yang FD, Bao YP, Shi J, Sun HQ, Lu L. Incubation of alcohol craving during abstinence in patients with alcohol dependence. Addict Biol 2015;20(3):513–22.

[172] Wang G, Shi J, Chen N, Xu L, Li J, Li P, Sun Y, Lu L. Effects of length of abstinence on decision-making and craving in methamphetamine abusers. PLoS One 2013;8(7):e68791.

[173] Grimm JW, Hope BT, Wise RA, Shaham Y. Neuroadaptation. Incubation of cocaine craving after withdrawal. Nature 2001;412(6843):141–2.

[174] Sorge RE, Stewart J. The contribution of drug history and time since termination of drug taking to footshock stress-induced cocaine seeking in rats. Psychopharmacology (Berl) 2005;183(2):210–7.

[175] Abdolahi A, Acosta G, Breslin FJ, Hemby SE, Lynch WJ. Incubation of nicotine seeking is associated with enhanced protein kinase A-regulated signaling of dopamine- and cAMP-regulated phosphoprotein of 32 kDa in the insular cortex. Eur J Neurosci 2010;31(4):733–41.

[176] Kirschmann EK, Pollock MW, Nagarajan V, Torregrossa MM. Effects of adolescent cannabinoid self-administration in rats on addiction-related behaviors and working memory. Neuropsychopharmacology 2017;42(5):989–1000.

[177] Kuntz KL, Twining RC, Baldwin AE, Vrana KE, Grigson PS. Heroin self-administration: I. Incubation of goal-directed behavior in rats. Pharmacol Biochem Behav 2008;90(3):344–8.

[178] Shalev U, Morales M, Hope B, Yap J, Shaham Y. Time-dependent changes in extinction behavior and stress-induced reinstatement of drug seeking following withdrawal from heroin in rats. Psychopharmacology (Berl) 2001;156(1):98–107.

[179] Tran-Nguyen LT, Fuchs RA, Coffey GP, Baker DA, O'Dell LE, Neisewander JL. Time-dependent changes in cocaine-seeking behavior and extracellular dopamine levels in the amygdala during cocaine withdrawal. Neuropsychopharmacology 1998;19(1):48–59.

[180] Grimm JW, Fyall AM, Osincup DP. Incubation of sucrose craving: effects of reduced training and sucrose pre-loading. Physiol Behav 2005;84(1):73–9.

[181] Li X, Caprioli D, Marchant NJ. Recent updates on incubation of drug craving: a mini-review. Addict Biol 2015;20(5):872–6.

[182] Guillem K, Ahmed SH, Peoples LL. Escalation of cocaine intake and incubation of cocaine seeking are correlated with dissociable neuronal processes in different accumbens subregions. Biol Psychiatry 2014;76(1):31–9.

[183] Lu L, Grimm JW, Dempsey J, Shaham Y. Cocaine seeking over extended withdrawal periods in rats: different time courses of responding induced by cocaine cues versus cocaine priming over the first 6 months. Psychopharmacology (Berl) 2004;176(1):101–8.

[184] Nic Dhonnchadha BA, Szalay JJ, Achat-Mendes C, Platt DM, Otto MW, Spealman RD, Kantak KM. D-cycloserine deters reacquisition of cocaine self-administration by augmenting extinction learning. Neuropsychopharmacology 2010;35(2):357–67.

[185] Perry CJ, McNally GP. Naloxone prevents the rapid reacquisition but not acquisition of alcohol seeking. Behav Neurosci 2012;126(4):599–604.

[186] Ahmed SH, Walker JR, Koob GF. Persistent increase in the motivation to take heroin in rats with a history of drug escalation. Neuropsychopharmacology 2000;22(4):413–21.

[187] Asin KE, Wirtshafter D, Tabakoff B. Failure to establish a conditioned place preference with ethanol in rats. Pharmacol Biochem Behav 1985;22(2):169–73.

[188] Bespalov AY, Tokarz ME, Bowen SE, Balster RL, Beardsley PM. Effects of test conditions on the outcome of place conditioning with morphine and naltrexone in mice. Psychopharmacology (Berl) 1999;141(2):118–22.

[189] Blatt SL, Takahashi RN. Experimental anxiety and the reinforcing effects of ethanol in rats. Braz J Med Biol Res 1999;32(4):457–61.

[190] Cohen A, Koob GF, George O. Robust escalation of nicotine intake with extended access to nicotine self-administration and intermittent periods of abstinence. Neuropsychopharmacology 2012;37(9):2153–60.

[191] Cunningham CL. Genetic relationship between ethanol-induced conditioned place preference and other ethanol phenotypes in 15 inbred mouse strains. Behav Neurosci 2014;128(4):430–45.

[192] Cunningham CL, Noble D. Methamphetamine-induced conditioned place preference or aversion depending on dose and presence of drug. Ann N Y Acad Sci 1992;654:431–3.

[193] de Guglielmo G, Kallupi M, Cole MD, George O. Voluntary induction and maintenance of alcohol dependence in rats using alcohol vapor self-administration. Psychopharmacology (Berl) 2017;234(13):2009–18.

[194] He X, Bao Y, Li Y, Sui N. The effects of morphine at different embryonic ages on memory consolidation and rewarding properties of morphine in day-old chicks. Neurosci Lett 2010;482(1):12–6.

[195] Morgan AD, Campbell UC, Fons RD, Carroll ME. Effects of agmatine on the escalation of intravenous cocaine and fentanyl self-administration in rats. Pharmacol Biochem Behav 2002;72(4):873–80.

[196] Pchelintsev MV, Gorbacheva EN, Zvartau EE. Simple methodology of assessment of analgesics' addictive potential in mice. Pharmacol Biochem Behav 1991;39(4):873–6.

[197] Phillips AG, LePiane FG. Reinforcing effects of morphine microinjection into the ventral tegmental area. Pharmacol Biochem Behav 1980;12(6):965–8.

[198] Phillips TJ, Broadbent J, Burkhart-Kasch S, Henderson C, Wenger CD, McMullin C, McKinnon CS, Cunningham CL. Genetic correlational analyses of ethanol reward and aversion phenotypes in short-term selected mouse lines bred for ethanol drinking or ethanol-induced conditioned taste aversion. Behav Neurosci 2005;119(4):892–910.

[199] Spyraki C, Fibiger HC, Phillips AG. Attenuation of heroin reward in rats by disruption of the mesolimbic dopamine system. Psychopharmacology (Berl) 1983;79(2-3):278–83.

[200] Stewart RB, Murphy JM, McBride WJ, Lumeng Ls, Li TK. Place conditioning with alcohol in alcohol-preferring and -nonpreferring rats. Pharmacol Biochem Behav 1996;53(3):487–91.

[201] Tuazon DB, Suzuki T, Misawa M, Watanabe S. Methylxanthines (caffeine and theophylline) blocked methamphetamine-induced conditioned place preference in mice but enhanced that induced by cocaine. Ann N Y Acad Sci 1992;654:531–3.

Chapter 3

Transcriptional and Epigenetic Regulation of Reward Circuitry in Drug Addiction

Andrew Eagle, Basma Al Masraf, and A.J. Robison
Department of Physiology, Michigan State University, East Lansing, MI, United States

INTRODUCTION

Drug addiction is characterized by compulsive drug seeking and drug intake despite adverse consequences. Although abuse of both licit and illicit substances is estimated to cost over $700 billion annually in health care, crime, and lost productivity in the United States alone [1], there are currently few pharmacological treatments for drug addiction, and those that do exist have low efficacy in many individuals. This paucity of effective treatment options may be due to inextirpable drug-induced changes in brain function that drive some or all of the addiction phenotype. Chronic drug exposure associated with addiction induces persistent changes in the structure and function of brain cells and circuits that underlie addictive behavior, for example, drug seeking and relapse [2]. Therefore, identifying target mechanisms that drive long-term functional changes in the brain is a critical step in illuminating disease etiology and developing new treatments. This will require a comprehensive understanding of the neurobiology of addiction, including the role of gene expression, and its regulation, in drug-induced changes in neuronal structure and function. We will summarize in the following chapter the transcriptional and epigenetic mechanisms of addiction and, furthermore, we will discuss the current implications of these findings in terms of prevention of drug addiction and drug discovery for the treatment of addiction.

MODELS OF ADDICTION

Various animal models have been integral to investigating the neurobiological mechanisms underlying addiction. Critical information about addiction biology has been derived from invertebrates, primates, and studies of human patients, but the majority of data available for causal links between cellular and molecular pathways and specific aspects of addiction behavior have been derived from rodent models. Broadly, the common rodent models consist of: self-administration, conditioned place preference (CPP), and sensitization, with each providing information about common and unique aspects of addictive phenotypes.

Drug self-administration is a translationally relevant paradigm that models addiction processes including: development, withdrawal, and relapse to addiction (See Chapter 2 for a complete description of models of addiction.). The basic premise of the model is training animals to respond on an operanda (e.g., nosepoke or lever) for administration of a drug, typically via intravenous injection. There are also a number of modifications to this paradigm that are used to examine different aspects of addiction behaviors, including extended vs limited drug access, abstinence vs extinction of the trained response, drug motivational states (breakpoints), and various models of drug seeking and relapse. These models inform studies investigating the contributions to human addiction relapse, such as re-introduction to environmental cues after abstinence, the influence of life stressors, etc. These include withdrawal in home cage (as a model for involuntary abstinence) and a variety of drug seeking reinstatement paradigms: cues (contextual and discrete stimuli previously associated with drug exposure), stress, and drug priming (e.g., low-dose drug exposure), key models of discrete aspects and stages of drug addiction. Furthermore, these models suggest common neurobiological and gene expression mechanisms that may underlie aspects of the disease [3], notably the motivation or drive to seek out drugs.

CPP is another common behavioral model to assess contextually associated expression of drug reward [4,5]. Typical CPP involves repeated pairings of a particular context with a drug to determine, in a later drug-free state, whether animals will "choose" to spend time in the environment paired with the drug over a vehicle-paired environment. This provides a measure of the contextual association to the drug, that is, learning, as well as the expression of the reward conferred by a

Neural Mechanisms of Addiction. https://doi.org/10.1016/B978-0-12-812202-0.00003-8
Copyright © 2019 Elsevier Inc. All rights reserved.

particular drug [6]. This is useful in determining the molecular and cellular neurobiology underlying the hedonic aspects of drug seeking.

Finally, in rodents, repeated drug administration can produce the phenomenon of sensitization, which is an enhanced or "sensitized" response to subsequent drug administration. Sensitization can be observed behaviorally, for example, increased drug-induced locomotion after repeated injections, and in neural function, for example, increased neuronal excitability. Drug sensitization occurs with most forms of administration, and so allows for the differentiation of contingent (self-administered) vs noncontingent (experimenter-administered) effects.

These paradigms (self-administration, CPP, and sensitization) model key aspects of addiction and have uncovered many critical neurobiological underpinnings of addiction, notably in the reward pathway. This modeling allows determination of patterns of gene expression and transcriptional and epigenetic machinery in the reward circuit and other brain regions that may drive some addiction phenotypes. Furthermore, they provide testable paradigms for assessing novel therapeutics for the treatment of addiction.

THE REWARD CIRCUITRY

There are two major nuclei of dopamine (DA) neurons in the brain, the substantia nigra (SN), which regulates initiation of movement and degenerates in Parkinson's disease, and ventral tegmental area (VTA), which is critical for motivated behavior. VTA DA neurons have major projections to the prefrontal cortex (PFC; the mesocortical pathway) and to the nucleus accumbens (NAc; the mesolimbic pathway), but also project to hippocampus, amygdala, and several other forebrain regions (Fig. 1). VTA DA neurons are regulated by both excitatory glutamatergic and inhibitory GABAergic signaling from local cells and afferents from other brain regions [6]. Glutamatergic inputs onto VTA DA neurons come from anterior cortex (including PFC) and several brainstem regions, including the lateral dorsal tegmental nucleus, subthalamic nucleus (STN), and bed nucleus of the stria terminalis, and many of these inputs directly regulate reward-related behaviors [7]. In addition, the VTA contains GABAergic neurons that directly regulate the DA neurons as well as a small number of glutamatergic neurons [6]. Thus, the VTA is a highly heterogeneous structure [8], and the complex local and extrinsic regulation of VTA DA output continues to be explored.

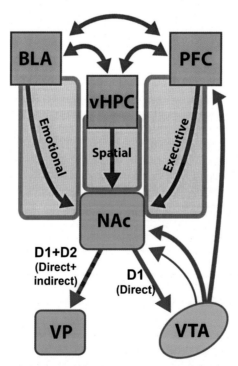

FIG. 1 Schematic of brain reward circuitry implicated in addiction. Dopaminergic (green) and glutamatergic (red) inputs converge on γ-aminobutyric acid (GABA)ergic (blue) medium spiny neurons in the nucleus accumbens (NAc). These inputs coordinate and regulate direct and indirect outputs that differentially contribute to drug-related behaviors. BLA=basolateral amygdala; D1=dopamine type 1 receptor; D2=dopamine type 2 receptor; PFC=prefrontal cortex; vHPC=ventral hippocampus; VP=ventral pallidum; VTA=ventral tegmental area.

Mesocortical DA is thought to be involved in emotional responses and control of cognition [9], but the reciprocal glutamatergic connections of PFC to VTA and NAc make this difficult to parse. Mesolimbic DA has traditionally been the pathway most directly linked to motivated behaviors, and the central role of mesolimbic DA neuron activity in reward-related processes has been well established [10,11]. Mesolimbic DA release exerts its effects primarily through activation of dopamine receptors (DRs) on NAc medium spiny neurons (MSNs). These MSNs are GABAergic, and—as with homologous cells in dorsal striatum—are composed of two largely separate populations that express predominantly either D1- or D2-type DRs [12,13]. D1 MSNs in the dorsal striatum project directly to the globus pallidus interna (GPi) and SN/VTA, inhibiting subsequent GABAergic output from these regions to the thalamus, and thus increasing thalamocortical drive—the direct pathway. Dorsal striatum D2 MSNs project to globus pallidus externa, which itself inhibits STN excitation of GPi and SN/VTA, ultimately reducing thalamocortical drive—the indirect pathway. In the NAc, the pathways are slightly different, and both D1 and D2 NAc MSNs project directly to ventral pallidum, while NAc D1 neurons have additional projections to the VTA and other regions. However, as with the dorsal striatum, the cellular and behavioral sequelae of NAc direct and indirect pathway activity are divergent and often opposite. Because D1 DRs increase responsiveness to glutamatergic excitation while D2 DRs decrease this glutamate excitability, VTA DA release facilitates the direct pathway while putting a brake on the indirect pathway, with the combined effect frequently increasing behavioral drive.

NAc MSNs receive glutamatergic inputs from several cortical and limbic structures, including medial and lateral divisions of the PFC, ventral hippocampus (vHPC), basolateral amygdala (BLA), and medial thalamus [14,15]. Cortical inputs onto the NAc likely regulate goal-directed behaviors, such as the seeking and approaching of substances/activities associated with reward, like palatable food, sexual partners, or drugs [16,17]. Thus, PFC input onto NAc seems to provide "executive control" required for planning actions to obtain rewards. Inputs from vHPC related to affective valence of locations in space can directly determine NAc outputs that influence goal-directed behavior. This applies to both reward- and aversion-driven behaviors, including feeding, context-dependent fear conditioning, and locomotor responses to drugs [18–20]. The amygdala, which is critical for expression of emotion and emotional learning [21], also sends glutamatergic inputs onto NAc MSNs, including D1-type cells [22]. BLA excitation of NAc MSNs facilitates reward-seeking behavior and is sufficient to support positive reinforcement [23,24]. Critically, BLA projects to several other brain regions, and it appears that these other projections are essential for fear- and anxiety-related learning and behaviors, while BLA-NAc projections drive reward and reinforcement [21].

The NAc thus acts as a "Grand Central Station" for factors driving motivated behaviors like drug seeking, integrating glutamatergic inputs that relate planning (PFC inputs), context and experience (HPC inputs), and emotional state (BLA inputs). The major function of mesolimbic DA is to modulate the impact of these excitatory inputs onto NAc, suppressing some and augmenting others, ultimately regulating the integration of incoming cortical/limbic signals and controlling differential thalamocortical outputs that drive behavior.

TRANSCRIPTION FACTORS INVOLVED IN ADDICTION

CREB

Cyclic adenosine monophosphate (cAMP) response element binding (CREB) protein is a ubiquitously expressed transcription factor that plays a critical role throughout the reward circuitry and may underlie long-term neuronal adaptations associated with drug addiction. CREB phosphorylation prevents proteolytic degradation and facilitates DNA binding, and thus tightly regulates transcription of target genes [25,26]. Many drugs of abuse activate calcium- and cAMP-dependent signaling in NAc, including MAP kinase cascades, regulating CREB activation and orchestrating downstream transcription [27,28] (Fig. 2).

The roles of CREB activity in the cellular and behavioral responses to drugs are diverse. Studies using viral overexpression or inhibition of CREB in mouse NAc demonstrate that CREB activity opposes the behavioral effects of drugs, blunting the rewarding effects of opiates or cocaine [29,30]. However, this reduction in the rewarding effects of the drug can actually drive increased self-administration [31], and thus may act as a mechanism for escalation of drug abuse, driving addiction. This hypothesis is corroborated on the cellular level, as chronic cocaine decreases the excitability of NAc MSNs while CREB overexpression increases the excitability of the same cells, perhaps providing a negative feedback response in the reward circuitry that results in behavioral tolerance driving escalation [32]. This CREB-mediated increase in membrane excitability was also reported in PFC pyramidal neurons [33], although the behavioral effects remain to be determined. CREB also enhances NMDA-type glutamate receptor (NMDAR)-mediated synaptic transmission and surface expression of NMDARs in NAc, and mimicking these effects using pharmacological enhancement of NMDAR function in NAc suppresses cocaine sensitivity [34], providing a potential synaptic mechanism for CREB effects on cell function. It is

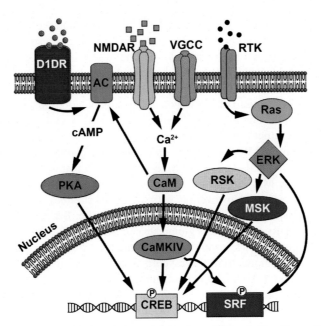

FIG. 2 Common signaling cascades leading to activity-dependent transcription factor (TF) activation. Extracellular signals, such as dopamine acting through the D1-type dopamine receptor (D1DR), and changes in membrane potential can lead to increases in second-messenger (cAMP and Ca^{2+}) or MAP kinase cascade signaling that converge on phosphorylation and activation of CREB and SRF complexes. These bind to specific elements in promoter regions to regulate transcription of a variety of genes, including the *FosB*. AC = adenylyl cyclase; VGCC = voltage gated calcium channel; RTK = receptor tyrosine kinase; ERK = extracellular signal-regulated kinase; PKA = protein kinase A; CaM = calmodulin; RSK = ribosomal S6 kinase; MSK = mitogen and stress-activated kinase; CaMKIV = calcium/calmodulin-dependent protein kinase IV.

unclear if this is associated with a corresponding increase in self-administration. NMDA antagonism reduces cocaine self-administration [35], yet enhancement of NMDAR function facilitates the extinction of cocaine self-administration [36]. However, in clinical trials, it was found that D-cycloserine, a co-agonist of NMDARs, increased craving with no facilitating effect on extinction learning, suggesting that enhanced NMDAR activity in NAc could produce increased cocaine seeking in addicts.

cAMP signaling drives formation of new dendritic spines, and this is mediated by phosphorylation of CREB [37,38]. New, or immature, dendritic spines are associated with glutamatergic synapses that have N-methyl-D-aspartate (NMDA)- but not α-amino-3-hydroxy-5-methyl-4-isoxazolepropionic acid (AMPA)-mediated transmission, often termed "silent synapses" [39]. CREB drives the generation of silent synapses [40], which are critical to the synaptic development associated with enhanced drug seeking underlying addiction [39].

Presumably, the cellular and behavioral effects of CREB arise from its regulation of gene transcription, and indeed considerable progress has been made in identifying gene targets mediating these effects using both candidate [30] and genome-wide [41,42] approaches. In addition, the mechanism of CREB transcriptional regulation crosses over with epigenetic mechanisms discussed in detail below. For example, CREB binding protein (CBP) regulates cocaine sensitivity by the acetylation of histones on the *FosB* promoter [43].

Taken together, these studies indicate that CREB may provide valuable inroads for the treatment of addiction. For example, environmental enrichment leads to decreased CREB levels and reduced cocaine self-administration [44]. Because CREB is a critical mediator of many processes throughout the brain both during development and in adult behaviors unrelated to addiction, direct systemic targeting of CREB is unlikely to be an ideal addiction treatment. However, the search for CREB's target genes and up- and downstream signaling partners may uncover druggable targets that could indeed prove useful in the treatment of substance abuse disorders.

SRF

Serum-response factor (SRF) is a member of the MADS-box (minichromosome maintenance factor 1 (MCM1), Agamous, Deficiens, SRF) family of transcription factors that share a conserved sequence motif and tend to recruit other transcription factors in multiregulatory complexes but have different DNA binding properties and heterodimer partners [45]. SRF binds serum-response elements (SREs) in many genes, including the *c-fos* [46], and DNA binding is dependent on SRF phosphorylation, which is downstream of some of the same signaling cascades that activate CREB [47]. In NAc, SRF is a critical mediator of the rewarding properties of cocaine, at least in part through its regulation of expression of the transcription factor ΔFosB [48], discussed in detail below.

SRF activity is regulated by a number of cofactors, and these interactions appear to be critical for many drug-related behaviors. For instance, inhibiting the cofactor Elk1 impairs cocaine sensitization and CPP and reduces cocaine induction of dendritic spines [49]. In a more complex example, Rho GTPases induce actin polymerization, freeing the SRF cofactor myelin and lymphocyte protein (MAL) from its interaction with unpolymerized actin such that it can stimulate SRF transcriptional regulation [50]. Importantly, the MAL-SRF complex is necessary for cocaine induction of NAc MSN dendritic spines and behavioral sensitization to cocaine [51]. MAL and SRF expression are increased after early withdrawal from cocaine self-administration, and SRF mediates cocaine-induced spine morphogenesis during this same time point, potentially via transcriptional activation of genes such as Arc and Rock1 [51]. Cocaine enhances the binding of the MAL-SRF by activation of Rho GTPases, such as Rap1b, which underlies this plasticity [52]. This suggests that cocaine activation of SRF regulates actin cytoskeletal remodeling underlying new spine formation, and that this may be a critical form of plasticity underlying addiction. However, there is also evidence that spinogenesis is a compensatory mechanism that helps to limit cocaine-self-administration [53], and parsing the causal role of changes in NAc spines in various motivated behaviors remains a challenge for the field.

NF-κB

Nuclear factor-kappa B (NF-κB) is an ubiquitously expressed transcriptional activator of inflammatory signaling that also regulates synaptic plasticity in brain [54]. NF-κB is upregulated in the NAc by chronic cocaine administration in mice [55] and is also elevated in postmortem NAc samples from chronic cocaine and heroin abusers [56,57]. Furthermore, NF-κB is important for the expression of cocaine [58] and morphine reward [59], and is critical for cocaine-mediated induction of dendritic spines in NAc MSNs [58]. These findings suggest that NF-κB or its target genes may represent novel targets for the treatment of addiction. This may be particularly important for chronic intravenous cocaine users who are at high risk of HIV infection, as cocaine enhances HIV replication, primarily by activating NF-κB [60].

MEF2

Multiple myocyte-specific enhancer factor 2 (MEF2) transcription factors (MEF2A-D) complex with cofactors, including histone deacetylases (HDACs) (see below), to activate or repress transcription, primarily of genes involved in cell survival and excitatory synapse regulation [61,62]. Neuronal activity regulates active transcription via MEF2 binding to MEF2 response elements (MREs) located on proximal promoters as well as distal enhancers [63]. Chronic cocaine decreases MEF2A/two-dimensional (2D) activity in NAc via a D1 DR activation and cAMP-dependent inhibition of calcineurin, a calcium-dependent phosphatase that activates MEF2 by dephosphorylating the inhibitory serine 444 [64]. Suppression of MEF2 is critical for cocaine-induced increases in dendritic spines in NAc MSNs, and increased MEF2 activity blocks cocaine-induced increase in spines and inhibits sensitization to chronic cocaine [64]. The activation and repression of MEF2 may be isoform specific, as cocaine also increases the expression of the MEF2C isoform in NAc via mammalian target of rapamycin (mTOR) signaling [65]. In addition, MEF2 expression in hippocampus and amygdala is important for repression of memory consolidation [66,67], perhaps playing a repressive role in the learned behaviors and habits critical in the addiction phenotype. Thus, MEF2 may play a unique role as a negative regulator of structural plasticity that may underlie resilience to addiction.

ΔFosB

ΔFosB is a member of the Fos family of activity-dependent transcription factors. It is a splice variant produced from the *FosB* gene, and the resulting ΔFosB protein lacks two c-terminal degron domains, rendering it exceedingly stable [68]. ΔFosB has a half-life of 8 days in vivo, in part due to its phosphorylation by calmodulin-dependent protein kinase II α (CaMKIIα) [69–71]. Because it is induced throughout the brain by chronic exposure to virtually all drugs of abuse [72], ΔFosB has been proposed as a molecular switch for addiction [73]. The expression of ΔFosB is regulated by both CREB and SRF [74], and its expression is regulated by cocaine via D1-cAMP signaling in a manner similar to regulation of CREB [75]. Indeed, ΔFosB is specifically induced in D1-type MSNs by most drugs of abuse [76], and this may be critical for its cellular and behavioral effects.

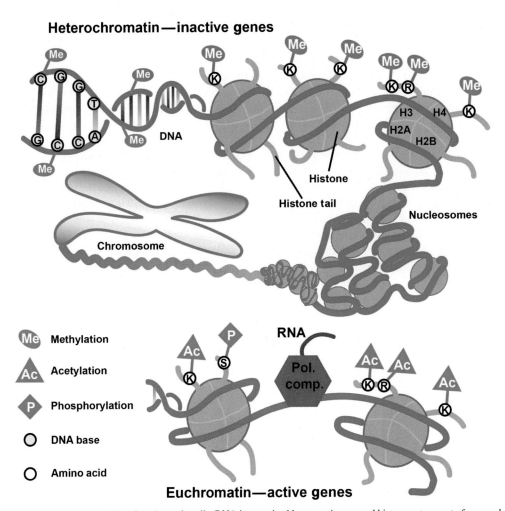

FIG. 3 Mechanisms of epigenetic regulation. In eukaryotic cells, DNA is organized by wrapping around histone octomers to form nucleosomes, which are further organized and condensed to form chromosomes. When a gene is inactive or repressed, DNA is tightly wrapped around histones (heterochromatin, top), due in part to strengthened interactions with methylated lysines (K) and argenines (R) on histone tails, and cytosine (C) bases found next to guanine (G) bases within the DNA strand are more often methylated. When a gene is active, histone tail acetylation reduces the strength of interaction with DNA, allowing for an open conformation (euchromatin, bottom) that permits binding of the RNA polymerase complex (purple) and active transcription.

ΔFosB expression in NAc appears to underlie multiple addiction-related behaviors. Silencing ΔFosB's transcriptional activity in NAc blocks drug reward and decreases drug self-administration [71,77–79] while ΔFosB overexpression in NAc enhances drug reward and increases drug self-administration [71,78,80]. ΔFosB is necessary and sufficient for cocaine induction of immature dendritic spine formation in NAc MSNs [81]. Moreover, ΔFosB overexpression decreases excitatory synaptic strength and increases dendritic spines expressing silent synapses in D1 NAc MSNs, but conversely decreases silent synapses in D2 NAc MSNs [82]. This correlates with behavioral effects, as ΔFosB increases the rewarding effects of cocaine when expressed in D1 but not D2 MSNs [82,83].

ΔFosB expression leads to NAc gene expression patterns related to those induced by chronic cocaine or CREB overexpression [41]. Many of its targets are critical for glutamatergic synapse structure and function, including AMPA receptor subunits [83,84], CaMKIIα [71], and cyclin-dependent kinase 5 (Cdk5) [85]. Thus, in light of its effects on MSN synaptic structure and function, ΔFosB may be critical for mediating the effects of drug exposure on the ability of NAc to integrate glutamatergic inputs from vHPC, PFC, and amygdala. Moreover, ΔFosB plays critical roles in the function of HPC [86] and PFC [87,88], and so may be a master regulator of drug effects on the entire reward circuitry (Fig. 3).

Additional Transcription Factors

There are many additional transcription factor mechanisms that may contribute to drug responses, and the field is expanding rapidly. For example, the transcription factor mothers against DPP homolog 3 (SMAD3) is implicated in cocaine-induced plasticity in NAc MSNs and mediates cocaine seeking after self-administration [89,90], through an activin-SMAD3-brahma-related gene 1 (BRG1) process. Members of the transforming growth factor beta (TGF-B) family, including activin, are ligands for activin receptors, which then bind SMAD3 and BRG1, promoting chromatin remodeling [91]. Cocaine activates activin receptors leading to phosphorylation of SMAD3 promoting its association with BRG1 orchestrating transcriptional changes in NAc that drive cocaine craving after withdrawal [90]. In addition, multiple Nur (nuclear receptor subfamily 4; NR4A) family members, which act as immediate early genes and transcription factors [92], are involved in dopaminergic neurotransmission in the striatum and are dysregulated in addiction, and may underlie a number of addictive phenotypes [93,94]. Finally, NPAS4 is linked to distinct programs of late-phase activity-dependent transcription leading to the enhancement of inhibitory signaling, possibly rewiring the connections between neurons to promote an inhibitory response [95–97]. Critically, NPAS4 is induced in NAc by cocaine and morphine [98–100], making this a likely candidate for the study of novel transcriptional programs that may underlie addiction. All of these regulators of transcription likely influence, and are influenced by, the structure of the genes they regulate. Epigenetic changes to gene structure are emerging as key regulators of drug effects and responses, and may be crucial mediators of transcriptional mechanisms underlying addiction phenotypes.

EPIGENETIC MECHANISMS

Posttranslational Histone Modifications

Many residues in the tails of histone proteins undergo a variety of covalent modifications, resulting in a complex "code" that is thought to control the accessibility of the genome to the transcriptional machinery [101,102]. Histone acetylation negates the positive charge of lysine residues in the histone tail, presumably reducing the strength of interaction with negatively charged DNA and thus increasing transcriptional activation. This process is controlled by histone acetyltransferases (HATs) and HDACs, each of which are made up of multiple enzyme classes whose expression and activity are exquisitely regulated within the brain and the rest of the body [103]. In addition, histone methylation has been associated with both transcriptional activation and repression depending on the particular residue and the extent of methylation [104,105]. Both lysines and arginines can be methylated by several families of histone methyltransferases (HMTs), and this reaction can be reversed by equally diverse histone demethylases. Histone tail modifications also include phosphorylation, ubiquitination, sumoylation, adenosine diphosphate (ADP) ribosylation, among many others [103], whose role in drug addiction remains to be clearly characterized.

Histone acetylation in the brain is regulated by multiple drugs of abuse, and this acetylation and the machinery that controls it have been linked to many of the physiological and behavioral phenotypes found in rodent models of drug abuse [2]. Cocaine induces differential and promoter-specific histone acetylation depending on exposure regimen. For instance, in striatum, acute but not chronic cocaine causes transiently increased H4 acetylation at the *cFos* promoter [106], consistent with cocaine's ability to induce the *cFos* acutely but desensitize it chronically [107]. In keeping with this pattern, H3 acetylation is persistently (at least 1 week after withdrawal) increased by chronic but not acute cocaine at the brain-derived neurotrophic factor *(BDNF)* promoter in NAc [106], and *BDNF* expression is induced by chronic but not acute cocaine [108]. These observed changes are consistent with the patterns of gene induction by cocaine, and a great deal of recent effort has established a causal link between histone acetylation machinery and the cellular and behavioral effects of cocaine, particularly in NAc.

In general, histone acetylation in NAc appears to drive increased responsiveness to the rewarding effects of drugs of abuse. Acute systemic or intra-NAc administration of HDAC inhibitors potentiates CPP and locomotor responses to psychostimulants and to opiates [106,109,110]. In contrast, long-term HDAC inhibition reduces drug responses [111,112], perhaps through homeostatic adaptations to initial enzyme inhibition. In keeping with a positive role for NAc histone acetylation in drug responses, overexpression of HDAC4 or HDAC5 decreases behavioral responses to cocaine [106,110], whereas genetic deletion of HDAC5 hypersensitizes mice to the chronic (but not acute) effects of the drug [110]. In addition, mutant mice with reduced expression of a major brain HAT, CBP, exhibit decreased sensitivity to chronic cocaine [43]. In contrast, withdrawal from chronic ethanol exposure increases HDAC activity and reduces histone acetylation in mouse brain [113], suggesting that specific acetylation effects are drug dependent or are modulated by withdrawal state. These studies imply that histone acetylation may be an avenue for intervention in addiction. However, the crucial role of brain histone acetylation in a variety of other behaviors, including general learning and cognition [114], means that effective therapy is unlikely to be as simple as systemic treatment with HAT inhibitors.

Histone tail methylation is complex, as both lysines and arginines can be mono-, di-, or tri-methylated, and the effects of these modifications on gene expression are dependent on the residue, the number of methyl groups, and the gene.

One well-studied modification in mouse brain is histone 3 lysine 9 dimethylation (H3K9me2), which is reduced in the NAc after chronic cocaine [81], and a genome-wide screen revealed cocaine-dependent increases and decreases in H3K9me2 association with the promoters of numerous genes in this brain region [42]. Cocaine causes a downregulation of HMTs that catalyze H3K9me2 in NAc, including G9a and G9a-like protein (GLP), and inhibition of these enzymes promotes cocaine-induced behaviors [81]. This, in turn, may orchestrate a variety of critical downstream transcriptional effects of cocaine, including regulation of the *FosB* gene: G9a binds to and represses the *FosB* [81] through induction of H3K9me2 on histones associated with the promoter [115], and thus G9a downregulation promotes the accumulation of ΔFosB observed after chronic cocaine. Moreover, in an apparent feedback loop, ΔFosB represses G9a expression and global H3K9me2 in NAc [81], perhaps contributing to the buildup of ΔFosB and subsequent transcriptional changes that occur in NAc with chronic cocaine exposure. Moreover, a mechanism for interregulation of H3K9me2 and histone acetylation may exist, as G9a is induced in NAc upon prolonged HDAC inhibition [116], suggesting that cocaine's ultimate effects on transcription may depend on synergistic alterations of the histone code.

Histone phosphorylation may also be robustly regulated in NAc by cocaine exposure. Cocaine activates the MAP kinase signaling cascade, presumably through D1-type dopamine receptors (D1DRs), which results in phosphorylation of H3 on serine 10 by mitogen- and stress-activated protein kinase 1 (MSK-1), which promotes the binding of CREB and other transcription factors to a number of immediate early genes, including the *c-Fos* [117]. This is further potentiated by cocaine-dependent activation of 32 kDa dopamine and cAMP-regulated phosphoprotein (DARPP-32), which inhibits protein phosphatase 1 preventing the dephosphorylation of H3S10 [118]. This signaling pathway appears to be critical for long-term effects of cocaine exposure, including locomotor sensitization [117], and since many other drugs of abuse also regulate MAP kinase signaling and activate the c-fos expression [119], it is likely to be a key regulator of the rewarding effects of many substances.

DNA methylation

Gene expression is also regulated epigenetically through direct methylation of DNA, wherein a methyl group is transferred from *S*-adenosylmethionine (SAM) to the $5'$ position on the cytosine pyrimidine ring (5-mC), most often in sequences containing a cytosine adjacent to a guanine [120]. These CpG sequences are clustered throughout the genome in "islands," predominantly in the promoter regions of genes, and methylation in these regions is generally associated with gene repression through preventing binding of transcription factors to their target sequences or recruiting corepressor complexes that may include HDACs or HMTs [2]. These effects are often mediated by methyl binding proteins (MBPs), which directly bind to methylated CpGs and recruit corepressor complexes. Mammalian DNA methyltransferases (DNMTs) are divided into the DNMT1 family, which primarily maintain DNA methylation patterns in proliferating cells but can establish some new patterns [121], and the DNMT3 family, which primarily perform novel de novo DNA methylation [122]. DNA demethylation involves a series of chemical reactions that produce multiple intermediaries, sometimes including hydroxymethylcytosine by ten-eleven translocation (TET) enzymes or conversion to thymine, and results in resynthesis of nonmethylated cytosine [123].

In NAc, DNA methylation is dynamically regulated by multiple drugs of abuse depending on method and length of exposure, time of measurement, and gene locus. Acute cocaine increases expression of DNMT3a in NAc while chronic cocaine decreases expression, although DNMT3a mRNA is again increased 1 month after withdrawal from chronic cocaine [124,125]. These effects are contrasted by cell-type-specific studies, which demonstrate that chronic, but not acute, cocaine increases DNMT3a expression in D1 MSNs and not D2 MSNs [126]. Moreover, local inhibition or knockout of DNMT3a in NAc attenuates cocaine reward [124] and prevention of DNMT activity in HPC prior to training impairs acquisition of cocaine CPP [127]. In keeping with this, systemic injection of SAM to boost overall DNA methylation increases sensitization to cocaine [128]. In contrast, systemic methyl supplementation using methionine reverses cocaine CPP, reduces locomotor sensitization, and attenuates drug seeking behavior during reinstatement to self-administration [129,130], suggesting that the effects of DNA methylation may be dependent on the method of manipulation. This is further complicated by the cocaine-specific nature of these effects, as systemic methionine had no effect on morphine or food CPP [130].

MBPs are also dynamically regulated by cocaine. Extended cocaine self-administration in rats induces the methyl CpG binding protein 2 (MeCP2) throughout the striatum, and striatal knockdown of MeCP2 reduces cocaine seeking [131]. Cocaine also regulates the inhibitory phosphorylation of MeCP2 after even acute exposure in both mice and rats [132,133], implying a potential negative feedback that can only be overcome by chronic exposure to the drug. Finally, removal of cytosine methylation may also be an important player in cocaine effects, as Tet1 expression is downregulated in NAc after chronic cocaine in both mice and human addicts [134]. Further, inhibition of Tet1 in NAc increases cocaine

CPP while overexpression inhibits the same behavior [134], demonstrating a key role for reversal of CpG methylation in behavioral responses to cocaine.

Determining the specific gene loci differentially methylated throughout the reward circuitry, and any subsequent behaviorally relevant effects on expression, is a challenge the field is only now beginning to address. Subunits of protein phosphatase 1, the MeCP2-associated protein Cdk5, and BDNF have all been linked to altered expression and promoter methylation after cocaine exposure [125,135,136], but causal links between the gene-specific methylation and cocaine-related behaviors have yet to be established. Gene-specific tools for manipulation of epigenetic marks are now becoming available, and such advances will be critical for determining that role of discreet DNA methylation in drug addiction.

CONCLUSIONS

Here, we present some of the most well-studied and clearly defined transcriptional and epigenetic mechanisms by which drugs of abuse modulate gene transcription within the reward circuitry to alter cellular and behavioral responses underlying some aspects of drug dependence. This overview is by no means comprehensive, and many additional mechanisms have already been described and/or are sure to be uncovered in future experiments. However, we have attempted to highlight examples that demonstrate the interconnected signaling and the cross-talk between mechanisms that drive drug-induced alterations in gene expression. For instance, convergence of cocaine-induced CREB and SRF signaling on the *FosB* gene regulates the expression of ΔFosB, which then orchestrates the expression of additional genes critical for drug-associated behaviors. Alternatively, drug-dependent reduction in DNA methylation can reduce the recruitment of HDAC complexes, increasing histone acetylation and permitting increased target gene expression. And, these particular mechanisms are tied together by CREB itself, which recruits the HAT CBP to drive histone acetylation and ΔFosB expression.

It is therefore hoped that as research progresses, new ties will be found between these mechanisms and others yet to be uncovered, eventually making clear the complex effects of chronic drug exposure on reward system gene transcription. A comprehensive picture of these processes may allow us to target region-, cell type-, and even circuit-specific mechanisms for novel forms of therapeutic intervention or prevention of drug addiction.

REFERENCES

[1] Singer BF, et al. Transient viral-mediated overexpression of alpha-calcium/calmodulin-dependent protein kinase II in the nucleus accumbens shell leads to long-lasting functional upregulation of alpha-amino-3-hydroxyl-5-methyl-4-isoxazole-propionate receptors: dopamine type-1 receptor and protein kinase a dependence. Eur J Neurosci 2010;31(7):1243–51.

[2] Robison AJ, Nestler EJ. Transcriptional and epigenetic mechanisms of addiction. Nat Rev Neurosci 2011;12(11):623–37.

[3] Wolf ME. The Bermuda triangle of cocaine-induced neuroadaptations. Trends Neurosci 2010;33(9):391–8.

[4] Tzschentke TM. C.P.P. REVIEW ON. Measuring reward with the conditioned place preference (CPP) paradigm: update of the last decade. Addict Biol 2007;12(3–4):227–462.

[5] Lüscher C, Malenka RC. Drug-evoked synaptic plasticity in addiction: from molecular changes to circuit remodeling. Neuron 2011;69(4):650–63.

[6] Nair-Roberts RG, et al. Stereological estimates of dopaminergic, GABAergic and glutamatergic neurons in the ventral tegmental area, substantia nigra and retrorubral field in the rat. Neuroscience 2008;152(4):1024–31.

[7] Bariselli S, et al. Ventral tegmental area subcircuits process rewarding and aversive experiences. J Neurochem 2016;139(6):1071–80.

[8] Margolis EB, et al. Identification of rat ventral tegmental area GABAergic neurons. PLoS One 2012;7(7).

[9] Nestler EJ, et al. Molecular neuropharmacology : a foundation for clinical neuroscience. 3rd ed. Columbus, OH: McGraw-Hill Education; 2015. p. xiv. 528 pp.

[10] Salamone JD, et al. Mesolimbic dopamine and the regulation of motivated behavior. Curr Top Behav Neurosci 2016;27:231–57.

[11] Thomas MJ, Kalivas PW, Shaham Y. Neuroplasticity in the mesolimbic dopamine system and cocaine addiction. Br J Pharmacol 2008;154 (2):327–42.

[12] Lobo MK. Molecular profiling of striatonigral and striatopallidal medium spiny neurons past, present, and future. Int Rev Neurobiol 2009;89:1–35.

[13] Surmeier DJ, et al. D1 and D2 dopamine-receptor modulation of striatal glutamatergic signaling in striatal medium spiny neurons. Trends Neurosci 2007;30(5):228–35.

[14] Sesack SR, Grace AA. Cortico-basal ganglia reward network: microcircuitry. Neuropsychopharmacology 2010;35(1):27–47.

[15] Floresco SB. The nucleus accumbens: an interface between cognition, emotion, and action. Annu Rev Psychol 2015;66:25–52.

[16] Kalivas PW, Volkow N, Seamans J. Unmanageable motivation in addiction: a pathology in prefrontal-accumbens glutamate transmission. Neuron 2005;45(5):647–50.

[17] Gruber AJ, Hussain RJ, O'Donnell P. The nucleus accumbens: a switchboard for goal-directed behaviors. PLoS One 2009;4(4).

[18] Kanoski SE, Grill HJ. Hippocampus contributions to food intake control: mnemonic, neuroanatomical, and endocrine mechanisms. Biol Psychiatry 2015.

[19] Fanselow MS. Contextual fear, gestalt memories, and the hippocampus. Behav Brain Res 2000;110(1–2):73–81.

[20] Vezina P, et al. Environment-specific cross-sensitization between the locomotor activating effects of morphine and amphetamine. Pharmacol Biochem Behav 1989;32(2):581–4.

[21] Janak PH, Tye KM. From circuits to behaviour in the amygdala. Nature 2015;517(7534):284–92.

[22] Charara A, Grace AA. Dopamine receptor subtypes selectively modulate excitatory afferents from the hippocampus and amygdala to rat nucleus accumbens neurons. Neuropsychopharmacology 2003;28(8):1412–21.

[23] Ambroggi F, et al. Basolateral amygdala neurons facilitate reward-seeking behavior by exciting nucleus accumbens neurons. Neuron 2008;59(4):648–61.

[24] Stuber GD, et al. Excitatory transmission from the amygdala to nucleus accumbens facilitates reward seeking. Nature 2011;475(7356):377–80.

[25] Bito H, Deisseroth K, Tsien RW. CREB phosphorylation and dephosphorylation: a Ca(2+)- and stimulus duration-dependent switch for hippocampal gene expression. Cell 1996;87(7):1203–14.

[26] Mayr B, Montminy M. Transcriptional regulation by the phosphorylation-dependent factor CREB. Nat Rev Mol Cell Biol 2001;2(8):599–609.

[27] Shaw-Lutchman TZ, et al. Regional and cellular mapping of cAMP response element-mediated transcription during naltrexone-precipitated morphine withdrawal. J Neurosci 2002;22(9):3663–72.

[28] Shaw-Lutchman TZ, et al. Regulation of CRE-mediated transcription in mouse brain by amphetamine. Synapse 2003;48(1):10–7.

[29] Barrot M, et al. CREB activity in the nucleus accumbens shell controls gating of behavioral responses to emotional stimuli. Proc Natl Acad Sci U S A 2002;99(17):11435–40.

[30] Carlezon Jr. WA, et al. Regulation of cocaine reward by CREB. Science 1998;282(5397):2272–5.

[31] Larson EB, et al. Over-expression of CREB in the nucleus accumbens shell increases cocaine reinforcement in self-administering rats. J Neurosci 2011;31(45):16447–57.

[32] Dong Y, et al. CREB modulates excitability of nucleus accumbens neurons. Nat Neurosci 2006;9(4):475–7.

[33] Dong Y, et al. Cocaine-induced plasticity of intrinsic membrane properties in prefrontal cortex pyramidal neurons: adaptations in potassium currents. J Neurosci 2005;25(4):936–40.

[34] Huang YH, et al. CREB modulates the functional output of nucleus Accumbens neurons: a critical role of N-methyl-d-aspartate glutamate receptor (NMDAR) receptors. J Biol Chem 2008;283(5):2751–60.

[35] Pulvirenti L, Maldonado-Lopez R, Koob GF. NMDA receptors in the nucleus accumbens modulate intravenous cocaine but not heroin self-administration in the rat. Brain Res 1992;594(2):327–30.

[36] Thanos PK, et al. D-cycloserine facilitates extinction of cocaine self-administration in rats. Synapse 2011;65(9):938–44.

[37] Murphy DD, Segal M. Morphological plasticity of dendritic spines in central neurons is mediated by activation of cAMP response element binding protein. Proc Natl Acad Sci 1997;94(4):1482–7.

[38] Segal M, Murphy DD. CREB activation mediates plasticity in cultured hippocampal neurons. Neural Plast 1998;6(3).

[39] Huang YH, Schluter OM, Dong Y. Silent synapses speak up: updates of the neural rejuvenation hypothesis of drug addiction. Neuroscientist 2015;21(5):451–9.

[40] Marie H, et al. Generation of silent synapses by acute in vivo expression of CaMKIV and CREB. Neuron 2005;45(5):741–52.

[41] McClung CA, Nestler EJ. Regulation of gene expression and cocaine reward by CREB and DeltaFosB. Nat Neurosci 2003;6(11):1208–15.

[42] Renthal W, et al. Genome-wide analysis of chromatin regulation by cocaine reveals a role for sirtuins. Neuron 2009;62(3):335–48.

[43] Levine AA, et al. CREB-binding protein controls response to cocaine by acetylating histones at the fosB promoter in the mouse striatum. Proc Natl Acad Sci U S A 2005;102(52):19186–91.

[44] Green TA, et al. Environmental enrichment produces a behavioral phenotype mediated by low cyclic adenosine monophosphate response element binding (CREB) activity in the nucleus Accumbens. Biol Psychiatry 2010;67(1):28–35.

[45] Shore P, Sharrocks AD. The MADS-box family of transcription factors. Eur J Biochem 1995;229(1):1–13.

[46] Treisman R. Identification and purification of a polypeptide that binds to the c-fos serum response element. EMBO J 1987;6(9):2711–7.

[47] Xia Z, et al. Calcium influx via the NMDA receptor induces immediate early gene transcription by a MAP kinase/ERK-dependent mechanism. J Neurosci 1996;16(17):5425–36.

[48] Vialou V, et al. Serum response factor and cAMP response element binding protein are both required for cocaine induction of {Delta}FosB. J Neurosci 2012;32(22):7577–84.

[49] Besnard A, et al. Alterations of molecular and behavioral responses to cocaine by selective inhibition of Elk-1 phosphorylation. J Neurosci 2011;31(40):14296–307.

[50] Miralles F, et al. Actin dynamics control SRF activity by regulation of its coactivator MAL. Cell 2003;113(3):329–42.

[51] Cahill ME, et al. The dendritic spine morphogenic effects of repeated cocaine use occur through the regulation of serum response factor signaling. Mol Psychiatry 2017;.

[52] Cahill ME, et al. Bidirectional synaptic structural plasticity after chronic cocaine administration occurs through Rap1 small GTPase signaling. Neuron 2016;89(3):566–82.

[53] Gourley SL, Taylor JR. Going and stopping: dichotomies in behavioral control by the prefrontal cortex. Nat Neurosci 2016;19(6):656–64.

[54] Engelmann C, Haenold R. Transcriptional control of synaptic plasticity by transcription factor NF-kappaB. Neural Plast 2016;2016. p. 7027949.

[55] Ang E, et al. Induction of nuclear factor-κB in nucleus accumbens by chronic cocaine administration. J Neurochem 2001;79(1):221–4.

[56] Bannon MJ, et al. A molecular profile of cocaine abuse includes the differential expression of genes that regulate transcription, chromatin, and dopamine cell phenotype. Neuropsychopharmacology 2014;39(9):2191–9.

[57] Levran O, et al. Synaptic plasticity and signal transduction gene polymorphisms and vulnerability to drug addictions in populations of European or African ancestry. CNS Neurosci Ther 2015;21(11):898–904.

[58] Russo SJ, et al. Nuclear factor κB signaling regulates neuronal morphology and cocaine reward. J Neurosci 2009;29(11):3529–37.

[59] Zhang X, et al. Involvement of p38/NF-κB signaling pathway in the nucleus accumbens in the rewarding effects of morphine in rats. Behav Brain Res 2011;218(1):184–9.

[60] Sahu G, et al. Cocaine promotes both initiation and elongation phase of HIV-1 transcription by activating NF-κB and MSK1 and inducing selective epigenetic modifications at HIV-1 LTR. Virology 2015;483:185–202.

[61] Flavell SW, et al. Activity-dependent regulation of MEF2 transcription factors suppresses excitatory synapse number. Science 2006;311 (5763):1008–12.

[62] Mao Z, et al. Neuronal activity-dependent cell survival mediated by transcription factor MEF2. Science 1999;286(5440):785–90.

[63] Flavell SW, et al. Genome-wide analysis of MEF2 transcriptional program reveals synaptic target genes and neuronal activity-dependent polyadenylation site selection. Neuron 2008;60(6):1022–38.

[64] Pulipparacharuvil S, et al. Cocaine regulates MEF2 to control synaptic and behavioral plasticity. Neuron 2008;59(4):621–33.

[65] Dietrich J-B, et al. Cocaine induces the expression of MEF2C transcription factor in rat striatum through activation of SIK1 and phosphorylation of the histone deacetylase HDAC5. Synapse 2012;66(1):61–70.

[66] Rashid AJ, Cole CJ, Josselyn SA. Emerging roles for MEF2 transcription factors in memory. Genes Brain Behav 2014;13(1):118–25.

[67] Cole CJ, et al. MEF2 negatively regulates learning-induced structural plasticity and memory formation. Nat Neurosci 2012;15(9):1255–64.

[68] Carle TL, et al. Proteasome-dependent and -independent mechanisms for FosB destabilization: Identification of FosB degron domains and implications for DeltaFosB stability. Eur J Neurosci 2007;25(10):3009–19.

[69] Ulery PG, Rudenko G, Nestler EJ. Regulation of ΔFosB stability by phosphorylation. J Neurosci 2006;26(19):5131–42.

[70] Ulery-Reynolds PG, et al. Phosphorylation of DeltaFosB mediates its stability in vivo. Neuroscience 2009;158(2):369–72.

[71] Robison AJ, et al. Behavioral and structural responses to chronic cocaine require a feedforward loop involving DeltaFosB and calcium/calmodulin-dependent protein kinase II in the nucleus Accumbens Shell. J Neurosci 2013;33(10):4295–307.

[72] Perrotti LI, et al. Distinct patterns of DeltaFosB induction in brain by drugs of abuse. Synapse 2008;62(5):358–69.

[73] Nestler EJ. Transcriptional mechanisms of addiction: role of ΔFosB. Philos Trans R Soc Lond B Biol Sci 2008;363:3245–55.

[74] Vialou V, et al. Serum response factor and cAMP response element binding protein are both required for cocaine induction of DeltaFosB. J Neurosci 2012;32(22):7577–84.

[75] Nye HE, et al. Pharmacological studies of the regulation of chronic FOS-related antigen induction by cocaine in the striatum and nucleus accumbens. J Pharmacol Exp Ther 1995;275(3):1671–80.

[76] Lobo MK, et al. DeltaFosB induction in striatal medium spiny neuron subtypes in response to chronic pharmacological, emotional, and optogenetic stimuli. J Neurosci 2013;33(47):18381–95.

[77] Kelz MB, et al. Expression of the transcription factor [Delta]FosB in the brain controls sensitivity to cocaine. Nature 1999;401(6750):272–6.

[78] Zachariou V, et al. An essential role for [Delta]FosB in the nucleus accumbens in morphine action. Nat Neurosci 2006;9(2):205–11.

[79] Muschamp JW, et al. DeltaFosB enhances the rewarding effects of cocaine while reducing the pro-depressive effects of the kappa-opioid receptor agonist U50488. Biol Psychiatry 2012;71(1):44–50.

[80] Colby CR, et al. Striatal cell type-specific overexpression of ΔFosB enhances incentive for cocaine. J Neurosci 2003;23(6):2488–93.

[81] Maze I, et al. Essential role of the histone methyltransferase G9a in cocaine-induced plasticity. Science 2010;327(5962):213–6.

[82] Grueter BA, et al. FosB differentially modulates nucleus accumbens direct and indirect pathway function. Proc Natl Acad Sci U S A 2013;110 (5):1923–8.

[83] Kelz MB, et al. Expression of the transcription factor deltaFosB in the brain controls sensitivity to cocaine. Nature 1999;401(6750):272–6.

[84] Vialou V, et al. DeltaFosB in brain reward circuits mediates resilience to stress and antidepressant responses. Nat Neurosci 2010;13(6):745–52.

[85] Chen J, et al. Induction of cyclin-dependent kinase 5 in the hippocampus by chronic electroconvulsive seizures: role of [Delta]FosB. J Neurosci 2000;20(24):8965–71.

[86] Eagle AL, et al. Experience-dependent induction of hippocampal DeltaFosB controls learning. J Neurosci 2015;35(40):13773–83.

[87] Dietz DM, et al. DeltaFosB induction in prefrontal cortex by antipsychotic drugs is associated with negative behavioral outcomes. Neuropsychopharmacology 2014;39(3):538–44.

[88] Vialou V, et al. Prefrontal cortical circuit for depression- and anxiety-related behaviors mediated by cholecystokinin: role of DeltaFosB. J Neurosci 2014;34(11):3878–87.

[89] Gancarz AM, et al. Activin receptor signaling regulates cocaine-primed behavioral and morphological plasticity. Nat Neurosci 2015;18(7):959–61.

[90] Wang Z-J, et al. BRG1 in the nucleus Accumbens regulates cocaine-seeking behavior. Biol Psychiatry 2016;80(9):652–60.

[91] Trotter KW, Archer TK. The BRG1 transcriptional coregulator. Nucl Recept Signal 2008;6:.

[92] Maxwell MA, Muscat GEO. The NR4A subgroup: Immediate early response genes with pleiotropic physiological roles. Nucl Recept Signal 2006;4.

[93] Bannon MJ, et al. Decreased expression of the transcription factor NURR1 in dopamine neurons of cocaine abusers. Proc Natl Acad Sci 2002;99 (9):6382–5.

[94] Campos-Melo D, et al. Nur transcription factors in stress and addiction. Front Mol Neurosci 2013;6:44.

[95] Lin Y, et al. Activity-dependent regulation of inhibitory synapse development by Npas4. Nature 2008;455(7217):1198–204.

[96] Bloodgood BL, et al. The activity-dependent transcription factor NPAS4 regulates domain-specific inhibition. Nature 2013;503(7474):121–5.

[97] Spiegel I, et al. Npas4 regulates excitatory-inhibitory balance within neural circuits through cell-type-specific gene programs. Cell 2014;157 (5):1216–29.

[98] Guo M-L, et al. Upregulation of Npas4 protein expression by chronic administration of amphetamine in rat nucleus accumbens in vivo. Neurosci Lett 2012;528(2):210–4.

[99] Martin TA, et al. Methamphetamine causes differential alterations in gene expression and patterns of histone acetylation/Hypoacetylation in the rat nucleus Accumbens. PLoS One 2012;7(3).

[100] Piechota M, et al. The dissection of transcriptional modules regulated by various drugs of abuse in the mouse striatum. Genome Biol 2010;11 (5):R48.

[101] Jenuwein T, Allis CD. Translating the histone code. Science 2001;293(5532):1074–80.

[102] LaPlant Q, Nestler EJ. CRACKing the histone code: cocaine's effects on chromatin structure and function. Horm Behav 2011;59(3):321–30.

[103] Borrelli E, et al. Decoding the epigenetic language of neuronal plasticity. Neuron 2008;60(6):961–74.

[104] Maze I, Nestler EJ. The epigenetic landscape of addiction. Ann N Y Acad Sci 2011;1216:99–113.

[105] Su IH, Tarakhovsky A. Lysine methylation and 'signaling memory. Curr Opin Immunol 2006;18(2):152–7.

[106] Kumar A, et al. Chromatin remodeling is a key mechanism underlying cocaine-induced plasticity in striatum. Neuron 2005;48(2):303–14.

[107] Renthal W, et al. Delta FosB mediates epigenetic desensitization of the c-fos gene after chronic amphetamine exposure. J Neurosci 2008;28 (29):7344–9.

[108] Li X, Wolf ME. Multiple faces of BDNF in cocaine addiction. Behav Brain Res 2015;279:240–54.

[109] McQuown SC, Wood MA. Epigenetic regulation in substance use disorders. Curr Psychiatry Rep 2010;12(2):145–53.

[110] Renthal W, et al. Histone deacetylase 5 epigenetically controls behavioral adaptations to chronic emotional stimuli. Neuron 2007;56(3):517–29.

[111] Romieu P, et al. Histone deacetylase inhibitors decrease cocaine but not sucrose self-administration in rats. J Neurosci 2008;28(38):9342–8.

[112] Kim WY, Kim S, Kim JH. Chronic microinjection of valproic acid into the nucleus accumbens attenuates amphetamine-induced locomotor activity. Neurosci Lett 2008;432(1):54–7.

[113] Pandey SC, et al. Brain chromatin remodeling: a novel mechanism of alcoholism. J Neurosci 2008;28(14):3729–37.

[114] Jarome TJ, Lubin FD. Epigenetic mechanisms of memory formation and reconsolidation. Neurobiol Learn Mem 2014;115:116–27.

[115] Heller EA, et al. Locus-specific epigenetic remodeling controls addiction- and depression-related behaviors. Nat Neurosci 2014;17(12):1720–7.

[116] Kennedy PJ, et al. Class I HDAC inhibition blocks cocaine-induced plasticity by targeted changes in histone methylation. Nat Neurosci 2013;16 (4):434–40.

[117] Brami-Cherrier K, et al. Role of the ERK/MSK1 signalling pathway in chromatin remodelling and brain responses to drugs of abuse. J Neurochem 2009;108(6):1323–35.

[118] Stipanovich A, et al. A phosphatase cascade by which rewarding stimuli control nucleosomal response. Nature 2008;453(7197):879–84.

[119] Sun WL, Quizon PM, Zhu J. Molecular mechanism: ERK signaling, drug addiction, and behavioral effects. Prog Mol Biol Transl Sci 2016;137:1–40.

[120] Klose RJ, Bird AP. Genomic DNA methylation: the mark and its mediators. Trends Biochem Sci 2006;31(2):89–97.

[121] Bestor TH. The DNA methyltransferases of mammals. Hum Mol Genet 2000;9(16):2395–402.

[122] Okano M, et al. DNA methyltransferases Dnmt3a and Dnmt3b are essential for de novo methylation and mammalian development. Cell 1999;99 (3):247–57.

[123] Day JJ, Sweatt JD. DNA methylation and memory formation. Nat Neurosci 2010;13(11):1319–23.

[124] LaPlant Q, et al. Dnmt3a regulates emotional behavior and spine plasticity in the nucleus accumbens. Nat Neurosci 2010;13(9):1137–43.

[125] Anier K, et al. DNA methylation regulates cocaine-induced behavioral sensitization in mice. Neuropsychopharmacology 2010;35(12):2450–61.

[126] Heiman M, et al. A translational profiling approach for the molecular characterization of CNS cell types. Cell 2008;135(4):738–48.

[127] Han J, et al. Effect of 5-aza-2-deoxycytidine microinjecting into hippocampus and prelimbic cortex on acquisition and retrieval of cocaine-induced place preference in C57BL/6 mice. Eur J Pharmacol 2010;642(1–3):93–8.

[128] Anier K, Zharkovsky A, Kalda A. S-adenosylmethionine modifies cocaine-induced DNA methylation and increases locomotor sensitization in mice. Int J Neuropsychopharmacol 2013;16(9):2053–66.

[129] Wright KN, et al. Methyl supplementation attenuates cocaine-seeking behaviors and cocaine-induced c-Fos activation in a DNA methylation-dependent manner. J Neurosci 2015;35(23):8948–58.

[130] Tian W, et al. Reversal of cocaine-conditioned place preference through methyl supplementation in mice: Altering global DNA methylation in the prefrontal cortex. PLoS One 2012;7(3).

[131] Im HI, et al. MeCP2 controls BDNF expression and cocaine intake through homeostatic interactions with microRNA-212. Nat Neurosci 2010;13 (9):1120–7.

[132] Mao LM, et al. Cocaine increases phosphorylation of MeCP2 in the rat striatum in vivo: a differential role of NMDA receptors. Neurochem Int 2011;59(5):610–7.

[133] Deng JV, et al. MeCP2 in the nucleus accumbens contributes to neural and behavioral responses to psychostimulants. Nat Neurosci 2010;13 (9):1128–36.

[134] Feng J, et al. Role of Tet1 and 5-hydroxymethylcytosine in cocaine action. Nat Neurosci 2015;18(4):536–44.

[135] Carouge D, et al. CDKL5 is a brain MeCP2 target gene regulated by DNA methylation. Neurobiol Dis 2010;38(3):414–24.

[136] Tian W, et al. Demethylation of c-MYB binding site mediates upregulation of Bdnf IV in cocaine-conditioned place preference. Sci Rep 2016;6:22087.

Chapter 4

Neural Circuit Plasticity in Addiction

Eden Anderson and Matthew Hearing

Department of Biomedical Sciences, Marquette University, Milwaukee, WI, United States

INTRODUCTION

Theories in neuropsychology increasingly define addiction as a progressive disorder that involves a transition from social use into a compulsive relapsing state. This transition in behavior results from genetic, developmental, and sociological vulnerabilities, that are compounded by pharmacodynamically induced adaptations in neurochemistry and cell physiology within the mesocorticolimbic dopamine (DA) system (i.e., the "reward circuit") responsible for reinforcement, reward, and pleasure [1–6]. While these adaptations are often transient during early intermittent use, repeated drug exposure eventually leads to more enduring changes in the synaptic physiology of brain circuits regulating decision making, learning, and responsivity to motivationally relevant environmental stimuli that strengthen drug-associated behaviors [7,8]. With chronic drug use, this may ultimately reduce the function of neural circuits responsible for cognitive regulation of decision making while increasing activation of neural networks responsible for habit-like behavior, leading to uncontrollable drug use [9–11]. This chapter will describe circuit-based neural plasticity believed to occur throughout the stages of addiction. Examination of drug-induced plasticity will primarily revolve around changes in three major reward-circuit substrates—the ventral tegmental area (VTA), nucleus accumbens (NAc), and prefrontal cortex (PFC)—and highlight when possible, relevant pathway- and cell-type specific adaptations.

VENTRAL TEGMENTAL AREA

A large body of work has established that despite their very different pharmacodynamic profiles, the rewarding properties of addictive drugs, as well as learning behaviors that result in drug consumption and development of conditioned associations between drugs and craving-inducing stimuli, depend on their ability to increase DA in brain regions like the NAc, despite their very different pharmacodynamic profiles [12–15] but see Badiani et al. [16].

Psychostimulants increase DA by blocking reuptake or inverting DA transport [17], whereas opiates indirectly activate dopaminergic neurons in the VTA through inhibition of local γ-aminobutyric acid (GABA) interneurons [18,19]. Mechanisms by which other drugs of abuse increase DA in the NAc are less clear. Nicotine appears to augment NAc DA transmission through a number of mechanisms, including direct activation of DA neurons, as well as indirectly through actions on VTA glutamate terminals and GABA neurons [20]. Ethanol on the other hand may be mediated through ethanol metabolites or indirect actions on GABA signaling [21–23].

Dopaminergic innervation of forebrain regions such as the NAc, amygdala, hippocampus, ventral pallidum (VP), and frontal cortex, comprise in part, the mesocorticolimbic system, and are critical constituents of the brain reward circuitry that mediate responses to rewards and reward-predicting stimuli (Fig. 1) [7,24–28]. Accordingly, leading addiction theories posit that the development of addiction begins with drug-induced adaptations in the VTA that alter DA transmission in downstream brain regions [29–32]. Overtime, DA can influence neural circuit modifications that promote hierarchical transfer of drug-related information from the VTA to these brain regions to develop addiction-related behavior and form links between the drug and environmental cues present during drug use that can trigger feelings of craving that promote relapse [24,28,29,32].

Structure, Function, and Anatomical Connectivity

The VTA and substantia nigra (SN) comprise two major subregions of the ventral midbrain (VM). Although originally perceived as a dopaminergic structure, the VTA is in fact a heterogeneous region consisting of DA (~65%), GABA

Neural Mechanisms of Addiction. https://doi.org/10.1016/B978-0-12-812202-0.00004-X
Copyright © 2019 Elsevier Inc. All rights reserved.

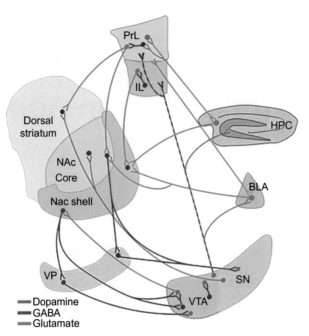

FIG. 1 Relevant circuitry regulating reward, decision-making, and motivated behavior for dopaminergic (blue) projections, GABAergic (red) projections, and glutamate (green) projections. Abbreviations: prelimbic cortex (PrL); infralimbic cortex (ILC); hippocampus (HPC); dorsal striatum (DS); nucleus accumbens (NAc); ventral pallidum (VP); basolateral amygdala (BLA); substantia nigra (SN); ventral tegmental area (VTA).

(30%), and glutamate neurons ($\sim$5%) [19,33–36]. Although still a debated and developing area of research, a large body of evidence indicates that in addition to cell-type differences based on neurotransmitter, DA neurons themselves display a large degree of heterogeneity—differing in their molecular and physiological properties, as well as their responses to rewarding and aversive stimuli—largely based on the downstream target of a given DA neuron [37–40].

Output of VTA DA neurons is dependent on a balance of excitatory and inhibitory input [33,41]. Glutamatergic afferents arising from a number of cortical and subcortical structures play an important role in the regulation of cellular activity and function of the VTA [42]. Negative autocrine input arises from presynaptic D2 receptors localized to DA axon terminals, as well as activation of D2 receptors localized on the soma and dendrites of DA neurons [43]—both of which serve as an autoreceptors to reduce the transmitter release and firing rates, respectively [44,45]. GABA is also an important source of inhibitory signaling in the VTA that arises from the VP, NAc, and rostromedial tegmental nucleus (RMTg; i.e., tail of the VTA [46,47]). Input from the NAc synapses on DA and local GABA neurons and exerts their effects through activation of GABA$_B$ and GABA$_A$ receptors, respectively [48]. This GABA$_B$-dependent signaling plays an important role in behavioral responses to cocaine, but not morphine [48]. DA neuron spontaneous activity and responsivity to drugs of abuse also appears to be regulated by direct GABAergic inputs arising from the RMTg [49,50]. This input is tightly regulated by glutamatergic input arising from the lateral habenula—and plays a role in aversion learning [51,52].

Excitatory Plasticity

Timeline of VTA Excitatory Plasticity

Induction of behaviorally relevant burst firing of VTA DA neurons largely relies on glutamatergic inputs [42]. Glutamate transmission is also critical to the rewarding properties and behavioral effects of drugs of abuse [42,53,54]. The first direct demonstration of synaptic plasticity induced by in vivo drug exposure came from Ungless et al. [55], showing that a single cocaine injection augmented glutamatergic synaptic strength [as measured by an increase in the AMPA (α-amino-3-hydroxy-5-methyl-4-isoxazole propionic acid) receptor (AMPAR) to NMDA (N-methyl-D-aspartate) receptor (NMDAR) ratio; Box 1] up to 5, but not 10 days after drug exposure (Fig. 2). Subsequent examination showed this adaptation occurs as early as 3 h following drug exposure [56]. Interestingly, while experimenter-administered cocaine, amphetamine [57], ethanol [58], and nicotine [59] all enhance the strength of DA neuron glutamate synapses, studies employing repeated cocaine and nicotine administration suggest that increasing the number of drug exposures does not extend the duration of these changes, as an increase in AMPA/NMDA ratio is no longer present 10 days

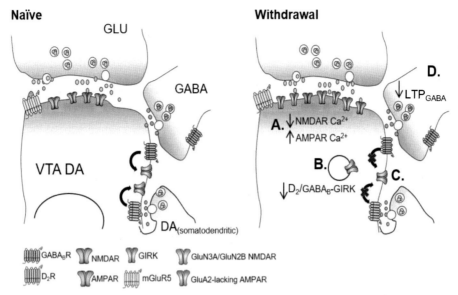

FIG. 2 Effects of acute and repeated drug exposure on VTA DA neuron excitatory and inhibitory signaling during withdrawal. During early withdrawal (1–7 days) VTA DA neuron excitatory synapses show increased AMPAR/NMDAR ratios due to (A) increased insertion of Ca^{2+}-permeable GluA2-lacking AMPARs and Ca^{2+}-impermeable GluN3a/GluN2B NMDARs, leading to increased burst firing of DA neurons. Parallel reductions in inhibitory include, (B) reduced GIRK channel surface expression, (C) reduced $GABA_B/D_2$ receptor-mediating currents, and (D) impaired long-term potentiation (LTP) at GABAergic synapses, contributing to increased tonic firing and membrane excitability of DA neurons.

BOX 1 Measuring excitatory plasticity

Long-Term Potentiation/Long-Term Depression

Most of the mechanistic work on long-term synaptic plasticity in the brain over the past few decades has focused on forms of long-term potentiation (LTP) and long-term depression (LTD). LTP and LTD refers to the persistent potentiation or suppression of postsynaptic responses (e.g., potentials or currents) to a presynaptic action potential, respectively [54]. To be considered long term, changes must persist for at least 1 h following induction and can occur at excitatory or inhibitory synapses. Functionally, LTP and LTD plays an important role in fine-tuning neural circuits, and have been posited as cellular correlates of learning and memory. Experimentally, modifications in strength are produced by electrically or optically stimulating afferent terminals at high frequencies (e.g., 100 Hz) to achieve LTP, and lower frequencies (1–10 Hz) to promote LTD. Importantly, the frequency can determine the mechanism by which plasticity is induced. For example, stimulating at 1 Hz leads to NMDA receptor (NMDAR)-dependent LTD, while stimulating at 10–13 Hz, promotes an mGluR5-dependent LTD [54,193].

Postsynaptic Plasticity

Ex vivo whole-cell patch clamp electrophysiology is a commonly used tool to assess pre- and postsynaptic modifications in excitatory synaptic strength following exposure to drugs of abuse. Assessments of synaptic strength often begin by measuring synaptic currents mediated predominantly by AMPA- and NMDA-type glutamate receptors. This is done by electrically or optically evoking glutamate release from presynaptic terminals that binds to and activates postsynaptic receptors—causing an influx of ions through AMPA receptors (AMPARs) and NMDARs that can be measured using a recording electrode as excitatory postsynaptic currents (EPSCs). Similar approaches can be used to evaluate alterations in inhibitory synaptic signaling by measuring alterations in evoked inhibitory postsynaptic currents (IPSCs) or activity-independent (miniature) IPSCs. As both types of currents are often present in a given cell, antagonists can be applied to bath recording solutions to isolate IPSCs or EPSCs.

Presynaptic Plasticity

Alterations in presynaptic transmission can be assessed electrophysiologically by examining quantal synaptic events, namely, the frequency of miniature EPSCs (mEPSCs) to assay plasticity in pooled inputs [292,293]. Changes in event frequency can be attributed to a number of things, including presynaptic transmitter release, increased numbers of postsynaptic AMPARs, or transient postsynaptic modifications (i.e., generation of silent synapses) [294,295]. Therefore, it is often advantageous to perform additional studies to determine whether alterations in mEPSC frequency align with presynaptic modifications. This is often done by using a paired pulse ratio (PPR), which measures synaptic efficacy that is inversely proportional to the presynaptic probability of release [177,296]. Briefly, this involves comparing the amplitude of two electrically or optically evoked EPSCs at predetermined interstimulus intervals (e.g., 50, 100, and 200ms), with the ratio defined as the amplitude of the second EPSC divided by the amplitude of the first EPSC.

following the last exposure [55,60]. On the other hand, when drugs like cocaine are intravenously self-administered, modifications in excitatory signaling have been observed after 90 days of withdrawal [61], suggesting that learned and voluntary drug intake and the pharmacodynamic effects of noncontingent exposure can induce differential changes in DA signaling.

Mechanisms Underlying VTA Excitatory Plasticity

Modifications in excitatory synaptic strength following acute cocaine require activation of D1/5 DA receptors and NMDAR-dependent activation of protein kinase C (PKC), but not Ca^{2+}/calmodulin-dependent protein kinase II (CaMKII) in DA neurons [55,56,62,63]. These adaptations do not appear to reflect alterations in presynaptic plasticity as measured by paired pulse ratios (PPRs), but rather reorganization of postsynaptic glutamatergic transmission onto DA neurons involving increased synaptic incorporation of Ca^{2+}-permeable GluA2-lacking AMPARs (Box 1) and simultaneously reducing NMDAR-mediated signaling [55,56,62,64–67].

Acute cocaine-induced reductions in NMDAR-mediated transmission are due to increased insertion of a noncanonical NMDAR subtype that expresses GluN3A/GluN2B subunits [68]. These receptors exhibit reduced Mg^{2+} sensitivity and reduced conductance, as they are less permeable to Ca^{2+} [69,70]. Importantly, insertion of these receptors appears to be a necessary step for the expression of cocaine-evoked increases in GluA2-lacking AMPAR [68]. This shift in AMPAR and NMDAR subunit composition likely represents a form of metaplasticity that alters the source of depolarizing synaptic Ca^{2+} through typical NMDARs to that of the Ca^{2+}-permeable AMPARs, leading to an impairment of NMDAR-dependent long-term potentiation (LTP, Box 1) of AMPAR signaling and endorsing further strengthening of drug-potentiated synapses through a specific form of synaptic plasticity that relies on GluA2-lacking AMPARs (Fig. 2) [56,65].

These alterations in NMDAR and AMPAR signaling are also responsible for increases in VTA DA neuron firing frequency and bursting activity that are observed up to 7 days following a single drug exposure [68]. Such increases in DA neuron activity, as well as the underlying alterations in excitatory synaptic strength appear to reflect an overall increase in DA neuron intrinsic excitability due to a reduced ability to activate small-conductance potassium channels (as well as other inhibitory adaptations discussed below), and are believed to contribute to the development of cocaine-seeking behavior [61,68]. Similar increases in synaptic strength and AMPAR signaling have been observed at VTA synapses following a single exposure to amphetamine, morphine, nicotine, and ethanol [59,71,72], suggesting that adaptations in VTA DA neurons reflect a common mechanism by which drugs of abuse initiate plasticity within the mesocorticolimbic system; ultimately shaping addiction-related behavior.

Notably, 1 week following a single cocaine exposure, increases in GluA2-lacking AMPAR transmission and reductions in NMDAR function return to baseline. This recovery is driven by endogenous activation of VTA group I metabotropic glutamate receptors (mGluRs) via Ca^{2+}-dependent local protein synthesis [62,68]. As such, endogenous mGluR activation may serve as an adaptive barrier that initially acts as a protective mechanism to counteract drug exposure. Repeated exposure may in essence overcome the critical window in which mGluR can control cocaine-evoked plasticity (by repeated drug exposure)—a step that may represent the first move toward recruitment of downstream plasticity in areas like the NAc and medial PFC (mPFC) [28,61,73,74]. Thus, it is tempting to speculate that individuals with deficient mGluR1-dependent long-term depression (LTD) mechanisms may exhibit a higher propensity to develop addiction, and that clinical screening of genes controlling mGluR function may provide a useful clinical tool in the fight to prevent addiction [65].

Inhibitory Plasticity

While many studies have showcased the importance of excitatory plasticity in the VTA in the development and persistence of addiction-related behavior, alterations in inhibitory transmission, largely mediated by GABA and DA, are also a critical determinant of cocaine actions [45]. Several lines of evidence indicate that inhibitory neurotransmission mediated by metabotropic (G protein-coupled) receptors for GABA ($GABA_BR$) and DA (D_2R) are also a critical determinant of cellular and behavioral responses to acute and repeated drug exposure [45]. Similarly, long-term adaptations in the synaptic strength of GABAergic synapses also appears to play a prominent role in behavior related to a number of abused drugs [75–77].

Timeline of VTA Inhibitory Plasticity

Although the timeline of known drug-induced effects on inhibitory signaling pales in comparison to the mapped effects on excitatory synaptic plasticity, evidence suggests that inhibitory adaptations may follow a similar time course, however, comparisons between experimenter- and self-administration models are still needed. Acute exposure to ethanol, morphine, cocaine, and nicotine reduces the ability of inhibitory synapses on VTA DA neurons to express LTP of GABA. Acute cocaine has been also shown to reduce D2 and GABA$_B$-mediated signaling [78]. These effects are observed as early as 2 h following morphine and nicotine, and 24 h after cocaine and ethanol, but similar to excitatory plasticity in the VTA dissipate by day 5 of withdrawal [75,78].

Alterations in VTA inhibitory signaling following acute drug exposure have been also shown to extend to local GABA neurons, with acute methamphetamine or cocaine reducing synaptic and somatodendritic GABA$_B$ signaling for as long as 7 days [79]. The functional implications of this plasticity are unclear, as downregulation would presumably increase activation of GABAergic signaling within the VTA, however, DA neurons are more excitable at this time point. As such, it is conceivable that impaired GABA signaling removes an intrinsic brake on GABA neuron spiking, which may in turn augment GABA transmission in the mesocorticolimbic system.

Reductions in VTA D2 and GABA$_B$ signaling are also observed following repeated cocaine, an effect that is evident following 4 days withdrawal, but again appears to dissipate within 10 days (Fig. 2) [78,80,81]. This reduction in inhibitory signaling is paralleled by a reduction in the ability of D2R to inhibit firing and an increase in the basal number of spontaneously active DA neurons and burst firing [82]. These data suggest that reductions in inhibitory signaling contribute to augmented DA transmission following repeated drug exposure. However, reductions in inhibitory signaling are paralleled by enhanced AMPAR signaling in the same neurons [78], and match well with increases in synaptic strength—making it difficult to parse out a role for one or the other. One possibility is that increases in DA neuron burst firing during exposure to drugs and drug-related stimuli may be due largely to enhanced glutamate transmission [83,84], while increases in spontaneous firing and intrinsic excitability are driven by reductions in inhibitory signaling.

Mechanisms Underlying VTA Inhibitory Plasticity

Mechanisms underlying drug-induced reductions in VTA inhibitory signaling are far less characterized than those of excitatory plasticity. Regarding metabotropic signaling, evidence points to a significant role for reductions in G protein-modulated effectors, such as G protein inwardly rectifying potassium (K$^+$) (GIRK) channels, which promote a hyperpolarizing efflux of K$^+$ that reduces excitability and spike firing [45,85,86]. Mice lacking GIRK channel expression display increases in motor activating responses to morphine and cocaine, increased responding for and intake of self-administered cocaine [87], and exhibit reduced D2R modulation of VTA DA neuron excitability [85,87,88]. Drug-induced reductions in GIRK signaling and associated drug behaviors have largely been attributed to a phosphorylation-dependent reduction in membrane expression of GIRKs and/or GABA$_B$/D2 receptors (Fig. 2) [86]. Increases in DA neuron excitability and firing may also be attributed to reductions in small conductance K$^+$ channels [68], or downregulation of cAMP/PKA (3',5'-cyclic adenosine monophosphate/protein kinase A) signaling [89]. Alternatively, drug-induced suppression of LTP of GABA$_A$-mediated transmission appears to reflect a disruption in the LTP induction mechanism, rather than an occlusion of this plasticity [75,76].

NUCLEUS ACCUMBENS

Structure, Function, and Anatomical Connectivity

The NAc region of the ventral striatum is critically involved in the attribution of incentive value to drug-paired cues that provides a motivational force behind drug-seeking behavior [90]. The NAc is a heterogeneous structure that can be divided into a core and shell subregion based on their connectivity and reward-related responses [91–93]. The NAc core plays a role in evaluation of reward and initializing reward-related motor activity [94–96], whereas the shell is highly connected with limbic and autonomic brain regions and heavily involved in motivation and establishing learned associations [94,97]. The NAc core is innervated by glutamatergic afferents arising from the dorsomedial [prelimbic cortex (PrL)] and dorsolateral (anterior insular cortex), parahippocampal formation, and nuclei of the basolateral amygdala (BLA) (Fig. 1) [98,99]. Alternatively, the shell is highly innervated by projections from the ventromedial PFC [infralimbic cortex (ILC)], the ventral hippocampus (vHPC), and aspects of the BLA (Fig. 1).

GABAergic medium-spiny projection neurons (MSNs) comprise >90% of cells in the NAc. Canonical understanding is that MSNs can largely be divided into two subpopulations based on the expression of peptides, DA receptors (D1 or D2), and downstream projection targets [47,100,101], with a small fraction (∼2%–17%) expressing both D1- and D2-receptors [102–104]. These two subpopulations of MSNs are significant, as emerging evidence supports the notion that D1- and D2-MSN populations often display different physiological characteristics [105,106] and exhibit different forms of drug-induced plasticity, and generally exert antagonistic effects in drug-associated behaviors [107–111].

Increasing evidence points to motivated behavior being encoded through the NAc in parallel pathways based not only on the subregion and MSN type, but also on the downstream projection of a given MSN. In more dorsal regions of the striatum, information is processed through discrete direct and indirect pathways, where D1-expressing MSNs (D1-MSNs) convey their information directly to the output nuclei of the basal ganglia via the globus pallidus and SN pars reticulata and D2-expressing MSNs (D2-MSNs) do so indirectly through synapses in the external segment of the globus pallidus which project to the subthalamic nucleus [112]. In contrast, recent data indicate that information processing within the NAc involves significantly more overlap between D1- and D2-MSN projections than previously believed [47,103,113–115]. For example, D1-MSNs project to the VM and VP, whereas D2-MSNs almost exclusively project to the VP [47,100,101]. Within the NAc core circuitry, D1- and D2-MSN innervate VP neurons that subsequently project to either the VM or medial dorsal thalamus [103].

Time-Dependent Plasticity in the NAc: An Overview

Acute Exposure and Withdrawal

A single exposure to psychostimulants, opioids, and nicotine is sufficient to produce relatively enduring increases in excitatory synaptic strength in the VTA [72]. In the NAc, psychostimulant-evoked plasticity also occurs, but generally on a more variable timescale and with a steeper induction threshold. Early electrophysiological and biochemical characterization of acute drug effects indicated that noncontingent acute exposure to cocaine, amphetamine, or morphine was not sufficient to augment synaptic strength (measured by changes in the AMPA/NMDA ratio) or AMPAR-specific transmission in the NAc shell during early (24–48 h) withdrawal [30,74,116–118]. Alternatively, in the NAc core, AMPAR signaling is elevated in pooled MSNs shortly after noncontingent acute amphetamine or cocaine exposure [119,120], suggesting that during the immediate periods following acute drug exposure, plasticity may be more prominent within the motor-centric core region of the NAc, whereas more enduring pharmacological effects responsible for long-term development of addicted behavior persist in the shell—a region associated with developing conditioned associations.

In support of this, repeated amphetamine has been shown to augment AMPAR signaling in pooled NAc shell MSNs 10–14 days but not 24 h following repeated amphetamine [120]. The mechanisms that account for the temporally specific regional differences in plasticity between NAc core and shell may lie in differences in the ability of amphetamine to increase extracellular DA. For example, in drug-naïve animals, the core is more sensitive to acute amphetamine than the shell, producing a larger increase in extracellular DA at lower concentrations of the drug [121], whereas following repeated amphetamine treatment, increases in extracellular DA and the associated behavioral sensitization, appear to be more robustly associated with the shell [122,123]. Alternatively, it is possible that adaptations following an acute drug exposure may simply require time to express in the NAc shell, as increases in AMPAR have been observed 7 days following acute exposure in the NAc shell, but only in D1-MSNs [111].

Repeated Exposure and Withdrawal

In contrast to acute drug exposure, early withdrawal from repeated or chronic cocaine, morphine, or ethanol, appears to be associated with a suppression of synaptic strength, measured as decreased AMPA/NMDA ratio and an impaired ability to induce NMDA- or mGluR2/3-dependent forms of LTD (Box 1) at pooled NAc shell MSN synapses when measured within the first 72 h [117,124–130]. Although not entirely clear, suppression of excitatory signaling likely reflects a combination of mechanisms including reduced AMPAR signaling and/or an increase in the expression of newly generated "silent synapses" that are enriched with functional NMDA-, but not AMPA-type receptors, and are believed to act as intermediary structures for subsequent increases in synaptic strength (Fig. 3) [131,132].

As with acute exposure, prolonging the withdrawal time after repeated exposure leads to potentiated excitatory signaling, whether it be following a single cocaine injection [107,111] or repeated cocaine, nicotine, morphine, or ethanol [110,117,118,120,129,133,134], an effect that can last 6–7 weeks [135,136]. With psychostimulants and opioids, increases in excitatory signaling are paralleled by an impaired ability to experimentally induce LTP of synaptic strength using high-frequency stimulation of pooled afferents in the NAc [107,118,137,138]. As with impaired LTD during early withdrawal, this likely reflects an occlusion due to the already potentiated state of synapses following withdrawal.

Several mechanisms have been proposed to underlie enhanced synaptic strength at NAc synapses, including, but not limited to, increased presynaptic glutamate release and insertion of postsynaptic AMPARs in both the NAc core and shell (Fig. 3) [107,108,110,117,133–135,139–144]. While the majority of AMPAR transmission under drug naïve conditions is mediated by receptors expressing the GluA2 subunit (i.e., GluA1/GluA2), drug exposure appears to favor formation of GluA1-homomeric AMPARs (GluA2-lacking; Fig. 3) [74,110,117,135,145]. Increased insertion of these receptor subtypes appears to depend on continuous protein translation [146] and has been shown numerous times to play an important role in addiction-related behavior, as antagonism or elimination of these receptors in the NAc prevents elevations in

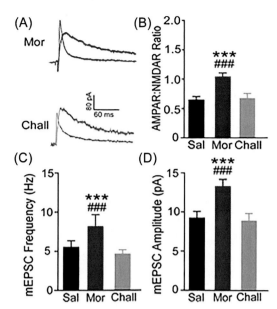

FIG. 3 Re-exposure to morphine during withdrawal synaptic strength (AMPAR:NMDAR ratio) and AMPAR-mediated currents in the NAc shell. Similar to cocaine [117,200], AMPAR/NMDAR ratios (A–B), and AMPAR-mediated mEPSCs (C–D) are elevated at NAc shell MSN synapses (pooled) following 10–14 days of withdrawal from repeated morphine. Increases in AMPAR/NMDAR ratios (B) and AMPAR signaling (C–D) are depotentiated 24 h following an acute challenge injection of morphine (gray, 10 mg/kg). Sal = saline, Mor = morphine + no challenge injection, Chall = morphine + challenge injection, mEPSC = miniature excitatory postsynaptic current; ms = milliseconds, pA = picoAmps, Hz = Hertz (frequency). ***$P < 0.001$ vs Sal, ###$P < 0.001$ vs. Chall.

cocaine-seeking following prolonged abstinence [74,108,135], disrupts the ability of a drug challenge to reinstate morphine place preference, and attenuates naloxone-induced affective and motivational components of withdrawal [147].

The apparent bimodal effects of repeated drug exposure during early withdrawal and protracted withdrawal have led some to believe that increases in excitatory signaling are attributable to a compensatory upregulation of synaptic strength due to a drug-induced suppression of signaling immediately following repeated exposure [116,117]; otherwise known as synaptic scaling [148]. Alternatively, an emerging concept in the addiction field is that exposure to drugs of abuse can promote the generation of excitatory glutamatergic synapses that express functional NMDARs, but lack functional AMPARs due to their absence from the synapse [149,150]. These so-called "silent synapses" are thought to occur through synaptogenesis-like events or by "silencing" preexisting synapses [151].

With regards to addiction, silent synapses likely represent a common substrate for experience-dependent synaptic plasticity produced by all drugs of abuse, as generation of these synapses has been directly observed in both the NAc shell and core following repeated cocaine and morphine [109,131,132,152–154], and may also be generated by repeated heroin, nicotine, and ethanol [155,156]. The expression of functional NMDAR, but not AMPARs suggests that reductions in AMPAR/NMDAR ratios during early withdrawal from cocaine reflect an increase in silent synapses [117,141]. Believed to act as intermediary structures—these synapses are thought to evolve into fully functional synapses or be pruned with the progression of withdrawal to ultimately enhance or weaken excitatory synaptic strength within select circuits [157–161]. However, as discussed below, the mechanisms and functional relevance of these adaptations may differ across drug class and cell type.

Short-Term Synaptic Plasticity Associated With Relapse

It is important to note that although LTP may be impaired, synapses are still capable of undergoing change following withdrawal from drug exposure. Further, the adaptations discussed above are all induced by past exposure to drugs and require a period of abstinence to take hold—suggesting a potential role in susceptibility to relapse. However, it is possible that engagement of drug-seeking behavior during a relapse event may involve additional rapid-onset forms of synaptic plasticity. It is worth noting that relapse-related data are often variable and difficult to interpret, however, we will discuss several mechanisms currently predominating the field below.

Beginning in the mid-1990s, a string of exciting papers showed that direct activation of AMPARs in the NAc core in chronic drug-treated animals led to expression of psychomotor sensitization and other addiction-related behaviors, such as drug seeking in reinstatement models [161a,161b,161c,161d]. More recently, it has been shown that re-exposure to the drug or to a drug-associated cue induces a rapid potentiation of NAc core glutamatergic signaling in cocaine, nicotine, or heroin withdrawn rats [133,134,138,144]. One straightforward interpretation of these findings is that a progressive enhancement of AMPAR signaling during withdrawal serves as a common mechanism for driving addiction-related behavior [162,163].

On the other hand, there is an emerging literature that supports a seemingly opposite hypothesis—that decreases in excitatory drive onto NAc MSNs, particularly in the NAc shell, may promote reinstatement to drug seeking. For example, viral-mediated overexpression of AMPARs in the shell attenuates cocaine-primed [164] and stress-induced reinstatement [164a]. Further, reinstatement is facilitated by NAc expression of a mutated, dominant negative AMPAR that degrades basal AMPAR function, while extinction and reinstatement of cocaine seeking is reduced by overexpression of wild-type GluA1R [164]. This fits well with electrophysiological and biochemical evidence that re-exposure to drug following withdrawal from experimenter-administered amphetamine, cocaine, and morphine, as well as cocaine self-administration causes a reduction in synaptic strength and/or AMPAR signaling in the NAc shell and core [117,130,165–168, 168a,120] (Hearing et al., unpublished—Fig. 4).

Reductions in AMPAR signaling appears to be a rapid and transient event, as it is observed as early as 20 min and 2 h following an ex vivo or in vivo cocaine challenge, respectively, but returns to a potentiated state within 5 days [169]. This plasticity appears to transcend reinstatement modalities, as it is also observed immediately following reinstatement induced by drug-associated cues and a pharmacological stressor, further supporting a role in driving relapse behavior [168]. Although caution should always be exercised when comparing plasticity across experimental approaches employing extinction/reinstatement versus abstinence without extinction, it is unlikely that data from [164] purely reflects instrumental extinction training, as enhanced synaptic strength persists in unidentified MSN populations in the NAc shell following extinction from repeated noncontingent cocaine [170] and cocaine self-administration in mice [168].

With cocaine (and amphetamine), short-term reductions in synaptic strength in the NAc shell involves mGluR5-mediated PKC- and protein synthesis-dependent internalization of AMPAR signaling—a phenomenon that may also involve activation of presynaptic endocannabinoid CB1 receptors [120,165–167,169,170]. Reductions in NAc shell AMPAR do appear to play a causal role in driving reward behavior, as blockade of mGluR5 and endocytosis of AMPARs inhibits reinstatement of place preference, while activation of mGluR5 or patterned optical stimulation known to promote mGluR5-dependent LTD is sufficient to trigger depotentiation of synaptic strength and promote reinstatement in the absence of drug [170]. Whether similar causal links are observed in the NAc core or following self-administration remains to be seen.

Afferent- and Cell-Specific Plasticity

Converging input from cortical and limbic brain regions onto a single MSN form a separate but unified system of information that can promote heterosynaptic plasticity and associative learning—a common feature of addiction phenotypes [108,171–173]. How this information is interpreted depends on which subtype of MSN (i.e., D1- vs D2-MSN) and where that neuron projects to (i.e., VP or SN) [103,113,174–176].

Examination of long-term changes in glutamate transmission originally focused on postsynaptic mechanisms, however, it is now clear that modifications of synaptic strength may occur at either side of the synapse, with the same synapse often undergoing various forms of plasticity. While postsynaptic plasticity generally involves changes in postsynaptic receptor number or properties, presynaptic plasticity involves an increase or decrease in neurotransmitter release [177]. With the use of optogenetics and transgenic mice that express fluorescent proteins in subpopulations of neurons, neuroscientists can now measure pathway- and cell-specific measurements of presynaptic and postsynaptic glutamate signaling following drug exposure. These approaches have begun to provide unprecedented insight into how drugs of abuse alter selective NAc neural circuits.

Presynaptic Plasticity

NAc Shell

Examination of pooled afferents indicates that psychostimulants and opioids may exert temporally differential effects on presynaptic glutamate release in the NAc shell. Examination of release probability using whole-cell ex vivo electrophysiology indicates that release of glutamate is generally reduced during early withdrawal from cocaine self-administration, but no longer reduced after 3–4 weeks of withdrawal [141]. A likely explanation for the lack of effect following more prolonged withdrawal is that enduring increases in release are reserved to specific afferent pathways, as release probability is increased at ILC-to-shell and paraventricular nucleus (PVN)-to-shell, but not BLA-to-shell projections 1 and 45 days

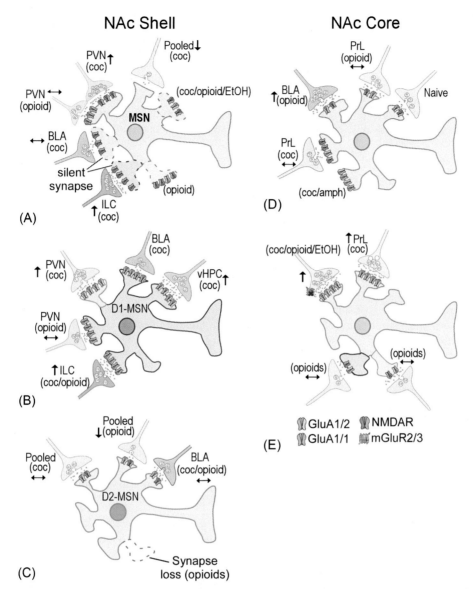

FIG. 4 Alterations in presynaptic glutamate release probability and postsynaptic glutamate receptor signaling in the NAc. The field lacks comparable data regarding how individual drug classes alter plasticity in the NAc with regards to time as well as afferent- and cell-type specificity. Examination of pooled MSN populations (tan synapse) have shown that early withdrawal (A, D) from repeated drug exposure is generally associated with a reduction in NAc shell (A) and core (A) AMPAR:NMDAR signaling that is thought to reflect an increase in silent synapse (dotted synapse) production within ILC/PVN and PrL pathways, respectively. Examination of pathway- and cell-type specific plasticity in the NAc shell (A) shows that presynaptic glutamate release is increased (upward arrow) during early (and prolonged) withdrawal from cocaine at ILC/PVN terminals, but not PVN or BLA terminals following repeated opioid or cocaine, respectively. In the NAc core (B), psychostimulants appear to upregulate AMPAR signaling during early withdrawal, while opioids increase release probability at BLA but not PrL inputs. With progression of withdrawal, synapses on D1-MSNs that were previously silent the NAc core and shell appear to mature through a progressive insertion of GluA2-containing and GLuA2-lacking AMPARs, respectively, following cocaine, whereas silent synapses on D2-MSNs following early opioid withdrawal are lost (green dotted synapse; C). Examination of specific afferent pathways to MSNs has begun to show both overlapping and divergent plasticity, but depict an overall effect of increased glutamate release at MSNs in the NAc shell (B) and core, primarily at D1-MSNs (B, C), that is also paralleled by a reduction in release at D2-MSNs following opioid withdrawal (C). Afferents: Infralimbic cortical (ILC, blue), paraventricular nucleus (PVN, yellow), ventral hippocampus (vHPC, pink), basolateral amygdala (BLA, purple), prelimbic cortex (PrC, green). Arrows designate whether release probability is increased, decreased or unchanged versus controls.

following noncontingent and self-administered cocaine [143]. The lack of effect at BLA-to-shell synapses, however, may reflect collection of data from pooled MSN subtypes, as the number of BLA-to-shell connections has been shown to increase selectively in D1-MSNs [178]. Interestingly, while regions like the NAc shell may receive the highest degree of innervation from the vHPC [139], glutamate release within the vHPC-to-shell pathway has not been directly examined, however, increased frequency of miniature excitatory postsynaptic currents (mEPSCs) have been observed [108].

Much like psychostimulants, opioids have also been shown to alter glutamate release at MSNs in the NAc shell, however, the temporal and anatomical dynamics of these changes are less clear. Examination of pooled inputs indicates that glutamate release is unaltered within the first 24 h of repeated noncontingent morphine or remifentanil self-administration, even when examining MSN subtypes or specific afferents to the NAc shell, such as those arising from the PVN or BLA [118,179,180]. On the other hand, more prolonged withdrawal (10–14 days) from morphine is associated with increased presynaptic release selectively at inputs to D1-MSNs, while excitatory drive at D2-MSNs is reduced (Fig. 3) [110,118]. These data suggest that although different drug classes may produce differing effects on specific inputs to the NAc shell, they share a common ability to promote progressive plasticity that requires time to develop.

NAc core

Compared with the NAc shell, direct evidence for withdrawal-dependent alterations in transmitter release in the NAc core is substantially lacking, however, a considerable amount of indirect evidence has demonstrated that exposure to numerous drugs of abuse alters presynaptic regulatory mechanisms which normally serve to inhibit synaptic transmission at the pre-synaptic level. For example, under baseline conditions, activation of presynaptic metabotropic group II/III receptors (mGluR2/3R) is accompanied by reductions in release probability and frequency of mEPSCs [127], whereas withdrawal from chronic cocaine or heroin causes a decrease in mGluR2/3-mediated inhibition of AMPAR-dependent transmission (Fig. 3) [127,137,181] but see Kasanetz et al. [182].

This reduction in mGluR2/3-dependent regulation of glutamate release appears to play a key role in modifying NAc core MSN postsynaptic signaling responsible for driving drug-related behavior across a number of drugs, including cocaine, heroin, morphine, nicotine, and ethanol [162,183–185]. Although unclear if multiple afferents exhibit tempered reductions in machinery regulating glutamate release in the NAc core, PrL-to-core synapses exhibit an impaired ability to undergo mGluR2/3-dependent LTD of synaptic strength [137]. Removal of this presynaptic "brake" may account for observed increases in synaptic glutamate in the NAc core that occurs during cue, stress-, and/or drug-induced reinstatement of cocaine and/or heroin seeking [133,142,144].

Interestingly, the possibility of increased glutamate release occurring selectively at PrL afferents during reinstatement of heroin seeking may help to explain the lack of effect on glutamate release from pooled inputs at D1- or D2-MSNs in the NAc core following a similar period of drug abstinence from repeated morphine [110]. Alternatively, it is possible that a lack of plasticity within the motor-centric NAc core reflects use of a model employing experimenter- versus self-administered drug. Regardless, the above data overwhelmingly suggest that drugs of abuse collectively facilitate increases in glutamate release at MSN synapses in the NAc that plays a critical role in drug-seeking and/or reward behavior.

Postsynaptic Plasticity

Rapid glutamatergic transmission in NAc MSNs is primarily mediated through activation of the AMPA- and NMDA-type ionotropic glutamate receptors. Alterations in expression, localization, and ion conductance of these receptors have profound effects on synaptic transmission and cellular physiology [54]. Drug-induced alterations in synaptic transmission represents a driving force behind the development of addicted behavior and the persistent risk for relapse; even following prolonged periods of abstinence [54,163,186–190]. Below we will discuss the progressive development of postsynaptic modifications in AMPA and NMDAR signaling, and their behavioral relevance to addiction-related behavior.

Cortico-Accumbens Plasticity

The mPFC-to-NAc pathway has long been proposed to undergo drug-induced synaptic plasticity that is important for driving opioid and psychostimulant behavior [154,162,191]. This pathway can be divided into two primary projections consisting of the ILC projection to the NAc shell and the PrL projection to the NAc core (Fig. 1). Optogenetic exploration of mPFC-to-NAc afferents has shown that AMPAR signaling is increased at ILC-to-shell and PrL-to-core synapses following repeated cocaine or morphine [107,110,111,154]. Although it is unclear exactly how long these changes persist, increased signaling has been observed up to 45 days postdrug exposure, and may contribute to escalations in drug seeking that occur with time [154,192].

Increased strength of both mPFC-to-NAc pathways appears to reflect a progressive maturation of silent synapses generated during early drug withdrawal through insertion of GluA2-lacking AMPARs in the ILC-to-shell pathway, and GluA2-containing AMPARs in the PrL-to-core pathway [153,154,192]. While cocaine-induced generation of these synapses reflects spinogenesis preferentially on D1R-MSNs, morphine-induced silent synapses result from the removal of AMPARs from synapses on D2-MSNs during the early stages of withdrawal [109]. Electrophysiological evidence suggests that overtime, "weakened" glutamatergic synapses on D2R-MSNs are removed—precipitating a loss of dendritic spines and reduced excitatory drive within the D2-striatal pathway [109,110]. Thus, both morphine and cocaine produce an increase in the ratio of D1R- to D2R-mediated signaling, although through different mechanisms.

Allocortical-Accumbens Plasticity

Neighboring synapses in the NAc shell receiving input from the vHPC or BLA also display increased synaptic strength [108,111,139,153]. Similar to PrL inputs, enhanced glutamate signaling at vHPC synapses appears to involve insertion of GluA2-containing AMPARs (Fig. 3) [108,139]. Alternatively, data reporting effects on postsynaptic signaling within BLA projections appear to be more susceptible to variations in the model used, however increases in AMPAR signaling likely reflects increased insertion of GluA2-lacking receptors selectively at BLA-to-shell synapses on D2-MSNs [108,111,153].

Relevance of Long- and Short-Term Accumbens Plasticity

Enduring pathway-specific increases in AMPARs have important implications for a variety of addiction-related behaviors. For example, optical stimulation of the ILC-to-shell (or PVN-to-shell) pathway at a low frequency (1 Hz), restores basal AMPAR transmission during withdrawal, disrupts subsequent expression of locomotor sensitization, and rescues signs of withdrawal [107,180]. Low-frequency (1–5 Hz) stimulation of BLA-to-shell afferents or ILC-to-shell and PrL-to-core pyramidal neurons promotes internalization of GluA2-lacking (shell) and GluA2-containing (core) AMPARs and disrupts withdrawal-induced increases in cocaine seeking [153,154]. Stimulation at a higher frequency (10–13 Hz), known to promote mGluR5-dependent LTD [193], reverses inward rectification of AMPARs at D1-MSNs and blocks cocaine relapse and reinstatement of morphine place preference [107,110,194].

Cortico-Accumbens Plasticity

The PrL-to-core pathway appears to play an important role in driving drug-seeking behavior across drug classes, whereas the ILC-to-shell projection may fundamentally serve different roles depending on the drug class—acting as a neural off switch for cocaine seeking, but an on switch for opioid seeking [191]. Data support the notion that repeated cocaine and opioid exposure and withdrawal enhance synaptic strength within both of these pathways [110,154,162,191], begging the question: does this plasticity plays functionally distinct roles?

In agreement with a role of the PrL-to-core pathway driving drug seeking, prolonged withdrawal from cocaine increase AMPAR signaling in this pathway, while cue-induced reinstatement involves potentiation of glutamate transmission in the NAc core [134,144,154,195]. Further, optical stimulation of this pathway using a pattern known to promote LTD (1–5 Hz), prevents subsequent maturation of generated silent synapses and attenuates subsequent relapse [154].

Alternatively, aligning with the purported role of the ILC-to-shell pathway as a neural off-switch for cocaine seeking (in conjunction with instrumental extinction training), optically induced reversal of synapse-based remodeling in the ILC-to-shell pathway prior to testing potentiates subsequent incubation of cocaine craving [154]. As synaptic strength is rapidly reduced in the NAc shell by exposure to drug or drug-associated stimuli—an effect that has been more recently shown to occur within the ILC-to-shell pathway [117,120,166,170], it is possible that rapid removal of AMPARs represents a synaptic trigger for relapse when this AMPAR-mediated "brake" is removed.

Accumbens Microcircuit Plasticity

Although the mechanism may differ across drug classes, the above data highlight a number of features of circuit level drug-induced plasticity that have important implications for future therapy development. First, while pathway-specific plasticity varies, enhanced glutamatergic transmission within the PFC-to-NAc pathway is shared across drug classes and models [107,108,110,143]. Moreover, similar to cocaine and morphine, withdrawal from nicotine self-administration potentiates NAc AMPAR signaling [133]. Further, withdrawal from chronic ethanol exposure is associated with increased amplitude of AMPAR signaling and suppressed synaptic strength at D2-MSNs [125,129]. Although the mechanism may differ across drug classes, the above data highlight a potential feature of drug-induced plasticity at the circuit level, a progressive increase in the ratio between excitatory synaptic drive at D1- over D2-MSN synapses during withdrawal. Regardless of the specific change and the specific model used, these data support the perspective that the enduring symptoms of drug addiction may be encoded by synaptic changes in the NAc.

Potential Targets for Opioid Dependence and Relapse

Although not prevalent with psychostimulants, preclinical and clinical data indicate that abrupt cessation of opioids can lead to precipitation of physical withdrawal and negative emotional states [147,196–200]. Manifestation of opioid-induced withdrawal symptoms align temporally with reductions in NAc core and shell GluA1 surface expression 1 and 72 h following repeated or acute morphine, respectively, while naloxone-precipitated withdrawal from chronic morphine is associated with reductions in NAc GluA2-lacking AMPARs [109,147,197,201–203]. Alternatively, AMPAR-mediated signaling and surface expression is elevated in the NAc shell and core immediately (3 h) following acute morphine (Hearing et al., unpublished; [147]), indicating that increases and decreases in AMPAR signaling play opposing roles in reward and withdrawal-associated aversive states, respectively, thus highlighting a potential target in treating opioid withdrawal, which can often drive misuse of prescription medications and lead to increased use of nontherapeutic opioids.

Although further research is needed, it is also possible that short-term relapse vulnerability driven by early withdrawal states versus that observed following more protracted withdrawal (postdetoxification) are driven by differing forms of pre- and postsynaptic glutamate plasticity. For example, alterations in presynaptic release probability appear to be manifest only following longer periods of withdrawal [110,180]. Further, biochemical and electrophysiological data suggest there morphine produces a time-dependent shift in NMDAR subunit expression, where increased GluN2A-expressing receptors during early withdrawal gives way to elevations specifically in GluN2B-expressing receptors with prolonged withdrawal [109,204,205]. Notably, this shift appears to be specific for opioids, as GluN2B expression appears to be upregulated during early and more protracted withdrawal—indicating a role for GluN2A receptors in development of dependence or expression of withdrawal, respectively [206]. This contention is further supported by data from transgenic GluN2A knockout mice showing marked loss of typical withdrawal signs, which can be restored by subsequent rescue of GluN2A selectively in the NAc [207,208]. Overall, these data highlight unique adaptations produced by opioids may help to mitigate relapse at varying stages of addiction.

PREFRONTAL CORTEX

Research over the past two decades has made it increasingly clear that the enduring vulnerability to relapse lies within long-lasting neural adaptations in corticostriatal glutamate circuits that are responsible for generating well-learned behaviors, but also cognitive flexibility that allows individuals to adapt their behavior in response to alterations in their environment [25,29,209]. In addiction, an individual's ability to inhibit drug seeking and drug taking is thought to reflect a pathological strengthening of drug-seeking behaviors or impairments in the capacity to control maladaptive behavior. These theories are not mutually exclusive and would likely involve modifications within prefrontal cortical circuits, as substantial clinical and preclinical evidence points to the PFC as a critical substrate for reinforcement learning, acquisition of drug self-administration, and relapse to drug seeking [8,16,29,162,210,211].

Clinical and preclinical evidence indicates that chronic drug exposure promotes a persistent reduction in the activity of the dorsal PFC [210,212–214]. In contrast, intoxication and exposure to drug-associated information (e.g., cues) but not naturally rewarding stimuli, is associated with increased activation of these regions [214–223]. More ventral PFC regions on the other hand display reduced activation during exposure to drug-associated stimuli [224]. Preclinical evidence suggests that activation of dorsal PFC is likely required for facilitating acquisition and relapse of drug seeking, while activation of the more ventral PFC exerts an inhibitory effect on drug seeking [16,191,225–227]. Taken together, these changes are thought to contribute to a reduced incentive to seek out previously rewarding stimuli (e.g., time with family and friends), an increased responsivity to drug-conditioned cues, and reduced inhibitory control over addicted behavior.

While clinical and preclinical data emphasize the importance of studying the maladaptive functioning of mPFC neurons, relatively few studies have reported drug-evoked synaptic plasticity in this region compared with regions like the VTA and NAc. In this section, we will provide an overview of inhibitory and excitatory neural plasticity in the mPFC, and discuss how these adaptations equate to current theories of addiction. We will primarily focus on the more medial PrL and ILC cortices, as these regions of the mPFC are a major source of NAc glutamate, are reciprocally connected with limbic structures such as the VTA and BLA [99,228–230], and are known substrates for relapse to drug seeking across numerous drug classes [90,210,226].

Structure, Function, and Physiology

The frontal cortex is an aggregate of several cortical regions, including the cingulate, PrL, ILC, and orbitofrontal cortices, which provide widespread glutamatergic input to cortical and subcortical structures. Principal glutamatergic (excitatory) pyramidal neurons and local-circuit GABA (inhibitory) neurons are the building blocks of mPFC microcircuits, with pyramidal neurons located in layers 2/3 and 5/6 making up 85% of neurons [231,232]. Collateral connections between pyramidal neurons as well as reciprocal connections between pyramidal and GABA interneurons work in conjunction to provide a dynamic balance of excitation and inhibition that is critical for cortical information processing and thus, function of this region (Fig. 1) [233].

These microcircuits are further modulated by a number of afferent projections, including DA and GABA arising from the VTA, and glutamatergic input from areas like the BLA, HPC, and thalamus (Fig. 1) [45]. This input can exert effects on pyramidal neurons through direct synaptic connections or indirectly through modulation of local GABAergic interneurons [234–237]. In return, layer 5/6 pyramidal neurons represent a major source of excitatory input to regions such as the VTA, NAc, BLA, and HPC that often permits direct and indirect regulation of these structures [99,228–230].

Drug-Induced Adaptations in the mPFC

Effects on Pyramidal Neuron Activity

The predominating focus in addiction has initially been on the role of the mesolimbic DA system. However, during the last two decades, it has become increasingly clear that the mPFC plays a key role in the attribution of salience to rewarding stimuli, compulsive drug taking, the expression of drug-associated memories, and relapse to drug seeking [172,191,211]. Overtime, repeated exposure to drugs of abuse can alter glutamatergic output from this region that is essential for engaging in drug-seeking behavior [8,29,162,238]. Studies on the acute effects of drugs on pyramidal neuron activity are relatively few in number, but in vivo studies show that cocaine, nicotine, and ethanol administration produces a transient depolarization and/or firing of layer 5/6 pyramidal neurons [144,239–241], followed by an overall reduction in firing rates in anesthetized and self-administering rats [242,243]. Similarly, in vitro slice electrophysiology studies have shown that intrinsic excitability of these neurons is not altered 24 h following a single cocaine exposure [85,244], suggesting that the threshold for enduring plasticity in this region is greater than areas like the VTA.

With repeated drug exposure, mPFC pyramidal neurons display an enhanced responsivity to cocaine and cocaine-related stimuli, however, these same neurons have reduced basal activity, indicative of a hyperfrontal state during drug seeking and taking and a hypofrontal basal state [242]. in vitro slice work has shown that repeated experimenter-administered cocaine triggers an enduring increase in pyramidal neuron membrane excitability in PrL, but not ILC layer 5/6 pyramidal neurons that persists up to 45 days postdrug exposure [85,244–247].

Alternatively, following more prolonged cocaine self-administration, there is a reduction in excitability of PrL pyramidal neurons that is significantly more pronounced in a subpopulation of rats that exhibit compulsive-like drug seeking [248]. It is possible that during early drug exposure, increased excitability of mPFC pyramidal neurons accounts for increased responding to cocaine-related information during withdrawal [216,242], and may permit these neurons to more readily undergo synaptic plasticity responsible for reinforcement learning related to development of drug-associated stimuli and drug-seeking behavior. This process may be further fueled by reductions in basal activity that amplify responsiveness of the mPFC to drug and drug-conditioned cues, thus allowing these cues to exert greater control over behavior through enhanced glutamate release in the mesocorticolimbic system [162,242].

In agreement with this possibility, Bossert et al. [249] showed that context-induced relapse to heroin seeking is driven by activation of sparsely distributed neurons in the ventral mPFC that appear to encode learned associations between heroin reward and heroin-associated stimuli that when inactivated, disrupt recall and relapse [249,250]. Alternatively, reductions in excitability following chronic cocaine use match well with reduced basal activation of similar regions in humans addicted to cocaine, suggesting that with more extensive drug use, the mPFC may no longer be required for engaging in drug seeking, but rather require activation of corticostriatal circuits involved in habit-like behavior [29,248,251–254]. Regardless of how these data are interpreted, they clearly identify that repeated drug exposure produces alterations in pyramidal neuron responsivity, reflecting a combination of modified synaptic transmission, as well as intrinsic synaptic structure and physiology.

Excitatory and Inhibitory Plasticity

A number of studies have provided evidence that repeated exposure to drugs of abuse produce enduring changes in synaptic reorganization of pyramidal neurons in the mPFC. With the exception of heroin, most drugs (amphetamine, cocaine, ethanol, and nicotine) promote increases in dendritic arborization and spine density [255–258] but see [259–261]. Cocaine and ethanol-induced alterations in spine density are associated with increased expression and transmission of NMDA (GluN) and AMPA (GluA) receptors, as well as activity-dependent synaptic plasticity [257,258,262,263]. Further, as early as 72–96 h following cocaine, there is an increase in D1R-mediated induction of LTP [246,264]. This facilitation of LTP appears to be due in part to a reduction in GABAergic transmission mediated by the ionotropic, GABA$_A$ receptor [245,264].

Alternatively, withdrawal from heroin self-administration does not appear to alter pyramidal neuron excitatory synaptic strength in the mPFC, however, it is worth noting that this may reflect a lack of parsing out plasticity within the PrL versus ILC pyramidal neurons [265]. These modifications, in conjunction with observed increases in excitability of mPFC pyramidal neurons, particularly following cocaine, may indicate that glutamatergic input to pyramidal neurons is augmented, potentially favoring input that drives relapse. However, at present, direct measures of excitatory synaptic strength in mPFC pyramidal neurons are lacking.

Pathway-Specific Plasticity

As discussed earlier in this chapter, the mPFC-to-NAc pathway has long been proposed to undergo drug-induced synaptic plasticity that is important for drug-seeking behavior associated with a number of drugs. Behavioral and electrophysiological evidence suggests that with cocaine, increased synaptic strength and activation of the PrL-to-core pathways plays an essential role in cue- and drug-induced reinstatement of cocaine seeking [140,154,162,176,191,238,266]. Augmented glutamatergic transmission at PrL-to-core synapses is thought to reflect a number of factors, including reduced basal tone on presynaptic glutamate autoreceptors [162,267], increased postsynaptic expression of AMPARs [135,140,154], as well as increased excitability of Layer 5/6 pyramidal neurons selective for the PrL cortex [85].

Similar increases in synaptic strength have been observed within the ILC-to-shell pathway, however, reversible lesion studies suggest that activation of the ILC-to-shell projection may serve as a neural off switch for cocaine seeking [191]. Based on the proposed inhibitory role of the ILC-to-shell pathway, it is plausible to suggest that a progressive enhancement of synaptic strength within this pathway may reflect a neuroprotective mechanism of the brain in an attempt to mitigate future relapse [154,191]. On the other hand, experiences leading to a reduction in strength at ILC-to-shell MSN synapses, such as that observed following re-exposure to cocaine or other relapse-inducing stimuli [107,117,154,268], would be predicted to reduce the ability of this pathway to inhibit relapse-associated behavior and thus drive reinstatement. This pro- versus antirelapse circuitry is substantiated by evidence showing that decreasing basal glutamate transmission during extinction/abstinence, particularly in the NAc shell, enhances reinstatement of drug seeking [164,269], Furthermore, optically induced suppression of synaptic strength at ILC-to-shell synapses in an extinguished context previously associated with drug is sufficient to reinstate place preference in the absence of drug—an effect that can be blocked by optical inhibition of this same pathway [170]. Similarly, reversing increases in silent synapse expression during early withdrawal in the ILC-to-shell and PrL-to-core pathway prevents subsequent strengthening of both pathways, reversal of this plasticity exerts opposing behavioral effects, resulting in potentiated or inhibited relapse, respectively [154].

Plasticity within prefrontal-based circuits outside of the NAc is likely to play an important role in addiction as well, however, data are substantially lacking. The mPFC has dense reciprocal connections with areas such as the amygdala, a known substrate that controls cue-induced reinstatement of drug seeking [99,270–272]. The BLA has key neuronal mPFC efferents, particularly from the ILC, that are known to mediate extinction of fear-related memories by reducing output from the amygdala [273,274]. Thus, this pathway may have a similar function in extinction of conditioned drug-seeking across drugs of abuse [191,211,275]. On the other hand, activation of BLA-to-PrL projections is critically involved in reinstatement of cocaine seeking [276] and plasticity within the amygdala-to-PFC pathway may play a significant role in developing opiate addiction memory formation associated with dependence and withdrawal [277]. In addition, mPFC-to-VTA projections appear to gate systemic morphine-induced excitation of a subset of DA neurons [278], whereas VTA DA-dependent gating of inhibition in the PFC is facilitated by cocaine [279].

Alterations in Ionic Conductance in mPFC Neurons

Modifications in ion channel function can have a profound effect on the intrinsic physiology of neurons, as well as how a given cell population responds to excitatory and inhibitory input. Several lines of evidence, mostly stemming from cocaine-related research, indicate that reductions in mPFC pyramidal neuron basal transmission, as well as augmented responsivity and glutamate release in response to drug stimuli, may both be explained in part by a drug-induced shift toward DA preferentially activating D1R rather than D2R.

Reductions in D2R expression or binding have been observed in the frontal cortex of addicted individuals and the mPFC of self-administering rats, the latter of which correlates with deficits in mPFC-dependent cognitive function [280–282]. in vivo electrophysiology and pharmacology data indicate that reductions in D2R-mediated inhibitory regulation of GABAergic signaling arising from local interneurons leaves an unopposed D1R-mediated enhancement of local GABA release and an overall reduction in evoked pyramidal neuron firing [283–286]. Notably, similar implications for

FIG. 5 Effects of repeated drug exposure on PFC neurotransmission in the dorsal mPFC. Drug-induced reductions D_2R binding/expression and increased D_1R activity (presumably in pyramidal neurons), promotes a shift towards D_1R-mediating signaling, leading to (A) increased activation of protein kinase A (PKA). Repeated cocaine also (B) promotes a phosphatase dependent internalization of GIRK, and a reduction in $GABA_BR$-mediated signaling that are paralleled by increased $GABA_A$ signaling (C). PKA-mediated reductions in VSSC/VGKC (D), as well as increased $GABA_AR$ transmission may underlie reductions in basal PFC activity, while increases in L-type Ca^{2+} (E) and glutamate receptor signaling (F) and reductions in $GABA_B$-GIRK signaling may contribute to increased responsivity of pyramidal neurons to drug-related stimuli. PP2A = protein phosphatase 2A, GIRK = G protein-inwardly rectifying K^+ channel, VGKC = voltage-gated potassium (K+) channels, VSSC = voltage-sensitive sodium channels.

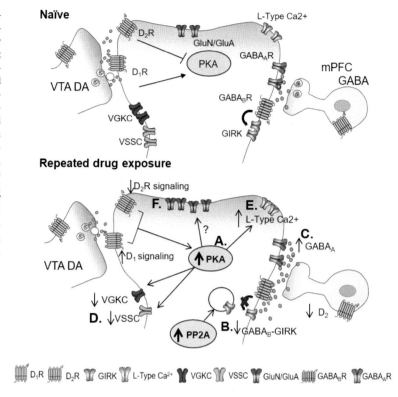

GABAergic transmission have also been observed with other drugs, such as nicotine and heroin [211,287]. This shift toward increased D1R activity may also underlie observed reductions in voltage-sensitive sodium channels (VSSCs) and decreased intracellular Ca^{2+} release in pyramidal neurons (Fig. 5) [288]. Together these adaptations would be predicted to reduce basal state activity and excitability of pyramidal neurons, and reduce behavioral flexibility, resulting in a state where only strong stimuli (e.g., drugs or conditioned cues) are capable of guiding behavior [289].

A cocaine-induced shift toward D1R tone may also underlie increased pyramidal neuron responsivity to drug-related information and downstream glutamate release. Abstinence from chronic cocaine treatment is associated with increased expression of proteins like activator of G-protein signaling 3 (AGS3) in the mPFC that negatively regulates signaling of inhibitory metabotropic G protein coupled receptors like D2 and $GABA_B$ [45,280]. This may in turn lead to an upregulation in signaling cascades that are usually inhibited by Gi/o-dependent signaling, such as the D1R/cAMP/PKA pathway [85,244,290], enhance Ca^{2+} influx, and reduce voltage-sensitive K^+ currents (Fig. 5) [85,244,247,291].

Increases in D1R activation appears to be an essential step in triggering reductions in mPFC pyramidal neuron $GABA_BR$ signaling that is responsible for regulating excitability of these cells [85]. This reduction occurs selectively in layer 5/6 (not layer 2/3) neurons in the PrL region of the mPFC, and reflects a phosphatase 2A (PP2A)-dependent endocytosis of GIRK channels (Fig. 5) [85]. Loss of this $GABA_BR$-GIRK pathway increases excitability and spike firing up to 40 days following repeated cocaine and promotes increased sensitivity to the behavioral effects of cocaine [297,298,85].

Short-Term Plasticity

Engaging in relapse behavior is thought to be initiated by transient changes in glutamate transmission in the VTA and NAc [117,165,299]; however, little is known regarding such changes in the mPFC. A study by Van den Oever et al. [265], provided the first direct evidence of such plasticity, showing that relapse to heroin seeking is mediated specifically by endocytosis of GluA2-containing AMPARs at pyramidal neuron synapses in the ILC [265]. It is possible that similar mechanisms promote relapse to other drugs of abuse, as re-exposure to heroin- as well as cocaine-conditioned cues both increase expression of proteins associated with endocytosis of GluA2-containing AMPARs in the mPFC, such as activity-regulated cytoskeleton (Arc) protein [219,220,300,301].

SUMMARY

In this chapter, we have highlighted how drugs of abuse alter the balance of excitatory and inhibitory signaling within key substrates of the mesocorticolimbic system—many of which are shared across drug types. However, different drugs of abuse undoubtedly produce distinct neurobiological, behavioral, and psychological effects that have implications for addiction modeling and treatment—thus generalizations across drugs of abuse should be made with extreme caution. While differences across drugs, cell types, and neural circuits were noted when pertinent, it is impossible to highlight all points of divergence in plasticity discussed here. Thus, it is important that the reader may not assume generalizations when reading this chapter, but rather explore the literature for themselves, as there are a number of fundamental gaps in how plasticity produced by drugs align temporally and anatomically. Accordingly, understanding divergences in plasticity can provide useful insight into potential therapies, it is likely that adaptations conserved across abused drugs represent a more targetable approach to treating addiction. As much of what we know regarding addiction has been derived from psychostimulant (i.e., cocaine) models, it is critical that future research utilizes new technologies to examine the overlap in relevance of plasticity across multiple drug types.

REFERENCES

[1] Nestler EJ. Molecular basis of long-term plasticity underlying addiction. Nat Rev Neurosci 2001;2:119–28.

[2] Nestler EJ. The neurobiology of cocaine addiction. Sci Pract Perspect 2005;3:4–10.

[3] Wise RA. Brain reward circuitry: insights from unsensed incentives. Neuron 2002;36:229–40.

[4] Wolf ME. Addiction: making the connection between behavioral changes and neuronal plasticity in specific pathways. Mol Interv 2002;2:146–57.

[5] Luscher C, Malenka RC. Drug-evoked synaptic plasticity in addiction: from molecular changes to circuit remodeling. Neuron 2011;69:650–63.

[6] Mameli M, Luscher C. Synaptic plasticity and addiction: learning mechanisms gone awry. Neuropharmacology 2011;61:1052–9.

[7] Hyman SE, Malenka RC, Nestler EJ. Neural mechanisms of addiction: the role of reward-related learning and memory. Annu Rev Neurosci 2006;29:565–98.

[8] Kalivas PW, O'Brien C. Drug addiction as a pathology of staged neuroplasticity. Neuropsychopharmacology 2008;33:166–80.

[9] Balleine BW, Delgado MR, Hikosaka O. The role of the dorsal striatum in reward and decision-making. J Neurosci 2007;27:8161–5.

[10] Jog MS. Building neural representations of habits. Science 1999;286:1745–9.

[11] Packard MG, Knowlton BJ. Learning and memory functions of the Basal Ganglia. Annu Rev Neurosci 2002;25:563–93.

[12] Di Chiara G, Imperato A. Drugs abused by humans preferentially increase synaptic dopamine concentrations in the mesolimbic system of freely moving rats. Proc Natl Acad Sci U S A 1988;85:5274–8.

[13] Pierce RC, Kumaresan V. The mesolimbic dopamine system: the final common pathway for the reinforcing effect of drugs of abuse? Neurosci Biobehav Rev 2006;30:215–38.

[14] Volkow ND, Fowler JS, Wang GJ, Swanson JM, Telang F. Dopamine in drug abuse and addiction: results of imaging studies and treatment implications. Arch Neurol 2007;64:1575–9.

[15] Wise RA. Drive, incentive, and reinforcement: the antecedents and consequences of motivation. Nebr Symp Motiv 2004;50:159–95.

[16] Badiani A, Belin D, Epstein D, Calu D, Shaham Y. Opiate versus psychostimulant addiction: the differences do matter. Nat Rev Neurosci 2011;12:685–700.

[17] Harris JE, Baldessarini RJ. Uptake of (3H)-catecholamines by homogenates of rat corpus striatum and cerebral cortex: effects of amphetamine analogues. Neuropharmacology 1973;12:669–79.

[18] Gysling K, Wang RY. Morphine-induced activation of A10 dopamine neurons in the rat. Brain Res 1983;277:119–27.

[19] Johnson SW, North RA. Opioids excite dopamine neurons by hyperpolarization of local interneurons. J Neurosci 1992;12:483–8.

[20] Fowler CD, Arends MA, Kenny PJ. Subtypes of nicotinic acetylcholine receptors in nicotine reward, dependence, and withdrawal: evidence from genetically modified mice. Behav Pharmacol 2008;19:461–84.

[21] Correa M, Salamone JD, Segovia KN, Pardo M, Longoni R, Spina L, Peana AT, Vinci S, Acquas E. Piecing together the puzzle of acetaldehyde as a neuroactive agent. Neurosci Biobehav Rev 2012;36:404–30.

[22] Guan YZ, Ye JH. Ethanol blocks long-term potentiation of GABAergic synapses in the ventral tegmental area involving mu-opioid receptors. Neuropsychopharmacology 2010;35:1841–9.

[23] Quertemont E, Tambour S, Tirelli E. The role of acetaldehyde in the neurobehavioral effects of ethanol: a comprehensive review of animal studies. Prog Neurobiol 2005;75:247–74.

[24] Bjorklund A, Dunnett SB. Fifty years of dopamine research. Trends Neurosci 2007;30:185–7.

[25] Kelley AE. Ventral striatal control of appetitive motivation: role in ingestive behavior and reward-related learning. Neurosci Biobehav Rev 2004;27:765–76.

[26] Kelley AE, Bakshi VP, Haber SN, Steininger TL, Will MJ, Zhang M. Opioid modulation of taste hedonics within the ventral striatum. Physiol Behav 2002;76:365–77.

[27] Salamone JD, Correa M. The mysterious motivational functions of mesolimbic dopamine. Neuron 2012;76:470–85.

[28] Wise RA. Rewards wanted: Molecular mechanisms of motivation. Discov Med 2004;4:180–6.

[29] Everitt BJ, Robbins TW. Neural systems of reinforcement for drug addiction: from actions to habits to compulsion. Nat Neurosci 2005;8:1481–9.

[30] Kim J, Park BH, Lee JH, Park SK, Kim JH. Cell type-specific alterations in the nucleus accumbens by repeated exposures to cocaine. Biol Psychiatry 2011;69:1026–34.

[31] Pascoli V, Terrier J, Hiver A, Luscher C. Sufficiency of mesolimbic dopamine neuron stimulation for the progression to addiction. Neuron 2015;88:1054–66.

[32] Volkow ND, Morales M. The brain on drugs: from reward to addiction. Cell 2015;162:712–25.

[33] Carr DB, Sesack SR. GABA-containing neurons in the rat ventral tegmental area project to the prefrontal cortex. Synapse 2000;38:114–23.

[34] Dobi A, Margolis EB, Wang HL, Harvey BK, Morales M. Glutamatergic and nonglutamatergic neurons of the ventral tegmental area establish local synaptic contacts with dopaminergic and nondopaminergic neurons. J Neurosci 2010;30:218–29.

[35] Van Bockstaele EJ, Pickel VM. GABA-containing neurons in the ventral tegmental area project to the nucleus accumbens in rat brain. Brain Res 1995;682:215–21.

[36] Yamaguchi T, Sheen W, Morales M. Glutamatergic neurons are present in the rat ventral tegmental area. Eur J Neurosci 2007;25:106–18.

[37] Ford CP, Mark GP, Williams JT. Properties and opioid inhibition of mesolimbic dopamine neurons vary according to target location. J Neurosci 2006;26:2788–97.

[38] Lammel S, Ion DI, Roeper J, Malenka RC. Projection-specific modulation of dopamine neuron synapses by aversive and rewarding stimuli. Neuron 2011;70:855–62.

[39] Lammel S, Hetzel A, Hackel O, Jones I, Liss B, Roeper J. Unique properties of mesoprefrontal neurons within a dual mesocorticolimbic dopamine system. Neuron 2008;57:760–73.

[40] Margolis EB, Mitchell JM, Ishikawa J, Hjelmstad GO, Fields HL. Midbrain dopamine neurons: projection target determines action potential duration and dopamine D(2) receptor inhibition. J Neurosci 2008;28:8908–13.

[41] White FJ. Synaptic regulation of mesocorticolimbic dopamine neurons. Annu Rev Neurosci 1996;19:405–36.

[42] Geisler S, Marinelli M, Degarmo B, Becker ML, Freiman AJ, Beales M, Meredith GE, Zahm DS. Prominent activation of brainstem and pallidal afferents of the ventral tegmental area by cocaine. Neuropsychopharmacology 2008;33:2688–700.

[43] Sesack SR, Aoki C, Pickel VM. Ultrastructural localization of D2 receptor-like immunoreactivity in midbrain dopamine neurons and their striatal targets. J Neurosci 1994;14:88–106.

[44] Beckstead MJ, Grandy DK, Wickman K, Williams JT. Vesicular dopamine release elicits an inhibitory postsynaptic current in midbrain dopamine neurons. Neuron 2004;42:939–46.

[45] Hearing MC, Zink AN, Wickman K. Cocaine-induced adaptations in metabotropic inhibitory signaling in the mesocorticolimbic system. Rev Neurosci 2012;23:325–51.

[46] Kaufling J, Veinante P, Pawlowski SA, Freund-Mercier MJ, Barrot M. Afferents to the GABAergic tail of the ventral tegmental area in the rat. J Comp Neurol 2009;513:597–621.

[47] Smith RJ, Lobo MK, Spencer S, Kalivas PW. Cocaine-induced adaptations in D1 and D2 accumbens projection neurons (a dichotomy not necessarily synonymous with direct and indirect pathways). Curr Opin Neurobiol 2013;23:546–52.

[48] Edwards NJ, Tejeda HA, Pignatelli M, Zhang S, McDevitt RA, Wu J, Bass CE, Bettler B, Morales M, Bonci A. Circuit specificity in the inhibitory architecture of the VTA regulates cocaine-induced behavior. Nat Neurosci 2017;20:438–48.

[49] Lecca S, Trusel M, Mameli M. Footshock-induced plasticity of GABAB signalling in the lateral habenula requires dopamine and glucocorticoid receptors. Synapse 2017;71:e21948.

[50] Lecca S, Melis M, Luchicchi A, Ennas MG, Castelli MP, Muntoni AL, Pistis M. Effects of drugs of abuse on putative rostromedial tegmental neurons, inhibitory afferents to midbrain dopamine cells. Neuropsychopharmacology 2011;36:589–602.

[51] Balcita-Pedicino JJ, Omelchenko N, Bell R, Sesack SR. The inhibitory influence of the lateral habenula on midbrain dopamine cells: ultrastructural evidence for indirect mediation via the rostromedial mesopontine tegmental nucleus. J Comp Neurol 2011;519:1143–64.

[52] Jhou TC, Geisler S, Marinelli M, Degarmo BA, Zahm DS. The mesopontine rostromedial tegmental nucleus: a structure targeted by the lateral habenula that projects to the ventral tegmental area of Tsai and substantia nigra compacta. J Comp Neurol 2009;513:566–96.

[53] Kauer JA. Learning mechanisms in addiction: synaptic plasticity in the ventral tegmental area as a result of exposure to drugs of abuse. Annu Rev Physiol 2004;66:447–75.

[54] Kauer JA, Malenka RC. Synaptic plasticity and addiction. Nat Rev Neurosci 2007;8:844–58.

[55] Ungless MA, Whistler JL, Malenka RC, Bonci A. Single cocaine exposure in vivo induces long-term potentiation in dopamine neurons. Nature 2001;411:583–7.

[56] Argilli E, Sibley DR, Malenka RC, England PM, Bonci A. Mechanism and time course of cocaine-induced long-term potentiation in the ventral tegmental area. J Neurosci 2008;28:9092–100.

[57] Ahn KC, Bernier BE, Harnett MT, Morikawa H. IP3 receptor sensitization during in vivo amphetamine experience enhances NMDA receptor plasticity in dopamine neurons of the ventral tegmental area. J Neurosci 2010;30:6689–99.

[58] Bernier BE, Whitaker LR, Morikawa H. Previous ethanol experience enhances synaptic plasticity of NMDA receptors in the ventral tegmental area. J Neurosci 2011;31:5205–12.

[59] Gao M, Jin Y, Yang K, Zhang D, Lukas RJ, Wu J. Mechanisms involved in systemic nicotine-induced glutamatergic synaptic plasticity on dopamine neurons in the ventral tegmental area. J Neurosci 2010;30:13814–25.

[60] Borgland SL, Malenka RC, Bonci A. Acute and chronic cocaine-induced potentiation of synaptic strength in the ventral tegmental area: electrophysiological and behavioral correlates in individual rats. J Neurosci 2004;24:7482–90.

[61] Chen BT, Bowers MS, Martin M, Hopf FW, Guillory AM, Carelli RM, Chou JK, Bonci A. Cocaine but not natural reward self-administration nor passive cocaine infusion produces persistent LTP in the VTA. Neuron 2008;59:288–97.

[62] Bellone C, Luscher C. Cocaine triggered AMPA receptor redistribution is reversed in vivo by mGluR-dependent long-term depression. Nat Neurosci 2006;9:636–41.

[63] Luu P, Malenka RC. Spike timing-dependent long-term potentiation in ventral tegmental area dopamine cells requires PKC. J Neurophysiol 2008;100:533–8.

[64] Good CH, Lupica CR. Afferent-specific AMPA receptor subunit composition and regulation of synaptic plasticity in midbrain dopamine neurons by abused drugs. J Neurosci 2010;30:7900–9.

[65] Mameli M, Bellone C, Brown MT, Luscher C. Cocaine inverts rules for synaptic plasticity of glutamate transmission in the ventral tegmental area. Nat Neurosci 2011;14:414–6.

[66] Pellegrini-Giampietro DE. An activity-dependent spermine-mediated mechanism that modulates glutamate transmission. Trends Neurosci 2003;26:9–11.

[67] Yuan T, Mameli M, O'Connor EC, Dey PN, Verpelli C, Sala C, Perez-Otano I, Luscher C, Bellone C. Expression of cocaine-evoked synaptic plasticity by GluN3A-containing NMDA receptors. Neuron 2013;80:1025–38.

[68] Creed M, Kaufling J, Fois GR, Jalabert M, Yuan T, Luscher C, Georges F, Bellone C. Cocaine exposure enhances the activity of ventral tegmental area dopamine neurons via calcium-impermeable NMDARs. J Neurosci 2016;36:10759–68.

[69] Roberts AC, Diez-Garcia J, Rodriguiz RM, Lopez IP, Lujan R, Martinez-Turrillas R, Pico E, Henson MA, Bernardo DR, Jarrett TM, Clendeninn DJ, Lopez-Mascaraque L, Feng G, Lo DC, Wesseling JF, Wetsel WC, Philpot BD, Perez-Otano I. Downregulation of NR3A-containing NMDARs is required for synapse maturation and memory consolidation. Neuron 2009;63:342–56.

[70] Tong G, Takahashi H, Tu S, Shin Y, Talantova M, Zago W, Xia P, Nie Z, Goetz T, Zhang D, Lipton SA, Nakanishi N. Modulation of NMDA receptor properties and synaptic transmission by the NR3A subunit in mouse hippocampal and cerebrocortical neurons. J Neurophysiol 2008;99:122–32.

[71] Brown MT, Bellone C, Mameli M, Labouebe G, Bocklisch C, Balland B, Dahan L, Lujan R, Deisseroth K, Luscher C. Drug-driven AMPA receptor redistribution mimicked by selective dopamine neuron stimulation. PLoS One 2010;5:e15870.

[72] Saal D, Dong Y, Bonci A, Malenka RC. Drugs of abuse and stress trigger a common synaptic adaptation in dopamine neurons. Neuron 2003;37:577–82.

[73] Berridge KC, Robinson TE. What is the role of dopamine in reward: hedonic impact, reward learning, or incentive salience? Brain Res Brain Res Rev 1998;28:309–69.

[74] Mameli M, Halbout B, Creton C, Engblom D, Parkitna JR, Spanagel R, Luscher C. Cocaine-evoked synaptic plasticity: persistence in the VTA triggers adaptations in the NAc. Nat Neurosci 2009;12:1036–41.

[75] Niehaus JL, Murali M, Kauer JA. Drugs of abuse and stress impair LTP at inhibitory synapses in the ventral tegmental area. Eur J Neurosci 2010;32:108–17.

[76] Nugent FS, Penick EC, Kauer JA. Opioids block long-term potentiation of inhibitory synapses. Nature 2007;446:1086–90.

[77] Polter AM, Kauer JA. Stress and VTA synapses: implications for addiction and depression. Eur J Neurosci 2014;39:1179–88.

[78] Arora D, Hearing M, Haluk DM, Mirkovic K, Fajardo-Serrano A, Wessendorf MW, Watanabe M, Lujan R, Wickman K. Acute cocaine exposure weakens GABA(B) receptor-dependent G-protein-gated inwardly rectifying K+ signaling in dopamine neurons of the ventral tegmental area. J Neurosci 2011;31:12251–7.

[79] Padgett CL, Lalive AL, Tan KR, Terunuma M, Munoz MB, Pangalos MN, Martinez-Hernandez J, Watanabe M, Moss SJ, Lujan R, Luscher C, Slesinger PA. Methamphetamine-evoked depression of GABA(B) receptor signaling in GABA neurons of the VTA. Neuron 2012;73:978–89.

[80] Ackerman JM, White FJ. A10 somatodendritic dopamine autoreceptor sensitivity following withdrawal from repeated cocaine treatment. Neurosci Lett 1990;117:181–7.

[81] Marinelli M, Cooper DC, Baker LK, White FJ. Impulse activity of midbrain dopamine neurons modulates drug-seeking behavior. Psychopharmacology 2003;168:84–98.

[82] Henry DJ, Greene MA, White FJ. Electrophysiological effects of cocaine in the mesoaccumbens dopamine system: repeated administration. J Pharmacol Exp Ther 1989;251:833–9.

[83] Chergui K, Charlety PJ, Akaoka H, Saunier CF, Brunet JL, Buda M, Svensson TH, Chouvet G. Tonic activation of NMDA receptors causes spontaneous burst discharge of rat midbrain dopamine neurons in vivo. Eur J Neurosci 1993;5:137–44.

[84] Tong ZY, Overton PG, Clark D. Antagonism of NMDA receptors but not AMPA/kainate receptors blocks bursting in dopaminergic neurons induced by electrical stimulation of the prefrontal cortex. J Neural Transm (Vienna) 1996;103:889–904.

[85] Hearing M, Kotecki L, Marron Fernandez de Velasco E, Fajardo-Serrano A, Chung HJ, Lujan R, Wickman K. Repeated cocaine weakens GABA(B)-Girk signaling in layer 5/6 pyramidal neurons in the prelimbic cortex. Neuron 2013;80:159–70.

[86] Rifkin RA, Moss SJ, Slesinger PA. G protein-gated potassium channels: a link to drug addiction. Trends Pharmacol Sci 2017;38:378–92.

[87] McCall NM, Kotecki L, Dominguez-Lopez S, Marron Fernandez de Velasco E, Carlblom N, Sharpe AL, Beckstead MJ, Wickman K. Selective ablation of GIRK channels in dopamine neurons alters behavioral effects of cocaine in mice. Neuropsychopharmacology 2017;42:707–15.

[88] Kotecki L, Hearing M, McCall NM, Marron Fernandez de Velasco E, Pravetoni M, Arora D, Victoria NC, Munoz MB, Xia Z, Slesinger PA, Weaver CD, Wickman K. GIRK channels modulate opioid-induced motor activity in a cell type- and subunit-dependent manner. J Neurosci 2015;35:7131–42.

[89] Stewart J, Vezina P. Microinjections of Sch-23390 into the ventral tegmental area and substantia nigra pars reticulata attenuate the development of sensitization to the locomotor activating effects of systemic amphetamine. Brain Res 1989;495:401–6.

[90] Kalivas PW, Volkow ND. The neural basis of addiction: a pathology of motivation and choice. Am J Psychiatry 2005;162:1403–13.

[91] Everitt BJ, Parkinson JA, Olmstead MC, Arroyo M, Robledo P, Robbins TW. Associative processes in addiction and reward. The role of amygdala-ventral striatal subsystems. Ann N Y Acad Sci 1999;877:412–38.

[92] Heimer L, Zahm DS, Churchill L, Kalivas PW, Wohltmann C. Specificity in the projection patterns of accumbal core and shell in the rat. Neuroscience 1991;41:89–125.

[93] Zahm DS, Brog JS. On the significance of subterritories in the "accumbens" part of the rat ventral striatum. Neuroscience 1992;50:751–67.

[94] Sesack SR, Grace AA. Cortico-Basal Ganglia reward network: microcircuitry. Neuropsychopharmacology 2010;35:27–47.

[95] Shiflett MW, Balleine BW. Contributions of ERK signaling in the striatum to instrumental learning and performance. Behav Brain Res 2011;218:240–7.

[96] Voorn P, Vanderschuren LJ, Groenewegen HJ, Robbins TW, Pennartz CM. Putting a spin on the dorsal-ventral divide of the striatum. Trends Neurosci 2004;27:468–74.

[97] Heimer L, Alheid GF, de Olmos JS, Groenewegen HJ, Haber SN, Harlan RE, Zahm DS. The accumbens: beyond the core-shell dichotomy. J Neuropsychiatry Clin Neurosci 1997;9:354–81.

[98] Brog JS, Salyapongse A, Deutch AY, Zahm DS. The patterns of afferent innervation of the core and shell in the "accumbens" part of the rat ventral striatum: immunohistochemical detection of retrogradely transported fluoro-gold. J Comp Neurol 1993;338:255–78.

[99] Sesack SR, Deutch AY, Roth RH, Bunney BS. Topographical organization of the efferent projections of the medial prefrontal cortex in the rat: an anterograde tract-tracing study with Phaseolus vulgaris leucoagglutinin. J Comp Neurol 1989;290:213–42.

[100] Le Moine C, Bloch B. D1 and D2 dopamine receptor gene expression in the rat striatum: sensitive cRNA probes demonstrate prominent segregation of D1 and D2 mRNAs in distinct neuronal populations of the dorsal and ventral striatum. J Comp Neurol 1995;355:418–26.

[101] Lobo MK, Karsten SL, Gray M, Geschwind DH, Yang XW. FACS-array profiling of striatal projection neuron subtypes in juvenile and adult mouse brains. Nat Neurosci 2006;9:443–52.

[102] Bertran-Gonzalez J, Bosch C, Maroteaux M, Matamales M, Herve D, Valjent E, Girault JA. Opposing patterns of signaling activation in dopamine D1 and D2 receptor-expressing striatal neurons in response to cocaine and haloperidol. J Neurosci 2008;28:5671–85.

[103] Kupchik YM, Brown RM, Heinsbroek JA, Lobo MK, Schwartz DJ, Kalivas PW. Coding the direct/indirect pathways by D1 and D2 receptors is not valid for accumbens projections. Nat Neurosci 2015;18:1230–2.

[104] Thibault D, Loustalot F, Fortin GM, Bourque MJ, Trudeau LE. Evaluation of D1 and D2 dopamine receptor segregation in the developing striatum using BAC transgenic mice. PLoS One 2013;8:e67219.

[105] Matamales M, Bertran-Gonzalez J, Salomon L, Degos B, Deniau JM, Valjent E, Herve D, Girault JA. Striatal medium-sized spiny neurons: identification by nuclear staining and study of neuronal subpopulations in BAC transgenic mice. PLoS One 2009;4:e4770.

[106] Valjent E, Bertran-Gonzalez J, Herve D, Fisone G, Girault JA. Looking BAC at striatal signaling: cell-specific analysis in new transgenic mice. Trends Neurosci 2009;32:538–47.

[107] Pascoli V, Turiault M, Luscher C. Reversal of cocaine-evoked synaptic potentiation resets drug-induced adaptive behaviour. Nature 2011;481:71–5.

[108] Pascoli V, Terrier J, Espallergues J, Valjent E, O'Connor EC, Luscher C. Contrasting forms of cocaine-evoked plasticity control components of relapse. Nature 2014;509:459–64.

[109] Graziane NM, Sun S, Wright WJ, Jang D, Liu Z, Huang YH, Nestler EJ, Wang YT, Schluter OM, Dong Y. Opposing mechanisms mediate morphine- and cocaine-induced generation of silent synapses. Nat Neurosci 2016;19:915–25.

[110] Hearing MC, Jedynak J, Ebner SR, Ingebretson A, Asp AJ, Fischer RA, Schmidt C, Larson EB, Thomas MJ. Reversal of morphine-induced cell-type-specific synaptic plasticity in the nucleus accumbens shell blocks reinstatement. Proc Natl Acad Sci U S A 2016;113:757–62.

[111] Terrier J, Luscher C, Pascoli V. Cell-Type Specific Insertion of GluA2-Lacking AMPARs with Cocaine Exposure Leading to Sensitization, Cue-Induced Seeking, and Incubation of Craving. Neuropsychopharmacology 2016;41:1779–89.

[112] Gerfen CR, Surmeier DJ. Modulation of striatal projection systems by dopamine. Annu Rev Neurosci 2011;34:441–66.

[113] Bock R, Shin JH, Kaplan AR, Dobi A, Markey E, Kramer PF, Gremel CM, Christensen CH, Adrover MF, Alvarez VA. Strengthening the accumbal indirect pathway promotes resilience to compulsive cocaine use. Nat Neurosci 2013;16:632–8.

[114] Kravitz AV, Tye LD, Kreitzer AC. Distinct roles for direct and indirect pathway striatal neurons in reinforcement. Nat Neurosci 2012;15:816–8.

[115] Lobo MK, Nestler EJ. The striatal balancing act in drug addiction: distinct roles of direct and indirect pathway medium spiny neurons. Front Neuroanat 2011;5:41.

[116] Boudreau AC, Wolf ME. Behavioral sensitization to cocaine is associated with increased AMPA receptor surface expression in the nucleus accumbens. J Neurosci 2005;25:9144–51.

[117] Kourrich S, Rothwell PE, Klug JR, Thomas MJ. Cocaine experience controls bidirectional synaptic plasticity in the nucleus accumbens. J Neurosci 2007;27:7921–8.

[118] Wu X, Shi M, Wei C, Yang M, Liu Y, Liu Z, Zhang X, Ren W. Potentiation of synaptic strength and intrinsic excitability in the nucleus accumbens after 10 days of morphine withdrawal. J Neurosci Res 2012;90:1270–83.

[119] Dobi A, Seabold GK, Christensen CH, Bock R, Alvarez VA. Cocaine-induced plasticity in the nucleus accumbens is cell specific and develops without prolonged withdrawal. J Neurosci 2011;31:1895–904.

[120] Jedynak J, Hearing M, Ingebretson A, Ebner SR, Kelly M, Fischer RA, Kourrich S, Thomas MJ. Cocaine and amphetamine induce overlapping but distinct patterns of AMPAR plasticity in nucleus accumbens medium spiny neurons. Neuropsychopharmacology 2016;41:464–76.

[121] Siciliano CA, Calipari ES, Jones SR. Amphetamine potency varies with dopamine uptake rate across striatal subregions. J Neurochem 2014;131:348–55.

[122] Giorgi O, Piras G, Lecca D, Corda MG. Differential activation of dopamine release in the nucleus accumbens core and shell after acute or repeated amphetamine injections: a comparative study in the Roman high- and low-avoidance rat lines. Neuroscience 2005;135:987–98.

[123] Pierce RC, Kalivas PW. Amphetamine produces sensitized increases in locomotion and extracellular dopamine preferentially in the nucleus accumbens shell of rats administered repeated cocaine. J Pharmacol Exp Ther 1995;275:1019–29.

[124] Jeanes ZM, Buske TR, Morrisett RA. In vivo chronic intermittent ethanol exposure reverses the polarity of synaptic plasticity in the nucleus accumbens shell. J Pharmacol Exp Ther 2011;336:155–64.

[125] Jeanes ZM, Buske TR, Morrisett RA. Cell type-specific synaptic encoding of ethanol exposure in the nucleus accumbens shell. Neuroscience 2014;277:184–95.

[126] Martin M, Chen BT, Hopf FW, Bowers MS, Bonci A. Cocaine self-administration selectively abolishes LTD in the core of the nucleus accumbens. Nat Neurosci 2006;9:868–9.

[127] Robbe D, Bockaert J, Manzoni OJ. Metabotropic glutamate receptor 2/3-dependent long-term depression in the nucleus accumbens is blocked in morphine withdrawn mice. Eur J Neurosci 2002;16:2231–5.

[128] Schramm-Sapyta NL, Olsen CM, Winder DG. Cocaine self-administration reduces excitatory responses in the mouse nucleus accumbens shell. Neuropsychopharmacology 2006;31:1444–51.

[129] Spiga S, Talani G, Mulas G, Licheri V, Fois GR, Muggironi G, Masala N, Cannizzaro C, Biggio G, Sanna E, Diana M. Hampered long-term depression and thin spine loss in the nucleus accumbens of ethanol-dependent rats. Proc Natl Acad Sci U S A 2014;111:E3745–3754.

[130] Thomas MJ, Beurrier C, Bonci A, Malenka RC. Long-term depression in the nucleus accumbens: a neural correlate of behavioral sensitization to cocaine. Nat Neurosci 2001;4:1217–23.

[131] Brown TE, Lee BR, Mu P, Ferguson D, Dietz D, Ohnishi YN, Lin Y, Suska A, Ishikawa M, Huang YH, Shen H, Kalivas PW, Sorg BA, Zukin RS, Nestler EJ, Dong Y, Schluter OM. A silent synapse-based mechanism for cocaine-induced locomotor sensitization. J Neurosci 2011;31:8163–74.

[132] Huang YH, Lin Y, Mu P, Lee BR, Brown TE, Wayman G, Marie H, Liu W, Yan Z, Sorg BA, Schluter OM, Zukin RS, Dong Y. In vivo cocaine experience generates silent synapses. Neuron 2009;63:40–7.

[133] Gipson CD, Kupchik YM, Shen H, Reissner KJ, Thomas CA, Kalivas PW. Relapse induced by cues predicting cocaine depends on rapid, transient synaptic potentiation. Neuron 2013;77:867–72.

[134] Gipson CD, Reissner KJ, Kupchik YM, Smith AC, Stankeviciute N, Hensley-Simon ME, Kalivas PW. Reinstatement of nicotine seeking is mediated by glutamatergic plasticity. Proc Natl Acad Sci U S A 2013;110:9124–9.

[135] Conrad KL, Tseng KY, Uejima JL, Reimers JM, Heng LJ, Shaham Y, Marinelli M, Wolf ME. Formation of accumbens GluR2-lacking AMPA receptors mediates incubation of cocaine craving. Nature 2008;454:118–21.

[136] Grimm JW, Hope BT, Wise RA, Shaham Y. Neuroadaptation. Incubation of cocaine craving after withdrawal. Nature 2001;412:141–2.

[137] Moussawi K, Pacchioni A, Moran M, Olive MF, Gass JT, Lavin A, Kalivas PW. N-Acetylcysteine reverses cocaine-induced metaplasticity. Nat Neurosci 2009;12:182–9.

[138] Shen H, Moussawi K, Zhou W, Toda S, Kalivas PW. Heroin relapse requires long-term potentiation-like plasticity mediated by NMDA2b-containing receptors. Proc Natl Acad Sci U S A 2011;108:19407–12.

[139] Britt JP, Benaliouad F, McDevitt RA, Stuber GD, Wise RA, Bonci A. Synaptic and behavioral profile of multiple glutamatergic inputs to the nucleus accumbens. Neuron 2012;76:790–803.

[140] McCutcheon JE, Wang X, Tseng KY, Wolf ME, Marinelli M. Calcium-permeable AMPA receptors are present in nucleus accumbens synapses after prolonged withdrawal from cocaine self-administration but not experimenter-administered cocaine. J Neurosci 2011;31:5737–43.

[141] Ortinski PI, Vassoler FM, Carlson GC, Pierce RC. Temporally dependent changes in cocaine-induced synaptic plasticity in the nucleus accumbens shell are reversed by D1-like dopamine receptor stimulation. Neuropsychopharmacology 2012;37:1671–82.

[142] Shen HW, Gipson CD, Huits M, Kalivas PW. Prelimbic cortex and ventral tegmental area modulate synaptic plasticity differentially in nucleus accumbens during cocaine-reinstated drug seeking. Neuropsychopharmacology 2014;39:1169–77.

[143] Suska A, Lee BR, Huang YH, Dong Y, Schluter OM. Selective presynaptic enhancement of the prefrontal cortex to nucleus accumbens pathway by cocaine. Proc Natl Acad Sci U S A 2013;110:713–8.

[144] Trantham-Davidson H, LaLumiere RT, Reissner KJ, Kalivas PW, Knackstedt LA. Ceftriaxone normalizes nucleus accumbens synaptic transmission, glutamate transport, and export following cocaine self-administration and extinction training. J Neurosci 2012;32:12406–10.

[145] Jia Z, Agopyan N, Miu P, Xiong Z, Henderson J, Gerlai R, Taverna FA, Velumian A, MacDonald J, Carlen P, Abramow-Newerly W, Roder J. Enhanced LTP in mice deficient in the AMPA receptor GluR2. Neuron 1996;17:945–56.

[146] Scheyer AF, Wolf ME, Tseng KY. A protein synthesis-dependent mechanism sustains calcium-permeable AMPA receptor transmission in nucleus accumbens synapses during withdrawal from cocaine self-administration. J Neurosci 2014;34:3095–100.

[147] Russell SE, Puttick DJ, Sawyer AM, Potter DN, Mague S, Carlezon Jr. WA, Chartoff EH. Nucleus accumbens AMPA receptors are necessary for morphine-withdrawal-induced negative-affective states in rats. J Neurosci 2016;36:5748–62.

[148] Turrigiano GG, Nelson SB. Hebb and homeostasis in neuronal plasticity. Curr Opin Neurobiol 2000;10:358–64.

[149] Dong Y, Nestler EJ. The neural rejuvenation hypothesis of cocaine addiction. Trends Pharmacol Sci 2014;35:374–83.

[150] Lee BR, Dong Y. Cocaine-induced metaplasticity in the nucleus accumbens: silent synapse and beyond. Neuropharmacology 2011;61:1060–9.

[151] Hanse E, Seth H, Riebe I. AMPA-silent synapses in brain development and pathology. Nat Rev Neurosci 2013;14:839–50.

[152] Koya E, Cruz FC, Ator R, Golden SA, Hoffman AF, Lupica CR, Hope BT. Silent synapses in selectively activated nucleus accumbens neurons following cocaine sensitization. Nat Neurosci 2012;15:1556–62.

[153] Lee BR, Ma YY, Huang YH, Wang X, Otaka M, Ishikawa M, Neumann PA, Graziane NM, Brown TE, Suska A, Guo C, Lobo MK, Sesack SR, Wolf ME, Nestler EJ, Shaham Y, Schluter OM, Dong Y. Maturation of silent synapses in amygdala-accumbens projection contributes to incubation of cocaine craving. Nat Neurosci 2013;16:1644–51.

[154] Ma YY, Lee BR, Wang X, Guo C, Liu L, Cui R, Lan Y, Balcita-Pedicino JJ, Wolf ME, Sesack SR, Shaham Y, Schluter OM, Huang YH, Dong Y. Bidirectional modulation of incubation of cocaine craving by silent synapse-based remodeling of prefrontal cortex to accumbens projections. Neuron 2014;83:1453–67.

[155] Bellinger FP, Davidson MS, Bedi KS, Wilce PA. Neonatal ethanol exposure reduces AMPA but not NMDA receptor levels in the rat neocortex. Brain Res Dev Brain Res 2002;136:77–84.

[156] Maggi L, Le Magueresse C, Changeux JP, Cherubini E. Nicotine activates immature "silent" connections in the developing hippocampus. Proc Natl Acad Sci U S A 2003;100:2059–64.

[157] Isaac JT, Nicoll RA, Malenka RC. Evidence for silent synapses: implications for the expression of LTP. Neuron 1995;15:427–34.

[158] Liao D, Hessler NA, Malinow R. Activation of postsynaptically silent synapses during pairing-induced LTP in CA1 region of hippocampal slice. Nature 1995;375:400–4.

[159] Malenka RC, Bear MF. LTP and LTD: an embarrassment of riches. Neuron 2004;44:5–21.

[160] Marie H, Morishita W, Yu X, Calakos N, Malenka RC. Generation of silent synapses by acute in vivo expression of CaMKIV and CREB. Neuron 2005;45:741–52.

[161] Petralia RS, Esteban JA, Wang YX, Partridge JG, Zhao HM, Wenthold RJ, Malinow R. Selective acquisition of AMPA receptors over postnatal development suggests a molecular basis for silent synapses. Nat Neurosci 1999;2:31–6.

[161a] Pierce RC, Bell K, Duffy P, Kalivas PW. Repeated cocaine augments excitatory amino acid transmission in the nucleus accumbens only in rats having developed behavioral sensitization. J Neurosci 1996;16:1550–60.

[161b] Cornish JL, Duffy P, Kalivas PW. A role for nucleus accumbens glutamate transmission in the relapse to cocaine-seeking behavior. Neuroscience 1999;93:1359–67.

[161c] Cornish JL, Kalivas PW. Glutamate transmission in the nucleus accumbens mediates relapse in cocaine addiction. J Neurosci 2000;20:Rc89.

[161d] Suto N, Tanabe LM, Austin JD, Creekmore E, Pham CT, Vezina P. Previous exposure to psychostimulants enhances the reinstatement of cocaine seeking by nucleus accumbens AMPA. Neuropsychopharmacology 2004;29:2149–59.

[162] Kalivas PW. The glutamate homeostasis hypothesis of addiction. Nat Rev Neurosci 2009;10:561–72.

[163] Kalivas PW, Hu XT. Exciting inhibition in psychostimulant addiction. Trends Neurosci 2006;29:610–6.

[164] Bachtell RK, Choi KH, Simmons DL, Falcon E, Monteggia LM, Neve RL, Self DW. Role of GluR1 expression in nucleus accumbens neurons in cocaine sensitization and cocaine-seeking behavior. Eur J Neurosci 2008;27:2229–40.

[164a] Sutton MA, Schmidt EF, Choi KH, Schad CA, Whisler K, Simmons D, Karanian DA, Monteggia LM, Neve RL, Self DW. Extinction-induced upregulation in AMPA receptors reduces cocaine-seeking behaviour. Nature 2003;421:70–5.

[165] Brebner K, Wong TP, Liu L, Liu Y, Campsall P, Gray S, Phelps L, Phillips AG, Wang YT. Nucleus accumbens long-term depression and the expression of behavioral sensitization. Science 2005;310:1340–3.

[166] Famous KR, Kumaresan V, Sadri-Vakili G, Schmidt HD, Mierke DF, Cha JH, Pierce RC. Phosphorylation-dependent trafficking of GluR2-containing AMPA receptors in the nucleus accumbens plays a critical role in the reinstatement of cocaine seeking. J Neurosci 2008;28:11061–70.

[167] Schmidt HD, Schassburger RL, Guercio LA, Pierce RC. Stimulation of mGluR5 in the accumbens shell promotes cocaine seeking by activating PKC gamma. J Neurosci 2013;33:14160–9.

[168] Ebner SR, Larson EB, Hearing MC, Ingebretson AE, Thomas MJ. Extinction and reinstatement of cocaine-seeking behavior in self-administering mice produces bidirectional effects on AMPAR-mediated synaptic plasticity in the nucleus accumbens shell. Psychopharmacology 2018; [In review].

[168a] Schmidt HD, Kimmey BA, Arreola AC, Pierce RC. Group I metabotropic glutamate receptor-mediated activation of PKC gamma in the nucleus accumbens core promotes the reinstatement of cocaine seeking. Addict Biol 2015;20:285–96.

[169] Ingebretson A, Hearing MC, Huffington ED, Thomas MJ. Endogenous dopamine and endocannabinoid signaling mediate cocaine-induced reversal of AMPAR synaptic potentiation in the nucleus accumbens shell. Neuropharmacology 2018;131:154–65.

[170] Benneyworth MA, Hearing MC, Asp AJ, Ingebretson AE, Schmidt CE, Silvis KA, Larson EB, Ebner SR, Thomas MJ. Synaptic depotentiation and mGluR5 activity in the nucleus accumbens drives cocaine-primed reinstatement. J Neurosci 2018; [in re-submission].

[171] Goto Y, Grace AA. Dopaminergic modulation of limbic and cortical drive of nucleus accumbens in goal-directed behavior. Nat Neurosci 2005;8:805–12.

[172] Hogarth L, Balleine BW, Corbit LH, Killcross S. Associative learning mechanisms underpinning the transition from recreational drug use to addiction. Ann N Y Acad Sci 2013;1282:12–24.

[173] Silva AJ, Zhou Y, Rogerson T, Shobe J, Balaji J. Molecular and cellular approaches to memory allocation in neural circuits. Science 2009;326:391–5.

[174] Hikida T, Kimura K, Wada N, Funabiki K, Nakanishi S. Distinct roles of synaptic transmission in direct and indirect striatal pathways to reward and aversive behavior. Neuron 2010;66:896–907.

[175] Lobo MK, Covington 3rd HE, Chaudhury D, Friedman AK, Sun H, Damez-Werno D, Dietz DM, Zaman S, Koo JW, Kennedy PJ, Mouzon E, Mogri M, Neve RL, Deisseroth K, Han MH, Nestler EJ. Cell type-specific loss of BDNF signaling mimics optogenetic control of cocaine reward. Science 2010;330:385–90.

[176] Stefanik MT, Kupchik YM, Brown RM, Kalivas PW. Optogenetic evidence that pallidal projections, not nigral projections, from the nucleus accumbens core are necessary for reinstating cocaine seeking. J Neurosci 2013;33:13654–62.

[177] Yang Y, Calakos N. Presynaptic long-term plasticity. Front Synaptic Neurosci 2013;5:8.

[178] MacAskill AF, Cassel JM, Carter AG. Cocaine exposure reorganizes cell type- and input-specific connectivity in the nucleus accumbens. Nat Neurosci 2014;17:1198–207.

[179] James AS, Chen JY, Cepeda C, Mittal N, Jentsch JD, Levine MS, Evans CJ, Walwyn W. Opioid self-administration results in cell-type specific adaptations of striatal medium spiny neurons. Behav Brain Res 2013;256:279–83.

[180] Zhu Y, Wienecke CF, Nachtrab G, Chen X. A thalamic input to the nucleus accumbens mediates opiate dependence. Nature 2016;530:219–22.

[181] Moussawi K, Zhou W, Shen H, Reichel CM, See RE, Carr DB, Kalivas PW. Reversing cocaine-induced synaptic potentiation provides enduring protection from relapse. Proc Natl Acad Sci U S A 2011;108:385–90.

[182] Kasanetz F, Deroche-Gamonet V, Berson N, Balado E, Lafourcade M, Manzoni O, Piazza PV. Transition to addiction is associated with a persistent impairment in synaptic plasticity. Science 2010;328:1709–12.

[183] Liechti ME, Lhuillier L, Kaupmann K, Markou A. Metabotropic glutamate 2/3 receptors in the ventral tegmental area and the nucleus accumbens shell are involved in behaviors relating to nicotine dependence. J Neurosci 2007;27:9077–85.

[184] Melendez RI, Hicks MP, Cagle SS, Kalivas PW. Ethanol exposure decreases glutamate uptake in the nucleus accumbens. Alcohol Clin Exp Res 2005;29:326–33.

[185] Reid MS, Mickalian JD, Delucchi KL, Berger SP. A nicotine antagonist, mecamylamine, reduces cue-induced cocaine craving in cocaine-dependent subjects. Neuropsychopharmacology 1999;20:297–307.

[186] Bowers MS, Chen BT, Bonci A. AMPA receptor synaptic plasticity induced by psychostimulants: the past, present, and therapeutic future. Neuron 2010;67:11–24.

[187] McGeehan AJ, Olive MF. The mGluR5 antagonist MPEP reduces the conditioned rewarding effects of cocaine but not other drugs of abuse. Synapse 2003;47:240–2.

[188] McGeehan AJ, Olive MF. Attenuation of cocaine-induced reinstatement of cocaine conditioned place preference by acamprosate. Behav Pharmacol 2006;17:363–7.

[189] Ortinski PI. Cocaine-induced changes in NMDA receptor signaling. Mol Neurobiol 2014;50:494–506.

[190] Pierce RC, Wolf ME. Psychostimulant-induced neuroadaptations in nucleus accumbens AMPA receptor transmission. Cold Spring Harb Perspect Med 2013;3:a012021.

[191] Peters J, Pattij T, De Vries TJ. Targeting cocaine versus heroin memories: divergent roles within ventromedial prefrontal cortex. Trends Pharmacol Sci 2013;34:689–95.

[192] Wolf ME, Tseng KY. Calcium-permeable AMPA receptors in the VTA and nucleus accumbens after cocaine exposure: when, how, and why? Front Mol Neurosci 2012;5:72.

[193] Grueter BA, Brasnjo G, Malenka RC. Postsynaptic TRPV1 triggers cell type-specific long-term depression in the nucleus accumbens. Nat Neurosci 2010;13:1519–25.

[194] Luscher C, Huber KM. Group 1 mGluR-dependent synaptic long-term depression: mechanisms and implications for circuitry and disease. Neuron 2010;65:445–59.

[195] Gipson CD, Kupchik YM, Kalivas PW. Rapid, transient synaptic plasticity in addiction. Neuropharmacology 2014;76(Pt B):276–86.

[196] Araki H, Kawakami KY, Jin C, Suemaru K, Kitamura Y, Nagata M, Futagami K, Shibata K, Kawasaki H, Gomita Y. Nicotine attenuates place aversion induced by naloxone in single-dose, morphine-treated rats. Psychopharmacology 2004;171:398–404.

[197] Chartoff EH, Connery HS. It's MORe exciting than mu: crosstalk between mu opioid receptors and glutamatergic transmission in the mesolimbic dopamine system. Front Pharmacol 2014;5:116.

[198] Eisenberg RM. Further studies on the acute dependence produced by morphine in opiate naive rats. Life Sci 1982;31:1531–40.

[199] Nutt JG, Jasinski DR. Methadone-naloxone mixtures for use in methadone maintenance programs. I. An evaluation in man of their pharmacological feasibility. II. Demonstration of acute physical dependence. Clin Pharmacol Ther 1974;15:156–66.

[200] Rothwell PE, Thomas MJ, Gewirtz JC. Protracted manifestations of acute dependence after a single morphine exposure. Psychopharmacology 2012;219:991–8.

[201] Glass MJ, Lane DA, Colago EE, Chan J, Schlussman SD, Zhou Y, Kreek MJ, Pickel VM. Chronic administration of morphine is associated with a decrease in surface AMPA GluR1 receptor subunit in dopamine D1 receptor expressing neurons in the shell and non-D1 receptor expressing neurons in the core of the rat nucleus accumbens. Exp Neurol 2008;210:750–61.

[202] Herrold AA, Persons AL, Napier TC. Cellular distribution of AMPA receptor subunits and mGlu5 following acute and repeated administration of morphine or methamphetamine. J Neurochem 2013;126:503–17.

[203] Jacobs EH, Wardeh G, Smit AB, Schoffelmeer AN. Morphine causes a delayed increase in glutamate receptor functioning in the nucleus accumbens core. Eur J Pharmacol 2005;511:27–30.

[204] Martin G, Guadano-Ferraz A, Morte B, Ahmed S, Koob GF, De Lecea L, Siggins GR. Chronic morphine treatment alters N-methyl-D-aspartate receptors in freshly isolated neurons from nucleus accumbens. J Pharmacol Exp Ther 2004;311:265–73.

[205] Murray F, Harrison NJ, Grimwood S, Bristow LJ, Hutson PH. Nucleus accumbens NMDA receptor subunit expression and function is enhanced in morphine-dependent rats. Eur J Pharmacol 2007;562:191–7.

[206] Hearing M, Graziane N, Dong Y, Thomas MJ. Opioid and psychostimulant plasticity: targeting overlap in nucleus accumbens glutamate signaling. Trends Pharmacol Sci 2018;39:276–94.

[207] Inoue M, Mishina M, Ueda H. Locus-specific rescue of GluRepsilon1 NMDA receptors in mutant mice identifies the brain regions important for morphine tolerance and dependence. J Neurosci 2003;23:6529–36.

[208] Miyamoto Y, Yamada K, Nagai T, Mori H, Mishina M, Furukawa H, Noda Y, Nabeshima T. Behavioural adaptations to addictive drugs in mice lacking the NMDA receptor epsilon1 subunit. Eur J Neurosci 2004;19:151–8.

[209] Barnes TD, Kubota Y, Hu D, Jin DZ, Graybiel AM. Activity of striatal neurons reflects dynamic encoding and recoding of procedural memories. Nature 2005;437:1158–61.

[210] Goldstein RZ, Volkow ND. Dysfunction of the prefrontal cortex in addiction: neuroimaging findings and clinical implications. Nat Rev Neurosci 2011;12:652–69.

[211] Van den Oever MC, Spijker S, Smit AB, De Vries TJ. Prefrontal cortex plasticity mechanisms in drug seeking and relapse. Neurosci Biobehav Rev 2010;35:276–84.

[212] Botelho MF, Relvas JS, Abrantes M, Cunha MJ, Marques TR, Rovira E, Fontes Ribeiro CA, Macedo T. Brain blood flow SPET imaging in heroin abusers. Ann N Y Acad Sci 2006;1074:466–77.

[213] McGinty JF, Whitfield Jr. TW, Berglind WJ. Brain-derived neurotrophic factor and cocaine addiction. Brain Res 2010;1314:183–93.

[214] Sun W, Rebec GV. The role of prefrontal cortex D1-like and D2-like receptors in cocaine-seeking behavior in rats. Psychopharmacology 2005;177:315–23.

[215] Chang JY, Janak PH, Woodward DJ. Comparison of mesocorticolimbic neuronal responses during cocaine and heroin self-administration in freely moving rats. J Neurosci 1998;18:3098–115.

[216] Childress AR, Mozley PD, McElgin W, Fitzgerald J, Reivich M, O'Brien CP. Limbic activation during cue-induced cocaine craving. Am J Psychiatry 1999;156:11–8.

[217] Garavan H, Pankiewicz J, Bloom A, Cho JK, Sperry L, Ross TJ, Salmeron BJ, Risinger R, Kelley D, Stein EA. Cue-induced cocaine craving: neuroanatomical specificity for drug users and drug stimuli. Am J Psychiatry 2000;157:1789–98.

[218] Grant S, London ED, Newlin DB, Villemagne VL, Liu X, Contoreggi C, Phillips RL, Kimes AS, Margolin A. Activation of memory circuits during cue-elicited cocaine craving. Proc Natl Acad Sci U S A 1996;93:12040–5.

[219] Hearing MC, Miller SW, See RE, McGinty JF. Relapse to cocaine seeking increases activity-regulated gene expression differentially in the prefrontal cortex of abstinent rats. Psychopharmacology 2008;198:77–91.

[220] Koya E, Spijker S, Voorn P, Binnekade R, Schmidt ED, Schoffelmeer AN, De Vries TJ, Smit AB. Enhanced cortical and accumbal molecular reactivity associated with conditioned heroin, but not sucrose-seeking behaviour. J Neurochem 2006;98:905–15.

[221] Neisewander JL, Baker DA, Fuchs RA, Tran-Nguyen LT, Palmer A, Marshall JF. Fos protein expression and cocaine-seeking behavior in rats after exposure to a cocaine self-administration environment. J Neurosci 2000;20:798–805.

[222] Schmidt ED, Voorn P, Binnekade R, Schoffelmeer AN, De Vries TJ. Differential involvement of the prelimbic cortex and striatum in conditioned heroin and sucrose seeking following long-term extinction. Eur J Neurosci 2005;22:2347–56.

[223] Zavala AR, Biswas S, Harlan RE, Neisewander JL. Fos and glutamate AMPA receptor subunit coexpression associated with cue-elicited cocaine-seeking behavior in abstinent rats. Neuroscience 2007;145:438–52.

[224] Bonson KR, Grant SJ, Contoreggi CS, Links JM, Metcalfe J, Weyl HL, Kurian V, Ernst M, London ED. Neural systems and cue-induced cocaine craving. Neuropsychopharmacology 2002;26:376–86.

[225] Rocha A, Kalivas PW. Role of the prefrontal cortex and nucleus accumbens in reinstating methamphetamine seeking. Eur J Neurosci 2010;31:903–9.

[226] Rogers JL, Ghee S, See RE. The neural circuitry underlying reinstatement of heroin-seeking behavior in an animal model of relapse. Neuroscience 2008;151:579–88.

[227] Weissenborn R, Robbins TW, Everitt BJ. Effects of medial prefrontal or anterior cingulate cortex lesions on responding for cocaine under fixed-ratio and second-order schedules of reinforcement in rats. Psychopharmacology 1997;134:242–57.

[228] Sesack SR, Pickel VM. Prefrontal cortical efferents in the rat synapse on unlabeled neuronal targets of catecholamine terminals in the nucleus accumbens septi and on dopamine neurons in the ventral tegmental area. J Comp Neurol 1992;320:145–60.

[229] Taber MT, Fibiger HC. Electrical stimulation of the prefrontal cortex increases dopamine release in the nucleus accumbens of the rat: modulation by metabotropic glutamate receptors. J Neurosci 1995;15:3896–904.

[230] Taber MT, Das S, Fibiger HC. Cortical regulation of subcortical dopamine release: mediation via the ventral tegmental area. J Neurochem 1995;65:1407–10.

[231] Kawaguchi Y. Groupings of nonpyramidal and pyramidal cells with specific physiological and morphological characteristics in rat frontal cortex. J Neurophysiol 1993;69:416–31.

[232] Kawaguchi Y, Kubota Y. Correlation of physiological subgroupings of nonpyramidal cells with parvalbumin- and calbindinD28k-immunoreactive neurons in layer V of rat frontal cortex. J Neurophysiol 1993;70:387–96.

[233] Isaacson JS, Scanziani M. How inhibition shapes cortical activity. Neuron 2011;72:231–43.

[234] Goldman-Rakic PS, Leranth C, Williams SM, Mons N, Geffard M. Dopamine synaptic complex with pyramidal neurons in primate cerebral cortex. Proc Natl Acad Sci U S A 1989;86:9015–9.

[235] Seamans JK, Gorelova N, Durstewitz D, Yang CR. Bidirectional dopamine modulation of GABAergic inhibition in prefrontal cortical pyramidal neurons. J Neurosci 2001;21:3628–38.

[236] Sesack SR, Snyder CL, Lewis DA. Axon terminals immunolabeled for dopamine or tyrosine hydroxylase synapse on GABA-immunoreactive dendrites in rat and monkey cortex. J Comp Neurol 1995;363:264–80.

[237] Sotres-Bayon F, Sierra-Mercado D, Pardilla-Delgado E, Quirk GJ. Gating of fear in prelimbic cortex by hippocampal and amygdala inputs. Neuron 2012;76:804–12.

[238] Scofield MD, Heinsbroek JA, Gipson CD, Kupchik YM, Spencer S, Smith AC, Roberts-Wolfe D, Kalivas PW. The nucleus accumbens: mechanisms of addiction across drug classes reflect the importance of glutamate homeostasis. Pharmacol Rev 2016;68:816–71.

[239] Lambe EK, Picciotto MR, Aghajanian GK. Nicotine induces glutamate release from thalamocortical terminals in prefrontal cortex. Neuropsychopharmacology 2003;28:216–25.

[240] Stein EA, Pankiewicz J, Harsch HH, Cho JK, Fuller SA, Hoffmann RG, Hawkins M, Rao SM, Bandettini PA, Bloom AS. Nicotine-induced limbic cortical activation in the human brain: a functional MRI study. Am J Psychiatry 1998;155:1009–15.

[241] Woodward JJ, Pava MJ. Effects of ethanol on persistent activity and up-States in excitatory and inhibitory neurons in prefrontal cortex. Alcohol Clin Exp Res 2009;33:2134–40.

[242] Sun W, Rebec GV. Repeated cocaine self-administration alters processing of cocaine-related information in rat prefrontal cortex. J Neurosci 2006;26:8004–8.

[243] Trantham-Davidson H, Lavin A. Acute cocaine administration depresses cortical activity. Neuropsychopharmacology 2004;29:2046–51.

[244] Dong Y, Nasif FJ, Tsui JJ, Ju WY, Cooper DC, Hu XT, Malenka RC, White FJ. Cocaine-induced plasticity of intrinsic membrane properties in prefrontal cortex pyramidal neurons: adaptations in potassium currents. J Neurosci 2005;25:936–40.

[245] Huang CC, Lin HJ, Hsu KS. Repeated cocaine administration promotes long-term potentiation induction in rat medial prefrontal cortex. Cereb Cortex 2007;17:1877–88.

[246] Huang CC, Yang PC, Lin HJ, Hsu KS. Repeated cocaine administration impairs group II metabotropic glutamate receptor-mediated long-term depression in rat medial prefrontal cortex. J Neurosci 2007;27:2958–68.

[247] Nasif FJ, Hu XT, White FJ. Repeated cocaine administration increases voltage-sensitive calcium currents in response to membrane depolarization in medial prefrontal cortex pyramidal neurons. J Neurosci 2005;25:3674–9.

[248] Chen BT, Yau HJ, Hatch C, Kusumoto-Yoshida I, Cho SL, Hopf FW, Bonci A. Rescuing cocaine-induced prefrontal cortex hypoactivity prevents compulsive cocaine seeking. Nature 2013;496:359–62.

[249] Bossert JM, Stern AL, Theberge FR, Marchant NJ, Wang HL, Morales M, Shaham Y. Role of projections from ventral medial prefrontal cortex to nucleus accumbens shell in context-induced reinstatement of heroin seeking. J Neurosci 2012;32:4982–91.

[250] Van den Oever MC, Rotaru DC, Heinsbroek JA, Gouwenberg Y, Deisseroth K, Stuber GD, Mansvelder HD, Smit AB. Ventromedial prefrontal cortex pyramidal cells have a temporal dynamic role in recall and extinction of cocaine-associated memory. J Neurosci 2013;33:18225–33.

[251] Fuchs RA, Branham RK, See RE. Different neural substrates mediate cocaine seeking after abstinence versus extinction training: a critical role for the dorsolateral caudate-putamen. J Neurosci 2006;26:3584–8.

[252] Jentsch JD, Taylor JR. Impulsivity resulting from frontostriatal dysfunction in drug abuse: implications for the control of behavior by reward-related stimuli. Psychopharmacology 1999;146:373–90.

[253] Pacchioni AM, Gabriele A, See RE. Dorsal striatum mediation of cocaine-seeking after withdrawal from short or long daily access cocaine self-administration in rats. Behav Brain Res 2011;218:296–300.

[254] See RE, Elliott JC, Feltenstein MW. The role of dorsal vs ventral striatal pathways in cocaine-seeking behavior after prolonged abstinence in rats. Psychopharmacology 2007;194:321–31.

[255] Brown RW, Kolb B. Nicotine sensitization increases dendritic length and spine density in the nucleus accumbens and cingulate cortex. Brain Res 2001;899:94–100.

[256] Crombag HS, Gorny G, Li Y, Kolb B, Robinson TE. Opposite effects of amphetamine self-administration experience on dendritic spines in the medial and orbital prefrontal cortex. Cereb Cortex 2005;15:341–8.

[257] Klenowski PM, Fogarty MJ, Shariff M, Belmer A, Bellingham MC, Bartlett SE. Increased synaptic excitation and abnormal dendritic structure of prefrontal cortex layer V pyramidal neurons following prolonged binge-like consumption of ethanol. eNeuro 2016;3.

[258] Kroener S, Mulholland PJ, New NN, Gass JT, Becker HC, Chandler LJ. Chronic alcohol exposure alters behavioral and synaptic plasticity of the rodent prefrontal cortex. PLoS One 2012;7:e37541.

[259] Radley JJ, Anderson RM, Cosme CV, Glanz RM, Miller MC, Romig-Martin SA, LaLumiere RT. The contingency of cocaine administration accounts for structural and functional medial prefrontal deficits and increased adrenocortical activation. J Neurosci 2015;35:11897–910.

[260] Robinson TE, Gorny G, Mitton E, Kolb B. Cocaine self-administration alters the morphology of dendrites and dendritic spines in the nucleus accumbens and neocortex. Synapse 2001;39:257–66.

[261] Robinson TE, Gorny G, Savage VR, Kolb B. Widespread but regionally specific effects of experimenter- versus self-administered morphine on dendritic spines in the nucleus accumbens, hippocampus, and neocortex of adult rats. Synapse 2002;46:271–9.

[262] Ben-Shahar O, Obara I, Ary AW, Ma N, Mangiardi MA, Medina RL, Szumlinski KK. Extended daily access to cocaine results in distinct alterations in Homer 1b/c and NMDA receptor subunit expression within the medial prefrontal cortex. Synapse 2009;63:598–609.

[263] Tang W, Wesley M, Freeman WM, Liang B, Hemby SE. Alterations in ionotropic glutamate receptor subunits during binge cocaine self-administration and withdrawal in rats. J Neurochem 2004;89:1021–33.

[264] Lu H, Cheng PL, Lim BK, Khoshnevisrad N, Poo MM. Elevated BDNF after cocaine withdrawal facilitates LTP in medial prefrontal cortex by suppressing GABA inhibition. Neuron 2010;67:821–33.

[265] Van den Oever MC, Goriounova NA, Li KW, Van der Schors RC, Binnekade R, Schoffelmeer AN, Mansvelder HD, Smit AB, Spijker S, De Vries TJ. Prefrontal cortex AMPA receptor plasticity is crucial for cue-induced relapse to heroin-seeking. Nat Neurosci 2008;11:1053–8.

[266] McFarland K, Kalivas PW. The circuitry mediating cocaine-induced reinstatement of drug-seeking behavior. J Neurosci 2001;21:8655–63.

[267] Baker DA, McFarland K, Lake RW, Shen H, Tang XC, Toda S, Kalivas PW. Neuroadaptations in cystine-glutamate exchange underlie cocaine relapse. Nat Neurosci 2003;6:743–9.

[268] Rothwell PE, Kourrich S, Thomas MJ. Synaptic adaptations in the nucleus accumbens caused by experiences linked to relapse. Biol Psychiatry 2011;69:1124–6.

[269] Self DW, Choi KH. Extinction-induced neuroplasticity attenuates stress-induced cocaine seeking: a state-dependent learning hypothesis. Stress 2004;7:145–55.

[270] Crombag HS, Bossert JM, Koya E, Shaham Y. Review. Context-induced relapse to drug seeking: a review. Philos Trans R Soc Lond Ser B Biol Sci 2008;363:3233–43.

[271] McDonald AJ. Organization of amygdaloid projections to the mediodorsal thalamus and prefrontal cortex: a fluorescence retrograde transport study in the rat. J Comp Neurol 1987;262:46–58.

[272] See RE, Fuchs RA, Ledford CC, McLaughlin J. Drug addiction, relapse, and the amygdala. Ann N Y Acad Sci 2003;985:294–307.

[273] Peters J, Vallone J, Laurendi K, Kalivas PW. Opposing roles for the ventral prefrontal cortex and the basolateral amygdala on the spontaneous recovery of cocaine-seeking in rats. Psychopharmacology 2008;197:319–26.

[274] Quirk GJ, Likhtik E, Pelletier JG, Pare D. Stimulation of medial prefrontal cortex decreases the responsiveness of central amygdala output neurons. J Neurosci 2003;23:8800–7.

[275] Ovari J, Leri F. Inactivation of the ventromedial prefrontal cortex mimics re-emergence of heroin seeking caused by heroin reconditioning. Neurosci Lett 2008;444:52–5.

[276] Stefanik MT, Kalivas PW. Optogenetic dissection of basolateral amygdala projections during cue-induced reinstatement of cocaine seeking. Front Behav Neurosci 2013;7:213.

[277] Rosen LG, Sun N, Rushlow W, Laviolette SR. Molecular and neuronal plasticity mechanisms in the amygdala-prefrontal cortical circuit: implications for opiate addiction memory formation. Front Neurosci 2015;9:399.

[278] Liu C, Fang X, Wu Q, Jin G, Zhen X. Prefrontal cortex gates acute morphine action on dopamine neurons in the ventral tegmental area. Neuropharmacology 2015;95:299–308.

[279] Buchta WC, Mahler SV, Harlan B, Aston-Jones GS, Riegel AC. Dopamine terminals from the ventral tegmental area gate intrinsic inhibition in the prefrontal cortex. Phys Rep 2017;5.

[280] Bowers MS, McFarland K, Lake RW, Peterson YK, Lapish CC, Gregory ML, Lanier SM, Kalivas PW. Activator of G protein signaling 3: a gatekeeper of cocaine sensitization and drug seeking. Neuron 2004;42:269–81.

[281] Briand LA, Flagel SB, Garcia-Fuster MJ, Watson SJ, Akil H, Sarter M, Robinson TE. Persistent alterations in cognitive function and prefrontal dopamine D2 receptors following extended, but not limited, access to self-administered cocaine. Neuropsychopharmacology 2008;33:2969–80.

[282] Volkow ND, Fowler JS. Addiction, a disease of compulsion and drive: involvement of the orbitofrontal cortex. Cereb Cortex 2000;10:318–25.

[283] Jayaram P, Steketee JD. Effects of cocaine-induced behavioural sensitization on GABA transmission within rat medial prefrontal cortex. Eur J Neurosci 2005;21:2035–9.

[284] Jayaram P, Steketee JD. Cocaine-induced increases in medial prefrontal cortical GABA transmission involves glutamatergic receptors. Eur J Pharmacol 2006;531:74–9.

[285] Kroener S, Lavin A. Altered dopamine modulation of inhibition in the prefrontal cortex of cocaine-sensitized rats. Neuropsychopharmacology 2010;35:2292–304.

[286] Nogueira L, Kalivas PW, Lavin A. Long-term neuroadaptations produced by withdrawal from repeated cocaine treatment: role of dopaminergic receptors in modulating cortical excitability. J Neurosci 2006;26:12308–13.

[287] Couey JJ, Meredith RM, Spijker S, Poorthuis RB, Smit AB, Brussaard AB, Mansvelder HD. Distributed network actions by nicotine increase the threshold for spike-timing-dependent plasticity in prefrontal cortex. Neuron 2007;54:73–87.

[288] Hu XT. Cocaine withdrawal and neuro-adaptations in ion channel function. Mol Neurobiol 2007;35:95–112.

[289] Seamans JK, Yang CR. The principal features and mechanisms of dopamine modulation in the prefrontal cortex. Prog Neurobiol 2004;74:1–58.

[290] Dong Y, White FJ. Dopamine D1-class receptors selectively modulate a slowly inactivating potassium current in rat medial prefrontal cortex pyramidal neurons. J Neurosci 2003;23:2686–95.

[291] Nasif FJ, Sidiropoulou K, Hu XT, White FJ. Repeated cocaine administration increases membrane excitability of pyramidal neurons in the rat medial prefrontal cortex. J Pharmacol Exp Ther 2005;312:1305–13.

[292] Oliet SH, Malenka RC, Nicoll RA. Bidirectional control of quantal size by synaptic activity in the hippocampus. Science 1996;271:1294–7.

[293] Choi S, Lovinger DM. Decreased frequency but not amplitude of quantal synaptic responses associated with expression of corticostriatal long-term depression. J Neurosci 1997;17:8613–20.

[294] Kerchner GA, Nicoll RA. Silent synapses and the emergence of a postsynaptic mechanism for LTP. Nat Rev Neurosci 2008;9:813–25.

[295] Graziane N, Dong Y. Electrophysiological analysis of synaptic transmission. New York, NY: Springer; 2015.

[296] Manabe T, Wyllie DJ, Perkel DJ, Nicoll RA. Modulation of synaptic transmission and long-term potentiation: effects on paired pulse facilitation and EPSC variance in the CA1 region of the hippocampus. J Neurophysiol 1993;70:1451–9.

[297] Witkowski G, Szulczyk B, Rola R, Szulczyk P. D(1) dopaminergic control of G protein-dependent inward rectifier K(+) (GIRK)-like channel current in pyramidal neurons of the medial prefrontal cortex. Neuroscience 2008;155:53–63.

[298] Wikowski G, Rola R, Szulcczyk P. Effect of cyclic adenosine monophosphate on the G protein-dependent inward rectifier K(+)-like channel current in medial prefrontal cortex pyramidal neurons. J Physiol Pharmacol 2012;63:457–62.

[299] Lu L, Grimm JW, Shaham Y, Hope BT. Molecular neuroadaptations in the accumbens and ventral tegmental area during the first 90 days of forced abstinence from cocaine self-administration in rats. J Neurochem 2003;85:1604–13.

[300] Chowdhury S, Sheperd JD, Okuno H, Lyford G, Petralia RS, Plath N, Kuhl D, Huganir RL, Worley PF. Arc/Arg3.1 interacts with the endocytic machinery to regulate AMPA receptor trafficking. Neuron 2006;52:445–59.

[301] Rial Verde EM, Lee-Osbourne J, Worley PF, Malinow R, Cline HT. Increased expression of the immediate-early gene arc/arg3.1 reduces AMPA receptor-mediated synaptic transmission. Neuron 2006;52:461–74.

Chapter 5

Glutamatergic Neuroplasticity in Addiction

Allison R. Bechard and Lori A. Knackstedt

Psychology Department, University of Florida, Gainesville, FL, United States

INTRODUCTION

Identifying the neurobiological substrates of drug seeking is an important precursor to permit the development of effective pharmacotherapies targeting drug-induced neuroadaptations. Animal models of addiction have been widely used to investigate the maladaptive neuroadaptations that mediate prolonged and persistent drug seeking. There are many advantages to using animal models, including the ability of more invasive and tightly controlled experiments that assess specific mechanisms of addiction. Many preclinical studies have implicated glutamatergic neuroplasticity in addiction (see reviews by Refs. [1–3]) with specific importance placed on the role of the nucleus accumbens (NA; [4–6]). The NA sits in a key position of the reward circuitry; glutamatergic projections from the prefrontal cortex (PFC), amygdala, and VTA converge here and NA output pathways project to the basal ganglia, which mediate habit versus goal-directed behavior. Alterations in the glutamate neurotransmitter system as a consequence of repeated drug administration have been strongly implicated in the protracted seeking of several drugs of abuse [5–8] including cocaine [9], nicotine [10], opiates [11], and ethanol [12].

Here we will focus on the significant body of work that examines the role of glutamatergic neuroadaptations in mediating drug seeking (relapse) after weeks to months of drug abstinence. In the sections below, we focus on the most commonly used animal paradigms that model relapse of drug seeking, highlighting findings that have the greatest translational potential for human addiction therapy. We will discuss how several drugs of abuse can affect mechanisms of glutamate homeostasis mechanisms, largely within the NA. Candidates for pharmacological manipulation will be discussed in light of their potential for translation. Finally, we will review the clinical research that investigates glutamatergic neuroplasticity in human addicts.

ANIMAL MODELS OF RELAPSE

The animal models that best recapitulate the components of relapse vulnerability in human drug addiction are those that use operant drug self-administration followed by a period of withdrawal and then an assessment of persistence in drug seeking. The self-administration paradigm can involve a short (e.g., 2 h) or extended (e.g., 6 h) period of drug access, in which the animal can self-administer the drug by making an operant response, such as a lever press or nose poke. Along with receiving the drug itself, the operant response is typically accompanied by drug-paired cues, such as a light and an auditory tone, that allow for later examination of cue-elicited responses during a test of persistent drug seeking. The number of operant responses made upon reexposure to stimuli known to cause relapse in humans, such as the drug-associated context, cues, or the drug itself, is considered a measure of relapse [13]. Stress-induced relapse can also be modeled in animals (e.g., Refs. [14,15]). Importantly, postdrug experience (e.g., extinction of the operant response made to acquire drug; [16]) and the duration of the period of withdrawal [17,18] can have differential effects on relapse. We will discuss three animal models of drug self-administration that vary in postdrug experience: the extinction-reinstatement paradigm, the abstinence-relapse paradigm, and the ABA renewal paradigm.

The *extinction-reinstatement paradigm* is one of the most commonly used in addiction modeling research. Here, the animal is trained to self-administer a drug in an operant chamber by performing a response that is, additionally paired with both an auditory and visual cue. After several weeks of drug self-administration, the animal enters an extinction phase, in which the operant response no longer yields delivery of the drug or the presentation of discrete cues. With time, this extinction training results in the attenuation of the drug-seeking behavior, which can then be *reinstated* by the drug-associated context, cues, stress, or the drug itself. The extinction-reinstatement paradigm is considered to model relapse behavior and possess both

Copyright © 2019 Elsevier Inc. All rights reserved.

construct and face validity [13]. However, human addicts are rarely subjected to a similar extinction phase, and thus the abstinence-relapse paradigm has become widely used due to greater face validity and potential translatability.

In the *abstinence-relapse paradigm* animals self-administer a drug in the operant chamber, and following several weeks of repeated drug use, the animals are forced into a period of abstinence during which they remain in their home cage (e.g., Ref. [19]). Although they receive daily handling, these animals do not experience extinction training. After several weeks, they are assessed for persistent drug-seeking behavior primed by reexposure to the drug-associated context (context-primed relapse) and/or the drug-associated cues (cue-primed relapse). When this test is done after 30–60 days of withdrawal it is often referred to as "incubated cocaine craving" (e.g., Ref. [17]). This experience is more akin to what happens to human addicts, in that they are often forced into a period of abstinence by being jailed, or through court-appointed drug treatment. When released from this period of forced abstinence, these people are likely to face their drug-associated context, cues, and stress, and be vulnerable to relapse.

Renewal is the term used to describe resurgence in performance of an extinguished behavior that occurs with a change of context [20]. This phenomenon highlights the importance of drug-taking contexts and their role in persistent drug-seeking and use. In the *ABA renewal paradigm*, animals are trained to earn reinforcement in context A, experience extinction of the operant response in a separate context B, and then tested for recovery of the extinguished behavior in context A. The renewal of reward-seeking when returned to the drug-taking context (context A) has been demonstrated for sucrose [21], alcohol [22], cocaine [20,23], and heroin [24]. Importantly, these studies show that extinction learning is not "unlearning" of the previously conditioned task, but rather acts as a mask resulting in the inhibition of a behavior [25].

Although differences exist in the neurobiology mediating relapse depending on postdrug experience, overlapping circuitry, and pharmacological candidates for therapy are numerous. As much of what we know from preclinical studies focuses on the nucleus accumbens core (NAc), this structure will be discussed in detail.

THE NEUROCIRCUITRY OF ADDICTION

The NA is an ideal structure for the investigation of drug-induced alterations due to the convergence of glutamatergic and dopaminergic inputs and its position as the major input structure to the basal ganglia. Glutamate efflux in the NAc accompanies drug-seeking behaviors [26]. Determining the alterations in circuitry as a consequence of drug self-administration is important for developing antirelapse prevention.

One of the major pathways mediating reinstatement of drug seeking is the PFC-NAc glutamatergic projection. The NA also receives glutamatergic projections from the amygdala, hippocampus, and thalamus. Glutamatergic and dopaminergic projections are also received from the ventral tegmental area (VTA). Dopamine and glutamate both play a role in drug addiction behaviors; however, glutamate is thought to have more involvement in the persistent seeking of drugs after a period of withdrawal, whereas dopamine is more highly implicated in reinforcement processes during acute drug exposure. Comprised of both D1- and D2-dopamine receptor-expressing medium spiny neurons (MSNs), the NA provides GABAergic innervation back to the VTA and the ventral pallidum [27]. There is also a small population (5%–10%) of NA interneurons (e.g., acetylcholine interneurons) that participate in stimulus detection and context recognition [28], which are likely to influence cue-induced reinstatement of drug seeking. These GABAergic interneurons are stimulated by projections from the VTA and can alter downstream signaling events that influence reward [29]; however, their role is still largely understudied. Also understudied is the influence of extracellular matrix proteins and interactions with zinc-dependent endopeptidases (e.g., MMP-2 and MMP-9) that are required for transient synaptic plasticity events mediating cue-induced reinstatement of cocaine, heroin, and nicotine seeking [30].

The NA itself is divided into two main functional areas: the NA core (NAc) and NA shell (NAs). Projections from the NAc remain within the basal ganglia, whereas those from the NAs extend outside the basal ganglia to the hypothalamus and amygdala. Although the NAs is involved in reward prediction and learning, it is the NAc that is, suggested to be essential in drug-taking behaviors and cue-elicited drug seeking (see review by Ref. [6]). Dissociable roles for the NAc and NAs are evidenced by pharmacological disconnection [31–33] and neurochemical studies. For example, whereas NAc infusions of AMPAR antagonists inhibit cocaine-primed reinstatement, dopamine antagonist infusions produce no effect [4]. In the NAs, however, inhibition of dopamine receptors prevents cocaine-primed reinstatement [34].

In order to determine the neurocircuitry of persistent drug seeking, researchers use methods of pharmacological disconnection and inactivation, electrophysiology, and optogenetics. Although there is much overlap, such methods have identified dissociable roles of specific brain regions for relapse primed by cues, context, drug, or postdrug experience. For example, the role of the hippocampus in learning and context-primed relapse may be greater than that of the basolateral amygdala (BLA; [35]). The BLA is suggested to be more involved with the formation of associations between the drug and conditioned cues following extinction training [36,37]. Dysfunction and hypoactivation of the PFC resulting from chronic drug use is implicated in the inability of the user to extinguish drug-seeking behavior [38]. In addition, drug class-specific alterations in the circuitry mediating relapse exist. For example, compared to cocaine-seeking, heroin-seeking may recruit more brain structures, and involve a greater role for the NAs [39]. The underlying circuitries can also differ due to postdrug experience, such as whether extinction procedures are carried out (and the context they are carried out in). In the case of cocaine, there is recruitment of the BLA and prelimbic cortex (PL) in context- and cue-primed tests of relapse following extinction training [23,40]. However, when extinction training occurs outside of the self-administration context (i.e., ABA renewal paradigm) there is recruitment of the dorsal hippocampus [23]. Circuitry mediating relapse following abstinence without extinction is less defined, and roles for several structures are yet to be determined. For example, the role of the HPC is undefined in context- and cue-primed relapse following abstinence without extinction. Even a role for the NA in relapse may be dependent on self-administration training contingent with discrete drug-associated cues, as when rats are trained to self-administer cocaine with a cue present, blocking the metabotropic glutamate receptor 5 (mGlu5) in the NAc attenuates context-primed relapse after abstinence [16] but when rats are trained in the absence of drug-paired cues, inactivation of this brain region does not alter context-primed relapse [41].

GLUTAMATERGIC ALTERATIONS INDUCED BY CHRONIC DRUG USE

Drugs of abuse induce lasting changes in the brain. These include alterations in synaptic plasticity that increases the motivation for drug seeking, resulting in a feed-forward mechanism. Occurring at the level of the NA, repeated drug use alters glutamate homeostasis mechanisms, mainly affecting interactions between neurons and glial cells. A tripartite synaptic architecture consisting of astrocytic processes adjacent to the synapse and the pre- and postsynaptic neurons regulates glutamatergic transmission. The role of glial cells in addiction focuses on astrocytes, as they control the uptake and nonsynaptic release of glutamate. Astrocytes are affected by repeated exposure to ethanol, cocaine, nicotine, and opioids [6,42,43].

Most glutamatergic pharmacotherapies are designed to act on receptor, transporter, or exchanger proteins that have undergone changes in expression level, distribution, or posttranslational modifications as a result of the chronic drug use. Therefore, the identification of glutamatergic changes induced by chronic drug use and their resulting function is of critical importance. In general, postsynaptic sensitivity to glutamate release depends on the number and type of postsynaptic receptors. The typical glutamatergic synapse has three different types of postsynaptic ionotropic glutamate receptors (iGluRs): alpha-amino-3-hydroxy-5-methyl-4-isoxazolepropionic acid (AMPA) receptors, N-methyl-D-aspartate (NMDA) receptors, and kainic acid (KA) receptors. Release of glutamate into the synapse primarily activates AMPA and NMDA receptors, and as a potential consequence, much less research has been done on KA receptors. Metabotropic glutamate (mGlu) receptors 1–8 are G protein-coupled receptors that mediate slower, modulatory glutamatergic transmission. They are divided into three main groups (Group 1, 2, and 3) based on pharmacological and signal transduction properties. The activation of ionotropic and metabotropic receptors initiates a host of intracellular signaling cascades implicated in a variety of processes underlying synaptic plasticity and learning (reviewed by Ref. [44]). As drug-induced changes in glutamatergic transmission are heavily implicated in persistent drug seeking, each of these receptors and downstream events is of investigative interest in the pursuit of pharmacological manipulations that prevent relapse.

Finally, relapse to drug seeking is accompanied by glutamate efflux in the NA, an effect that has been observed across multiple classes of drugs. Relapse to cocaine seeking induced by cocaine itself [9], cocaine-associated cues [9a], and reexposure to the self-administration context following abstinence [19] is accompanied by glutamate efflux in the NA core. This efflux has been determined to be synaptic in nature, arising from the dmPFC to NA core projection [9]. Similar observations have been made during relapse to heroin [45], nicotine [10], and ethanol seeking [46].

GLT-1

Following synaptic glutamate release, astrocytes remove glutamate from the synaptic cleft, thereby terminating glutamatergic signaling and preventing excitotoxicity. The main mechanism for the removal of glutamate is via the Na^+-dependent glial glutamate transporter (GLT)-1, which is responsible for more than 90% of glutamate reuptake [47]. Downregulation of GLT-1 is associated with repeated drug abuse, including cocaine [48], nicotine [49], ethanol [50], and opioids [45]. Decreased expression of GLT-1 is also associated with increased levels of extrasynaptic glutamate that promotes synaptic plasticity mediating relapse, such as greater activation of postsynaptic glutamate receptors [6]. GLT-1 is being investigated as a therapeutic target for multiple classes of substance use disorders [51]. Two compounds have been at the forefront of this treatment strategy for their ability to restore function of GLT-1: N-acetylcysteine and ceftriaxone [51].

Cocaine

Chronic self-administration of cocaine has repeatedly been shown to decrease GLT-1 in the NAc [48,52]. The upregulation of NAc GLT-1 by pharmacological intervention (e.g., ceftriaxone treatment) reliably attenuates cocaine reinstatement [18,48,52,53]. Reduced GLT-1 has been associated with the glutamate efflux in the NAc observed during cocaine reinstatement [26]. However, GLT-1 upregulation via ceftriaxone was not enough to prevent such glutamate efflux during relapse following abstinence without extinction training [19].

Nicotine

Increased extracellular glutamate efflux in the NAc also accompanies cue-induced reinstatement of nicotine seeking [10]. In addition, membrane content of GLT-1 in the NAc is reduced in nicotine self-administering animals [10,49]. Four weeks of daily N-acetylcysteine resulted in a reduction in the number of cigarettes smoked, when excluding individuals with high rates of alcohol use [49]. GLT-1 is also reduced in the striatum, but not the hippocampus, of mice exposed to e-cigarette vapor containing nicotine [54].

Ethanol

Studies investigating the effects of self-administration of ethanol on GLT-1 report mixed results. In response to ethanol, glutamate transporter binding and function in the cortex has been found to decrease [55]. Whereas one study found no differences in GLT-1 expression in the hippocampus, hypothalamus, or cortex [56], another found decreased GLT-1 in the NA and no changes in the PFC [57]. In continuously drinking rats, GLT-1 levels decreased in the striatum and hippocampus and were restored by ceftriaxone [58]. However, in rats given 5 weeks of intermittent access to alcohol self-administration, GLT-1 is not decreased in the NAc and ceftriaxone does not alter its expression [59]. Other studies using ceftriaxone have also reported an upregulation of GLT-1 in the NAc and PFC, and an associated attenuation in ethanol self-administration and reinstatement (see review by Ref. [51]).

Heroin

The self-administration of heroin reduces GLT-1 and glutamate uptake in the NAc [45]. Pharmacological upregulation of GLT-1 by ceftriaxone prevents glutamate spillover and the reinstatement of heroin seeking [45].

In summary, NAc GLT-1 is consistently found to be downregulated by the self-administration of multiple classes of drugs, with the exception of intermittently self-administered ethanol. The upregulation of GLT-1 in the NAc reduces seeking of many classes of drugs.

xCT

Glial cells also regulate extracellular glutamate levels through the cystine-glutamate exchanger (system xc-), which releases intracellular glutamate into the extracellular space in exchange for extracellular cystine in a 1:1 ratio [60]. The catalytic subunit of system xc-, xCT, is downregulated after repeated drug use [49]. System xc- accounts for more than 50% of the basal, extrasynaptic glutamate in the NAc, and its impairment results in the decreased basal levels of NAc glutamate associated with withdrawal from cocaine [61]. Moreover, the maintenance of extracellular glutamate via system xc- provides tone on the mGlu2/3 autoreceptors that regulate synaptic release of glutamate [62], providing a mechanism for synaptic glutamate spillover and potentiated synaptic plasticity mediating drug seeking [6]. Accordingly, activating system xc- via N-acetylcysteine or ceftriaxone has attenuating effects on drug reinstatement [12,61].

Cocaine

Following withdrawal from repeated cocaine self-administration, the function of system xc- in the NAc is decreased, and accompanied by reduced expression xCT [48]. The upregulation of xCT via ceftriaxone is associated with prevention of cocaine reinstatement [48], possibly via its ability to normalize AMPARs [63]. The latter effect was demonstrated by intra-NAc knockdown of xCT with an antisense oligonucleotide, which prevented ceftriaxone from attenuating reinstatement and from upregulating GLT-1 and resulted in increased surface expression of AMPAR subunits GluA1 and GluA2 [63].

Nicotine

Nicotine self-administering animals showed decreased xCT expression in the NA and VTA, accompanied by reduced somatic signs of withdrawal [49]. Chronic exposure to nicotine containing vapor produced by e-cigarettes resulted in decreased xCT in the striatum and hippocampus of mice [54].

Ethanol

Ethanol intake and relapse to ethanol-consumption following continuous access to ethanol are attenuated by ceftriaxone, an effect associated with upregulation of xCT in the PFC and NA [12]. Ceftriaxone was found to reduce ethanol intake after only 2 days of treatment, which corresponded with upregulated xCT and GLT-1 levels in the PFC, NA, and amygdala [64]. Oral gavage of ethanol did not change hippocampal and striatal xCT [58]. However, following intermittent access to ethanol, xCT in the NAc is not decreased relative to ethanol-naive controls but ceftriaxone attenuates ethanol consumption while increasing xCT expression [59].

Heroin

Unlike the other classes of drugs, chronic self-administration of heroin resulted in an increase in surface expression of NAc xCT 2–3 weeks later [45]. However, this increase in protein expression was not accompanied by an increase in system xc-activity.

In summary, NAc xCT is consistently found to be downregulated by the self-administration of multiple classes of drugs, with the exception of intermittently self-administered ethanol and heroin. The upregulation of xCT in the NAc reduces the seeking of many classes of drugs.

mGlu2/3

Presynaptic metabotropic glutamate receptors 2/3 (mGlu2/3) regulate glutamate and dopamine release in the VTA and NA, regions implicated in hosting mGlu2/3 alterations affecting drug seeking [65,66]. In the NA, mGlu2/3 is expressed on presynaptic terminals that synapse with MSNs, and their activation decreases extracellular glutamate levels [67]. The role of mGlu2/3 in drug seeking has primarily been tested using pharmacological manipulations.

Cocaine

A loss of tone on mGlu2/3 autoreceptors has been proposed to occur in response to decreased basal glutamate levels present in the Nac 2–3 weeks following cessation of cocaine-self-administration [6]. In the cocaine-reinstatement model, an antagonist of mGlu2/3 prevents N-acetylcysteine from inhibiting cocaine seeking and it was hypothesized that increasing NA basal levels of glutamate restored tone on the mGlu2/3 autoreceptors and dampened glutamate release [62]. A recent study using mGlu2 mutant rats showed that decreased mGlu2 expression reduced sensitivity to cocaine reward and relapse [66]. In line with this hypothesis, the mGlu2/3 agonist, LY379268, inhibited cocaine-seeking, but also food-seeking behaviors [68].

Nicotine

Withdrawal from self-administration of nicotine reduced mGlu2/3 function in the NAs, amygdala, PFC, hypothalamus, hippocampus, and VTA [65,69]. Accordingly, agonists at inhibitory presynaptic mGlu2/3 attenuated drug seeking, whereas antagonists ameliorated withdrawal-associated reward deficits [69]. Systemic, as well as intra-NA and intra-VTA, administration of LY379268 decreased self-administration of nicotine, but not food, whereas cue-induced reinstatement of both nicotine and food seeking were blocked [65].

Ethanol

While basal glutamate levels in early withdrawal from alcohol consumption are increased (rather than decreased as following cocaine), the release of glutamate in the NAc continues to mediate the reinstatement of alcohol seeking [46]. Thus, the administration of the mGlu2/3 agonist LY379268 attenuates both stress- and cue-induced ethanol seeking in the extinction-reinstatement model of ethanol relapse [70]. The increase in NAc basal extracellular glutamate that occurs after chronic ethanol consumption is independent of a decrease in mGlu2/3 function in this brain region [71]. However, in the NAs, the ability of mGlu2/3 stimulation to reduce glutamate levels is dampened in alcohol-dependent rats and viral vector-mediated overexpression of mGlu2 along the vmPFC-NAs pathway decreases alcohol consumption [72]. Thus, mGlu2/3 function is disrupted in the NAs and ameliorating this dysfunction attenuates alcohol seeking.

Heroin

At low doses (0.3 or 1.0 μg), infusing LY379268 into the NAs dose-dependently attenuated heroin reinstatement, but was ineffective in the NAc. However, a much larger dose (3.0 μg) into the NAc was able to attenuate reinstatement [73].

In summary, mGlu2/3 agonists administered systemically or directly into the NA core and shell attenuate drug seeking across many classes of drugs. Similarly, overexpression of mGlu2 in vmPFC neurons that project to the NAs decreases ethanol consumption. This effect is likely due to the ability of such manipulations to attenuate the synaptic release of glutamate that drives relapse to drug seeking. In fact, this effect has been explicitly demonstrated for the reinstatement of cocaine seeking induced by cocaine-associated cues [9a]. However, these compounds also attenuate food seeking, raising concern for their use in human addicts.

mGlu5

Stimulation of mGlu5 receptors in the NAc has the opposite effect as activating presynaptic mGlu2/3, and results in the reinstatement of drug seeking. Several studies have investigated the role of mGlu5 in addiction, and converging evidence shows increased mGlu5 function following repeated use of cocaine, nicotine, ethanol, and heroin. This effect has been mostly inferred by the attenuation of reinstatement following the use of antagonists.

Cocaine

mGlu5 has distinct roles in drug seeking, including in the reinforcement of cocaine, extinction learning, and cocaine reinstatement. Cocaine self-administration decreases the expression of mGlu5 in the NA, and to a more modest degree, in the hippocampus and PFC [74,75], while it increases levels of mGlu5 in the dorsal striatum [76]. However, cocaine-self administration followed by extinction training reduces mGlu5 expression in the dorsal striatum and PFC, and increases mGlu5 expression in the hippocampus and NA relative to yoked-saline controls [76]. Pharmacological antagonists of mGlu5, such as MPEP [77] and MTEP [16] infused into the NA attenuate cocaine relapse. MTEP attenuates context-induced relapse after abstinence and cue-induced relapse following extinction training when infused into the NAc [16]. Intra-dorsal lateral striatum (dlSTR) MTEP infusions do not attenuate relapse, but attenuate subsequent extinction learning [16], an effect associated with decreased mGlu5 surface expression [78]. Systemic administration of MTEP attenuates context-induced relapse after abstinence, without affecting extinction learning [78].

Nicotine

In the human brain, chronic exposure to tobacco smoke reduces the density of mGlu5 [79]. In animals, antagonism of mGlu5 inhibits nicotine self-administration and nicotine- [80] and cue-primed reinstatement [81]. Somatic signs of nicotine withdrawal are absent in mGlu5 (−/−) mice [82].

Ethanol

Increased signaling of mGlu5 in the NAc occurs following prolonged consumption of ethanol [83], dependent on time of withdrawal [84]. Antagonism of mGlu5 with MPEP prevents cue-induced reinstatement of ethanol seeking, an effect associated with decreased signaling of downstream pathways in the NAs and BLA [85].

Heroin

There is a greater effect of mGlu5 antagonists in the NAs versus the NAc on heroin seeking. For example, MPEP administered into the NAs dose-dependently attenuates relapse to heroin seeking, whereas infusions into the NAc and dorsal striatum are ineffective [86].

Thus, mGlu5 antagonism (or negative allosteric modulation) consistently decreases drug seeking across classes. As with mGlu2/3, there may be an important role of extinction training in mGlu5 function following self-administration of alcohol and cocaine. Unfortunately, mGlu5 antagonism is associated with deficits in some aspects of learning (for review, see Ref. [87]), and thus medications targeting mGlu5 do not hold translational potential for treating addiction.

AMPA RECEPTOR SUBUNIT 1 AND 2 (GluA1/2)

Activation of AMPARs in the NA is a precursor to MSN activation by glutamatergic inputs that results in drug seeking (for review, see Ref. [88]). Accordingly, infusing an AMPAR agonist into the NA induces cocaine reinstatement, whereas an antagonist blocks reinstatement [4]. This early work using pharmacological manipulations of AMPARs to bidirectionally affect cocaine reinstatement has now been followed up using several classes of drugs and methods of approach (for review, see Refs. [6,7]).

More recently, alterations in AMPA receptor subunit composition have been associated with persistent drug seeking. In adult drug-naïve rats, synaptic AMPARs in the NA are primarily GluA1A2 subtype, although a small population of subunit GluA2A3 is also present [89]. When the GluA2 subunit is lacking, the AMPAR is Ca^{2+} permeable and [90] has a higher conductance. The formation of these receptors is proposed to partially underlie the increased reactivity of NAc neurons to drug-associated cues. CP-AMPARs comprise only about 10% of the NA in drug-naive animals. As a result of repetitive drug abuse, changes in presynaptic glutamate levels increase GluA1 surface expression without affecting GluA2 surface expression, thus leading to the formation of greater numbers of GluA2-lacking CP-AMPA receptors [91].

Cocaine

AMPAR alterations in the NA following repeated cocaine use and its role in cocaine reinstatement is complex, and is dependent on contingency of exposure and training during self-administration, withdrawal duration, and post-drug experience (reviewed by Ref. [88]). In general, cocaine self-administration and withdrawal increase the surface expression of the GluA1 subunit of AMPA receptors in the NAc and potentiate postsynaptic mechanisms mediating relapse [92]. Use of viral vectors to impair the transport of NAc GluA1 attenuates the reinstatement of cocaine seeking [93]. Following cocaine-cue extinction learning, protein kinase A phosphorylates GluA1 receptors at site serine 845 that is, distributed primarily in the extrasynaptic space [89] and stimulates the formation of CP-AMPARs [17]. The regulation of CP-AMPARs is of particular interest as they are thought to mediate the intensification of cocaine craving associated with increased periods of withdrawal (i.e., the incubation of cocaine craving; [89]). Disruption of GluA2 trafficking in either the NAc or NAs attenuates cocaine-primed reinstatement [89a]. In line with a functional role for increased GluA1 surface expression and the formation of CP-AMPAs is the finding that intra-NA infusion of the CP-AMPA antagonist Naspm attenuates relapse to cocaine seeking following extinction [94] and abstinence [17].

Nicotine

Animals that show persistent nicotine-seeking behaviors have increased levels of GluA1, but not GluA2, in the NAc [10]. This suggests that the increased levels of AMPARs were largely comprised of GluA2-lacking receptors. Increased extracellular glutamate in the NAc during cue-induced reinstatement of nicotine seeking is associated with increased MSN spine head diameter and ratio of AMPA to NMDA currents [10].

Ethanol

Long-term ethanol consumption results in strengthened excitatory AMPAR signaling. Pati et al. [71] found increased surface expression of GluA1 in the NAc at 24 h withdrawal from alcohol consumption. A positive allosteric modulator of the AMPAR (aniracetam) increases alcohol-reinforced responses and cue-induced reinstatement of alcohol, without affecting sucrose-reinforced responses or locomotion [95]. In addition, an AMPAR antagonist blocks this aniracetam-induced increase in alcohol self-administration [95].

Heroin

Following self-administration of heroin, reexposure to drug-associated cues results in downregulation of the GluA2 subunit of the AMPAR in synaptic membranes of the mPFC [96]. Blocking GluA2 endocytosis in the mPFC attenuated cue-induced relapse to heroin seeking [96]. In addition, AMPAR antagonist (NBQX) administration into the NAc inhibits both cue- and drug-primed heroin seeking [11].

Thus, an increase in surface GluA1 expression is consistently found in the NAc following self-administration of addictive drugs and is likely a hallmark of increased excitability of the MSN that drives drug seeking. AMPA receptor antagonists consistently attenuate drug seeking but are unusable in the clinic as they would have widespread effects on neuronal function.

NMDA RECEPTORS

Cocaine

The effects of cocaine on NMDARs are not straightforward. NMDARs have extensive interactions with several membrane proteins and signaling cascades, allowing for heterogenous and thus mixed reports of cocaine effects. NMDAR expression can differ due to brain region, history of cocaine exposure, cocaine administration paradigm, and period of withdrawal (reviewed by Ref. [97]). For example, cocaine self-administration, but not noncontingent administration, increases hippocampal NMDAR subunit GluN1, and dorsal striatal NMDAR subunit GluN2A [76]. Both self-administered and noncontingent cocaine produce increases in PFC GluN2A expression and GluN1 expression in the hippocampus and NA. Furthermore, extinction training following cocaine self-administration increases PFC GluN2A, and GluN1 expression in the hippocampus and NA [76]. Without extinction training, animals that self-administer cocaine show increased GluN1 expression in the NA after 1, 30, or 90 days of withdrawal, and in the VTA after 1 and 90 days, compared to animals trained to self-administer sucrose [98]. Although upregulation of the GluN1 subunit of NMDARs has been reported in the VTA across several methodologies, and in the NA for some methodologies, however, changes in protein expression may not reflect changes in NMDAR functionality. It may be that NMDAR subunit composition is a better indicator for assessing cocaine induced-alterations that influence relapse behavior. The upregulation of GluN2B-containing NMDARs in the NA following cocaine has been inferred from increased sensitivity to GluN2B antagonists and electrophysiological data. Moreover, such increases are suggested to occur at "silent synapses," which are those newly generated in response to cocaine that contain NMDARs but not AMPARs [99].

Nicotine

Following nicotine self-administration, NMDA receptors are increased in the VTA, amygdala [99a], and NAc [10]. This is associated with increased membrane bound GluN2A- and GluN2B-containing NMDARs thought to mediate cue-induced nicotine reinstatement [10]. In support of this hypothesis, a GluN2A or GluN2B antagonist infused into the NAc attenuated cue-induced nicotine seeking, without affecting locomotion [10].

Ethanol

Ethanol consumption inhibits NMDAR-mediated calcium influx in the NAc, cortex, amygdala, hippocampus, and VTA [100]. Within the VTA, increased expression of the NR2A subunit increases ethanol reward [101]. Following intermittent access to alcohol and both 24h and 7 days withdrawal, GluN2A is increased while GluN2B is decreased, in the NAs [84]. However, in the NAc, increased GluN2A and GluN2B are only observed following 24h and not 7 days withdrawal from both continuous and intermittent access to alcohol consumption [84]. Withdrawal symptoms of ethanol, such as seizures, can be blocked using NMDAR antagonists, implicating an associated increase in excitatory transmission due to chronic alcohol use [100].

Heroin

Upregulated NAc surface expression of NMDARs, and specifically, the GluN2B-containing receptors, is associated with heroin seeking [102]. Occurring at prefrontal-accumbens synapses, stimulation of these NR2Bs is required for heroin relapse [102]. Spillover of synaptic glutamate to extrasynaptic GluN2B-containing NMDARs in the NAc is associated with long-lasting downregulation of surface expression of GLT-1 induced by chronic heroin use [45].

In summary, many classes of addictive drugs alter NMDA receptor subunit expression in the NA and other reward-related brain regions. In early withdrawal from cocaine self-administration, such changes account for the formation of "silent" AMPAR-lacking synapses in the NA. To date, the present of silent synapses have not been assessed following heroin, nicotine, or alcohol self-administration.

GLUTAMATERGIC PHARMACOTHERAPIES FOR RELAPSE PREVENTION

GLT and xCT Targets

Ceftriaxone

Ceftriaxone is a third-generation β-lactam (cephalosporin) antibiotic effective against Gram-positive and Gram-negative aerobic, and some anaerobic, bacteria. Already approved by the FDA as a treatment for a variety of infections and meningitis, ceftriaxone has strong potential to serve as an antirelapse treatment for several substance use disorders. This application stems from its ability to upregulate system xc- (xCT) and GLT-1, which are primary regulators of glutamate homeostasis. The convergence of evidence suggesting that impaired glutamate homeostasis drives relapse across multiple drugs classes has lead preclinical researchers to evaluate ceftriaxone's ability as a relapse prevention agent for cocaine, ethanol, nicotine, and heroin, with compelling results [51].

N-Acetylcysteine (NAC)

NAC is a precursor to glutathione and has been used since the 1960s to combat paracetamol overdose and loosen mucus. It can be taken intravenously, by mouth or inhaled as a mist, and is considered safe, effective, and low cost. NAC is thought to improve tone on presynaptic mGluRs via facilitation of glial glutamate release. In addition to its ability to upregulate GLT-1 and xCT expression levels, NAC holds promise as an antirelapse treatment and has been assessed in the prevention of relapse to cocaine, nicotine, ethanol, and opiates, largely with positive outcomes (for review, see Ref. [6]).

Augmentin

The combination of a β-lactam antibiotic, amoxicillin, and a β-lactamase inhibitor, clavulanate, is FDA-approved and marketed as Augmentin. In preclinical studies, augmentin attenuated ethanol consumption and increased GLT-1 and xCT expression in the NA and PFC [103]. Clavulanic acid has been used to treat symptoms of morphine and cocaine addiction. It decreases morphine's rewarding properties, locomotor sensitization, and hypothermia, at a 20-fold lower dose than necessary for ceftriaxone [104]. At 100-fold lower dose than necessary for ceftriaxone, clavulanic acid increased GLT-1 in the NA and decreased the rewarding efficacy of cocaine, without affecting acquisition of cocaine self-administration or retention of learning [105].

AMPAR and NMDAR Targets

Amantadine is an NMDAR antagonist developed originally as an antiviral medication. However, effects on cocaine addiction have been mixed. Another NMDAR antagonist approved for treatment of late-stage Alzheimer's disease, memantine, has shown positive outcomes for treatment of opioid addiction [106] but little effect in alcohol dependence [107]. An NMDAR coagonist, D-cycloserine, has mixed results in treatment of cue reactivity in cocaine [108], nicotine [109], and alcohol addicts [110], although preclinical models show enhanced extinction learning [111–113].

mGluR Targets

Based on the effects on glutamatergic receptors outlined above, it would be reasonable to pursue activation of presynaptic mGlu2/3 or the blocking of postsynaptic mGlu5 as potential pharmacological candidates to prevent drug seeking [2]. Modafinil, a treatment for narcolepsy, has a variety of targets, including dopamine, and glutamate. Modafinil was shown to increase extracellular glutamate levels in the NAc via activation of xC- and mGlu2/3 and inhibit cocaine-primed reinstatement [114] and morphine-primed reinstatement [115]. Due to potential side effects on natural reward (e.g., sucrose, food) and locomotion found with mGlu2/3 manipulation (e.g., [115a]), the pursuit of mGlu5 may be of greater advantage. However, some success using an mGluR2-specific positive allosteric modulator (AZD8529) has been found with nicotine addiction, in which doses effective in preventing nicotine seeking were low enough that they did not impact food seeking [116]. Moreover, antagonizing mGlu5 has the potential for negatively impacting learning, including impairing performance on spatial and working memory tasks [117], and extinction of both conditioned fear and conditioned cocaine cues [78,118]. Although preclinical results show negative allosteric modulators of mGlu5 decrease nicotine self-administration and cue reinstatement, potentiated symptoms of withdrawal and multiple unwanted side effects currently prevent their use clinically [79].

CLINICAL EVIDENCE OF GLUTAMATERGIC NEUROPLASTICITY IN ADDICTION

In a double-blind crossover study using N-acetylcysteine to treat cocaine craving, reactivity to cocaine cues, and subjective desire to use cocaine decreased [119,120]. A pilot study testing three different doses found N-acetylcysteine to be safe and tolerable, and the majority of subjects who completed the study either terminated or significantly reduced their cocaine use during treatment [121]. In a double-blind study, nicotine users were treated for 4 weeks with oral N-acetylcysteine or placebo and were evaluated for subjective cravings and carbon monoxide levels [49]. Self-reports indicated a reduction in cigarettes smoked in the N-acetylcysteine treatment group, with no other effects [49]. In addition, a genetic linkage in the NMDA 2A gene, which encodes the 2A subunit of NMDAR, was identified in susceptibility to heroin addiction [49a]. Finally, PET studies utilizing an mGlu5 ligand revealed that recently abstinent cocaine users have reduced mGlu5 binding in the ventral striatum [122,123], in agreement with the findings in rats extinguished from cocaine self-administration [48].

CONCLUSION

The literature reviewed here strongly indicates that addictive drugs share the ability to alter glutamatergic protein expression and signaling, especially in the NAc. Some differences exist in terms of the ability of heroin and alcohol to alter GLT-1 and/or xCT expression and function, but the majority of glutamatergic changes are similar across classes of drugs. In support of these similarities, two compounds which alter glutamate uptake and nonsynaptic release, ceftriaxone and N-acetylcysteine, consistently attenuate the seeking of nicotine, heroin, alcohol, and cocaine. N-acetylcysteine has shown promise in human trials for reducing cocaine and nicotine use. Taken together, these results indicate that targeting glutamatergic adaptations holds the potential for reversing drug-induced neuroadaptations and preventing relapse to drug seeking.

REFERENCES

[1] Kalivas PW, Lalumiere RT, Knackstedt L, Shen H. Glutamate transmission in addiction. Neuropharmacology 2009;56(Suppl 1):169–73. https://doi.org/10.1016/j.neuropharm.2008.07.011.

[2] Olive MF. Metabotropic glutamate receptor ligands as potential therapeutics for addiction. Curr Drug Abuse Rev 2009;2:83–98.

[3] Uys JD, LaLumiere RT. Glutamate: the new frontier in pharmacotherapy for cocaine addiction. CNS Neurol Disord Drug Targets 2008;7:482–91.

[4] Cornish JL, Kalivas PW. Glutamate transmission in the nucleus accumbens mediates relapse in cocaine addiction. J Neurosci 2000;20:RC89.

[5] Quintero GC. Role of nucleus accumbens glutamatergic plasticity in drug addiction. Neuropsychiatr Dis Treat 2013;9:1499–512. https://doi.org/10.2147/NDT.S45963.

[6] Scofield MD, Heinsbroek JA, Gipson CD, et al. The nucleus accumbens: mechanisms of addiction across drug classes reflect the importance of glutamate homeostasis. Pharmacol Rev 2016;68:816–71. https://doi.org/10.1124/pr.116.012484.

[7] Knackstedt LA, Kalivas PW. Glutamate and reinstatement. Curr Opin Pharmacol 2009;9:59–64. https://doi.org/10.1016/j.coph.2008.12.003.

[8] Spencer S, Scofield M, Kalivas PW. The good and bad news about glutamate in drug addiction. J Psychopharmacol (Oxford) 2016;30:1095–8. https://doi.org/10.1177/0269881116655248.

[9] McFarland K, Lapish CC, Kalivas PW. Prefrontal glutamate release into the core of the nucleus accumbens mediates cocaine-induced reinstatement of drug-seeking behavior. J Neurosci 2003;23:3531–7.

[9a] Smith A, Scofield M, Heinsbroek J, Gipson C, Neuhofer D, Roberts-Wolfe D, Spencer S, Garcia-Keller C, Stankeviciute N, Smith J, Allen N, Lorang M, Griffin III W, Boger B, Kalivas P. Accumbens nNOS interneurons regulate cocaine relapse. J Neurosci 2017;37(4):742–56. Behavioral/Cognitive https://doi.org/10.1523/JNEUROSCI.2673-16.2017.

[10] Gipson CD, Reissner KJ, Kupchik YM, et al. Reinstatement of nicotine seeking is mediated by glutamatergic plasticity. Proc Natl Acad Sci U S A 2013;110:9124–9. https://doi.org/10.1073/pnas.1220591110.

[11] LaLumiere RT, Kalivas PW. Glutamate release in the nucleus accumbens core is necessary for heroin seeking. J Neurosci 2008;28:3170–7. https://doi.org/10.1523/JNEUROSCI.5129-07.2008.

[12] Alhaddad H, Das SC, Sari Y. Effects of ceftriaxone on ethanol intake: a possible role for xCT and GLT-1 isoforms modulation of glutamate levels in P rats. Psychopharmacology (Berl) 2014;231:4049–57. https://doi.org/10.1007/s00213-014-3545-y.

[13] Epstein DH, Preston KL, Stewart J, Shaham Y. Toward a model of drug relapse: an assessment of the validity of the reinstatement procedure. Psychopharmacology (Berl) 2006;189:1–16. https://doi.org/10.1007/s00213-006-0529-6.

[14] Banna KM, Back SE, Do P, See RE. Yohimbine stress potentiates conditioned cue-induced reinstatement of heroin-seeking in rats. Behav Brain Res 2010;208:144–8. https://doi.org/10.1016/j.bbr.2009.11.030.

[15] Buffalari DM, See RE. Footshock stress potentiates cue-induced cocaine-seeking in an animal model of relapse. Physiol Behav 2009;98:614–7. https://doi.org/10.1016/j.physbeh.2009.09.013.

[16] Knackstedt LA, Trantham-Davidson HL, Schwendt M. The role of ventral and dorsal striatum mGluR5 in relapse to cocaine-seeking and extinction learning. Addict Biol 2014;19:87–101. https://doi.org/10.1111/adb.12061.

[17] Conrad KL, Tseng KY, Uejima JL, et al. Formation of accumbens GluR2-lacking AMPA receptors mediates incubation of cocaine craving. Nature 2008;454:118–21. https://doi.org/10.1038/nature06995.

[18] Fischer-Smith KD, ACW H, Rebec GV. Differential effects of cocaine access and withdrawal on glutamate type 1 transporter expression in rat nucleus accumbens core and shell. Neuroscience 2012;210:333–9. https://doi.org/10.1016/j.neuroscience.2012.02.049.

[19] LaCrosse AL, Hill K, Knackstedt LA. Ceftriaxone attenuates cocaine relapse after abstinence through modulation of nucleus accumbens AMPA subunit expression. Eur Neuropsychopharmacol 2016;26:186–94. https://doi.org/10.1016/j.euroneuro.2015.12.022.

[20] Hamlin AS, Clemens KJ, GP MN. Renewal of extinguished cocaine-seeking. Neuroscience 2008;151:659–70. https://doi.org/10.1016/j.neuroscience.2007.11.018.

[21] Hamlin AS, Blatchford KE, GP MN. Renewal of an extinguished instrumental response: neural correlates and the role of D1 dopamine receptors. Neuroscience 2006;143:25–38. https://doi.org/10.1016/j.neuroscience.2006.07.035.

[22] Hamlin AS, Newby J, GP MN. The neural correlates and role of D1 dopamine receptors in renewal of extinguished alcohol-seeking. Neuroscience 2007;146:525–36. https://doi.org/10.1016/j.neuroscience.2007.01.063.

[23] Fuchs RA, Evans KA, Ledford CC, et al. The role of the dorsomedial prefrontal cortex, basolateral amygdala, and dorsal hippocampus in contextual reinstatement of cocaine seeking in rats. Neuropsychopharmacology 2005;30:296–309. https://doi.org/10.1038/sj.npp.1300579.

[24] Bossert JM, Liu SY, Lu L, Shaham Y. A role of ventral tegmental area glutamate in contextual cue-induced relapse to heroin seeking. J Neurosci 2004;24:10726–30. https://doi.org/10.1523/JNEUROSCI.3207-04.2004.

[25] Bouton ME. Context, ambiguity, and unlearning: sources of relapse after behavioral extinction. Biol Psychiatry 2002;52:976–86.

[26] Trantham-Davidson H, LaLumiere RT, Reissner KJ, et al. Ceftriaxone normalizes nucleus accumbens synaptic transmission, glutamate transport, and export following cocaine self-administration and extinction training. J Neurosci 2012;32:12406–10. https://doi.org/10.1523/JNEUROSCI.1976-12.2012.

[27] Heinsbroek JA, Neuhofer DN, Griffin WC, et al. Loss of Plasticity in the D2-Accumbens Pallidal Pathway Promotes Cocaine Seeking. J Neurosci 2017;37:757–67. https://doi.org/10.1523/JNEUROSCI.2659-16.2017.

[28] Kimura M, Yamada H, Matsumoto N. Tonically active neurons in the striatum encode motivational contexts of action. Brain Dev 2003;25(Suppl 1): S20–3.

[29] Brown MTC, Tan KR, O'Connor EC, et al. Ventral tegmental area GABA projections pause accumbal cholinergic interneurons to enhance associative learning. Nature 2012;492:452–6. https://doi.org/10.1038/nature11657.

[30] Smith ACW, Kupchik YM, Scofield MD, et al. Synaptic plasticity mediating cocaine relapse requires matrix metalloproteinases. Nat Neurosci 2014;17:1655–7. https://doi.org/10.1038/nn.3846.

[31] Di Ciano P, Everitt BJ. Direct interactions between the basolateral amygdala and nucleus accumbens core underlie cocaine-seeking behavior by rats. J Neurosci 2004;24:7167–73. https://doi.org/10.1523/JNEUROSCI.1581-04.2004.

[32] Fuchs RA, Evans KA, Parker MC, See RE. Differential involvement of the core and shell subregions of the nucleus accumbens in conditioned cue-induced reinstatement of cocaine seeking in rats. Psychopharmacology (Berl) 2004;176:459–65. https://doi.org/10.1007/s00213-004-1895-6.

[33] Ito R, Robbins TW, Everitt BJ. Differential control over cocaine-seeking behavior by nucleus accumbens core and shell. Nat Neurosci 2004;7:389–97. https://doi.org/10.1038/nn1217.

[34] Anderson SM, Bari AA, Pierce RC. Administration of the D1-like dopamine receptor antagonist SCH-23390 into the medial nucleus accumbens shell attenuates cocaine priming-induced reinstatement of drug-seeking behavior in rats. Psychopharmacology (Berl) 2003;168:132–8. https://doi.org/10.1007/s00213-002-1298-5.

[35] Fuchs RA, Feltenstein MW, See RE. The role of the basolateral amygdala in stimulus-reward memory and extinction memory consolidation and in subsequent conditioned cued reinstatement of cocaine seeking. Eur J Neurosci 2006;23:2809–13. https://doi.org/10.1111/j.1460-9568.2006.04806.x.

[36] McLaughlin J, See RE. Selective inactivation of the dorsomedial prefrontal cortex and the basolateral amygdala attenuates conditioned-cued reinstatement of extinguished cocaine-seeking behavior in rats. Psychopharmacology (Berl) 2003;168:57–65. https://doi.org/10.1007/s00213-002-1196-x.

[37] Stefanik MT, Kalivas PW. Optogenetic dissection of basolateral amygdala projections during cue-induced reinstatement of cocaine seeking. Front Behav Neurosci 2013;7:213. https://doi.org/10.3389/fnbeh.2013.00213.

[38] Kalivas PW, O'Brien C. Drug addiction as a pathology of staged neuroplasticity. Neuropsychopharmacology 2008;33:166–80. https://doi.org/10.1038/sj.npp.1301564.

[39] Rogers JL, Ghee S, See RE. The neural circuitry underlying reinstatement of heroin-seeking behavior in an animal model of relapse. Neuroscience 2008;151:579–88. https://doi.org/10.1016/j.neuroscience.2007.10.012.

[40] Rogers JL, See RE. Selective inactivation of the ventral hippocampus attenuates cue-induced and cocaine-primed reinstatement of drug-seeking in rats. Neurobiol Learn Mem 2007;87:688–92. https://doi.org/10.1016/j.nlm.2007.01.003.

[41] See RE, Elliott JC, Feltenstein MW. The role of dorsal vs ventral striatal pathways in cocaine-seeking behavior after prolonged abstinence in rats. Psychopharmacology (Berl) 2007;194:321–31. https://doi.org/10.1007/s00213-007-0850-8.

[42] Adermark L, Bowers MS. Disentangling the role of astrocytes in alcohol use disorder. Alcohol Clin Exp Res 2016;40:1802–16. https://doi.org/10.1111/acer.13168.

[43] Sargeant TJ, Miller JH, Day DJ. Opioidergic regulation of astroglial/neuronal proliferation: where are we now? J Neurochem 2008;107:883–97. https://doi.org/10.1111/j.1471-4159.2008.05671.x.

[44] Cleva RM, Gass JT, Widholm JJ, Olive MF. Glutamatergic targets for enhancing extinction learning in drug addiction. Curr Neuropharmacol 2010;8:394–408. https://doi.org/10.2174/157015910793358169.

[45] Shen H, Scofield MD, Boger H, et al. Synaptic glutamate spillover due to impaired glutamate uptake mediates heroin relapse. J Neurosci 2014;34:5649–57. https://doi.org/10.1523/JNEUROSCI.4564-13.2014.

[46] Gass JT, Sinclair CM, Cleva RM, et al. Alcohol-seeking behavior is associated with increased glutamate transmission in basolateral amygdala and nucleus accumbens as measured by glutamate-oxidase-coated biosensors. Addict Biol 2011;16:215–28. https://doi.org/10.1111/j.1369-1600.2010.00262.x.

[47] Danbolt NC. Glutamate uptake. Prog Neurobiol 2001;65:1–105. https://doi.org/10.1016/S0301-0082(00)00067-8.

[48] Knackstedt LA, Melendez RI, Kalivas PW. Ceftriaxone restores glutamate homeostasis and prevents relapse to cocaine seeking. Biol Psychiatry 2010;67:81–4. https://doi.org/10.1016/j.biopsych.2009.07.018.

[49] Knackstedt LA, LaRowe S, Mardikian P, et al. The role of cystine-glutamate exchange in nicotine dependence in rats and humans. Biol Psychiatry 2009;65:841–5. https://doi.org/10.1016/j.biopsych.2008.10.040.

[49a] Zhong HJ, Huo ZH, Dang J, Chen J, Zhu YS, Liu JH. Functional polymorphisms of the glutamate receptor N-methyl D-aspartate 2A gene are associated with heroin addiction. Genet Mol Res 2014;13:8714–21. https://doi.org/10.4238/2014.October.27.12.

[50] Rao PSS, Sari Y. Glutamate transporter 1: target for the treatment of alcohol dependence. Curr Med Chem 2012;19:5148–56.

[51] Roberts-Wolfe DJ, Kalivas PW. Glutamate Transporter GLT-1 as a Therapeutic Target for Substance Use Disorders. CNS Neurol Disord Drug Targets 2015;14:745–56.

[52] Fischer KD, ACW H, Rebec GV. Role of the major glutamate transporter GLT1 in nucleus accumbens core versus shell in cue-induced cocaine-seeking behavior. J Neurosci 2013;33:9319–27. https://doi.org/10.1523/JNEUROSCI.3278-12.2013.

[53] Reissner KJ, Gipson CD, Tran PK, et al. Glutamate transporter GLT-1 mediates N-acetylcysteine inhibition of cocaine reinstatement. Addict Biol 2015;20:316–23. https://doi.org/10.1111/adb.12127.

[54] Alasmari F, Crotty Alexander LE, Nelson JA, et al. Effects of chronic inhalation of electronic cigarettes containing nicotine on glial glutamate transporters and α-7 nicotinic acetylcholine receptor in female CD-1 mice. Prog Neuropsychopharmacol Biol Psychiatry 2017;77:1–8. https://doi.org/10.1016/j.pnpbp.2017.03.017.

[55] Schreiber R, Freund WD. Glutamate transport is downregulated in the cerebral cortex of alcohol-preferring rats. Med Sci Monit 2000;6:649–52.

[56] Devaud LL. Ethanol dependence has limited effects on GABA or glutamate transporters in rat brain. Alcohol Clin Exp Res 2001;25:606–11.

[57] Sari Y, Sreemantula SN. Neuroimmunophilin GPI-1046 reduces ethanol consumption in part through activation of GLT1 in alcohol-preferring rats. Neuroscience 2012;227:327–35. https://doi.org/10.1016/j.neuroscience.2012.10.007.

[58] Alshehri FS, Althobaiti YS, Sari Y. Effects of administered ethanol and methamphetamine on glial glutamate transporters in rat striatum and hippocampus. J Mol Neurosci 2017;61:343–50. https://doi.org/10.1007/s12031-016-0859-8.

[59] Stennett BA, Frankowski JC, Peris J, Knackstedt LA. Ceftriaxone reduces alcohol intake in outbred rats while upregulating xCT in the nucleus accumbens core. Pharmacol Biochem Behav 2017;159:18–23. https://doi.org/10.1016/j.pbb.2017.07.001.

[60] McBean GJ. Cerebral cystine uptake: a tale of two transporters. Trends Pharmacol Sci 2002;23:299–302.

[61] Baker DA, McFarland K, Lake RW, et al. Neuroadaptations in cystine-glutamate exchange underlie cocaine relapse. Nat Neurosci 2003;6:743–9. https://doi.org/10.1038/nn1069.

[62] Moran MM, Mcfarland K, Melendez RI, et al. Cystine/glutamate exchange regulates metabotropic glutamate receptor presynaptic inhibition of excitatory transmission and vulnerability to cocaine seeking. J Neurosci 2005;25:6389–93. https://doi.org/10.1523/JNEUROSCI.1007-05.2005.

[63] LaCrosse AL, O'Donovan SM, Sepulveda-Orengo MT, et al. Contrasting the role of xCT and GLT-1 upregulation in the ability of ceftriaxone to attenuate the cue-induced reinstatement of cocaine-seeking and normalize AMPA receptor subunit expression. J Neurosci 2017; https://doi.org/10.1523/JNEUROSCI.3717-16.2017.

[64] Rao PSS, Sari Y. Effects of ceftriaxone on chronic ethanol consumption: a potential role for xCT and GLT1 modulation of glutamate levels in male P rats. J Mol Neurosci 2014;54:71–7. https://doi.org/10.1007/s12031-014-0251-5.

[65] Liechti ME, Lhuillier L, Kaupmann K, Markou A. Metabotropic glutamate 2/3 receptors in the ventral tegmental area and the nucleus accumbens shell are involved in behaviors relating to nicotine dependence. J Neurosci 2007;27:9077–85. https://doi.org/10.1523/JNEUROSCI.1766-07.2007.

[66] Yang H-J, Zhang H-Y, Bi G-H, et al. Deletion of type 2 metabotropic glutamate receptor decreases sensitivity to cocaine reward in rats. Cell Rep 2017;20:319–32. https://doi.org/10.1016/j.celrep.2017.06.046.

[67] Cartmell J, Schoepp DD. Regulation of neurotransmitter release by metabotropic glutamate receptors. J Neurochem 2000;75:889–907.

[68] Peters J, Kalivas PW. The group II metabotropic glutamate receptor agonist, LY379268, inhibits both cocaine- and food-seeking behavior in rats. Psychopharmacology (Berl) 2006;186:143–9. https://doi.org/10.1007/s00213-006-0372-9.

[69] Liechti ME, Markou A. Role of the glutamatergic system in nicotine dependence : implications for the discovery and development of new pharmacological smoking cessation therapies. CNS Drugs 2008;22:705–24.

[70] Zhao Y, Dayas CV, Aujla H, et al. Activation of group II metabotropic glutamate receptors attenuates both stress and cue-induced ethanol-seeking and modulates c-fos expression in the hippocampus and amygdala. J Neurosci 2006;26:9967–74. https://doi.org/10.1523/JNEUROSCI.2384-06.2006.

[71] Pati D, Kelly K, Stennett B, et al. Alcohol consumption increases basal extracellular glutamate in the nucleus accumbens core of Sprague-Dawley rats without increasing spontaneous glutamate release. Eur J Neurosci 2016;44:1896–905. https://doi.org/10.1111/ejn.13284.

[72] Meinhardt MW, Hansson AC, Perreau-Lenz S, Bauder-Wenz C, Stahlin O, Heilig M, Harper C, Drescher KU, Spanagel R, Sommer WH. Rescue of infralimbic mGluR2 deficit restores control over drug-seeking behavior in alcohol dependence. J Neurosci 2013;33:2794–806. https://doi.org/10.1523/JNEUROSCI.4062-12.2013.

[73] Bossert JM, Gray SM, Lu L, Shaham Y. Activation of group II metabotropic glutamate receptors in the nucleus accumbens shell attenuates context-induced relapse to heroin seeking. Neuropsychopharmacology 2006;31:2197–209. https://doi.org/10.1038/sj.npp.1300977.

[74] Hao Y, Martin-Fardon R, Weiss F. Behavioral and functional evidence of metabotropic glutamate receptor 2/3 and metabotropic glutamate receptor 5 dysregulation in cocaine-escalated rats: factor in the transition to dependence. Biol Psychiatry 2010;68:240–8. https://doi.org/10.1016/j.biopsych.2010.02.011.

[75] Pomierny-Chamiolo L, Miszkiel J, Frankowska M, et al. Cocaine self-administration, extinction training and drug-induced relapse change metabotropic glutamate mGlu5 receptors expression: evidence from radioligand binding and immunohistochemistry assays. Brain Res 2016; https://doi.org/10.1016/j.brainres.2016.11.014.

[76] Pomierny-Chamiolo L, Miszkiel J, Frankowska M, et al. Withdrawal from cocaine self-administration and yoked cocaine delivery dysregulates glutamatergic mGlu5 and NMDA receptors in the rat brain. Neurotox Res 2015;27:246–58. https://doi.org/10.1007/s12640-014-9502-z.

[77] Kumaresan V, Yuan M, Yee J, et al. Metabotropic glutamate receptor 5 (mGluR5) antagonists attenuate cocaine priming- and cue-induced reinstatement of cocaine seeking. Behav Brain Res 2009;202:238–44. https://doi.org/10.1016/j.bbr.2009.03.039.

[78] Knackstedt LA, Schwendt M. mGlu5 receptors and relapse to cocaine-seeking: the role of receptor trafficking in postrelapse extinction learning deficits, Neural Plast 2016;2016: 9312508. https://doi.org/10.1155/2016/9312508.

[79] Chiamulera C, Marzo CM, Balfour DJK. Metabotropic glutamate receptor 5 as a potential target for smoking cessation. Psychopharmacology (Berl) 2017;234:1357–70. https://doi.org/10.1007/s00213-016-4487-3.

[80] Tessari M, Pilla M, Andreoli M, et al. Antagonism at metabotropic glutamate 5 receptors inhibits nicotine- and cocaine-taking behaviours and prevents nicotine-triggered relapse to nicotine-seeking. Eur J Pharmacol 2004;499:121–33. https://doi.org/10.1016/j.ejphar.2004.07.056.

[81] Bespalov AY, Dravolina OA, Sukhanov I, et al. Metabotropic glutamate receptor (mGluR5) antagonist MPEP attenuated cue- and schedule-induced reinstatement of nicotine self-administration behavior in rats. Neuropharmacology 2005;49(Suppl 1):167–78. https://doi.org/10.1016/j.neuropharm.2005.06.007.

[82] Stoker AK, Olivier B, Markou A. Involvement of metabotropic glutamate receptor 5 in brain reward deficits associated with cocaine and nicotine withdrawal and somatic signs of nicotine withdrawal. Psychopharmacology (Berl) 2012;221:317–27. https://doi.org/10.1007/s00213-011-2578-8.

[83] Cozzoli DK, Goulding SP, Zhang PW, et al. Binge drinking upregulates accumbens mGluR5-Homer2-PI3K signaling: functional implications for alcoholism. J Neurosci 2009;29:8655–68. https://doi.org/10.1523/JNEUROSCI.5900-08.2009.

[84] Obara I, Bell RL, Goulding SP, et al. Differential effects of chronic ethanol consumption and withdrawal on homer/glutamate receptor expression in subregions of the accumbens and amygdala of P rats. Alcohol Clin Exp Res 2009;33:1924–34. https://doi.org/10.1111/j.1530-0277.2009.01030.x.

[85] Schroeder JP, Spanos M, Stevenson JR, et al. Cue-induced reinstatement of alcohol-seeking behavior is associated with increased ERK1/2 phosphorylation in specific limbic brain regions: blockade by the mGluR5 antagonist MPEP. Neuropharmacology 2008;55:546–54. https://doi.org/10.1016/j.neuropharm.2008.06.057.

[86] Lou Z, Chen L, Liu H, et al. Blockade of mGluR5 in the nucleus accumbens shell but not core attenuates heroin seeking behavior in rats. Acta Pharmacol Sin 2014;35:1485–92. https://doi.org/10.1038/aps.2014.93.

[87] Simonyi A, Schachtman TR, Christoffersen GR. Metabotropic glutamate receptor subtype 5 antagonism in learning and memory. Eur J Pharmacol 2010;639:17–25. https://doi.org/10.1016/j.ejphar.2009.12.039.

[88] Wolf ME, Ferrario CR. AMPA receptor plasticity in the nucleus accumbens after repeated exposure to cocaine. Neurosci Biobehav Rev 2010;35:185–211. https://doi.org/10.1016/j.neubiorev.2010.01.013.

[89] Ferrario CR, Loweth JA, Milovanovic M, et al. Distribution of AMPA receptor subunits and TARPs in synaptic and extrasynaptic membranes of the adult rat nucleus accumbens. Neurosci Lett 2011;490:180–4. https://doi.org/10.1016/j.neulet.2010.12.036.

[89a] Famous KR, Kumaresan V, Sadri-Vakili G, Schmidt HD, Mierke DF, Cha J-HJ, Pierce RC. Phosphorylation-dependent trafficking of GluR2-containing AMPA receptors in the nucleus accumbens plays a critical role in the reinstatement of cocaine seeking. J Neurosci 2008;28:11061–70. https://doi.org/10.1523/JNEUROSCI.1221-08.2008.

[90] Liu SJ, Zukin RS. Ca2+-permeable AMPA receptors in synaptic plasticity and neuronal death. Trends Neurosci 2007;30:126–34. https://doi.org/10.1016/j.tins.2007.01.006.

[91] Wolf ME. The Bermuda Triangle of cocaine-induced neuroadaptations. Trends Neurosci 2010;33:391–8. https://doi.org/10.1016/j.tins.2010.06.003.

[92] Wolf ME. Synaptic mechanisms underlying persistent cocaine craving. Nat Rev Neurosci 2016;17:351–65. https://doi.org/10.1038/nrn.2016.39.

[93] Anderson SM, Famous KR, Sadri-Vakili G, et al. CaMKII: a biochemical bridge linking accumbens dopamine and glutamate systems in cocaine seeking. Nat Neurosci 2008;11:344–53. https://doi.org/10.1038/nn2054.

[94] White SL, Ortinski PI, Friedman SH, et al. A critical role for the glua1 accessory protein, SAP97, in cocaine seeking. Neuropsychopharmacology 2016;41:736–50. https://doi.org/10.1038/npp.2015.199.

[95] Cannady R, Fisher KR, Durant B, et al. Enhanced AMPA receptor activity increases operant alcohol self-administration and cue-induced reinstatement. Addict Biol 2013;18:54–65. https://doi.org/10.1111/adb.12000.

[96] Van den Oever MC, Goriounova NA, Li KW, et al. Prefrontal cortex AMPA receptor plasticity is crucial for cue-induced relapse to heroin-seeking. Nat Neurosci 2008;11:1053–8. https://doi.org/10.1038/nn.2165.

[97] Ortinski PI. Cocaine-induced changes in NMDA receptor signaling. Mol Neurobiol 2014;50:494–506. https://doi.org/10.1007/s12035-014-8636-6.

[98] Lu L, Grimm JW, Shaham Y, Hope BT. Molecular neuroadaptations in the accumbens and ventral tegmental area during the first 90 days of forced abstinence from cocaine self-administration in rats. J Neurochem 2003;85:1604–13.

[99] Huang YH, Lin Y, Mu P, et al. In vivo cocaine experience generates silent synapses. Neuron 2009;63:40–7. https://doi.org/10.1016/j.neuron.2009.06.007.

[99a] Kenny PJ, Chartoff E, Roberto M, Carlezon WA, Markou A. NMDA receptors regulate nicotine-enhanced brain reward function and intravenous nicotine self-administration: role of the ventral tegmental area and central nucleus of the amygdala. Neuropsychopharmacology 2009;34:266–81. https://doi.org/10.1038/npp.2008.58.

[100] Rao PSS, Bell RL, Engleman EA, Sari Y. Targeting glutamate uptake to treat alcohol use disorders. Front Neurosci 2015;9:144. https://doi.org/10.3389/fnins.2015.00144.

[101] D'Souza MS. Glutamatergic transmission in drug reward: implications for drug addiction. Front Neurosci 2015;9:404. https://doi.org/10.3389/fnins.2015.00404.

[102] Shen H, Moussawi K, Zhou W, et al. Heroin relapse requires long-term potentiation-like plasticity mediated by NMDA2b-containing receptors. Proc Natl Acad Sci U S A 2011;108:19407–12. https://doi.org/10.1073/pnas.1112052108.

[103] Hakami AY, Alshehri FS, Althobaiti YS, Sari Y. Effects of orally administered Augmentin on glutamate transporter 1, cystine-glutamate exchanger expression and ethanol intake in alcohol-preferring rats. Behav Brain Res 2017;320:316–22. https://doi.org/10.1016/j.bbr.2016.12.016.

[104] Schroeder JA, Tolman NG, FF MK, et al. Clavulanic acid reduces rewarding, hyperthermic and locomotor-sensitizing effects of morphine in rats: a new indication for an old drug? Drug Alcohol Depend 2014;142:41–5. https://doi.org/10.1016/j.drugalcdep.2014.05.012.

[105] Kim J, John J, Langford D, et al. Clavulanic acid enhances glutamate transporter subtype I (GLT-1) expression and decreases reinforcing efficacy of cocaine in mice. Amino Acids 2016;48:689–96. https://doi.org/10.1007/s00726-015-2117-8.

[106] Bisaga A, Comer SD, Ward AS, et al. The NMDA antagonist memantine attenuates the expression of opioid physical dependence in humans. Psychopharmacology (Berl) 2001;157:1–10.

[107] Evans SM, Levin FR, Brooks DJ, Garawi F. A pilot double-blind treatment trial of memantine for alcohol dependence. Alcohol Clin Exp Res 2007;31:775–82. https://doi.org/10.1111/j.1530-0277.2007.00360.x.

[108] Price KL, Baker NL, McRae-Clark AL, et al. A randomized, placebo-controlled laboratory study of the effects of D-cycloserine on craving in cocaine-dependent individuals. Psychopharmacology (Berl) 2013;226:739–46. https://doi.org/10.1007/s00213-011-2592-x.

[109] Kamboj SK, Joye A, Das RK, et al. Cue exposure and response prevention with heavy smokers: a laboratory-based randomised placebo-controlled trial examining the effects of D-cycloserine on cue reactivity and attentional bias. Psychopharmacology (Berl) 2012;221:273–84. https://doi.org/10.1007/s00213-011-2571-2.

[110] Kamboj SK, Massey-Chase R, Rodney L, et al. Changes in cue reactivity and attentional bias following experimental cue exposure and response prevention: a laboratory study of the effects of D-cycloserine in heavy drinkers. Psychopharmacology (Berl) 2011;217:25–37. https://doi.org/10.1007/s00213-011-2254-z.

[111] Davis M, Ressler K, Rothbaum BO, Richardson R. Effects of D-cycloserine on extinction: translation from preclinical to clinical work. Biol Psychiatry 2006;60:369–75. https://doi.org/10.1016/j.biopsych.2006.03.084.

[112] Parnas AS, Weber M, Richardson R. Effects of multiple exposures to D-cycloserine on extinction of conditioned fear in rats. Neurobiol Learn Mem 2005;83:224–31. https://doi.org/10.1016/j.nlm.2005.01.001.

[113] Weber M, Hart J, Richardson R. Effects of D-cycloserine on extinction of learned fear to an olfactory cue. Neurobiol Learn Mem 2007;87:476–82. https://doi.org/10.1016/j.nlm.2006.12.010.

[114] Mahler SV, Hensley-Simon M, Tahsili-Fahadan P, et al. Modafinil attenuates reinstatement of cocaine seeking: role for cystine-glutamate exchange and metabotropic glutamate receptors. Addict Biol 2014;19:49–60. https://doi.org/10.1111/j.1369-1600.2012.00506.x.

[115] Tahsili-Fahadan P, Carr GV, Harris GC, Aston-Jones G. Modafinil blocks reinstatement of extinguished opiate-seeking in rats: mediation by a glutamate mechanism. Neuropsychopharmacology 2010;35:2203–10. https://doi.org/10.1038/npp.2010.94.

[115a] Peters J, Kalivas PW. The group II metabotropic glutamate receptor agonist, LY379268, inhibits both cocaine- and food-seeking behavior in rats. Psychopharmacology (Berl) 2006;186:143–9. https://doi.org/10.1007/s00213-006-0372-9.

[116] Justinova Z, Panlilio LV, Secci ME, et al. The Novel Metabotropic Glutamate Receptor 2 Positive Allosteric Modulator, AZD8529, Decreases Nicotine Self-Administration and Relapse in Squirrel Monkeys. Biol Psychiatry 2015;78:452–62. https://doi.org/10.1016/j.biopsych.2015.01.014.

[117] Tan SZK, Ganella DE, Dick ALW, et al. Spatial learning requires mGlu5 signalling in the dorsal hippocampus. Neurochem Res 2015;40:1303–10. https://doi.org/10.1007/s11064-015-1595-0.

[118] Bird MK, Lohmann P, West B, et al. The mGlu5 receptor regulates extinction of cocaine-driven behaviours. Drug Alcohol Depend 2014;137:83–9. https://doi.org/10.1016/j.drugalcdep.2014.01.017.

[119] LaRowe SD, Mardikian P, Malcolm R, et al. Safety and tolerability of N-acetylcysteine in cocaine-dependent individuals. Am J Addict 2006;15:105–10. https://doi.org/10.1080/10550490500419169.

[120] LaRowe SD, Myrick H, Hedden S, et al. Is cocaine desire reduced by N-acetylcysteine? Am J Psychiatry 2007;164:1115–7. https://doi.org/10.1176/ajp.2007.164.7.1115.

[121] Mardikian PN, LaRowe SD, Hedden S, et al. An open-label trial of N-acetylcysteine for the treatment of cocaine dependence: a pilot study. Prog Neuropsychopharmacol Biol Psychiatry 2007;31:389–94. https://doi.org/10.1016/j.pnpbp.2006.10.001.

[122] Martinez D, Slifstein M, Nabulsi N, et al. Imaging glutamate homeostasis in cocaine addiction with the metabotropic glutamate receptor 5 positron emission tomography radiotracer [(11)C]ABP688 and magnetic resonance spectroscopy. Biol Psychiatry 2014;75:165–71. https://doi.org/10.1016/j.biopsych.2013.06.026.

[123] Milella MS, Marengo L, Larcher K, et al. Limbic system mGluR5 availability in cocaine dependent subjects: a high-resolution PET [(11)C]ABP688 study. Neuroimage 2014;98:195–202. https://doi.org/10.1016/j.neuroimage.2014.04.061.

Chapter 6

Fos-Expressing Neuronal Ensembles in Addiction Research

Bruce T. Hope

Behavioral Neuroscience Research Branch, Intramural Research Program, National Institute on Drug Abuse-National Institutes of Health, Baltimore, MD, United States

NEURONAL ENSEMBLES IN ADDICTION

Maladaptive learned behaviors are at the core of addiction. During drug use, addicts learn to associate their drug-taking activities with drug reward and environmental stimuli that eventually become cues that contribute to drug relapse [1–7]. One important characteristic is that these drug-related cues and behaviors are nearly always complex and highly specific, such as recognizing specific drug-associated friends or environments. The second characteristic is that addicts still perform many other nondrug-related learned behaviors in their daily life. Animal models have been good at modeling these characteristics [7–9]. As with addicts, drug-related cues and behaviors in these animal models can be complex and highly specific, and the animals can still perform many other nondrug-related learned behaviors. The key point is that that both humans and animals discriminate between many different drug- and nondrug-related learned behaviors and their associated cues with a high degree of resolution [10].

Any neural mechanism that mediates these drug-related learned behaviors must be capable of encoding learned associations with a comparably high degree of resolution while allowing for the encoding all other behaviors and learned associations—we call this factor "mechanistic resolution" [10]. While many popular hypotheses help explain why the performance of a learned behavior, or its formation, is increased or decreased, few account for the high degree of behavioral specificity or cue selectivity in the behavior. The relevant questions are why does an addict, or an animal, exhibit drug-related behaviors in response to one drug-paired cue and not to other nondrug-paired stimuli? How are drug-related behaviors stored in the brain without significantly affecting other behaviors? These characteristics are all clues to determining what neural mechanisms are capable of encoding the maladaptive learned behaviors in human addiction.

To date, the neuronal ensemble hypothesis is the only proposed neural mechanism with enough mechanistic resolution to encode complex highly specific learned behaviors and associations in human addiction and related animal models [11–17] (Fig. 1). In 1949, Hebb proposed that memories underlying complex learned behaviors are stored within specific patterns of neurons that he called "cell assemblies" [18]. The specific cell assembly, now called neuronal ensemble, is selected by a specific pattern of excitatory input coming from afferents that carry specific information about the animal's environment, their interoceptive (internal) experience of the drug, and their past experiences [13,19–26]. The small number of sparsely distributed neurons that receive the highest level of integrated excitatory input become part of the neuronal ensemble. In contrast, different patterns of excitatory input carrying information about different stimuli or experiences select different neuronal ensembles. Based on recent studies in the addiction field [27–35], 1% or less of neurons in a brain area receive enough integrated excitatory input to be part of a neuronal ensemble encoding a specific learned association. In brain areas with millions of neurons, this provides an immense number of possible neuronal ensemble patterns, especially since different ensembles can overlap to some extent. This provides a very high degree of mechanistic resolution capable of encoding an enormous number of different memories and behaviors, including complex highly specific learned associations underlying drug addiction [10].

This chapter covers what we currently know about neuronal ensembles in drug addiction research. We begin with a brief background about relevant animal models and techniques used to study them; more detailed descriptions of animal models can be found in other chapters. We then describe the evidence supporting a causal role for neuronal ensembles encoding drug-related memories and behaviors in rat and mouse models; we sometimes supplement these studies with similar findings from food reward studies. We then describe unique molecular and cellular alterations found within neuronal

Neural Mechanisms of Addiction. https://doi.org/10.1016/B978-0-12-812202-0.00006-3
Copyright © 2019 Elsevier Inc. All rights reserved.

FIG. 1 Activation of a specific neuronal ensemble is determined by afferent information to the Target brain area. Specific exteroceptive information about the animal's environment is transferred via specific patterns of neural activity in sensory brain areas to activate specific neuronal ensembles in Afferent brain areas. Specific interoceptive information about the drug's internal stimuli properties is transferred via specific patterns of neural activity in sensory brain areas to activate specific neuronal ensembles in Afferent brain areas. Specific information about the animal's past experiences are transferred from association cortices to activate specific neuronal ensembles in Afferent brain areas (or directly to the Target area). Neurons within the Target brain area that receive the highest levels of persistent integrated excitatory input from the Afferent brain areas are selected to be part of the activated neuronal ensemble.

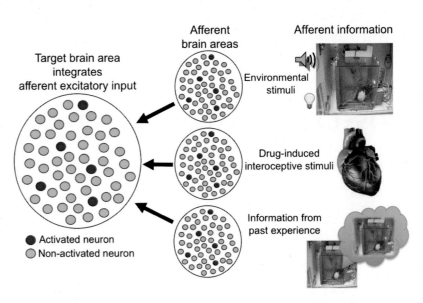

BEHAVIORAL MODELS USED TO STUDY ENSEMBLES IN ADDICTION RESEARCH

Most neuronal ensemble studies in addiction research used a form of operant conditioning called drug self-administration where a rat learns to associate lever pressing responses with drug reward (or reinforcer) and reward-associated cues [5,9,36–38]. The motivation to lever press can be further modulated by surrounding cues and contexts that indicate when drug reward is available or not. Then in the absence of drug reward on test day, one of the cues is assessed for its ability to renew lever pressing, or to reinstate lever pressing after prior extinction training.

Some studies discussed below use a form of Pavlovian conditioning called conditioned place preference (CPP) where the rat or mouse learns to associate drug reward with a specific drug-paired environment (or context), but not with a distinct unpaired environment [35,39–42]. Then in the absence of the drug reward on test day, the preference of the rat or mouse for the drug-paired versus drug-unpaired context is assessed. A few studies below use a form of Pavlovian conditioning called "context-specific" sensitization where rats or mice learn to associate repeated drug administration with one drug-paired context, but not with another unpaired control context [27,43–45]. Then on test day, the animals are assessed for drug-induced locomotor responses that are enhanced (or sensitized) only in the drug-paired context, but not in the drug-unpaired context. One additional study used a similar repeated drug administration procedure, but conditioned locomotion was assessed on test day in the absence of the drug reward [46].

In all of these models, specific neuronal ensembles are thought to be formed during training to encode the learned associations between drug-seeking behavior, the drug reward and related cues. Then on test day, the drug-related cues are thought to reactivate the neuronal ensemble encoding the learned association to induce the learned drug-related behavior.

TECHNIQUES USED TO STUDY NEURONAL ENSEMBLES IN ADDICTION RESEARCH

Shortly after Hebb described his "cell assembly" hypothesis [18], electrodes were placed in the brains of animals to record single neurons during behavior [15,47–51]. These electrodes provide excellent temporal resolution to examine correlations between neural activity and behavior and have been used successfully for over 60 years. In addiction studies, nearly all of these in vivo electrophysiology studies assessed the activity of one neuron at a time (single unit recording) [51–71], which can support a role for these individual neurons in behavior, but does not directly support a role for neuronal ensembles in behavior—the latter requires recording co-activation of multiple neurons during behavior. However very few studies used multiple electrodes to assess activity of multiple neurons firing together during behavior or during cue presentations [72], and none of these were used in addiction research models.

Alternatively, the expression of immediate early genes (IEGs), such as Fos, Arc, and Zif, has been used many times over 30 years to identify brain areas that were activated in many different behavioral models used in addiction research; early examples include [73–76]; and for more detail and examples, see [38]. IEGs are induced within minutes to a few hours in neurons that were strongly activated during behavior or cue presentations. The most commonly used IEG is Fos, which is generally activated in 1% or less of neurons in a given brain area (see below but also [*38*]). The percentages discussed in this review are percentages of Fos-expressing neurons in a three-dimensional brain volume that were calculated from percentages of Fos-expressing neurons observed in two-dimensional brain slices. While increased Fos expression is often interpreted as representing increased activity of a whole brain area, Fos expression most likely represents increased activity of only the 1% of neurons that are expressing the gene and indicates little about activity in the surrounding 99% of neurons in the brain area.

Since most early studies of Fos induction came from molecular biology labs, most explanations for how Fos was induced were based on known molecular mechanisms. Over the years, however, more physiologically relevant studies indicate that electrophysiological mechanisms better explain Fos induction in the brains of behaving animals (Fig. 2); see also [38] for more details. Briefly, the idea is that some neurons in a brain area receive more activated excitatory glutamate afferent input than other neurons; the specific set of activated afferent inputs is determined by the information conveyed to the brain area. Neurons integrate this excitatory input via electrophysiological mechanisms (which can also be modulated by GABA and other neurotransmitters) to depolarize the cell membrane. When the neurons are sufficiently depolarized, NMDA-type glutamate receptors and voltage-sensitive calcium channels (VSCCs) open and allow calcium influx into the neurons. High levels of neural activity-dependent calcium influx activate the MAP kinase intracellular signaling pathway that phosphorylates and activates transcription factor complexes on the Fos promoter to induce Fos expression. Under physiological conditions, dopamine modulates the function of other electrophysiological channels on the cell membrane, which further enhances neural activity of the most active neurons while decreasing neural activity of the less activated neurons. In general, Fos expression can be considered an indicator of high levels of calcium influx in the most persistently and strongly activated neurons.

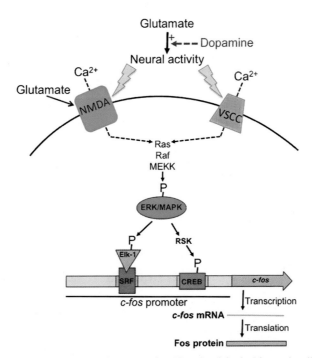

FIG. 2 Neural activity activates the *Fos* promoter to induce Fos expression. Neural activity is driven primarily by excitatory glutamate neurotransmission. Dopamine further enhances activity of the most strongly activated neurons within neuronal ensembles, while attenuating activity of the surrounding less activated majority of neurons. Strong neural activity allows calcium influx through NMDA glutamate receptor channels and VSCCs. When neurons are persistently and strongly activated enough, intracellular calcium levels increase enough to activate the MAP kinase (ERK/MAPK) signaling pathway that produces phosphorylation of the transcription factors Elk-1/SRF and CREB on the *Fos* promoter. Promoter activation induces transcription of *Fos* mRNA that is then translated to Fos protein. Thus, *Fos* promoter activation, *Fos* mRNA and Fos protein are all indicators of strong persistent neural activity.

The study of neuronal ensembles in addiction research has been focused on Fos-expressing ensembles, due in large part to recent technologies that exploit the Fos promoter to identify and manipulate neuronal ensembles activated during behavior and memory recall; for more detail, see [77]. Daun02 inactivation was the first method used to demonstrate a causal role for Fos-expressing neuronal ensembles in addiction research [27] (Fig. 3). In this procedure, Fos-LacZ transgenic rats [78] co-express beta-galactosidase (βGal) with Fos in neurons that were strongly activated during behavior or memory recall. The prodrug Daun02 is then injected into the brain area of interest 90–120 min later, when βGal levels are maximal. Daun02 is catalyzed into daunorubicin by βGal in only the Fos-expressing neurons, which inactivates [79] and kills [32,80,81] only those neurons that co-expressed βGal and Fos. Several days later, the experimenter tries to reactivate the Fos-expressing ensemble of interest to determine whether the drug- or cue-paired ensemble mediates the learned behavior or memory of interest. In more mechanistic studies, endogenous Fos and green fluorescent protein (GFP) in transgenic Fos-GFP rats have been used to identify unique molecular and cellular alterations within Fos-expressing ensembles [43,82–84]. These reagents and techniques are the basis for most of this discussion of neuronal ensembles in addiction research.

STRIATAL NEURONAL ENSEMBLES MEDIATE LEARNED BEHAVIORS IN ADDICTION RESEARCH MODELS

The nucleus accumbens in the ventral striatum plays a central role in the neural circuitry of drug addiction by integrating highly specific information about the environment and past experience [85,86]. It receives high-resolution cue-related information via excitatory glutamatergic inputs from the prefrontal cortex, basolateral amygdala, ventral subiculum of the hippocampus, and the dorsomedial nucleus of the thalamus [86,87]. Neurons within the nucleus accumbens that receive the highest levels of excitatory input will be in an activated state that is further enhanced by dopaminergic input from the ventral tegmental area that conveys information about salience and reward value [15,88–92]. The nucleus accumbens is thought to integrate information for action selection and motivation to act [91–93].

Most in vivo electrophysiology studies in addiction research examined correlations between single unit recordings in the nucleus accumbens and either lever pressing or cue presentations following self-administration training [51–55]. Neural activity increased or decreased in a time-locked fashion with lever pressing for cocaine or heroin, or with presentation of the drug-associated cues in the absence of the drugs [52,56–61]. Importantly, the specific neurons activated by different rewards, such as heroin, cocaine, sucrose or water, during self-administration are largely different from each other [62–70]. The specific neurons activated by cocaine- versus food-related cues are also different from each other [64,71]. This provides correlative evidence for different neuronal ensembles being responsive to different drug rewards and related cues.

Fos-expressing neuronal ensembles in the nucleus accumbens have been shown to mediate context-specific sensitization of cocaine-induced locomotion [27] and context-induced reinstatement of cocaine seeking in rats [31]. In both models, Daun02 inactivation of ensembles (previously activated by the specific drug-associated cues or context) subsequently reduced the drug cue and context-induced behaviors. Daun02 inactivation of ensembles that were previously activated by the nondrug-associated cues or contexts did not affect the drug cue and context-induced behaviors. Overall, these

FIG. 3 Daun02 inactivation procedure selectively ablates strongly activated Fos-expressing neuronal ensembles in *Fos-lacZ* transgenic rats. Persistently strong neural activity activates the *Fos* promoter in the *Fos-lacZ* transgene to induce transcription of the *lacZ* coding sequence to *lacZ* mRNA. *LacZ* mRNA is then translated to β-galactosidase (βgal) protein. The endogenous non-transgenic *Fos* gene is also activated to co-express Fos protein with βgal protein in the same persistently strongly activated neurons. When βgal levels are maximal (90–120 min post-stimulation), the inert prodrug Daun02 is injected into the brain area. Daun02 is catalyzed by βgal to the active cytotoxin Daunorubicin, which initially inactivates neurons, followed by apoptosis and death of the βgal-expressing neurons. Since βgal is induced in only the strongly activated Fos-expressing neurons, Daun02 injections selectively ablate only the Fos-expressing neuronal ensemble and not the surrounding majority of Fos-negative neurons.

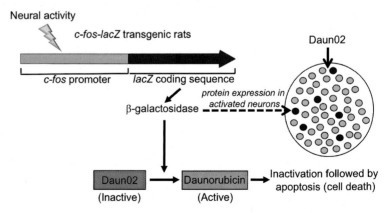

data indicated that specific drug-paired cues select specific neuronal ensembles in the nucleus accumbens to mediate learned cue-induced behaviors.

The dorsal striatum is also thought to play a role in mediating cue-induced drug seeking [86,87]. This brain area receives high-resolution cue-related information via excitatory glutamatergic inputs from the neocortex and thalamus [86,87]. Neurons within the dorsal striatum that receives the highest levels of excitatory input will be in an activated state that is further enhanced by dopaminergic input from the substantia nigra that conveys information about salience and reward value [86,89]. The dorsal striatum is also thought to integrate this information for action selection.

Fos-expressing neuronal ensembles in the dorsomedial striatum have been shown to mediate cue-induced methamphet-amine seeking following voluntary abstinence [30]. Daun02 inactivation of ensembles previously activated by the specific methamphetamine-associated cue subsequently reduced cue-induced drug seeking. Daun02 inactivation of these methamphetamine cue-associated ensembles did not alter cue-induced food seeking even though the food cues also induced Fos within the dorsomedial striatum. Overall, these data indicated that specific methamphetamine-paired cues select specific Fos-expressing neuronal ensembles in the dorsomedial striatum to mediate the learned cue-induced behavior.

In all three of these striatal ensemble studies, Fos-expressing neurons within the drug cue-associated ensembles make up less than 1% of neurons in the brain areas [27,30,31]. This suggests that the 99% or more Fos-negative neurons that surround the drug-paired Fos-expressing neuronal ensemble play a relatively insignificant role in mediating the learned behaviors. The Fos-expressing neurons in these studies were composed of multiple cell types. The majority of Fos-expressing neurons were medium spiny neurons of approximately equal numbers of the D1 receptor and D2 receptor cell types [27,30,31]. The small numbers of remaining Fos-expressing neurons were different types of interneurons. Overall the proportions of cell types within the Fos-expressing ensembles approximated the proportions found in previous anatomical studies. Thus, these learned behaviors were not mediated by a whole brain area or by any one cell type.

PREFRONTAL CORTEX NEURONAL ENSEMBLES MEDIATE LEARNED BEHAVIORS IN ADDICTION RESEARCH MODELS

The prefrontal cortex is composed of many different subregions that play different roles in addiction research behavior models. The prefrontal cortex integrates high-resolution information about the environment and past experience via excitatory glutamatergic inputs between different areas of the prefrontal cortex, from the basolateral amygdala, hippocampus, and thalamus. The medial prefrontal cortex is thought to integrate this information and provides input about the value of ongoing information to the nucleus accumbens and contributes to action selection and motivation to act.

In vivo electrophysiology studies also found correlations between single-unit recordings in the prefrontal cortex and either lever pressing or cue presentations following self-administration training. Neural activity increased or decreased in a time-locked fashion with lever pressing for cocaine or heroin, or with presentation of the drug-associated cues in the absence of the drugs [54,56,67,94–96]. Importantly, the specific neurons activated by the different rewards, heroin, cocaine, saccharin, or water, during self-administration are different from each other. This provides correlative evidence for different neuronal ensembles being responsive to different drug rewards and related cues.

Fos-expressing neuronal ensembles in the ventral prefrontal cortex, the infralimbic cortex [28,32,97–99] and the orbitofrontal cortex [29] have been shown to mediate cue or context-induced drug (and food) seeking in rats. In all cases, Daun02 inactivation of ensembles previously activated by the specific drug-associated cues or context subsequently reduced these cue- and context-induced behaviors. Daun02 inactivation of ensembles that were previously activated by nondrug-associated cues or context did not affect the cue- or context-induced behaviors. Overall, these data indicated that specific drug-paired cues select specific neuronal ensembles in the prefrontal cortex to mediate the learned cue-induced behaviors.

Three of these medial prefrontal cortex studies expose the difference between assessing behavioral roles of whole brain areas versus neuronal ensembles. In each of the first two studies, two different Fos-expressing neuronal ensembles in the same infralimbic cortex brain area were shown to mediate either the initial cocaine (or food) self-administration memory or the behaviorally opposing extinction memory [97,98]. Daun02 inactivation of Fos-expressing ensembles activated during recall of cocaine (or food) self-administration reduced subsequent recall of this self-administration behavior. Daun02 inactivation of extinction-related ensembles activated during extinction of the self-administration behavior had the opposite effect of increasing subsequent recall of this self-administration behavior. This suggests that the extinction-related ensemble actively suppresses the initial self-administration memory, which is also activated by the same set of cues. It is not known how the brain "remembers" whether the extinction or self-administration experience was more recent, to activate or not the extinction memory. Overall, these data indicate that different ensembles can intermingle with the same brain area to mediate opposing effects on the same behavior. The third of these medial prefrontal cortex studies used discriminative cues to indicate availability of a food reward: DS+ cues indicated availability of the food reward while DS− cues indicated lack of availability of the food reward [99]. Daun02 inactivation of Fos-expressing ensembles activated by the DS+ cue reduced subsequent food-seeking behavior. Daun02 inactivation of ensembles activated by the DS− cue had the opposite effect of increasing subsequent food-seeking behavior. These data are the strongest indication that different ensembles can intermingle with the same brain area to mediate opposing effects on the same behavior, and could only be shown using ensemble-specific methods.

Three of the studies directly compared the effects of whole brain area versus ensemble-specific manipulations on drug (and food) seeking. The first of these studies examined neuronal ensembles in cue-induced alcohol seeking in rats [32]. Daun02 inactivation of Fos-expressing ensembles in the infralimbic cortex, which were previously activated by the alcohol-related cue, subsequently increased cue-induced alcohol seeking on test day. This experiment by itself indicates that alcohol training can produce inhibitory Fos-expressing ensembles in the medial prefrontal cortex. More importantly, they also used Daun02 inactivation of all infralimbic neurons using a control transgenic CAG-lacZ rat that expresses betaGal in all neurons. Inhibiting all neurons in the infralimbic cortex, rather than just the cue-activated ensemble, had no effect on behavior. This experiment raises a lot of questions about the interpretations of previous studies that used whole brain area (as well as cell type) manipulations to assess their role in learned behaviors in addiction research. The other two studies compared the effects of whole brain inactivation (with baclofen+muscimol) of the entire infralimbic cortex versus ensemble-specific inactivation (Daun02 inactivation) on subsequent recall of cocaine (and food) self-administration and extinction memories [97,98]. Baclofen+ muscimol inactivation had a moderate but significant effect on the self-administration memory but no effect on the extinction memory. While the effect on the self-administration memory was qualitatively similar to that ensemble-specific manipulation with Daun02 inactivation, only the ensemble-specific manipulation could detect the extinction memory in this brain area. Whole brain manipulations or cell-type manipulations affect all neuronal ensembles, including ensembles that have opposite effects on the learned behavior. Thus the net effect of these latter manipulations is very difficult to interpret, and raises questions about how to interpret whole brain or cell-type manipulations.

In all seven of these prefrontal cortex studies, Fos-expressing neurons within the cue-associated ensembles make up 2% or less of all neurons in the medial prefrontal cortex [28,29,32,97–99]. We estimate that Fos-expressing ensembles in the Suto et al. study used as few as 0.01%–0.1% of all neurons in the infralimbic cortex [99]. This suggests that at least 98% of the Fos-negative neurons that surround the Fos-expressing neuronal ensemble play a relatively insignificant role in mediating the learned behaviors. The Fos-expressing neurons were also composed of multiple cell types. The majority (75%–90%) of Fos-expressing neurons were glutamatergic while the remaining neurons were mostly GABAergic interneurons, proportions similar to that found in anatomical studies [28,29,32,97–99].

AMYGDALA NEURONAL ENSEMBLES MEDIATE LEARNED BEHAVIORS IN ADDICTION RESEARCH MODELS

The amygdala is involved in encoding associations between sensory cues and rewarding or aversive stimuli [100–103]. Fos-expressing neuronal ensembles in the *basolateral nucleus* of the amygdala have been shown to mediate cue or context-induced drug seeking in rats. Daun02 inactivation of neuronal ensembles in the basolateral nucleus reduced nicotine-induced CPP and reinstatement of nicotine seeking [35]. Optogenetic regulation of Fos-expressing neuronal ensembles in the basolateral nucleus indicate that the acute reward information properties of nicotine [41] or cocaine alone can be encoded within this nucleus and paired with previously neutral stimuli to guide behavior.

Fos-expressing neuronal ensembles in the *central nucleus* of the amygdala have been shown to encode emotional states associated with alcohol dependence [33]. Daun02 inactivation of neuronal ensembles activated during abstinence in alcohol-dependent rats produced a long-term decrease of alcohol seeking, but only a transient decrease was observed in nondependent rats. In a separate study, Daun02 inactivation of neuronal ensembles activated during cue-induced nicotine seeking reduced subsequent cue-induced nicotine seeking, but not following Daun02 inactivation of ensembles in a distinct non-paired context [34].

Fos-expressing neuronal ensembles in the *lateral nucleus* of the amygdala have been suggested to mediate cocaine CPP in mice [42]. CREB overexpression was induced in a small percentage of randomly selected neurons in the lateral amygdala *prior* to CPP training. CREB overexpression increases the likelihood of a neuron being incorporated into a neuronal ensemble during subsequent learning. When these CREB-overexpressing neurons were selectively inactivated after CPP training, cocaine CPP was reduced. However, the relationship of these CREB-overexpressing neurons that were selected *prior* to learning, to Fos-expressing neuronal ensembles that were selected *during* learning is not known.

In these amygdala studies, Fos-expressing neurons within the cue-associated ensembles make up 3% or less of all neurons in the relevant amygdala nucleus [33–35,40–42]. This suggests that at least 97% of the Fos-negative neurons that surround the Fos-expressing neuronal ensemble play a relatively insignificant role in mediating the learned behaviors. Cell types of the Fos-expressing neuronal ensembles were not assessed.

GENERAL CONCLUSIONS ABOUT FOS-EXPRESSING NEURONAL ENSEMBLES IN ADDICTION RESEARCH

At the very least, the above studies demonstrate that Fos-expressing neuronal ensembles mediate information flow necessary for cue-specific behavior and memory recall in many different well-established operant and Pavlovian models in addiction research. These ensembles were found in most of the important forebrain areas in the neural circuitry of addiction, including nucleus accumbens, dorsal striatum, prefrontal cortex, and amygdala. Highly specific and complex information about the drug-related cues and past experience determine the specific drug-paired ensemble. Information about different cues and past experience activate different ensembles. Thus Fos-expressing ensembles are capable of the necessary high degree of mechanistic resolution for encoding drug-related learned associations and still allow for encoding of all other behaviors and learned associations used by humans and animals.

Neuronal ensemble mechanisms are more congruent with the high degree of cue- and behavior-specificity of learned behaviors in addiction. In contrast, hypotheses about alterations induced throughout a brain area, regardless of whether neurons are activated during the behavior, are not congruent with these characteristics of addiction [10,104–110]. These whole brain region hypotheses do not have the same mechanistic resolution and cannot explain the ability to store different memories and behaviors [10]. Furthermore, Fos-expressing ensembles were consistently shown to be comprised of all cell types in proportions predicted from anatomical studies, and using only small percentages of each neuron cell type in a brain area. Thus both whole brain area and cell-type-specific hypotheses about learned behaviors in addiction are not congruent with the characteristics of learned behaviors in addiction research models. However, these whole brain or cell-type hypotheses can indirectly affect the general molecular and cellular background upon which the relevant learning-specific mechanisms operate within neuronal ensembles.

Further work is necessary to demonstrate that permanent alterations are induced within these neurons to produce the long-lasting trace or *engram* that encodes memories. The work to identify unique alterations within Fos-expressing ensembles is described below.

UNIQUE GENE EXPRESSION ALTERATIONS WITHIN FOS-EXPRESSING NEURONAL ENSEMBLES

Fos-expressing neuronal ensembles also undergo unique molecular alterations during learning and recall of memories in addiction research [38,77]. To date, all studies have used fluorescence-activated cell sorting (FACS) and in situ hybridization [111–115]. For FACS, brain areas are extracted from brains 2–3 h after the behavior of interest or following memory recall, and the cells are separated and labeled with antibodies that identify Fos and neuronal cells. FACS separates Fos-positive neurons from Fos-negative neurons into separate tubes for subsequent molecular analyses [116,117]. This technique is useful for assessing many gene products. In situ hybridization is a more focused technique for identifying specific gene targets. Brains are extracted 30–120 min after the behavior of interest or following memory recall, and then the activated Fos-expressing neurons are fluorescently labeled for Fos mRNA and the gene of interest.

To date, only three studies have examined molecular alterations within Fos-expressing neuronal ensembles activated during an operant behavior model. FACS analyses indicated that context- and cue-induced methamphetamine seeking [114,115] or cue-induced heroin seeking [112] after self-administration training induced a number of unique gene alterations within Fos-expressing neurons in dorsal striatum that were not significantly induced in the majority of neurons that were Fos-negative. In other words, only the small number of Fos-expressing neurons were altered, and not the bulk of neurons that were Fos-negative. Many of these uniquely altered genes were IEGs, known to be rapidly induced in activated neurons, such as Fos, FosB, and Arc [111–115]. Other uniquely altered genes were homer-2 as well as glutamate receptor subunits that encode proteins capable of altering synaptic efficacy in these neurons [114]. In situ hybridization analyses in separate experiments verified regulation of these genes in only the Fos-expressing neurons [111–115]. As mentioned above, extinction memories are mediated by different Fos-expressing neurons [97,98]. These extinction-related neurons also induced many of the same sets of gene alterations that were observed with recall of the self-administration memories [114].

Only one study had examined molecular alterations within Fos-expressing neuronal ensembles activated during a passive learned behavior model [113]. FACS analyses indicated that context-specific sensitization of cocaine-induced locomotion induced many unique gene alterations within Fos-expressing neurons in striatum that were not significantly induced in the majority of neurons that were Fos-negative. Again, it appeared that the bulk of altered gene expression was induced only in the small number of Fos-expressing neurons, and not in the bulk of neurons that were Fos-negative. Many of these uniquely altered genes were IEGs, as well as a large number of other genes.

It is significant that the behavioral manipulations induced gene expression alterations primarily in the small number of Fos-expressing neurons. We directly compared acute methamphetamine-induced gene expression in Fos-expressing striatal neurons (from FACS), with gene expression from the more commonly used homogenate preparations of whole striatum [113]. In general, methamphetamine-induced expression of genes such as Fos and FosB were approximately 10-fold greater in the FACS-sorted Fos-positive neurons than in homogenates, while increased expression of homer-2 and arc could only be observed in the FACS-sorted Fos-positive neurons and not in homogenates. Overall, large alterations of gene expression were induced only in Fos-expressing ensembles activated by memory recall in an addiction-related behavioral model.

UNIQUE ELECTROPHYSIOLOGICAL ALTERATIONS WITHIN FOS-EXPRESSING NEURONAL ENSEMBLES

Fos-expressing neuronal ensembles also undergo unique electrophysiological alterations during learning and recall of memories in addiction research [43,46,83,84,118,119]. In these studies, Fos-expressing neurons are identified in slice preparations taken from transgenic Fos-GFP mice or rats that co-express GFP with the endogenous Fos. The electrophysiological characteristics of Fos/GFP-positive neurons are then compared to nearby Fos/GFP-negative neurons. In general, these alterations can be divided into two categories. Synaptic alterations include pre- and post-synaptic connections between neurons [43,83,84]. Intrinsic excitability alterations [46,118,119] are more concerned with how these synaptic inputs are integrated primarily by the cell body [120]. These two types of alterations work together to alter the selection and strength of activation of ensemble neurons.

Most studies examined these electrophysiological alterations within Fos-expressing neuronal ensembles in the nucleus accumbens that were activated during passive conditioning behavior models in mice. Following context-specific sensitization of cocaine-induced locomotion, only the GFP-positive neurons exhibited synaptic alterations, including decreased ratios of AMPA/NMDA glutamate receptor activation (an indicator of postsynaptic alterations) and decreased frequency of spontaneous excitatory postsynaptic currents (EPSCs) and altered paired pulse ratios (indicators of presynaptic vesicle release) [43,84]. The most notable synaptic alterations on these GFP-positive neurons were increased silent synapses. None of these synaptic alterations were induced in the surrounding GFP-negative neurons that make up approximately 99% of neurons. If the mice were placed in a different nondrug-paired context to induce GFP in a new set of neurons, then none of the synaptic alterations were observed [84].

Intrinsic excitability alterations were assessed following conditioned locomotion for cocaine reward and conditioned approach for sucrose reward in mice [46,119]. GFP-positive neurons in the nucleus accumbens core were more excitable than the surrounding GFP-negative neurons due to increased membrane resistance and decreased medium afterhyperpolarizations (mAHP). Following subsequent extinction training, these intrinsic excitability differences between GFP-positive and GFP-negative neurons were absent. It is not known if the same sets of neurons were GFP-positive following conditioning and extinction; the two Warren et al. studies [97,98] described above suggest that these may be different sets of neurons. Overall, synaptic connections and intrinsic excitability are uniquely altered in GFP and Fos-expressing neurons in the nucleus accumbens following training to associate the rewards with related cues.

Similar electrophysiological alterations have been observed within Fos-expressing neuronal ensembles in the medial prefrontal cortex that were activated following operant conditioning in rats. For synaptic alterations, yohimbine-induced reinstatement of food seeking decreased postsynaptic AMPA/NMDA ratios and altered presynaptic paired pulse ratios only for GFP-positive neurons [83], similar to the postsynaptic and presynaptic alterations described above for the passive conditioning behavioral models in mice. For intrinsic excitability alterations, food self-administration training increased excitability of only the GFP-positive neurons, primarily by increasing membrane resistance, but decreased excitability of the surrounding GFP-negative neurons primarily by increasing mAHP [118]. Daun02 inactivation of Fos-expressing neuronal ensembles in medial prefrontal cortex of Fos-lacZ rats following food self-administration indicates that these Fos-expressing neuronal ensembles also directly mediate the self-administration behavior [118]. Overall, unique synaptic and intrinsic excitability alterations are induced only in GFP- and Fos-expressing neurons in rat medial prefrontal cortex that mediate the association between food rewards and associated cues. These alterations are similar to those observed in the GFP- and Fos-expressing neurons in nucleus accumbens of mice following passive conditioning training.

UNIQUE DENDRITIC SPINE ALTERATIONS ON FOS-EXPRESSING NEURONAL ENSEMBLES

Dendritic spines on Fos-expressing neuronal ensembles can also undergo unique morphological alterations during learning and recall of memories in addiction research [45]. Dendritic spines receive excitatory input from glutamatergic afferents onto the spine head as well as dopamine and other inputs onto the spine neck. Spine alterations suggest that long-term alterations of afferent input can play a role in formation and maintenance of neuronal ensembles. In the one related study, spines on Fos-expressing neurons were identified by injecting a fluorescent-dye DiI into brain sections that were already immunolabeled for Fos expression [45]. The density of dendritic spines and morphological alterations of Fos-positive and Fos-negative neurons were assessed using confocal microscopy.

Following context-specific sensitization of amphetamine-induced locomotion, dendritic spine density and the number of medium-size spines increased on Fos-positive neurons in the nucleus accumbens shell subregion [45]. These spine alterations were not observed on the surrounding Fos-negative neurons that make up the majority of neurons in this brain area. When amphetamine was repeatedly injected into the nondrug-paired environment or under conditions that reduced drug-context conditioning, spine alterations were also not observed. Overall, unique spine density and morphological alterations were induced only in Fos-expressing ensembles activated by memory recall in addiction-related behavioral model.

FOS-EXPRESSING NEURONAL ENSEMBLES ARE LIKELY A SUBSET OF ALL NEURONS ACTIVATED DURING BEHAVIOR OR MEMORY RECALL

The exact relationship between neural activity assessed using in vivo electrophysiology versus that assessed using Fos promoter activation is not known. Previous attempts incorrectly compared the number of Fos-positive neurons to the number of electrophysiological activity spikes recorded from randomly selected neurons, which we now know that 95% or more of them would have been Fos-negative [121,122]. in vivo electrophysiology studies often identify up to 20%–40% of neurons that are activated during lever responding for drug reward or when responding to drug-related cues [51–56,67,94–96], while normally 1% and sometimes far less neurons (e.g., 0.01%–0.1% estimated in the Suto et al. study [99]) are activated enough to induce Fos expression [38,77]. We hypothesize that Fos is expressed in only the most active neurons observed by in vivo electrophysiology methods (for detailed description and references, see [38]). When a neuron receives enough excitatory input, calcium enters the cells through NMDA glutamate receptor and VSCCs. Low to moderate levels of calcium influx are rapidly sequestered by strong calcium buffering mechanisms. However, strong and persistent neural activity can induce higher levels of calcium influx that overwhelm the calcium buffering mechanism to activate calcium-dependent intracellular signaling pathways. One of these pathways is the MAP kinase pathway which activates the Elk-1/SRF and CREB transcription factors on the Fos promoter. Thus, we hypothesize that Fos is induced in only the most strongly and most persistently activated neurons observed by in vivo electrophysiology during behavior and memory recall.

We suggest that many, if not all, of the molecular, electrophysiological, and morphological alterations induced only in Fos-expressing neuronal ensembles are dependent on this, or a similar calcium-dependent mechanism, for determining which neurons are altered during learning and memory recall. As described above, Daun02 inactivation of these Fos-expressing neurons indicates that these uniquely altered neurons are selected during the learning of associations between behaviors and associated cues, contexts, and rewards. Put together, these data suggest that the unique set of alterations induced in Fos-expressing neurons play an important role in the formation of long-lasting neuronal ensembles during learning and in maintenance of these ensembles during memory recall. The selection of only a small percentage of neurons in an ensemble, perhaps 0.1% or lower [99], increases the number of possible of neuronal ensembles for enhanced mechanistic resolution and flexibility of learned responses in an animal's lifetime. It also enhances discrimination between distinct ensembles underlying distinct memories. However, further research is necessary to support these hypotheses. It should be noted that we never imply that Fos is necessary for these processes. The Daun02 inactivation studies suggest only that the expression of Fos happens to correspond reasonably close to the threshold for selection of neuronal ensembles that encode the learned behaviors and memories.

MANIPULATING FOS-EXPRESSING NEURONAL ENSEMBLES TO TREAT HUMAN ADDICTION?

Although currently pure speculation, the Fos-expressing neuronal ensemble hypothesis could be applied to treating human addiction. Previous pharmacological interventions target receptors or channels on all neurons or a specific cell type [123–127], regardless of whether the neurons are activated or involved in mediating the behavior of interest. In contrast, the unique physical characteristics of Fos-expressing ensembles that are strongly activated during behavior or memory recall could provide unique targets for pharmacological intervention in maladaptive learning disorders such as human addiction. For example, NMDA receptors on strongly depolarized membranes enter into an "open" state that can preferentially bind drugs such as ketamine or memantine [128–130]. However, the most effective targets may be cell body-based mechanisms that are selectively activated at thresholds that correspond with Fos promoter activation, such as MAP kinase and other calcium-dependent signaling pathways. Recalling behaviors or memories related to an addict's specific experiences can put the relevant neuronal ensembles into a unique activated state, and then activity-dependent pharmacological agents can selectively modulate only these ensembles and thus only the relevant addiction-related memories. Again, only more research will demonstrate whether this is truly possible, but it provides a powerful avenue for specifically targeting the memories that underlie drug relapse and addiction.

CONFLICT OF INTEREST

All authors have no conflicts of interest to disclose.

ACKNOWLEDGMENTS

This work was supported by the Intramural Research Program/National Institute on Drug Abuse/National Institutes of Health Grant# DA000467.

REFERENCES

[1] Goldberg SR. Stimuli associated with drug injections as events that control behavior. Pharmacol Rev 1976;27:325–40.

[2] O'Brien CP, Ehrman RN. J. W. Ternes. In: Goldberg S, Stolerman I, editors. Behavioral analysis of drug dependence. Orlando: Academic Press; 1986. p. 329–56.

[3] Stewart J, de Wit H, Eikelboom R. Role of unconditioned and conditioned drug effects in the self-administration of opiates and stimulants. Psychol Rev 1984;91:251–68.

[4] Wikler A. Dynamics of drug dependence. Implications of a conditioning theory for research and treatment. Arch Gen Psychiatry 1973;28:611–6.

[5] Crombag HS, Bossert JM, Koya E, Shaham Y. Review. Context-induced relapse to drug seeking: a review. Philos Trans R Soc Lond B Biol Sci 2008;363:3233–43.

[6] Shalev U, Grimm JW, Shaham Y. Neurobiology of relapse to heroin and cocaine seeking: a review. Pharmacol Rev 2002;54:1–42.

[7] Robinson TE, Berridge KC. The neural basis of drug craving: an incentive-sensitization theory of addiction. Brain Res Rev 1993;18:247–91.

[8] Bossert JM, Marchant NJ, Calu DJ, Shaham Y. The reinstatement model of drug relapse: recent neurobiological findings, emerging research topics, and translational research. Psychopharmacology 2013;229:453–76.

[9] Marchant NJ, Li X, Shaham Y. Recent developments in animal models of drug relapse. Curr Opin Neurobiol 2013;23:675–83.

[10] Warren BL, Suto N, Hope BT. Mechanistic resolution required to mediate operant learned behaviors: insights from neuronal ensemble-specific inactivation. Front Neural Circ 2017;11:28.

[11] Buzsaki G, Moser EI. Memory, navigation and theta rhythm in the hippocampal-entorhinal system. Nat Neurosci 2013;16:130–8.

[12] Knierim JJ, Zhang K. Attractor dynamics of spatially correlated neural activity in the limbic system. Annu Rev Neurosci 2012;35:267–85.

[13] Nicolelis MA, Fanselow EE, Ghazanfar AA. Hebb's dream: the resurgence of cell assemblies. Neuron 1997;19:219–21.

[14] C. M. Pennartz, J. D. Berke, A. M. Graybiel, R. Ito, C. S. Lansink, M. van der Meer, A. D. Redish, K. S. Smith, P. Voorn, Corticostriatal interactions during learning, memory processing, and decision making. J Neurosci 29, 12831–12838 (2009).

[15] Pennartz CM, Groenewegen HJ, Lopes da Silva FH. The nucleus accumbens as a complex of functionally distinct neuronal ensembles: an integration of behavioural, electrophysiological and anatomical data. Prog Neurobiol 1994;42:719–61.

[16] Penner MR, Mizumori SJ. Neural systems analysis of decision making during goal-directed navigation. Prog Neurobiol 2012;96:96–135.

[17] Schwindel CD, McNaughton BL. Hippocampal-cortical interactions and the dynamics of memory trace reactivation. Prog Brain Res 2011;193:163–77.

[18] Hebb DO. The organization of behavior; a neuropsychological theory. A Wiley book in clinical psychology, New York: Wiley; 1949. p. xix [335 p].

[19] Best PJ, White AM, Minai A. Spatial processing in the brain: the activity of hippocampal place cells. Annu Rev Neurosci 2001;24:459–86.

[20] Deadwyler SA, Hampson RE. Ensemble activity and behavior: what's the code? Science 1995;270:1316–8.

[21] Deadwyler SA, Hampson RE. The significance of neural ensemble codes during behavior and cognition. Annu Rev Neurosci 1997;20:217–44.

[22] Doetsch GS. Patterns in the brain. Neuronal population coding in the somatosensory system. Physiol Behav 2000;69:187–201.

[23] Guzowski JF, Knierim JJ, Moser EI. Ensemble dynamics of hippocampal regions CA3 and CA1. Neuron 2004;44:581–4.

[24] Johnson A, Fenton AA, Kentros C, Redish AD. Looking for cognition in the structure within the noise. Trends Cogn Sci 2009;13:55–64.

[25] Leutgeb S, Leutgeb JK, Moser MB, Moser EI. Place cells, spatial maps and the population code for memory. Curr Opin Neurobiol 2005;15:738–46.

[26] Mittler T, Cho J, Peoples LL, West MO. Representation of the body in the lateral striatum of the freely moving rat: single neurons related to licking. Exp Brain Res Exp Hirnforsch Exp Cereb 1994;98:163–7.

[27] Koya E, Golden SA, Harvey BK, Guez-Barber DH, Berkow A, Simmons DE, Bossert JM, Nair SG, Uejima JL, Marin MT, Mitchell TB, Farquhar D, Ghosh SC, Mattson BJ, Hope BT. Targeted disruption of cocaine-activated nucleus accumbens neurons prevents context-specific sensitization. Nat Neurosci 2009;12:1069–73.

[28] Bossert JM, Stern AL, Theberge FR, Cifani C, Koya E, Hope BT, Shaham Y. Ventral medial prefrontal cortex neuronal ensembles mediate context-induced relapse to heroin. Nat Neurosci 2011;.

[29] Fanous S, Goldart EM, Theberge FR, Bossert JM, Shaham Y, Hope BT. Role of orbitofrontal cortex neuronal ensembles in the expression of incubation of heroin craving. J Neurosci 2012;32:11600–9.

[30] Caprioli D, Venniro M, Zhang M, Bossert JM, Warren BL, Hope BT, Shaham Y. Role of Dorsomedial striatum neuronal ensembles in incubation of methamphetamine craving after voluntary abstinence. J Neurosci 2017;37:1014–27.

[31] Cruz FC, Babin KR, Leao RM, Goldart EM, Bossert JM, Shaham Y, Hope BT. Role of nucleus accumbens shell neuronal ensembles in context-induced reinstatement of cocaine-seeking. J Neurosci 2014;34:7437–46.

[32] Pfarr S, Meinhardt MW, Klee ML, Hansson AC, Vengeliene V, Schonig K, Bartsch D, Hope BT, Spanagel R, Sommer WH. Losing control: excessive alcohol seeking after selective inactivation of cue-responsive neurons in the Infralimbic cortex. J Neurosci 2015;35:10750–61.

[33] de Guglielmo G, Crawford E, Kim S, Vendruscolo LF, Hope BT, Brennan M, Cole M, Koob GF, George O. Recruitment of a neuronal Ensemble in the Central Nucleus of the amygdala is required for alcohol dependence. J Neurosci 2016;36:9446–53.

[34] Funk D, Coen K, Tamadon S, Hope BT, Shaham Y, Le AD. Role of central amygdala neuronal ensembles in incubation of nicotine craving. J Neurosci 2016;36:8612–23.

[35] Xue YX, Chen YY, Zhang LB, Zhang LQ, Huang GD, Sun SC, Deng JH, Luo YX, Bao YP, Wu P, Han Y, Hope BT, Shaham Y, Shi J, Lu L. Selective inhibition of amygdala neuronal ensembles encoding nicotine-associated memories inhibits nicotine preference and relapse. Biol Psychiatry 2017;82:781–93.

[36] Roberts DC, Corcoran ME, Fibiger HC. On the role of ascending catecholaminergic systems in intravenous self-administration of cocaine. Pharmacol Biochem Behav 1977;6:615–20.

[37] Pickens CL, Airavaara M, Theberge F, Fanous S, Hope BT, Shaham Y. Neurobiology of the incubation of drug craving. Trends Neurosci 2011;34:411–20.

[38] Cruz FC, Javier Rubio F, Hope BT. Using c-fos to study neuronal ensembles in corticostriatal circuitry of addiction. Brain Res 2015;1628:157–73.

[39] Ramirez S, Liu X, MacDonald CJ, Moffa A, Zhou J, Redondo RL, Tonegawa S. Activating positive memory engrams suppresses depression-like behaviour. Nature 2015;522:335–9.

[40] Redondo RL, Kim J, Arons AL, Ramirez S, Liu X, Tonegawa S. Bidirectional switch of the valence associated with a hippocampal contextual memory engram. Nature 2014;513:426–30.

[41] Gore F, Schwartz EC, Brangers BC, Aladi S, Stujenske JM, Likhtik E, Russo MJ, Gordon JA, Salzman CD, Axel R. Neural representations of unconditioned stimuli in Basolateral amygdala mediate innate and learned responses. Cell 2015;162:134–45.

[42] Hsiang HL, Epp JR, van den Oever MC, Yan C, Rashid AJ, Insel N, Ye L, Niibori Y, Deisseroth K, Frankland PW, Josselyn SA. Manipulating a "cocaine engram" in mice. J Neurosci 2014;34:14115–27.

[43] Koya E, Cruz FC, Ator R, Golden SA, Hoffman AF, Lupica CR, Hope BT. Silent synapses in selectively activated nucleus accumbens neurons following cocaine sensitization. Nat Neurosci 2012;15:1556–62.

[44] Whitaker LR, Carneiro de Oliveira PE, McPherson KB, Fallon RV, Planeta CS, Bonci A, Hope BT. Associative learning drives the formation of silent synapses in neuronal ensembles of the nucleus accumbens. Biol Psychiatry 2016;80:246–56.

[45] Singer BF, Bubula N, Li D, Przybycien-Szymanska MM, Bindokas VP, Vezina P. Drug-paired contextual stimuli increase dendritic spine dynamics in select nucleus Accumbens neurons. Neuropsychopharmacology 2016;41:2178–87.

[46] Ziminski JJ, Sieburg MC, Margetts-Smith G, Crombag HS, Koya E. Regional differences in striatal neuronal ensemble excitability following cocaine and extinction memory retrieval in Fos-GFP mice. Neuropsychopharmacology 2018;43:718–27.

[47] Buzsaki G. Large-scale recording of neuronal ensembles. Nat Neurosci 2004;7:446–51.

[48] Mountcastle VB. Modality and topographic properties of single neurons of cat's somatic sensory cortex. J Neurophysiol 1957;20:408–34.

[49] John ER, Schwartz EL. The neurophysiology of information processing and cognition. Annu Rev Psychol 1978;29:1–29.

[50] O'Keefe J. A review of the hippocampal place cells. Prog Neurobiol 1979;13:419–39.

[51] Carelli RM, Deadwyler SA. Cellular mechanisms underlying reinforcement-related processing in the nucleus accumbens: electrophysiological studies in behaving animals. Pharmacol Biochem Behav 1997;57:495–504.

[52] Carelli RM, King VC, Hampson RE, Deadwyler SA. Firing patterns of nucleus accumbens neurons during cocaine self-administration in rats. Brain Res 1993;626:14–22.

[53] Chang JY, Paris JM, Sawyer SF, Kirillov AB, Woodward DJ. Neuronal spike activity in rat nucleus accumbens during cocaine self-administration under different fixed-ratio schedules. Neuroscience 1996;74:483–97.

[54] Chang JY, Sawyer SF, Lee RS, Woodward DJ. Electrophysiological and pharmacological evidence for the role of the nucleus accumbens in cocaine self-administration in freely moving rats. J Neurosci 1994;14:1224–44.

[55] Kiyatkin EA, Brown PL. I.V. Cocaine induces rapid, transient excitation of striatal neurons via its action on peripheral neural elements: single-cell, iontophoretic study in awake and anesthetized rats. Neuroscience 2007;148:978–95.

[56] Chang JY, Janak PH, Woodward DJ. Neuronal and behavioral correlations in the medial prefrontal cortex and nucleus accumbens during cocaine self-administration by rats. Neuroscience 2000;99:433–43.

[57] Haracz JL, Tschanz JT, Wang Z, Griffith KE, Rebec GV. Amphetamine effects on striatal neurons: implications for models of dopamine function. Neurosci Biobehav Rev 1998;22:613–22.

[58] Kiyatkin EA. Dopamine in the nucleus accumbens: cellular actions, drug- and behavior-associated fluctuations, and a possible role in an organism's adaptive activity. Behav Brain Res 2002;137:27–46.

[59] Peoples LL, Uzwiak AJ, Gee F, West MO. Operant behavior during sessions of intravenous cocaine infusion is necessary and sufficient for phasic firing of single nucleus accumbens neurons. Brain Res 1997;757:280–4.

[60] Guillem K, Ahmed SH. Incubation of Accumbal neuronal reactivity to cocaine cues during abstinence predicts individual vulnerability to relapse. Neuropsychopharmacology 2017;.

[61] Guillem K, Ahmed SH, Peoples LL. Escalation of cocaine intake and incubation of cocaine seeking are correlated with dissociable neuronal processes in different accumbens subregions. Biol Psychiatry 2014;76:31–9.

[62] Cameron CM, Carelli RM. Cocaine abstinence alters nucleus accumbens firing dynamics during goal-directed behaviors for cocaine and sucrose. Eur J Neurosci 2012;35:940–51.

[63] Carelli RM, Deadwyler SA. A comparison of nucleus accumbens neuronal firing patterns during cocaine self-administration and water reinforcement in rats. J Neurosci 1994;14:7735–46.

[64] Carelli RM. The nucleus accumbens and reward: neurophysiological investigations in behaving animals. Behav Cogn Neurosci Rev 2002;1:281–96.

[65] Carelli RM. Nucleus accumbens cell firing during goal-directed behaviors for cocaine vs. 'natural' reinforcement. Physiol Behav 2002;76:379–87.

[66] Carelli RM, Wondolowski J. Selective encoding of cocaine versus natural rewards by nucleus accumbens neurons is not related to chronic drug exposure. J Neurosci 2003;23:11214–23.

[67] Chang JY, Janak PH, Woodward DJ. Comparison of mesocorticolimbic neuronal responses during cocaine and heroin self-administration in freely moving rats. J Neurosci 1998;18:3098–115.

[68] Deadwyler SA, Hayashizaki S, Cheer J, Hampson RE. Reward, memory and substance abuse: functional neuronal circuits in the nucleus accumbens. Neurosci Biobehav Rev 2004;27:703–11.

[69] Opris I, Hampson RE, Deadwyler SA. The encoding of cocaine vs. natural rewards in the striatum of nonhuman primates: categories with different activations. Neuroscience 2009;163:40–54.

[70] Roop RG, Hollander JA, Carelli RM. Accumbens activity during a multiple schedule for water and sucrose reinforcement in rats. Synapse 2002;43:223–6.

[71] Carelli RM, Ijames SG. Selective activation of accumbens neurons by cocaine-associated stimuli during a water/cocaine multiple schedule. Brain Res 2001;907:156–61.

[72] Lansink CS, Jackson JC, Lankelma JV, Ito R, Robbins TW, Everitt BJ, Pennartz CM. Reward cues in space: commonalities and differences in neural coding by hippocampal and ventral striatal ensembles. J Neurosci 2012;32:12444–59.

[73] Robertson HA, Peterson MR, Murphy K, Robertson GS. D1-dopamine receptor agonists selectively activate striatal c-fos independent of rotational behaviour. Brain Res 1989;503:346–9.

[74] Young ST, Porrino LJ, Iadarola MJ. Cocaine induces striatal c-fos-immunoreactive proteins via dopaminergic D1 receptors. Proc Natl Acad Sci U S A 1991;88:1291–5.

[75] Graybiel AM, Moratalla R, Robertson HA. Amphetamine and cocaine induce drug-specific activation of the c-fos gene in striosome-matrix compartments and limbic subdivisions of the striatum. Proc Natl Acad Sci U S A 1990;87:6912–6.

[76] Hope B, Kosofsky B, Hyman SE, Nestler EJ. Regulation of immediate early gene expression and AP-1 binding in the rat nucleus accumbens by chronic cocaine. Proc Natl Acad Sci U S A 1992;89:5764–8.

[77] Cruz FC, Koya E, Guez-Barber DH, Bossert JM, Lupica CR, Shaham Y, Hope BT. New technologies for examining the role of neuronal ensembles in drug addiction and fear. Nat Rev Neurosci 2013;14:743–54.

[78] Kasof GM, Mandelzys A, Maika SD, Hammer RE, Curran T, Morgan JI. Kainic acid-induced neuronal death is associated with DNA damage and a unique immediate-early gene response in c-fos-lacZ transgenic rats. J Neurosci 1995;15:4238–49.

[79] Barker GR, Banks PJ, Scott H, Ralph GS, Mitrophanous KA, Wong LF, Bashir ZI, Uney JB, Warburton EC. Separate elements of episodic memory subserved by distinct hippocampal-prefrontal connections. Nat Neurosci 2017;20:242–50.

[80] Ghosh AK, Khan S, Marini J, Nelson JC, Farquhar D. A daunorubicin b-galactoside prodrug for use in conjunction with gene-directed enzyme prodrug therapy. Tetrahedron Lett 2000;41:4871–4.

[81] Farquhar D, Pan BF, Sakurai M, Ghosh A, Mullen CA, Nelson JA. Suicide gene therapy using E. coli beta-galactosidase. Cancer Chemother Pharmacol 2002;50:65–70.

[82] Barth AL, Gerkin RC, Dean KL. Alteration of neuronal firing properties after in vivo experience in a FosGFP transgenic mouse. J Neurosci 2004;24:6466–75.

[83] Cifani C, Koya E, Navarre BM, Calu DJ, Baumann MH, Marchant NJ, Liu QR, Khuc T, Pickel J, Lupica CR, Shaham Y, Hope BT. Medial prefrontal cortex neuronal activation and synaptic alterations after stress-induced reinstatement of palatable food seeking: a study using c-fos-GFP transgenic female rats. J Neurosci 2012;32:8480–90.

[84] Whitaker LR, Carneiro de Oliveira PE, McPherson KB, Fallon RV, Planeta CS, Bonci A, Hope BT. Associative learning drives the formation of silent synapses in neuronal ensembles of the nucleus Accumbens. Biol Psychiatry 2015;.

[85] Bolam JP, Hanley JJ, Booth PA, Bevan MD. Synaptic organisation of the basal ganglia. J Anat 2000;196(Pt 4):527–42.

[86] Voorn P, Vanderschuren LJ, Groenewegen HJ, Robbins TW, Pennartz CM. Putting a spin on the dorsal-ventral divide of the striatum. Trends Neurosci 2004;27:468–74.

[87] Gerfen CR. The neostriatal mosaic: multiple levels of compartmental organization in the basal ganglia. Annu Rev Neurosci 1992;15:285–320.

[88] Wilson CJ, Kawaguchi Y. The origins of two-state spontaneous membrane potential fluctuations of neostriatal spiny neurons. J Neurosci 1996;16:2397–410.

[89] Surmeier DJ, Ding J, Day M, Wang Z, Shen W. D1 and D2 dopamine-receptor modulation of striatal glutamatergic signaling in striatal medium spiny neurons. Trends Neurosci 2007;30:228–35.

[90] O'Donnell P. Dopamine gating of forebrain neural ensembles. Eur J Neurosci 2003;17:429–35.

[91] Nicola SM. The nucleus accumbens as part of a basal ganglia action selection circuit. Psychopharmacology 2007;191:521–50.

[92] van der Meer MA, Redish AD. Ventral striatum: a critical look at models of learning and evaluation. Curr Opin Neurobiol 2011;21:387–92.

[93] Mogenson GJ, Yang CR. The contribution of basal forebrain to limbic-motor integration and the mediation of motivation to action. Adv Exp Med Biol 1991;295:267–90.

[94] Chang JY, Zhang L, Janak PH, Woodward DJ. Neuronal responses in prefrontal cortex and nucleus accumbens during heroin self-administration in freely moving rats. Brain Res 1997;754:12–20.

[95] Guillem K, Kravitz AV, Moorman DE, Peoples LL. Orbitofrontal and insular cortex: neural responses to cocaine-associated cues and cocaine self-administration. Synapse 2010;64:1–13.

[96] West EA, Saddoris MP, Kerfoot EC, Carelli RM. Prelimbic and infralimbic cortical regions differentially encode cocaine-associated stimuli and cocaine-seeking before and following abstinence. Eur J Neurosci 2014;39:1891–902.

[97] Warren BL, Mendoza MP, Cruz FC, Leao RM, Caprioli D, Rubio FJ, Whitaker LR, McPherson KB, Bossert JM, Shaham Y, Hope BT. Distinct Fos-expressing neuronal ensembles in the ventromedial prefrontal cortex mediate food reward and extinction memories. J Neurosci 2016;36:6691–703.

[98] B. L. Warren, V. Selvam, M. Venniro, M. P. Mendoza, L. Kane, R. Quintano-Feliciano, R. Madangopal, L. Komer, L. R. Whitaker, F. J. Rubio, J. M. Bossert, D. Caprioli, Y. Shaham, B. T. Hope, Fos-expressing neuronal ensembles in the ventromedial prefrontal cortex mediate cocaine self-administration and extinction memories. J Neurosci, (submitted).

[99] Suto N, Laque A, De Ness GL, Wagner GE, Watry D, Kerr T, Koya E, Mayford MR, Hope BT, Weiss F. Distinct memory engrams in the infralimbic cortex of rats control opposing environmental actions on a learned behavior. elife 2016;5.

[100] Davis M. The role of the amygdala in fear and anxiety. Annu Rev Neurosci 1992;15:353–75.

[101] LeDoux J. The amygdala. Curr Biol 2007;17:R868–874.

[102] Gallagher M, Chiba AA. The amygdala and emotion. Curr Opin Neurobiol 1996;6:221–7.

[103] Holland PC, Gallagher M. Amygdala circuitry in attentional and representational processes. Trends Cogn Sci 1999;3:65–73.

[104] Kalivas PW. The glutamate homeostasis hypothesis of addiction. Nat Rev Neurosci 2009;10:561–72.

[105] Koob GF. The neurobiology of addiction: a neuroadaptational view relevant for diagnosis. Addiction 2006;101(Suppl. 1):23–30.

[106] Bowers MS, Chen BT, Bonci A. AMPA receptor synaptic plasticity induced by psychostimulants: the past, present, and therapeutic future. Neuron 2010;67:11–24.

[107] Mameli M, Luscher C. Synaptic plasticity and addiction: learning mechanisms gone awry. Neuropharmacology 2011;.

[108] Nestler EJ. Molecular basis of long-term plasticity underlying addiction. Nat Rev Neurosci 2001;2:119–28.

[109] Russo SJ, Dietz DM, Dumitriu D, Morrison JH, Malenka RC, Nestler EJ. The addicted synapse: mechanisms of synaptic and structural plasticity in nucleus accumbens. Trends Neurosci 2010;33:267–76.

[110] Wolf ME, Ferrario CR. AMPA receptor plasticity in the nucleus accumbens after repeated exposure to cocaine. Neurosci Biobehav Rev 2010;35:185–211.

[111] Guez-Barber D, Fanous S, Golden SA, Schrama R, Koya E, Stern AL, Bossert JM, Harvey BK, Picciotto MR, Hope BT. FACS identifies unique cocaine-induced gene regulation in selectively activated adult striatal neurons. J Neurosci 2011;31:4251–9.

[112] Fanous S, Guez-Barber DH, Goldart EM, Schrama R, Theberge FR, Shaham Y, Hope BT. Unique gene alterations are induced in FACS-purified Fos-positive neurons activated during cue-induced relapse to heroin seeking. J Neurochem 2013;124:100–8.

[113] Liu QR, Rubio FJ, Bossert JM, Marchant NJ, Fanous S, Hou X, Shaham Y, Hope BT. Detection of molecular alterations in methamphetamine-activated Fos-expressing neurons from a single rat dorsal striatum using fluorescence-activated cell sorting (FACS). J Neurochem 2014;128:173–85.

[114] Rubio FJ, Liu QR, Li X, Cruz FC, Leao RM, Warren BL, Kambhampati S, Babin KR, McPherson KB, Cimbro R, Bossert JM, Shaham Y, Hope BT. Context-induced reinstatement of methamphetamine seeking is associated with unique molecular alterations in Fos-expressing dorsolateral striatum neurons. J Neurosci 2015;35:5625–39.

[115] Li X, Rubio FJ, Zeric T, Bossert JM, Kambhampati S, Cates HM, Kennedy PJ, Liu QR, Cimbro R, Hope BT, Nestler EJ, Shaham Y. Incubation of methamphetamine craving is associated with selective increases in expression of Bdnf and trkb, glutamate receptors, and epigenetic enzymes in cue-activated fos-expressing dorsal striatal neurons. J Neurosci 2015;35:8232–44.

[116] Guez-Barber D, Fanous S, Harvey BK, Zhang Y, Lehrmann E, Becker KG, Picciotto MR, Hope BT. FACS purification of immunolabeled cell types from adult rat brain. J Neurosci Methods 2012;203:10–8.

[117] Rubio FJ, Li X, Liu QR, Cimbro R, Hope BT. Fluorescence activated cell sorting (FACS) and gene expression analysis of Fos-expressing neurons from fresh and frozen rat brain tissue. J Vis Exp 2016;114, https://doi.org/10.3791/54358. PMID: 27685012.

[118] Whitaker LR, Warren BL, Venniro M, Harte TC, McPherson KB, Beidel J, Bossert JM, Shaham Y, Bonci A, Hope BT. Bidirectional modulation of intrinsic excitability in rat Prelimbic cortex neuronal ensembles and non-ensembles after operant learning. J Neurosci 2017;37:8845–56.

[119] Ziminski JJ, Hessler S, Margetts-Smith G, Sieburg MC, Crombag HS, Koya E. Changes in appetitive associative strength modulates nucleus Accumbens, but not orbitofrontal cortex neuronal ensemble excitability. J Neurosci 2017;37:3160–70.

[120] Kourrich S, Calu DJ, Bonci A. Intrinsic plasticity: an emerging player in addiction. Nat Rev Neurosci 2015;16:173–84.

[121] Labiner DM, Butler LS, Cao Z, Hosford DA, Shin C, McNamara JO. Induction of c-fos mRNA by kindled seizures: complex relationship with neuronal burst firing. J Neurosci 1993;13:744–51.

[122] Sgambato V, Abo V, Rogard M, Besson MJ, Deniau JM. Effect of electrical stimulation of the cerebral cortex on the expression of the Fos protein in the basal ganglia. Neuroscience 1997;81:93–112.

[123] Heilig M, Koob GF. A key role for corticotropin-releasing factor in alcohol dependence. Trends Neurosci 2007;30:399–406.

[124] Heilig M, Goldman D, Berrettini W, O'Brien CP. Pharmacogenetic approaches to the treatment of alcohol addiction. Nat Rev Neurosci 2011;12:670–84.

[125] Garcia-Pardo MP, Roger-Sanchez C, Rodriguez-Arias M, Minarro J, Aguilar MA. Pharmacological modulation of protein kinases as a new approach to treat addiction to cocaine and opiates. Eur J Pharmacol 2016;781:10–24.

[126] Javitt DC, Schoepp D, Kalivas PW, Volkow ND, Zarate C, Merchant K, Bear MF, Umbricht D, Hajos M, Potter WZ, Lee CM. Translating glutamate: from pathophysiology to treatment. Sci Transl Med 2011;3.

[127] Pierce RC, O'Brien CP, Kenny PJ, Vanderschuren LJ. Rational development of addiction pharmacotherapies: successes, failures, and prospects. Cold Spring Harb Perspect Med 2012;2.

[128] Johnson JW, Kotermanski SE. Mechanism of action of memantine. Curr Opin Pharmacol 2006;6:61–7.

[129] Sinner B, Graf BM. Ketamine. Handb Exp Pharmacol 2008;313–33.

[130] Wood PL. The NMDA receptor complex: a long and winding road to therapeutics. IDrugs: Investig Drugs J 2005;8:229–35.

Chapter 7

Interoceptive Stimulus Effects of Drugs of Abuse

Brady M. Thompson*, Rick A. Bevins*, and Jennifer E. Murray[†]
*Department of Psychology, University of Nebraska-Lincoln, Lincoln, NE, United States; [†]Department of Psychology, University of Guelph, Guelph, ON, Canada

BODY PERCEPTION AND EMOTIONAL PROCESSING

Interoception is broadly considered to be the perception—whether implicit or explicit—of the internal state of the self [1–3]. Various forms of interoception include proprioception, the orientation of the body in space, physiological states such as osmolarity or hunger, and nociception of harmed or damaged body areas. These interoceptive signals then must be interpreted as 'feelings' to evoke motivational states such as movement, drinking, feeding, or avoidance. This interpretation is a form of emotional processing. Two influential theories of emotional processing include the James-Lange theory and the Schachter-Singer theory. In the James-Lange theory, physiological changes in the body give rise to emotion [4]. The Schachter-Singer theory, however, recognized that the same physiological state might indicate different emotional states. For instance, a rapid heartbeat will emerge during a long, rejuvenating run or while watching a horror film, and therefore, physiological changes in the body are cognitively processed and assigned a 'label' based on the context in which they arise [5]. In both of these theories, emotions rely on the interpretation of interoceptive physiological changes. That is, the physiological changes came first. The somatic marker hypothesis is a theory of decision making that relies on emotion-evoking 'markers' that represent interoceptive physiological states. This theory adds a new dimension to emotional processing—a dimension of neural representation of the body based on learning experience. These 'somatic markers' can modulate that interpretation of physiological changes within the body [6]. These implicit interpretations are then made explicit in the form of emotions that guide decision making.

The interactions of interoception, motivation, and emotional regulation are multifaceted. On one of these dimensions, the interpretation of physiological changes as motivational states can be implicit or explicit. Homeostatic regulation is initially implicit (e.g., subconscious) but can then transition to explicit (conscious) recognition of the interoceptive state when a behavioral response may be required [7–10]. As an example, internal temperature changes can evoke goose bumps and shivering when chilled; or flushing and perspiration when warmed. Higher-level processing of this temperature change can evoke goal-directed seeking of an additional outcome. In these instances, the addition of a jacket for warmth or the removal of a cardigan helps to cool the body. Implicit versus explicit processing in temperature regulation have been demonstrated to have distinct neural pathways. For example, in a study investigating the neural basis of temperature regulation, rats were trained to press a lever to turn on a heat lamp. Rats that received subsequent lesions of the hypothalamus increased their lever pressing for heat access because they were no longer able to reflexively regulate their body temperature [11–13]. Therefore, in rodents, the motivation for homeostatic regulation of temperature can be processed on multiple levels, from the hypothalamus—a region known to monitor basic body processes and maintain homeostasis—as well as from the limbic system, implicated in emotional processing, learning Pavlovian associations between stimuli, and maintaining representations of related outcomes [9,14].

Olds and Milner's [15] classic study on the mechanisms of reward using alternating currents in the rat brain septal region found that limbic system activation is integral to reinforcement mechanisms. This limbic system is inextricably bound with substance use disorders. From early observations that drugs that are abused by humans evoke increased dopamine levels in the nucleus accumbens [16] to the development and refinement of elegant theories devoted to understanding the neurological basis of drug addiction [17–28], there is no question that drugs of abuse exogenously activate an endogenous reward and reinforcement system. Isolating the particular circuits responsible for distinct stages of drug-associated behaviors is an

Copyright © 2019 Elsevier Inc. All rights reserved.

ongoing endeavor. In the case of interoception, this self-perception and assignment of emotional value to a physiological state can then be integrated with cognitive factors and result in a motivational guide to direct behavior.

The ability to appropriately process (or act on) the state of the internal milieu is impaired in addiction. The mechanism behind this impairment has been suggested to be dysregulated alliesthesia—the notion that hedonic judgments depend on the individual's internal state [24,29]. The inability to appropriately assess the internal state results in aberrant reward valuation of drug-related outcomes and decreased valuation of nondrug-related outcomes. In rats, this can mean resistance to punished drug self-administration [30,31] or an escalation of drug taking [32,33]. In humans, this compromised insight can contribute to many of the potential behavioral criteria for a substance use disorder as determined by the Diagnostic and Statistical Manual of Mental *Disorders* (DSM-V) including those that indicate a loss of control over intake (criteria 1 and 2) or continued use despite personal (criterion 6), professional, (criteria 5 and 7), and physical (criteria 8 and 9) costs [34,35].

INTEROCEPTION, DRUG CONSUMPTION, AND THE INSULA

In order for the higher-order processing of body states to occur, there needs to be a way for the brain to access information regarding these physiological states. An afferent neural system that processes the physiological condition of the body has been identified and described in detail [1,2]. Briefly, sympathetic afferents innervate lamina I of the spinal cord which then projects to the ventromedial posterior nucleus of the thalamus. Conversely, parasympathetic afferents provide input to the nucleus of the solitary track, which then innervates the parvocellular portion of the ventromedial posterior nucleus. These pathways are then projected on to the insula to begin the processing of the interoceptive state. The insula has been associated with combining a perception of the internal state of the body with information from sensory, motor, homeostatic, and limbic reward/reinforcement pathways [7,10,36].

The role of interoceptive factors in addiction has been frequently discussed, and there are a number of excellent reviews on the subject [3,24,35,37–39]. Within the context of drug dependence, aberrant reward valuation of drug outcomes is related to dysregulated drug intake. Indeed, insula activity measured by the BOLD signal is increased in smokers exposed to visual smoking cues [40] and in drinkers exposed to visual alcohol cues who transitioned, or escalated, from moderate to heavy alcohol use over the course of a year [41]. In contrast, insula activity evoked by an aversive respiratory stimulus is attenuated in problem stimulant users compared with controls [42]. These findings indicate that under some circumstances, there may be an interoceptive bias for drug-related stimuli and a dampened response to nondrug stimuli. Therefore, insula hyperactivity evoked by drug-associated stimuli has been proposed as a causal factor in impaired inhibitory control over cue-driven drug seeking [38]. However, chronic heavy alcohol users showed higher insula activity in response to non-alcohol-related images than alcohol-related images [41], supporting the notion that the influence of the insula on drug-related behaviors is not unidirectional. Nevertheless, damage to the insula decreases smoking [43–45] and symptoms of withdrawal [46] in smokers who have experienced a stroke, suggesting restoration of control over drug intake.

This restored control of escalated drug intake following insula damage was replicated in a recent rodent study [47]. Drug users, whether human or nonhuman, generally titrate their drug consumption to achieve individually ideal, relatively stable levels of drug [48–51]. Under certain circumstances such as extended access to limitless drug, however, this titration goes awry, and control over drug intake is lost [52,53]. This lost control over drug intake has also been related to insula processing [47]. There, rats that had escalated their cocaine intake re-regulated that intake following post-escalation insula lesions as shown by stable self-administration levels during subsequent repeated self-administration sessions [47]. Conversely, insula lesions before training exacerbate cocaine escalation and relapse following punishment-induced suppression [47,54]. Given that the insula has been shown to be active during context-induced relapse following punishment-induced suppression of cocaine self-administration [55] and that insula inactivation prevented context-induced reinstatement of cocaine self-administration [56], the role of the insula in this task may be to moderate the level of relapse rather than simply to allow—or to prevent—emergence of that relapse behavior. On the other hand, inactivation of the insula also reduced nicotine self-administration, motivation, and reinstatement following extinction but had no effect on food-seeking behaviors [57]. Further, blocking dopamine D1 receptors [58], blocking orexin receptors [59], and deep brain stimulation [60] in insula disrupt nicotine self-administration. Combined, these seemingly contradictory findings indicate the need for interoceptive awareness and assessment in maintaining appetitive drug-seeking behavior, even though that interoception has been dysregulated through drug use [24]. Even more work is required to disambiguate the exact nature of the role of the insula in interoceptive processing for distinct types of addiction-related behaviors from early stages through to the perpetuation of compulsive drug taking.

DRUG CRAVING AS INTEROCEPTION

Drug craving is central to theories of addiction [22,25,61–63]. Various operationalizations of craving have been offered and contested over the years. Early versions included pharmacological factors of tolerance and obfuscated notions of craving for a particular drug or craving for a feeling of an altered state of mind regardless of drug class [64]. However, context specificity of drug craving was observed, noting that '[an alcoholic] may stay in an institution for many months without showing any distress, even if no therapy is being carried out. Here one certainly cannot speak of an irresistible craving or a physical demand for alcohol, even though on leaving the institution he may immediately turn to alcohol' [64], illustrating that craving and relapse are not necessarily dependent on direct physiological effects of the drug. More recently, craving is described as an internally perceived subjective state [65] that can speed initiation of drug use [66,67] and is intensified by an absence of drug [68]. Craving can be evoked by a drug-associated stimulus [69] as well as by interoceptive states such as perceived stress [70] or negative mood [71].

Further, the direct interoceptive effects of a drug can, in some individuals, trigger a loss of control over drug intake, resulting in a binge. This drug-induced drug seeking has been observed in drug-addicted humans [64] and modeled in drug-elicited reinstatement tasks in rodents [72]. This reinstatement effect, acute exposure to the interoceptive drug stimulus after a period of abstinence, can lead to escalated drug taking and has the ability to both strengthen nucleus accumbens reward-related synaptic connections and weaken top-down prefrontal cortex control-related synaptic connections, making the addictive behavior more likely [73]. Drug-evoked craving has also been incorporated in a limited number of cue-exposure therapy studies in humans [74,75] described later. Interestingly, in rats that have a cocaine self-administration history, an injection of cocaine methiodide—a cocaine analog that does not cross the blood-brain barrier—evokes a meso-limbic glutamate and dopamine release that is sufficient to reinstate extinguished cocaine seeking. In rats with no cocaine self-administration history, cocaine methiodide has no effects on the brain or behavior [76,77]. These fascinating studies demonstrate that abused drugs can evoke peripheral processes that become conditioned stimuli (CSs) for the impending central effects of the drug, providing another mechanism whereby the interoceptive drug effects can evoke subsequent drug craving.

OPERANT CONDITIONING OF INTEROCEPTION

Typically, drugs of abuse have been conceptualized as the reinforcers or rewards in addiction models. However, the effects of drugs can also be internally perceived by an individual and used as a cue to guide appropriate behavior. Drug-discrimination studies that involve operant learning mechanisms have been a staple for behavioral pharmacologists to investigate the nature of interoceptive effects of drugs of abuse in both human and nonhuman subjects [78–81]. The purpose of these studies is to determine the discriminability of one drug stimulus from another stimulus, and to establish to what extent subsequently tested compounds are similar—or generalize—to the drug under investigation. The four main components necessary for drug discrimination studies are the subject, the drug, the response, and the reinforcer [82]. In its most common form, the two-lever drug-discrimination task in rats uses a standard operant chamber in which experimenter-administered drug is differentially reinforced from a baseline non-pharmacological state, placebo (in humans), or another drug stimulus [82–84]. Briefly, on sessions in which the training drug under investigation is administered by the experimenter, responses on a specific lever (e.g., right) are reinforced according to the schedule set by the researcher (e.g., FR25), typically with food, sugar pellets, or sweetened solutions, whereas responding on the opposite lever has no scheduled consequence. On interspersed sessions in which the subject is given vehicle, the contingency is reversed such that responding on the other lever (i.e., left) is now reinforced. Therefore, for each session, the interoceptive effects of the drug under investigation guide appropriate lever pressing. Variations on this procedure include use of nose pokes into one of two apertures instead of lever presses, drug-drug discriminations instead of drug-no drug discriminations, and more complex operants. For instance, a three-lever drug-discrimination procedure has been used to compare different drugs or different doses of the same drug within the same training phase [85,86].

Since the fundamental nature of the drug stimulus can change with drug dose, the training drug dose can affect discriminability and subsequent generalizability [80]. A dose with a just noticeable difference above a non-pharmacological state may have unique qualities that differ from a no-drug state and higher doses. However, higher drug doses tend to be easier to discriminate from no-drug than lower drug doses. If the dose is too high, it can cause a general locomotor or cognitive impairment [87]. Interestingly, subjects that train with low doses may perform better on the discrimination when tested later with higher doses because of the individual's prior need to attend to small changes in their interoceptive state [88,89]. Acquiring a drug discrimination also becomes increasingly challenging when testing two qualitatively similar drugs rather than two distinct classes of drugs or drug versus no-drug. For instance, in humans, the stimulants methamphetamine and d-amphetamine are much harder to discriminate from one another than either stimulant compared with placebo or the opiate hydromorphone [90].

Some work has investigated the neural basis of drug discriminative stimuli using intracerebral microinjections [91]. For instance, direct intracranial infusions of nicotine into the dorsal hippocampus generalized to moderate [92] and large [93] systemic training doses of nicotine. However, intracranial infusions of nicotine into the nucleus accumbens either did not [92] or only partially generalized to the systemic nicotine discriminative stimulus [94]. Further, medial prefrontal cortex infusions of nicotine generalized to the systemic nicotine [95], a finding that is particularly interesting in light of the role of the medial prefrontal cortex in learning about stimulus associations [96]. Importantly, these sets of findings support the notion that the pathways mediating the interoceptive discriminative stimulus effects of nicotine are at least somewhat distinct from those mediating its reinforcing effects [97,98].

PAVLOVIAN CONDITIONING OF INTEROCEPTION

The drug discrimination models that use classical (Pavlovian) conditioning procedures, albeit less common than the operant models, have allowed researchers to study distinct associative-learning factors affecting the influence of interoceptive drug stimuli on approach behavior that may impact the development and perpetuation of addiction more directly. In one such model of drug-discriminated goal tracking, drugs of abuse function as occasion setters, modulating the association between a light CS and a sucrose unconditioned stimulus (US) [99,100]. In brief, on experimenter-administered drug sessions, the interoceptive effects of the drug can indicate that a short light CS presentation is followed by a 4-s access to sucrose. On intermixed nondrug sessions, the light CSs are still presented, but sucrose is never available. Rats in this positive feature occasion setting task increase goal-tracking during the light stimulus *only* on drug sessions [101–104]. Conversely, rats can learn that an interoceptive drug stimulus is a negative feature—indicating that the light will *not* be followed by sucrose. In this situation, the light CS will evoke sucrose seeking only on those sessions in which the drug is *not* administered [103,105]. Notably, nicotine trained as a negative feature has been shown to pass the summation and retardation-of-acquisition tests of a conditioned inhibitor [106]. In these tests, a stimulus that has been trained as a negative feature indicating when a CS will *not* be followed by a US is subsequently tested for conditioned inhibitory characteristics [107]. Specifically, the suspected conditioned inhibitor must reduce the amount of responding evoked by a separately trained excitatory CS when they are presented together (summation) as well as display slowed subsequent excitatory conditioning (retardation-of-acquisition). These exciting findings using nicotine demonstrate that the interoceptive stimulus effects of a drug can indeed be imbued with negative value through conditioning as an occasion setter that then transfers inhibitory properties across learning contexts. Further, and of particular relevance to drug addiction research, positive feature conditioning with a drug *enhanced* conditioned responding evoked by an excitatory CS never previously experienced in that drug state [108]. This transfer of occasion setting was specific to CSs that were already learned to be excitatory (although this learning took place in a different drug state), and a novel drug without that conditioning history had no enhancing effect. That an interoceptive drug stimulus can impact *other* associations by virtue of an acquired value through conditioning history has profound implications for addiction theory and treatment that will be discussed later.

In another version of a drug-discriminated goal tracking task, the interoceptive stimulus effects of the drug *directly* indicate when the appetitive US will be available in the environment. This preparation is distinct from the occasion setting task described above in that there is no discrete light stimulus CS—the drug itself serves as the purported CS. By far, nicotine has been the most studied drug in this preparation [109]. Briefly, on sessions in which nicotine has been experimenter-administered subcutaneously, rats are noncontingently provided with intermittent access to sucrose in the chamber (typically 36 4-s deliveries). On interspersed saline sessions, the sucrose is withheld. Anticipatory sucrose-seeking behavior rapidly emerges on nicotine sessions (calculated using the time interval before sucrose is delivered) compared with saline sessions (calculated using the equivalent time interval) [110,111]. This drug-discriminated goal-tracking behavior [112] is centrally, but not peripherally, mediated [113], sensitive to drug stimulus salience [110,114], sucrose stimulus salience [111,115], and extinction processes [110,113]. Importantly, this discrimination is specific to the neuropharmacological processes affected by nicotine (i.e., not simply a drug versus no-drug discrimination [113]. Further, control of goal tracking by nicotine reflects a direct learned association between nicotine and sucrose [116] and not recall of sucrose availability dependent on being under the influence of the drug [i.e., not state-dependent learning; [117,118]].

Investigations are currently underway to identify the brain structures involved in the development and expression of nicotine as a CS. Thus far, in rats that have this appetitive conditioning history with nicotine show enhanced c-Fos expression in the dorsomedial striatum compared with pseudoconditioned control rats equally exposed to nicotine, chambers, and sucrose [119]. Follow-up investigations revealed that pre- and post-training lesions of the posterior dorsomedial stratum disrupted acquisition and expression, respectively, of the nicotine-evoked conditioned response [120].

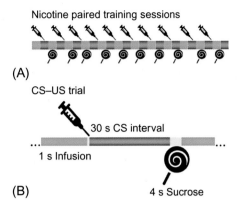

FIG. 1 Panel A shows the training schematic for the intravenous nicotine conditioned stimulus (CS). There were 10 nicotine-sucrose pairings within every 2-h session, each spaced ~11 min apart. Panel B depicts the timing of a single nicotine-sucrose pairing trial within the training sessions. Each 1-s infusion was followed 30 s later with 4-s access to liquid sucrose. Rats readily learned to selectively goal track for sucrose during that 30-s interval [121].

In contrast, post-training inactivation of the anterior portion of the dorsomedial striatum before a saline-no sucrose session disinhibited sucrose seeking that would otherwise have been low [120]. Combined, these results point to some intriguing complexities regarding the nature of conditioned responding evoked by the interoceptive drug stimulus that are specific to the learning history rather than to simple unconditioned effects of nicotine.

Studies investigating the nature of the interoceptive nicotine CS have also been extended to intravenous infusions instead of the presession subcutaneous injections [121]. The use of multiple, intra-session, noncontingent nicotine infusions allow for conditioning to be observed with each nicotine-sucrose pairing rather than once every 36 pairings (see above). In this variation of the model, nicotine infusions are spaced approximately 11 min apart across the 2-h session to allow for enhanced stimulus detection (Fig. 1A), and each 4-s sucrose delivery occurs 30 s after each infusion (Fig. 1B). Rats readily came to selectively increase goal tracking during those 30-s intervals between the nicotine infusion and sucrose. Again, the intravenous nicotine stimulus was centrally mediated and sensitive to extinction processes [121]. This model was further used to investigate theoretical notions of cue competition between the interoceptive nicotine stimulus and an exteroceptive light stimulus on control over the sucrose-seeking conditioned response. In the overshadowing study, rats trained with the compound nicotine + light stimulus (Fig. 2A) had noncontingent intravenous infusions of nicotine that occurred at the onset of a 30-s light stimulus signaling a brief sucrose delivery (Fig. 2B) [106]. Following acquisition of conditioned responding to this compound CS, intermittent tests of responding to each of the elements (i.e., nicotine or light) were conducted to assess control of each stimulus over the sucrose-seeking conditioned response (Fig. 2C). Compound training with the nicotine + light CS resulted in decreased responding to the light element of the stimulus relative to a group trained with the light CS paired with sucrose alone and controlled unpaired exposure to nicotine [106]. In another cue competition study investigating the blocking effect, groups were initially trained under one of three conditions—nicotine paired with sucrose, nicotine unpaired with sucrose, or equal exposure to conditioning chamber without nicotine or sucrose (Fig. 3A–C, respectively) [122]. Following acquisition of the sucrose-seeking conditioned response in the nicotine paired group, all groups received training with the nicotine + light compound CS paired with sucrose (cf. Fig. 2A). Subsequent tests of element-evoked responding (cf. Fig. 2C) showed that only the group that previously had nicotine paired with the sucrose showed a greater proportion of responding to the nicotine element than the light following compound stimulus training [122]. Combined, these studies demonstrate clear competition of the interoceptive drug CS with the exteroceptive visual light CS that resulted in decreased control of the light over appetitive conditioned responding. This outcome reveals a gaping hole in what we understand regarding how the interoceptive environment evoked by drugs of abuse may impact what is learned about the external drug-consumption environment.

INTERACTIONS OF INTEROCEPTION AND SELF-ADMINISTRATION

Given the evidence of strong drug-evoked interoceptive control of learning and behavior described above, it stands to reason that an interaction between multiple stimulus effects may be obfuscating the line between stimulus and reinforcer. For example, nicotine has been shown to be a self-administered reinforcer [123–125], but its reinforcing effects are sensitive to other factors such as presence of environmental stimuli [126]. The contingent presentation of these stimuli along with the nicotine infusion drives higher rates of nicotine self-administration than nicotine alone, and Caggiula and

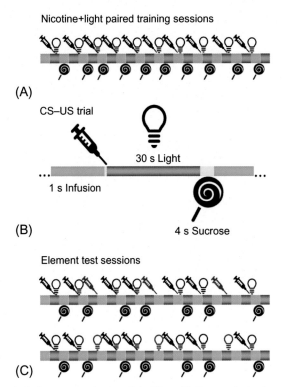

FIG. 2 Panel A shows the training schematic for intravenous nicotine + light CS used in the overshadowing experiment. There were 10 nicotine + light–sucrose pairings within every 2-h session, each spaced ~11 min apart. Panel B depicts the timing of a single nicotine + light–sucrose pairing trial within the training sessions. Each 1-s infusion was followed 30 s later with 4-s access to liquid sucrose; the 30-s interval was illuminated with the houselight. Panel C depicts the element tests of the overshadowing experiment. On these tests, the session was conducted as in training, except that three of the stimulus presentations were one of the two CS elements (nicotine or light) alone, not followed by sucrose. The order of testing nicotine or light was counterbalanced for each rat. The exteroceptive light stimulus controlled less of the conditioned response when trained in compound with nicotine than when trained in compound with saline [106].

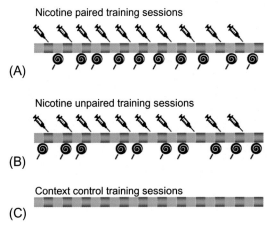

FIG. 3 This figure depicts the three initial training groups for the blocking experiment. Panel A shows the standard 10 nicotine-sucrose pairings within every 2-h session, each spaced ~11 min apart for the paired group. Panel B shows training for the unpaired group. Sucrose was presented at least 4 min away from any nicotine infusion. Panel C depicts that the context alone condition received no nicotine or sucrose deliveries. All groups then received nicotine + light–sucrose training as in the overshadowing experiment (cf. Fig. 2A) followed by element tests (cf. Fig. 2C). An appetitive conditioning history with nicotine blocked subsequent control of the light over conditioned responding compared with the unpaired and chamber control groups [122].

colleagues argue that the reinforcing effects of nicotine in animal models and in humans are enhanced by the presence of the conditioned exteroceptive cues [127,128]. Conversely, nicotine can also enhance the reinforcing value of other stimuli in the environment such as lights or sucrose [129–132]. Therefore, there is a drug-related positive feedback loop that may be driving a perpetuation of enhanced reinforcing value of nicotine and nicotine-associated stimuli. That is, environmental stimuli may be imbuing nicotine with enhanced value, and, at the same time, nicotine may be imbuing these other stimuli with enhanced value.

Nicotine readily serves as an interoceptive CS prompting sucrose-seeking behavior [114,121]. Further, nicotine can also be imbued with excitatory [108] or inhibitory [118] effects, impacting subsequent learning. These findings, combined with the positive feedback loop described above, lead to the hypothesis that, by virtue of being predictive of sucrose, the appetitive *value* of nicotine is increased, and therefore, rats with that appetitive nicotine conditioning history will self-administer more nicotine. This sort of acquired value enhancement has been demonstrated experimentally in a human Pavlovian conditioning task [133]. In that study, participants who did not initially prefer the interoceptive stimulus evoked by diazepam were provided a noncontingent high monetary reward (US) in a computer-based task whenever they were in the diazepam state; when in the nondrug state, these participants were provided a low monetary US. Notably, the US was not contingent on behavior. With repeated pairings, these participants shifted their preference to the diazepam stimulus and reported increased liking of that stimulus. This important finding directly demonstrates that drug preference—and even drug liking—can develop solely as a result of conditioning history.

The feasibility of enhanced drug value in self-administration models is also supported by studies using the acquisition of a new response (ANR) procedure. ANR is a theoretical concept that tests the acquired conditioned reinforcing properties of an appetitive CS. For instance, rats trained with an auditory CS explicitly paired with a food US subsequently showed much higher rates of lever pressing reinforced by that auditory CS than controls that had the auditory stimulus and food deliveries explicitly unpaired [134]. Similar procedures have been used to determine that a drug-associated light CS can support ANR [135–137]. Although these studies have used exteroceptive auditory or light CSs, the nicotine CS has been shown to operate in a similar manner to more traditional exteroceptive stimuli. The learned conditioned response is sensitive to nicotine CS salience [110,121], behavioral extinction [110], and cue competition including overshadowing and blocking [106,122]. In this way, the interoceptive drug cue is 'behaving' similar to how exteroceptive lights and sounds typically do in animal learning research [99].

It stands to reason that the nicotine stimulus may support ANR through conditioned reinforcing effects acquired by appetitive conditioning with a sucrose US [37,99,109,116]. Indeed, recent research from our group has shown this direct interaction between these two stimulus effects of nicotine. Specifically, an appetitive conditioning history with nicotine as a CS *enhanced* subsequent nicotine self-administration when assessed in repeated progressive ratio schedule sessions [138]. In that study, rats were initially trained to lever press to self-administer nicotine under one of two conditions, either standard infusion or the standard infusion followed 30 s later with brief access to liquid sucrose [121]. Intake at that stage was capped at 10 infusions per session. In the following stage, all rats could nose poke in a novel operandum for unlimited nicotine infusions within 2 h on a progressive ratio schedule. Rats that had the nicotine paired with sucrose in the first stage took more nicotine than the rats that were in the standard condition. This finding exemplifies the notion that a drug stimulus can be imbued with additional appetitive qualities that will directly impact the development of drug addiction.

CLOSING REFLECTIONS: INTEROCEPTION AND INTERVENTION

The complexity of the stimulus effects of drugs of abuse as demonstrated by acquired excitatory and inhibitory properties [108,139], competition with exteroceptive appetitive stimuli [106,122], and enhanced appetitive value [138] following appetitive associative conditioning, broadly highlights the pressing need to understand how drug stimuli impact what is learned about the exteroceptive environment. From the clinical perspective, incorporating an understanding of this impact may partially explain why cue-exposure therapies for drug dependence typically show disappointing outcomes. Cue-exposure therapies are designed to exploit Pavlovian conditioning processes. In practice, individuals are exposed to drug-related stimuli, and the drug-evoked conditioned responses are experienced and presumably extinguished when drug ingestion does not follow this exposure. This approach is meant to decrease the drive or craving for a drug after exposure to the drug-related stimulus, and the hope is that this extinguished conditioned response evoked by drug-related stimuli will generalize to the real world. Unfortunately, this practice has continued to show little promise for drug addiction [140–142].

Conversely, exposure therapy techniques have been successful to treat some specific phobias [143]. In this type of anxiety disorder, the stimulus evoking the fear response is discrete (e.g., spiders or flying) and can be reliably presented *in vivo* or through intense guided imagery. However, drugs of abuse have been paired with a multitude of environments, people, temporal contexts, paraphernalia, and levels of internal arousal or stress. In short, there are innumerable specific stimuli that need to be extinguished. As such, researchers have begun investigating personalized cues [144], the use of evoking craving conditioned responses in multiple contexts [145], guided imagery [146], targeting memory reconsolidation mechanisms [147], and incorporating virtual reality [148]. A more recent effort has been to develop a smartphone application based on established cue-exposure techniques [149], but its efficacy is currently under evaluation.

Further complicating the situation is that drug-related associations are formed outside of the clinic, and there is evidence that the cues used to evoke craving in the clinical setting are not even correlated with those experienced in the real world [150]. Although conducting exposure therapy in the actual drug-taking environments would address some of these issues, it is largely untenable due to the multitude of possible contexts associated with drug use, the increased expense of clinician time and travel, proprietors being unwilling to have drug treatment occurring in their establishments, and the potential dangers associated with the contexts in which illegal substances may be in use.

The one stimulus, however, that is consistent across all of these cues, contexts, people, and situations is the drug itself. Even further, the cue competition effects of overshadowing and blocking [106,122] described earlier suggest that if nicotine is serving as both a stimulus and as a reinforcer in human smokers, the exteroceptive cues associated with the rewarding effects of tobacco may control less behavior than often assumed. Therefore, in substance-dependent humans, drug exposure may be competing with the stimuli generally thought to evoke drug seeking and thus possibly reducing the impact of cue-exposure therapies [109]. These notions combine to form the argument that controlled exposure to the drug stimulus and extinction of the subsequent cravings could be used as a method of treatment. Incorporating the drug cue into therapy is not a new idea. Controlled exposure to an interoceptive ethanol cue has been shown to reduce frequency and volume of alcohol consumed in nondependent drinkers [75] and reduced desire to drink and slowed the rate of alcohol consumption in dependent drinkers [74]. Therefore, depending on the needs and situation of the particular individual [38], addiction prevention methods may not want to focus on abstinence but rather focus on decreased consumption. Harm reduction methods using nicotine replacement or methadone maintenance are already in place for tobacco and heroin consumption, respectively [151,152], and there has been some recent headway investigating individual differences in the mechanisms that predict successful transitions from heavy to moderated drinking [153–155]. These include levels of motivation and confidence, but also cognitive factors such as self-efficacy. Cognitive appraisal of interoceptive states and subsequent behavioral inhibition has been suggested to be impaired in drug dependence, leading to impulsive drug seeking evoked by implicit recognition of drug-associated stimuli [38]. This model could be extended even further to recognize the interoceptive stimulus effects of drug. Interoceptive drug exposure could therefore be used to carefully incorporate cognitive therapy methods of recognizing and coping with the interoceptive state, as well as being mindful of other rewards that are products of decreased drug consumption such as positive social interactions, lack of embarrassment, improved health, and making responsible and nonlife-threatening choices.

ACKNOWLEDGMENTS

The writing of this work was supported in part by NIH research grants DA034389 and DA039356 awarded to RAB. NIH had no other involvement other than financial support. The authors declare no financial conflicts of interest.

REFERENCES

[1] Craig AD. How do you feel? Interoception: the sense of the physiological condition of the body. Nat Rev Neurosci 2002;3(8):655–66. https://doi.org/10.1038/nrn894.

[2] Craig AD (Bud). Interoception: The sense of the physiological condition of the body. Curr Opin Neurobiol 2003;13(4):500–5. https://doi.org/10.1016/S0959-4388(03)00090-4.

[3] Paulus MP, Stewart JL. Interoception and drug addiction. Neuropharmacology 2014;76:342–50. https://doi.org/10.1016/j.neuropharm.2013.07.002.

[4] James W. II.—What is an emotion? Mind 1884;9(34):188–205. https://doi.org/10.1093/mind/os-IX.34.188.

[5] Schachter S, Singer J. Cognitive, social, and physiological determinants of emotional state. Psychol Rev 1962;69(5):379–99. https://doi.org/10.1037/h0046234.

[6] Damasio AR. The somatic marker hypothesis and the possible functions of the prefrontal cortex. Phil Trans Roy Soc B: Biol Sci 1996;351 (1346):1413–20. https://doi.org/10.1098/rstb.1996.0125.

[7] Critchley HD, Wiens S, Rotshtein P, Ohman A, Dolan RJ. Neural systems supporting interoceptive awareness. Nat Neurosci 2004;7(2):189–95. https://doi.org/10.1038/nn1176.

[8] Lau H, Rosenthal D. Empirical support for higher-order theories of conscious awareness. Trends Cogn Sci 2011;15(8):365–73. https://doi.org/10.1016/j.tics.2011.05.009.

[9] LeDoux JE. Emotion circuits in the brain. Annu Rev Neurosci 2000;23(1):155–84. https://doi.org/10.1146/annurev.neuro.23.1.155.

[10] LeDoux JE, Pine DS. Using neuroscience to help understand fear and anxiety: a two-system framework. Am J Psychiatry 2016;173(11):1083–93. https://doi.org/10.1176/appi.ajp.2016.16030353.

[11] Refinetti R, Carlisle HJ. A reevaluation of the role of the lateral hypothalamus in behavioral temperature regulation. Physiol Behav 1987;40 (2):189–92. https://doi.org/10.1016/0031-9384(87)90206-X.

[12] Satinoff E, Rutstein J. Behavioral thermoregulation in rats with anterior hypothalamic lesions. J Comp Physiol Psychol 1970;71(1):77–82. http://www.ncbi.nlm.nih.gov/pubmed/5452104.

[13] Schulze G, Tetzner M, Topolinski H. Operant thermoregulation of rats with anterior hypothalamic lesions. Naunyn Schmiedeberg's Arch Pharmacol 1981;318(1):43–8. https://doi.org/10.1007/BF00503311.

[14] Morrison SF, Nakamura K. Central neural pathways for thermoregulation. Front Biosci (Landmark Edition) 2011;16:74–104. http://www.ncbi.nlm. nih.gov/pubmed/21196160.

[15] Olds J, Milner P. Positive reinforcement produced by electrical stimulation of septal area and other regions of rat brain. J Comp Physiol Psychol 1954;47(6):419–27. https://doi.org/10.1037/h0058775.

[16] Di Chiara G, Imperato A. Drugs abused by humans preferentially increase synaptic dopamine concentrations in the mesolimbic system of freely moving rats. Proc Natl Acad Sci U S A, 1988; 85(14): 5274–8. https://doi.org/10.1073/pnas.85.14.5274

[17] Belin D, Belin-Rauscent A, Murray JE, Everitt BJ. Addiction: failure of control over maladaptive incentive habits. Curr Opin Neurobiol 2013;23 (4):564–72. https://doi.org/10.1016/j.conb.2013.01.025.

[18] Berridge KC, Robinson TE, Aldridge JW. Dissecting components of reward: "liking", "wanting", and learning. Curr Opin Pharmacol 2009;9 (1):65–73. https://doi.org/10.1016/j.coph.2008.12.014.

[19] Everitt BJ, Robbins TW. Drug addiction: updating actions to habits to compulsions ten years on. Annu Rev Psychol 2016;67(1)https://doi.org/ 10.1146/annurev-psych-122414-033457.

[20] Kalivas PW, Volkow N, Seamans J. Unmanageable motivation in addiction: a pathology in prefrontal-accumbens glutamate transmission. Neuron 2005;45(5):647–50. https://doi.org/10.1016/j.neuron.2005.02.005.

[21] Koob GF, Le Moal M. Drug addiction, dysregulation of reward, and allostasis. Neuropsychopharmacology 2001;24(2):97–129. https://doi.org/ 10.1016/S0893-133X(00)00195-0.

[22] Koob GF, Volkow ND. Neurocircuitry of addiction. Neuropsychopharmacology 2010;35(1):217–38. https://doi.org/10.1038/npp.2009.110.

[23] Koob GF, Volkow ND. Neurobiology of addiction: a neurocircuitry analysis. Lancet Psych 2016;3(8):760–73. https://doi.org/10.1016/S2215-0366 (16)00104-8.

[24] Paulus MP, Tapert SF, Schulteis G. The role of interoception and alliesthesia in addiction. Pharmacol Biochem Behav 2009;94(1):1–7. https://doi. org/10.1016/j.pbb.2009.08.005.

[25] Robinson TE, Berridge KC. The neural basis of drug craving: an incentive-sensitization theory of addiction. Brain Res Rev 1993;18(1):247–91. https://doi.org/10.1016/0165-0173(93)90013-P.

[26] Schoenbaum G, Chang C-Y, Lucantonio F, Takahashi YK. Thinking outside the box: orbitofrontal cortex, imagination, and how we can treat addiction. Neuropsychopharmacology 2016;41(13):2966–76. https://doi.org/10.1038/npp.2016.147.

[27] Scofield MD, Heinsbroek JA, Gipson CD, Kupchik YM, Spencer S, Smith AC, et al. The nucleus Accumbens: mechanisms of addiction across drug classes reflect the importance of glutamate homeostasis. Pharmacol Rev 2016;68(3):816–71. https://doi.org/10.1124/pr.116.012484.

[28] Wise RA. Dopamine and reward: the anhedonia hypothesis 30 years on. Neurotox Res 2008;14(2–3):169–83. https://doi.org/10.1007/BF03033808.

[29] Cabanac, M. (1971). Physiological role of pleasure. Science (New York, NY), 173(4002), 1103–7. http://www.ncbi.nlm.nih.gov/pubmed/5098954.

[30] Pelloux Y, Everitt BJ, Dickinson A. Compulsive drug seeking by rats under punishment: effects of drug taking history. Psychopharmacology 2007;194(1):127–37. https://doi.org/10.1007/s00213-007-0805-0.

[31] Torres OV, Jayanthi S, Ladenheim B, McCoy MT, Krasnova IN, Cadet JL. Compulsive methamphetamine taking under punishment is associated with greater cue-induced drug seeking in rats. Behav Brain Res 2017; https://doi.org/10.1016/j.bbr.2017.03.009.

[32] Ahmed SH, & Koob GF. (1998). Transition from moderate to excessive drug intake: change in hedonic set point. Science (New York, N.Y.), 282 (5387), 298–300. http://www.ncbi.nlm.nih.gov/pubmed/9765157.

[33] Beckmann JS, Gipson CD, Marusich Ja, Bardo MT. Escalation of cocaine intake with extended access in rats: dysregulated addiction or regulated acquisition? Psychopharmacology 2012;222(2):257–67. https://doi.org/10.1007/s00213-012-2641-0.

[34] American Psychiatric Association. Diagnostic and statistical manual of mental disorders, fifth edition (DSM-5). 5th ed. Washington, DC: American Psychiatric Publishing; 2013.

[35] Goldstein RZ, Craig ADB, Bechara A, Garavan H, Childress AR, Paulus MP, et al. The neurocircuitry of impaired insight in drug addiction. Trends Cogn Sci 2009;13(9):372–80. https://doi.org/10.1016/j.tics.2009.06.004.

[36] Craig AD (Bud). How do you feel–now? The anterior insula and human awareness. Nat Rev Neurosci, 2009; 10(1): 59–70. https://doi.org/10.1038/ nrn2555

[37] Bevins RA, Besheer J. Interoception and learning: import to understanding and treating diseases and psychopathologies. ACS Chem Neurosci 2014; 5(8):624–31. https://doi.org/10.1021/cn5001028.

[38] Noël X, Brevers D, Bechara A. A triadic neurocognitive approach to addiction for clinical interventions. Front Psych 2013;4:179. https://doi.org/ 10.3389/fpsyt.2013.00179.

[39] Verdejo-Garcia A, Clark L, Dunn BD. The role of interoception in addiction: a critical review. Neurosci Biobehav Rev 2012;36(8):1857–69. https:// doi.org/10.1016/j.neubiorev.2012.05.007.

[40] McClernon FJ, Conklin CA, Kozink RV, Adcock RA, Sweitzer MM, Addicott MA, DeVito AM. Hippocampal and insular response to smoking-related environments: Neuroimaging evidence for drug-context effects in nicotine dependence. Neuropsychopharmacology 2015;41(3):877–85. https://doi.org/10.1038/npp.2015.214.

[41] Dager AD, Anderson BM, Rosen R, Khadka S, Sawyer B, Jiantonio-Kelly RE, et al. Functional magnetic resonance imaging (fMRI) response to alcohol pictures predicts subsequent transition to heavy drinking in college students. Addiction 2014;109(4):585–95. https://doi.org/10.1111/ add.12437.

[42] Stewart JL, Juavinett AL, May AC, Davenport PW, Paulus MP. Do you feel alright? Attenuated neural processing of aversive interoceptive stimuli in current stimulant users. Psychophysiology 2015;52(2):249–62. https://doi.org/10.1111/psyp.12303.

[43] Gaznick N, Tranel D, McNutt A, Bechara A. Basal ganglia plus insula damage yields stronger disruption of smoking addiction than basal ganglia damage alone. Nicot Tob Res 2014;16(4):445–53. https://doi.org/10.1093/ntr/ntt172.

[44] Naqvi NH, Rudrauf D, Damasio H, Bechara A. Damage to the insula disrupts addiction to cigarette smoking. Science (New York, NY) 2007;315 (5811):531–4. https://doi.org/10.1126/science.1135926.

[45] Suñer-Soler R, Grau A, Gras ME, Font-Mayolas S, Silva Y, Dávalos A, et al. Smoking cessation 1 year poststroke and damage to the insular cortex. Stroke 2012;43(1):131–6. https://doi.org/10.1161/STROKEAHA.111.630004.

[46] Abdolahi A, Williams GC, Benesch CG, Wang HZ, Spitzer EM, Scott BE, et al. Damage to the insula leads to decreased nicotine withdrawal during abstinence, Addiction 2015; (Abingdon, England) https://doi.org/10.1111/add.13061.

[47] Rotge J-Y, Cocker PJ, Daniel M-L, Belin-Rauscent A, Everitt BJ, Belin D. Bidirectional regulation over the development and expression of loss of control over cocaine intake by the anterior insula. Psychopharmacology 2017;234(9–10):1623–31. https://doi.org/10.1007/s00213-017-4593-x.

[48] Sughondhabirom A, Jain D, Gueorguieva R, Coric V, Berman R, Lynch WJ, et al. A paradigm to investigate the self-regulation of cocaine administration in humans. Psychopharmacology 2005;180(3):436–46. https://doi.org/10.1007/s00213-005-2192-8.

[49] Woodward M, Tunstall-Pedoe H. Self-titration of nicotine: evidence from the Scottish Heart Health Study. Addiction 1993;88(6):821–30. http://www.ncbi.nlm.nih.gov/pubmed/8329973.

[50] Zimmer BA, Dobrin CV, Roberts DCS. Examination of behavioral strategies regulating cocaine intake in rats. Psychopharmacology 2013;225 (4):935–44. https://doi.org/10.1007/s00213-012-2877-8.

[51] Zittel-Lazarini A, Cador M, Ahmed SH. A critical transition in cocaine self-administration: behavioral and neurobiological implications. Psychopharmacology 2007;192(3):337–46. https://doi.org/10.1007/s00213-007-0724-0.

[52] Tornatzky W, Miczek KA. Cocaine self-administration "binges": transition from behavioral and autonomic regulation toward homeostatic dysregulation in rats. Psychopharmacology 2000;148(3):289–98. http://www.ncbi.nlm.nih.gov/pubmed/10755742.

[53] Zimmer BA, Oleson EB, Roberts DC. The motivation to self-administer is increased after a history of spiking brain levels of cocaine. Neuropsychopharmacology 2012;37(8):1901–10. https://doi.org/10.1038/npp.2012.37.

[54] Pelloux Y, Murray JE, Everitt BJ. Differential roles of the prefrontal cortical subregions and basolateral amygdala in compulsive cocaine seeking and relapse after voluntary abstinence in rats. Eur J Neurosci 2013;38(January):3018–26. https://doi.org/10.1111/ejn.12289.

[55] Pelloux Y, Hoots KJ, Cifani C, Adhikary S, Martin J, Minier-Toribio A, et al. Context-induced relapse to cocaine seeking after punishment-imposed abstinence is associated with activation of cortical and subcortical brain regions. Addict Biol 2017; https://doi.org/10.1111/adb.12527.

[56] Arguello AA, Wang R, Lyons CM, Higginbotham JA, Hodges MA, Fuchs RA. Role of the agranular insular cortex in contextual control over cocaine-seeking behavior in rats. Psychopharmacology 2017;1–11. https://doi.org/10.1007/s00213-017-4632-7.

[57] Forget B, Pushparaj A, Le Foll B. Granular insular cortex inactivation as a novel therapeutic strategy for nicotine addiction. Biol Psychiatry 2010;68 (3):265–71. https://doi.org/10.1016/j.biopsych.2010.01.029.

[58] Kutlu MG, Burke D, Slade S, Hall BJ, Rose JE, Levin ED. Role of insular cortex D_1 and D_2 dopamine receptors in nicotine self-administration in rats. Behav Brain Res 2013;256:273–8. https://doi.org/10.1016/j.bbr.2013.08.005.

[59] Hollander JA, Lu Q, Cameron MD, Kamenecka TM, Kenny PJ. Insular hypocretin transmission regulates nicotine reward. Proc Natl Acad Sci U S A 2008;105(49):19480–5. https://doi.org/10.1073/pnas.0808023105.

[60] Pushparaj A, Hamani C, Yu W, Shin DS, Kang B, Nobrega JN, & Le Foll B. (2013). Electrical stimulation of the insular region attenuates nicotine-taking and nicotine-seeking behaviors. Neuropsychopharmacology, 38(4), 690–8. https://doi.org/10.1038/npp.2012.235

[61] Ferguson SG, Shiffman S. The relevance and treatment of cue-induced cravings in tobacco dependence. J Subst Abus Treat 2009;36(3):235–43. https://doi.org/10.1016/j.jsat.2008.06.005.

[62] Robbins TW, Ersche KD, Everitt BJ. Drug addiction and the memory systems of the brain. Ann N Y Acad Sci 2008;1141:1–21. https://doi.org/10.1196/annals.1441.020.

[63] Skinner MD, Aubin H-J. Craving's place in addiction theory: contributions of the major models. Neurosci Biobehav Rev 2010;34(4):606–23. https://doi.org/10.1016/j.neubiorev.2009.11.024.

[64] Jellinek EM. The disease concept of alcoholism. New Haven, CT: Hillhouse Press; 1960.

[65] Gray MA, Critchley HD. Interoceptive basis to craving. Neuron 2007;54(2):183–6. https://doi.org/10.1016/j.neuron.2007.03.024.

[66] Shiffman S, Dunbar M, Kirchner T, Li X, Tindle H, Anderson S, et al. Smoker reactivity to cues: Effects on craving and on smoking behavior. J Abnorm Psychol 2013;122(1):264–80. https://doi.org/10.1037/a0028339.

[67] Stevenson JG, Oliver JA, Hallyburton MB, Sweitzer MM, Conklin CA, McClernon FJ. Smoking environment cues reduce ability to resist smoking as measured by a delay to smoking task. Addict Behav 2017;67:49–52. https://doi.org/10.1016/j.addbeh.2016.12.007.

[68] Grimm JW, Hope BT, Wise RA, Shaham Y. Incubation of cocaine craving after withdrawal. Nature 2001;412(6843):141–2. https://doi.org/10.1038/35084134.

[69] Begh R, Smith M, Ferguson SG, Shiffman S, Munafò MR, Aveyard P. Association between smoking-related attentional bias and craving measured in the clinic and in the natural environment. Psychol Addict Behav 2016;30(8):868–75. https://doi.org/10.1037/adb0000231.

[70] Sinha R. Modeling stress and drug craving in the laboratory: implications for addiction treatment development. Addict Biol 2009;14(1):84–98. https://doi.org/10.1111/j.1369-1600.2008.00134.x.

[71] Perkins KA, Karelitz JL, Giedgowd GE, Conklin CA. Negative mood effects on craving to smoke in women versus men. Addict Behav 2013;38 (2):1527–31. https://doi.org/10.1016/j.addbeh.2012.06.002.

[72] Bossert JM, Marchant NJ, Calu DJ, Shaham Y. The reinstatement model of drug relapse: recent neurobiological findings, emerging research topics, and translational research. Psychopharmacology 2013;229(3):453–76. https://doi.org/10.1007/s00213-013-3120-y.

[73] Shen H, Gipson CD, Huits M, Kalivas PW. Prelimbic cortex and ventral tegmental area modulate synaptic plasticity differentially in nucleus Accumbens during cocaine-reinstated drug seeking. Neuropsychopharmacology 2014;39(5):1169–77. https://doi.org/10.1038/npp.2013.318.

[74] Rankin H, Hodgson R, Stockwell T. Cue exposure and response prevention with alcoholics: a controlled trial. Behav Res Ther 1983;21(4):435–46.

[75] Sitharthan T, Sitharthan G, Hough MJ, Kavanagh DJ. Cue exposure in moderation drinking: a comparison with cognitive-behavior therapy. J Consult Clin Psychol 1997;65(5):878–82. https://doi.org/10.1037//0022-006X.65.5.878.

[76] Wang B, You Z-B, Oleson EB, Cheer JF, Myal S, Wise RA. Conditioned contribution of peripheral cocaine actions to cocaine reward and cocaine-seeking. Neuropsychopharmacology 2013;38(9):1763–9. https://doi.org/10.1038/npp.2013.75.

[77] Wise RA, Wang B, You Z-B. Cocaine serves as a peripheral interoceptive conditioned stimulus for central glutamate and dopamine release. PLoS One 2008;3(8):e2846. https://doi.org/10.1371/journal.pone.0002846.

[78] Colpaert F. Drug discrimination in neurobiology. Pharmacol Biochem Behav 1999;64(2):337–45. https://doi.org/10.1016/S0091-3057(99)00047-7.

[79] Overton DA, Batta SK. Investigation of narcotics and antitussives using drug discrimination techniques. J Pharmacol Exp Ther 1979;211 (2):401–8.

[80] Stolerman IP, Childs E, Ford MM, Grant Ka. Role of training dose in drug discrimination: a review. Behav Pharmacol 2011;22(5–6):415–29. https://doi.org/10.1097/FBP.0b013e328349ab37.

[81] Wooters TE, Bevins RA, Bardo MT. Neuropharmacology of the interoceptive stimulus properties of nicotine. Curr Drug Abuse Rev 2009;2 (3):243–55. http://www.ncbi.nlm.nih.gov/pubmed/20443771.

[82] Young R. Drug discrimination. Methods of Behavior Analysis in Neuroscience. CRC Press/Taylor & Francis; 2009.

[83] Overton DA. Drug discrimination training with progressively lowered doses. Science (New York, NY) 1979;205(4407):720–1. https://doi.org/10.1126/science.462182.

[84] Schassburger RL, Levin ME, Weaver MT, Palmatier MI, Caggiula AR, Donny EC, et al. Differentiating the primary reinforcing and reinforcement-enhancing effects of varenicline. Psychopharmacology 2015;232(5):975–83. https://doi.org/10.1007/s00213-014-3732-x.

[85] Jones HE, Bigelow GE, Preston KL. Assessment of opioid partial agonist activity with a three-choice hydromorphone dose-discrimination procedure. J Pharmacol Exp Ther 1999;289(3):1350–61. https://www.ncbi.nlm.nih.gov/pubmed/10336526.

[86] Sannerud CA, Ator NA. Drug discrimination analysis of midazolam under a three-lever procedure: I. Dose-dependent differences in generalization and antagonism. J Pharmacol Exp Ther 1995;272(1):100–11. https://www.ncbi.nlm.nih.gov/pubmed/7815322.

[87] McGaughy J, Sarter M. Behavioral vigilance in rats: Task validation and effects of age, amphetamine, and benzodiazepine receptor ligands. Psychopharmacology 1995;117(3):340–57. https://www.ncbi.nlm.nih.gov/pubmed/7770610.

[88] Kantak KM, Riberdy A, Spealman RD. Cocaine-opioid interactions in groups of rats trained to discriminate different doses of cocaine. Psychopharmacology 1999;147(3):257–65. http://www.ncbi.nlm.nih.gov/pubmed/10639683.

[89] Stadler JR, Caul WF, Barrett RJ. Effects of training dose on amphetamine drug discrimination: dose-response functions and generalization to cocaine. Pharmacol Biochem Behav 2001;70(2–3):381–6. http://www.ncbi.nlm.nih.gov/pubmed/11701211.

[90] Lamb RJ, Henningfield JE. Human d-amphetamine drug discrimination: methamphetamine and hydromorphone. J Exp Anal Behav 1994;61 (2):169–80. https://doi.org/10.1901/jeab.1994.61-169.

[91] Holtzman SG, Locke KW. Neural mechanisms of drug stimuli: experimental approaches. Psychopharmacol Ser 1988;4:138–53. http://www.ncbi.nlm.nih.gov/pubmed/3293038.

[92] Shoaib M, Stolerman IP. Brain sites mediating the discriminative stimulus effects of nicotine in rats. Behav Brain Res 1996;78:183–8. https://www.ncbi.nlm.nih.gov/pubmed/8864050.

[93] Meltzer LT, Rosecrans JA. Investigations on the CNS sites of action of the discriminative stimulus effects of arecoline and nicotine. Pharmacol Biochem Behav 1981;15(1):21–6. http://www.ncbi.nlm.nih.gov/pubmed/7291225.

[94] Miyata H, Ando K, Yanagita T. Brain regions mediating the discriminative stimulus effects of nicotine in rats. Ann N Y Acad Sci 2006;965 (1):354–63. https://doi.org/10.1111/j.1749-6632.2002.tb04177.x.

[95] Miyata H, Ando K, Yanagita T. Medial prefrontal cortex is involved in the discriminative stimulus effects of nicotine in rats. Psychopharmacology 1999;145(2):234–6. https://doi.org/10.1007/s002130051054.

[96] Popescu AT, Zhou MR, Poo M. Phasic dopamine release in the medial prefrontal cortex enhances stimulus discrimination. Proc Natl Acad Sci 2016;113(22):E3169–76. https://doi.org/10.1073/pnas.1606098113.

[97] Corrigall WA, Franklin KB, Coen KM, Clarke PB. The mesolimbic dopaminergic system is implicated in the reinforcing effects of nicotine. Psychopharmacology 1992;107(2–3):285–9. http://www.ncbi.nlm.nih.gov/pubmed/1615127.

[98] Pich EM, Pagliusi SR, Tessari M, Talabot-Ayer D, Hooft van Huijsduijnen R, Chiamulera C. Common neural substrates for the addictive properties of nicotine and cocaine. Science (New York, NY) 1997;275(5296):83–6. http://www.ncbi.nlm.nih.gov/pubmed/8974398.

[99] Bevins RA, Murray JE. Internal stimuli generated by abused substances: role of Pavlovian conditioning and its implications for drug addiction. In: Schachtman T, Reilly S, editors. Associative learning and conditioning: Human and non-human applications. New York, NY: Oxford University Press; 2011. p. 270–89.

[100] Bevins RA, Palmatier MI. Extending the role of associative learning processes in nicotine addiction. Behav Cogn Neurosci Rev 2004;3(3):143–58. https://doi.org/10.1177/1534582304272005.

[101] Besheer J, Fisher KR, Durant B. Assessment of the interoceptive effects of alcohol in rats using short-term training procedures. Alcohol 2012;46 (8):747–55. https://doi.org/10.1016/j.alcohol.2012.08.003.

[102] Palmatier MI, Wilkinson JL, Metschke DM, Bevins RA. Stimulus properties of nicotine, amphetamine, and Chlordiazepoxide as positive features in a Pavlovian appetitive discrimination task in rats. Neuropsychopharmacology 2004;731–41. https://doi.org/10.1038/sj.npp.1300629.

[103] Reichel CM, Wilkinson JL, Bevins RA. Methamphetamine functions as a positive and negative drug feature in a Pavlovian appetitive discrimination task. Behav Pharmacol 2007;18(8):755–65. https://doi.org/10.1097/FBP.0b013e3282f14efc.

[104] Wilkinson JL, Li C, Bevins RA. Pavlovian drug discrimination with bupropion as a feature positive occasion setter: substitution by methamphetamine and nicotine, but not cocaine. Addict Biol 2009;14(2):165–73. https://doi.org/10.1111/j.1369-1600.2008.00141.x.

[105] Bevins RA, Wilkinson JL, Palmatier MI, Siebert HL, Wiltgen SM. Characterization of nicotine's ability to serve as a negative feature in a Pavlovian appetitive conditioning task in rats. Psychopharmacology 2006;184(3–4):470–81. https://doi.org/10.1007/s00213-005-0079-3.

[106] Murray JE, Wells NR, Bevins RA. Nicotine competes with a visual stimulus for control of conditioned responding. Addict Biol 2011;16(1):152–62. https://doi.org/10.1111/j.1369-1600.2010.00228.x.

[107] Rescorla RA. Pavlovian conditioned inhibition. Psychol Bull 1969;72(2):77–94. https://doi.org/10.1037/h0027760.

[108] Palmatier MI, Bevins RA. Occasion setting by drug states: functional equivalence following similar training history. Behav Brain Res 2008;195 (2):260–70. https://doi.org/10.1016/j.bbr.2008.09.009.

[109] Murray JE, Polewan RJ, Bevins RA. Rethinking the nicotine stimulus. In: Murray JE, editor. Exposure therapy: New developments. Hauppauge, NY: Nova Science Publishers; 2012. p. 95–120.

[110] Murray JE, Bevins RA. The conditional stimulus effects of nicotine vary as a function of training dose. Behav Pharmacol 2007;18(8):707–16. http://www.ncbi.nlm.nih.gov/pubmed/17989508.

[111] Wilkinson JL, Murray JE, Li C, Wiltgen SM, Penrod RD, Berg SA, et al. Interoceptive Pavlovian conditioning with nicotine as the conditional stimulus varies as a function of the number of conditioning trials and unpaired sucrose deliveries, Behav Pharmacol 2006;17(2):161–72. http://www.ncbi.nlm.nih.gov/pubmed/16495724.

[112] Farwell BJ, Ayres JJ. Stimulus-reinforcer and response-reinforcer relations in the control of conditioned appetitive headpoking ("goal tracking") in rats. Learn Motiv 1979;10(3):295–312. https://doi.org/10.1016/0023-9690(79)90035-3.

[113] Besheer J, Palmatier MI, Metschke DM, Bevins RA. Nicotine as a signal for the presence or absence of sucrose reward: a Pavlovian drug appetitive conditioning preparation in rats. Psychopharmacology 2004;172(1):108–17. https://doi.org/10.1007/s00213-003-1621-9.

[114] Murray JE, Bevins RA. Behavioral and neuropharmacological characterization of nicotine as a conditional stimulus. Eur J Pharmacol 2007;561 (1):91–104. http://www.ncbi.nlm.nih.gov/pubmed/17343849.

[115] Murray JE, Penrod RD, Bevins Ra. Nicotine-evoked conditioned responding is dependent on concentration of sucrose unconditioned stimulus. Behav Process 2009;81(1):136–9. https://doi.org/10.1016/j.beproc.2009.01.002.

[116] Pittenger ST, Bevins RA. Interoceptive conditioning with a nicotine stimulus is susceptible to reinforcer devaluation. Behav Neurosci 2013;127 (3):465–73. https://doi.org/10.1037/a0032691.

[117] Bevins RA, Penrod RD, Reichel CM. Nicotine does not produce state-dependent effects on learning in a Pavlovian appetitive goal tracking task with rats. Behav Brain Res 2007;177(1):134–41. https://doi.org/10.1016/j.bbr.2006.10.026.

[118] Murray JE, Walker AW, Polewan RJ, Bevins RA. An examination of NMDA receptor contribution to conditioned responding evoked by the conditional stimulus effects of nicotine. Psychopharmacology 2011;213(1):131–41. https://doi.org/10.1007/s00213-010-2022-5.

[119] Charntikov S, Tracy ME, Zhao C, Li M, Bevins RA. Conditioned response evoked by nicotine conditioned stimulus preferentially induces c-Fos expression in medial regions of caudate-putamen. Neuropsychopharmacology 2012;37(4):876–84. https://doi.org/10.1038/npp.2011.263.

[120] Charntikov S, Pittenger ST, Swalve N, Li M, Bevins RA. Double dissociation of the anterior and posterior dorsomedial caudate-putamen in the acquisition and expression of associative learning with the nicotine stimulus. Neuropharmacology 2017;121:111–9. https://doi.org/10.1016/j. neuropharm.2017.04.026.

[121] Murray JE, Bevins RA. Acquired appetitive responding to intravenous nicotine reflects a Pavlovian conditioned association. Behav Neurosci 2009;123(1):97–108. https://doi.org/10.1037/a0013735.

[122] Murray JE, Bevins RA. Excitatory conditioning to the interoceptive nicotine stimulus blocks subsequent conditioning to an exteroceptive light stimulus. Behav Brain Res 2011;221(1):314–9. https://doi.org/10.1016/j.bbr.2011.03.020.

[123] Chiamulera C, Borgo C, Falchetto S, Valerio E, Tessari M. (1996). Nicotine reinstatement of nicotine self-administration after long-term extinction. Psychopharmacology, 127(1–2), 102–107. https://doi.org/10.1007/BF02805981.

[124] LeSage MG, Burroughs D, Dufek M, Keyler DE, Pentel PR. Reinstatement of nicotine self-administration in rats by presentation of nicotine-paired stimuli, but not nicotine priming. Pharmacol Biochem Behav 2004;79(3):507–13. https://doi.org/10.1016/j.pbb.2004.09.002.

[125] Liu X, Caggiula AR, Palmatier MI, Donny EC, Sved AF. Cue-induced reinstatement of nicotine-seeking behavior in rats: effect of bupropion, persistence over repeated tests, and its dependence on training dose. Psychopharmacology 2008;196(3):365–75. https://doi.org/10.1007/s00213-007-0967-9.

[126] Caggiula AR, Donny EC, White AR, Chaudhri N, Booth S, Gharib Ma, et al. Environmental stimuli promote the acquisition of nicotine self-administration in rats. Psychopharmacology 2002;163(2):230–7. https://doi.org/10.1007/s00213-002-1156-5.

[127] Caggiula AR, Donny EC, White AR, Chaudhri N, Booth S, Gharib Ma, et al. Cue dependency of nicotine self-administration and smoking. Pharmacol Biochem Behav 2001;70(4):515–30. https://doi.org/10.1016/S0091-3057(01)00676-1.

[128] Rose JE, Salley A, Behm FM, Bates JE, Westman EC. Reinforcing effects of nicotine and non-nicotine components of cigarette smoke. Psychopharmacology 2010;210(1):1–12. https://doi.org/10.1007/s00213-010-1810-2.

[129] Barret ST, Bevins RA. Nicotine enhances operant responding for qualitatively distinct reinforcers under maintenance and extinction conditions. Pharmacol Biochem Behav 2013;114–115:9–15. http://www.ncbi.nlm.nih.gov/pubmed/24422211.

[130] Donny EC, Chaudhri N, Caggiula AR, Evans-Martin FF, Booth S, Gharib MA, et al. Operant responding for a visual reinforcer in rats is enhanced by noncontingent nicotine: implications for nicotine self-administration and reinforcement. Psychopharmacology 2003;169(1):68–76. https://doi.org/10.1007/s00213-003-1473-3.

[131] Guy EG, Fletcher PJ. The effects of nicotine exposure during Pavlovian conditioning in rats on several measures of incentive motivation for a conditioned stimulus paired with water. Psychopharmacology 2014;231(11):2261–71. https://doi.org/10.1007/s00213-013-3375-3.

[132] Perkins KA, Karelitz JL, Boldry MC. Nicotine acutely enhances reinforcement from non-drug rewards in humans. Front Psych 2017;8:65. https://doi.org/10.3389/fpsyt.2017.00065.

[133] Alessi SM, Roll JM, Reilly MP, Johanson C-E. Establishment of a diazepam preference in human volunteers following a differential-conditioning history of placebo versus diazepam choice. Exp Clin Psychopharmacol 2002;10(2):77–83. http://www.ncbi.nlm.nih.gov/pubmed/12022801.

[134] Hyde TS. The effect of Pavlovian stimuli on the acquisition of a new response. Learn Motiv 1976;7(2):223–39. https://doi.org/10.1016/0023-9690(76)90030-8.

[135] Di Ciano P, Everitt BJ. Conditioned reinforcing properties of stimuli paired with self-administered cocaine, heroin or sucrose: implications for the persistence of addictive behaviour. Neuropharmacology 2004;47(Suppl 1):202–13. https://doi.org/10.1016/j.neuropharm.2004.06.005.

[136] Flavell CR, Barber DJ, Lee JLC. Behavioural memory reconsolidation of food and fear memories. Nat Commun 2011;2:504. https://doi.org/10.1038/ncomms1515.

[137] Lee JLC, Di Ciano P, Thomas KL, Everitt BJ. Disrupting reconsolidation of drug memories reduces cocaine-seeking behavior. Neuron 2005;47(6):795–801. https://doi.org/10.1016/j.neuron.2005.08.007.

[138] Charntikov S, Pittenger S, Swalve N, Bevins RA. In: The effect of appetitive interoceptive conditioning on nicotine reinforcement: a novel nicotine self-administration model. 44th Annual Society for Neuroscience Conference (p. 231.07/V9). Washington, DC; 2014.

[139] Murray JE, Walker AW, Li C, Wells NR, Penrod RD, Bevins Ra. Nicotine trained as a negative feature passes the retardation-of-acquisition and summation tests of a conditioned inhibitor. Learn Memory (Cold Spring Harbor, NY) 2011;18(7):452–8. https://doi.org/10.1101/lm.2177411.

[140] Conklin CA, Tiffany ST. Applying extinction research and theory to cue-exposure addiction treatments. Addiction 2002;97(2):155–67. (Abingdon, England). http://www.ncbi.nlm.nih.gov/pubmed/11860387.

[141] Kaplan GB, Heinrichs SC, Carey RJ. Treatment of addiction and anxiety using extinction approaches: neural mechanisms and their treatment implications. Pharmacol Biochem Behav 2011;97(3):619–25. https://doi.org/10.1016/j.pbb.2010.08.004.

[142] Niaura R, Abrams DB, Shadel WG, Rohsenow DJ, Monti PM, Sirota AD. Cue exposure treatment for smoking relapse prevention: a controlled clinical trial. Addiction 1999;94(5):685–95. https://doi.org/10.1046/j.1360-0443.1999.9456856.x.

[143] Choy Y, Fyer AJ, Lipsitz JD. Treatment of specific phobia in adults. Clin Psychol Rev 2007;27(3):266–86. https://doi.org/10.1016/j.cpr.2006.10.002.

[144] Conklin CA, Perkins KA, Robin N, McClernon FJ, Salkeld RP. Bringing the real world into the laboratory: Personal smoking and nonsmoking environments. Drug Alcohol Depend 2010;111(1–2):58–63. https://doi.org/10.1016/j.drugalcdep.2010.03.017.

[145] Conklin CA. Environments as cues to smoke: Implications for human extinction-based research and treatment. Exp Clin Psychopharmacol 2006;14(1):12–9. https://doi.org/10.1037/1064-1297.14.1.12.

[146] Erblich J, Montgomery GH, Bovbjerg DH. Script-guided imagery of social drinking induces both alcohol and cigarette craving in a sample of nicotine-dependent smokers. Addict Behav 2009;34(2):164–70. https://doi.org/10.1016/j.addbeh.2008.10.007.

[147] Taylor JR, Olausson P, Quinn JJ, Torregrossa MM. Targeting extinction and reconsolidation mechanisms to combat the impact of drug cues on addiction. Neuropharmacology 2009;56(Suppl 1):186–95. https://doi.org/10.1016/j.neuropharm.2008.07.027.

[148] Hone-Blanchet A, Wensing T, Fecteau S. The use of virtual reality in craving assessment and cue-exposure therapy in substance use disorders. Front Hum Neurosci 2014;8:844. https://doi.org/10.3389/fnhum.2014.00844.

[149] Mellentin AI, Stenager E, Nielsen B, Nielsen AS, Yu F. A smarter pathway for delivering cue exposure therapy? The design and development of a smartphone app targeting alcohol use disorder. JMIR mHealth uHealth 2017;5(1):e5. https://doi.org/10.2196/mhealth.6500.

[150] Shiffman S, Li X, Dunbar MS, Tindle HA, Scholl SM, Ferguson SG. Does laboratory cue reactivity correlate with real-world craving and smoking responses to cues? Drug Alcohol Depend 2015;155:163–9. https://doi.org/10.1016/j.drugalcdep.2015.07.673.

[151] Amato L, Davoli M, Perucci CA, Ferri M, Faggiano F, Mattick RP. An overview of systematic reviews of the effectiveness of opiate maintenance therapies: available evidence to inform clinical practice and research. J Subst Abus Treat 2005;28(4):321–9. https://doi.org/10.1016/j.jsat.2005.02.007.

[152] Cahill K, Stevens S, Perera R, Lancaster T. Pharmacological interventions for smoking cessation: an overview and network meta-analysis. In: Cahill K, editor. Cochrane database of systematic reviews. Chichester, UK: John Wiley & Sons, Ltd.; 2013. p. CD009329. https://doi.org/10.1002/14651858.CD009329.pub2

[153] Hoffmann E, Davis AK, Ashrafioun L, Kraus SW, Rosenberg H, Bannon EE, et al. Evaluation of the criterion and predictive validity of the alcohol reduction strategies—Current confidence (ARS-CC) in a natural drinking environment. Addict Behav 2013;38(4):1940–3. https://doi.org/10.1016/j.addbeh.2012.12.021.

[154] Kuerbis A, Armeli S, Muench F, Morgenstern J. Motivation and self-efficacy in the context of moderated drinking: global self-report and ecological momentary assessment. Psychol Addict Behav 2013;27(4):934–43. https://doi.org/10.1037/a0031194.

[155] Kuerbis A, Armeli S, Muench F, Morgenstern J. Profiles of confidence and commitment to change as predictors of moderated drinking: a person-centered approach. Psychol Addict Behav 2014;28(4):1065–76. https://doi.org/10.1037/a0036812.

Chapter 8

Maladaptive Memory Mechanisms in Addiction and Relapse

Matthew T. Rich*,†,‡ and Mary M. Torregrossa*,†

*Department of Psychiatry, University of Pittsburgh, Pittsburgh, PA, United States, †Center for Neuroscience, University of Pittsburgh, Pittsburgh, PA, United States, ‡Center for the Neural Basis of Cognition, University of Pittsburgh, Pittsburgh, PA, United States

INTRODUCTION

Addiction is described as a repetitive sequence of behavior, consisting of periods of drug use, withdrawal, abstinence, and relapse. A defining characteristic of the disorder is that drug use continues despite the occurrence of negative consequences. Addicted individuals often struggle with the dilemma of either engaging in drug-seeking behavior, which has immediate short-term rewarding effects, but also the associated negative consequences, or maintaining abstinence, which has more long-term benefits. The National Household Survey on Drug Abuse by the National Institute of Health estimate relapse rates to be between 40% and 60%. This high tendency for relapse, sometimes even after extensive periods of abstinence, is a challenge for the successful treatment of addiction [1]. One explanation for high relapse rates is that chronic substance abuse induces long-lasting synaptic alterations within neural circuits regulating reward-associated memories, which can make avoiding thoughts and actions directed toward drug use extremely difficult. Indeed, it has been proposed that there is a progressive increase in craving during early abstinence, and that craving stays elevated thereafter, suggesting that memories or thoughts of drug use do not dissipate with time, potentially explaining the greater risk of relapse with increasing abstinence [2–4].

During drug-free abstinent periods, individuals often experience reexposure to the addictive substance, periods of stress, and/or interactions with drug-associated environmental stimuli (See [5,6]). These relapse triggers act on various, often overlapping, pathways in the brain, and are controlled by a diverse set of neural mechanisms. Specific drug-associated stimuli can vary widely from person to person, but can include friends and family members with whom the individual engaged in drug-related behaviors, a specific context or location in which drugs were formerly used, and drug paraphernalia [7–9]. Broadly there are two main categories of drug-associated stimuli studied in preclinical models of addiction: contextual and discrete cues. They each have distinct anatomical and molecular correlates and so, will be considered separately throughout this chapter.

Neural systems that govern reward, learning, memory, and executive function are particularly sensitive to drugs of abuse [10,11]. Thus, memories associated with drug use are thought to become stronger than those associated with natural rewards. Chronic drug use thus continually hijacks this circuitry further driving continued and increasing drug use. This escalation culminates in the compulsive behaviors that are characteristic of addiction [12]. Uncontrolled drug-seeking behavior increases the likelihood of relapse, making addiction extremely difficult to overcome. The modification of synaptic architecture by drugs of abuse has been directly linked to relapse-like behavior [13–15].

Understanding the neurobiological mechanisms that regulate the ability of memories to drive relapse has the potential to inform treatments that extend and maintain abstinence. In this chapter, we will briefly discuss animal models (see also Chapter 2) that have been utilized to determine the effects of drug-associated stimuli on relapse behavior. In this context, we will provide an update on the latest neuroscience research investigating the neuroanatomical circuitry that is altered following chronic drug use with a focus on the specific components and molecular mechanisms that are thought to underlie drug-associated memory. Finally, we will comment on recent translational attempts to combine behavioral therapeutic approaches with pharmacological interventions with the goal of enhancing the efficacy of memory-based relapse prevention treatments.

Neural Mechanisms of Addiction. https://doi.org/10.1016/B978-0-12-812202-0.00008-7
Copyright © 2019 Elsevier Inc. All rights reserved.

ANIMAL MODELS OF RELAPSE

Many paradigms have been developed to study drug-seeking behavior in a laboratory setting, but perhaps the most relevant preclinical model for studying the mechanisms by which memories drive drug seeking and relapse is the Reinstatement Model of Drug Seeking [16,17]. The reinstatement model is useful for several reasons. First, it can be utilized for multiple categories of drugs of abuse including stimulants (cocaine, amphetamine), opioids (heroin, morphine), nicotine, alcohol, and most recently cannabinoids [18–20,37]. Second, reinstatement can be reliably triggered by drug-associated stimuli as well as by stress and drug reexposure and enables measurement of the subject's behavioral and neural responses [21–25]. Third, this paradigm mirrors patterns of human drug use, in which the subject learns to administer a drug under their own volition (as opposed to experimenter-administered drugs) and after a period of abstinence, episodes of relapse are reliably observed [26].

The reinstatement model consists of several distinct phases. During the first stage, animals learn to make an operant response (i.e., nose poke or lever press) to self-administer a drug in the presence of environmental stimuli (cues) that become associated with the interoceptive effects of the drug. Next, the operant response is extinguished by allowing animals to make nonreinforced responses. Over time, the animals adapt their behavior to cease performing the operant response, as it no longer results in a rewarding drug infusion, or the cues associated with the drug. Finally, the now extinguished response is reinstated. In cue-induced reinstatement, the cues previously presented with drug infusion are again presented when the animal makes a response though no drug is available. A slight variation of cue-induced reinstatement involves performing extinction in a new context, and then returning the rat to the original drug-taking context, to determine how much the drug-taking environment itself contributes to relapse [27]. Reinstatement can also be triggered by stressful stimuli, which likely engages memories of drug-induced stress reduction, and by acute reexposure to the drug of abuse, which reactivates memories related to the interoceptive effects of the drug. Typically, the number of operant responses is much higher than that under extinction conditions, thus giving a quantitative measure of how much the rat is seeking out the drug relative to normal, exploration-based responding. Similar behavioral strategies are often used clinically in drug-dependent individuals to measure craving and relapse-like behavior, making the reinstatement paradigm a valid translational model to study relapse [28,29].

Another variant of the reinstatement model is referred to as the "incubation of craving" model [16,30], which involves drug self-administration followed by a given number of withdrawal days (typically ≥21 days), during which time the animals do not have access to the self-administration chambers or the drug-associated cues. After withdrawal, a typical cue-induced reinstatement test is performed. These studies have found that increasing the withdrawal period results in increased reinstatement relative to that observed after just 1 day of withdrawal [16]. Incubation of craving has been extensively characterized following withdrawal from cocaine [31,32], methamphetamine [33,34], heroin [35], nicotine [36], and for the cannabinoid receptor agonist, WIN55,212-2 [37].

Recent studies have utilized the reinstatement and incubation models in combination with the latest neuroscience techniques to shed light on the underlying anatomical circuitry and neural mechanisms that control the ability of memories to drive relapse, which will be discussed in the following sections.

NEUROANATOMICAL BASIS FOR MALADAPTIVE MEMORY AND RELAPSE

Neuroanatomical Basis of Reinstatement

The mesocorticolimbic system is a series of connected brain regions that guides the seeking of naturally occurring rewards that are important for survival. The circuit is also activated by pleasurable and reinforcing substances (i.e., drugs of abuse). By serving as a mediator of reward-motivated behaviors, the mesocorticolimbic system plays an important role in drug-seeking and relapse-like behavior. Midbrain dopamine (DA) neurons located in the ventral tegmental area (VTA) send projections to cortical and subcortical limbic structures, including the prefrontal cortex (PFC), anterior cingulate cortex (ACC), nucleus accumbens (NAc), hippocampus, hypothalamus, and amygdala [38–40]. These structures are coupled to each other by dense excitatory and inhibitory projections, forming a complex, interconnected network that functions to control adaptive behaviors [41–43].

Drugs of abuse, as with naturally occurring rewards, increase the firing rate of dopaminergic neurons, resulting in locally elevated levels of DA within limbic structures [44,45]. Drug-rewards increase DA levels more than physiological rewards, which may be responsible for the strong reinforcing effects of addictive substances and the ability of drugs of abuse to produce pathologically strong memories [46,47]. Indeed, human imaging studies have revealed that drug-induced increases in DA release are accompanied by a subjective feeling of euphoria [48], which creates an expectation for future

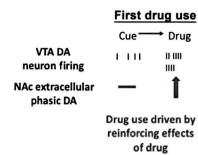

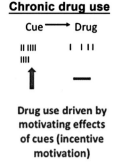

First drug use

Cue ⟶ Drug

Chronic drug use

Cue ⟶ Drug

VTA DA neuron firing

NAc extracellular phasic DA

Drug use driven by reinforcing effects of drug

Drug use driven by motivating effects of cues (incentive motivation)

FIG. 1 Illustration of adaptations in the mesolimbic dopamine system that occurs with chronic pairing of environmental cues with drug reward. Upon first exposure to drug, any environmental cues present have yet to be associated with the feelings of drug reward. As such, VTA DA neurons increase firing in response to the drug, not the cue, and this is associated with increased phasic DA release in the NAc. After chronic use of drugs paired with environmental cues, the cues themselves are able to elicit VTA DA neuron firing and increased phasic DA in the NAc, and less of a response occurs to the drug itself. This process of learning that the cue predicts drug can lead to strong brain activation and desire for drug (craving/relapse) when these cues are later encountered in the environment, even if the individual is attempting to maintain abstinence.

use, causing DA neurons to fire in anticipation of the reward [49]. Environmental contexts or stimuli associated with drug use can also become conditioned by the drug-induced release of DA. The stimuli become predictive of drug use, resulting in DA release in response to subsequent presentations of the stimuli, even if the drug itself is withheld [47,50]. Current research suggests that drug-paired stimuli enhance excitatory transmission onto VTA DA neurons, thus driving the corresponding increase in DA release (Fig. 1; [51]).

In addition, DA modulates glutamatergic and GABAergic synaptic activity within cortical and subcortical structures [52,53]. Long-lasting synaptic changes have been observed in regions, such as the amygdala, NAc, and PFC that ultimately encode drug-associated memories and drive drug-seeking behavior and relapse [38,54–57]. The remainder of this section will discuss the specific anatomical correlates that underlie drug-seeking behavior, and how maladaptive memories influence the activity of these circuits.

Acute drug infusions and reexposure to drug-associated stimuli induce Fos expression in several cortical and limbic brain regions [58–60], including the hippocampus, which is involved in contextual memory [61] and associative learning [62,63]. Indeed, both the dorsal hippocampus (DH) and ventral hippocampus (VH) have been associated with context-induced reinstatement [64,65].

In addition, Fos expression is observed in the amygdala, in particular the basolateral amygdala (BLA), which regulates emotionally salient memories, including those related to fear and drugs and is particularly involved in cue- and context-induced reinstatement [66,67]. The amygdala also has a series of reciprocal connections to areas of the PFC, including the orbitofrontal cortex (OFC), dorsomedial PFC (dmPFC, which consists of the prelimbic cortex [PL] and regions of the ACC), and ventromedial PFC (vmPFC), which includes the infralimbic cortex (IL). Specific interaction between the OFC and BLA, either through direct connections, or a common downstream target, is critical for context-induced reinstatement [68,69]. Furthermore, the BLA sends direct projections to the NAc, which is a critical regulator of reward-related behavior (discussed further below). Previous research has shown that optogenetic inhibition of BLA projections to either the NAc core or to the PL PFC reduces cue-induced reinstatement of cocaine seeking [70]. Similarly, ablation of NAc projecting BLA neurons via retrograde virus-based expression of a diphtheria toxin receptor showed attenuation of alcohol cue-induced reinstatement [71]. However, the effect was not as pronounced as for cocaine, which may be due to differences in the method, or could indicate that memories associated with different substances engage different circuits. Nevertheless, these studies support the idea that the BLA encodes memories associated with drug use that then drives downstream brain structures responsible for initiating drug seeking (Fig. 2; [72–74]).

Further evidence of a role of the NAc in regulating the output of maladaptive memories in addiction comes from studies showing that Fos expression is increased in the NAc shell following reinstatement of sucrose-, cocaine-, and alcohol-seeking [59,75,76] and that both the NAc core and shell are involved in context-, cue-, and stress-induced reinstatement [77–82]. Yet, there has been some controversy over the precise role of the NAc shell vs. NAc core in the regulation of drug seeking. For years, the prevailing belief was that the NAc core, via inputs from the dmPFC, promotes drug seeking, while the NAc shell, via inputs from the vmPFC, limits drug seeking [83]. However, anatomical studies demonstrate that there is not a clear dichotomy in these corticostriatal pathways, as there are also strong connections from vmPFC to NAc core and from dmPFC to NAc shell [84,85]. Indeed, recent behavioral evidence suggests that *both* the vmPFC-NAc and dmPFC-NAc pathways can promote relapse, and that some of the dichotomies observed could be due to the class of drug studied (e.g., stimulant vs. opiate). For example, activation of glutamatergic projections from vmPFC to NAc shell can induce context-induced reinstatement of heroin seeking [86], and inhibition of the vmPFC prevents cue- and heroin-primed reinstatement of heroin seeking [87]. In contrast, inhibiting the vmPFC promotes reinstatement of cocaine seeking [88].

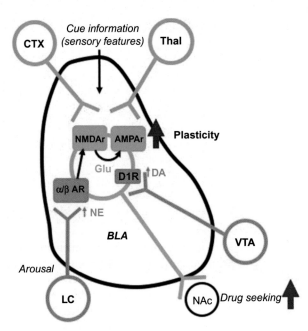

FIG. 2 Diagram of inputs and outputs to the BLA that regulates the formation of drug-cue associative memories. Sensory information (e.g., audio and visual) is conveyed to the BLA from sensory cortices (CTX) and thalamus (Thal). These glutamatergic projections converge on BLA principle neurons that also receive arousal information via locus coeruleus (LC), norepinephrine (NE), signaling onto adrenergic receptors (ARs), and motivational information from VTA DA neurons. These neuromodulatory inputs can enhance glutamatergic plasticity on BLA neurons that project to the NAc to drive cue-elicited drug-seeking behavior.

Therefore, the various pharmacological actions of drugs of abuse may induce differential plasticity in these PFC to NAc circuits. In addition, it is possible that the same brain region can both inhibit and activate drug seeking. For example, clusters or ensembles of neurons within the vmPFC (IL) respond both to cues predictive of reward availability to promote reward seeking, and to cues predictive of reward omission to inhibit behavior [89]. Thus, the learning history of the animal regarding drug availability may influence how memories drive drug seeking. Indeed, a recent study used retrograde expression of DREADDs in the IL-NAc shell pathway to show that activation of IL projections to the shell inhibits drug seeking. However, this was only observed if the rats had undergone instrumental extinction (i.e., extinction of lever pressing) prior to the drug-seeking test [90]. This manipulation had no effect in the absence of extinction learning, providing further evidence that the IL-NAc shell circuit can encode memories related to drug availability to influence both the drive to seek and inhibit drug use.

Neuroanatomical Basis of Incubation of Craving

As with reinstatement models, the neuroanatomical substrates important for incubation of craving have been typically identified either by analyzing markers of cellular activity [increased Fos expression or extracellular regulated kinase (ERK) phosphorylation] or by reversible inactivation of specific brain regions (via application of the GABA agonists muscimol and baclofen). Similar to reinstatement, reexposure to drug-associated cues after abstinence has been shown to increase Fos expression throughout the mesocorticolimbic system, including the OFC, ACC, NAc shell and core, VTA, BLA, and CeA [91–93]. In addition, time-dependent increases in ERK phosphorylation were observed in the vmPFC, but not dmPFC [94], and reversible inactivation of the OFC, vmPFC, and CeA has all been shown to prevent withdrawal-dependent increases in drug-seeking behavior [33,34,91,94]. Again, the involvement of the vmPFC in the drive to seek drugs after abstinence further supports the idea that the vmPFC can turn drug seeking on or off depending on whether or not extinction learning occurs, as extinction training is not normally included in incubation of craving studies. Finally, the NAc shell is also critical for the incubation of craving phenomenon. Numerous studies have found that plasticity within the NAc shell is required to observe increased cue-induced drug seeking after protracted abstinence [95,96], and these plasticity mechanisms are discussed in section "Neural Mechanisms Underlying Maladaptive Memory and Relapse".

Neuroanatomical Basis for Maladaptive Habit Memory in Addiction

Another theory of addiction posits that the transition from casual drug use to a substance use disorder involves a temporal shift in the regions of striatum that are activated in response to stimuli associated with the drug of abuse [97,98]. Evidence suggests that the ventral striatum (VS), which includes the NAc, and to an extent, the dorsomedial striatum (DMS) control the initial drug seeking and reinstatement responses discussed above, but with chronic drug use, the dorsolateral striatum (DLS) is preferentially activated [99,100]. For example, it is thought that when an individual is reexposed to a drug-associated context, the DLS, which controls habitual behaviors, is activated, triggering uncontrollable urges, and drug-seeking [101,102]. Consistent with this hypothesis, several studies have shown that a long history of cocaine or alcohol self-administration leads to the development of habit-like response patterns that is dependent on the integrity of the DLS [101,103,104]. In addition, inactivation of the DLS following extended access to cocaine, impairs the ability of cocaine-associated cues to evoke drug seeking [105]. Similarly, blockade of dopamine transmission in the DLS reduces cue-induced cocaine seeking, but only in rats trained under high response requirements [106]. Interestingly, the transition from VS to DLS control over drug seeking may be mediated by a shift in afferent input from the amygdala, with the initial recruitment of the DLS dependent on BLA activity, but the long-term maintenance of habit-like drug seeking recruiting input from the CeA (Fig. 3; [107]). Therefore, the mechanisms by which memories associated with drug use drive craving and relapse-like behavior may evolve with extended drug experience to activate habit-based response strategies.

NEURAL MECHANISMS UNDERLYING MALADAPTIVE MEMORY AND RELAPSE

Developing a pharmacological or behavioral therapy aimed at preventing relapse requires studying the cellular underpinnings that drive drug-seeking behavior. Just as there are unique anatomical correlates for the different phases of addiction and relapse, there are also differences in the neurotransmitters, receptors, and signaling molecules that promote the formation of maladaptive drug-associated memories that drive relapse. In the following sections, we will review research into the neural mechanisms responsible for relapse, while highlighting the contributions of recent technological advances that have led to new insights.

Molecular Mechanisms of Drug-Associated Memory Formation and Maintenance

Repetitive drug use generates powerful learned associations between environmental stimuli and the acute pharmacological and reinforcing properties of the drug [108–110]. Over time, reexposure to the drug and any associated stimuli (discrete cues, environmental context), can elicit conditioned physiological and behavioral responses, such as increased heart rate,

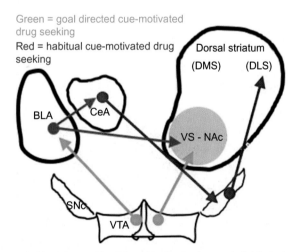

FIG. 3 Circuits driving goal-directed versus habitual drug seeking. Many theorize that in some individuals chronic drug use lead to a loss of control over intake that is compulsive and not mediated by the reinforcing effects of the drug, much like a habit. Animal studies have found that goal-directed, or controlled, drug intake largely involves VTA DA signaling to the BLA NAc (green pathways). On the other hand, conditions that lead to habitual drug seeking are thought to lead to a switch in the circuits that control behavior to one that transition from the BLA to central nucleus of the amygdala (CeA; red pathway). The CeA can then activate dopamine neurons in the substantia nigra pars compacta (SNc) that project to the dorsal lateral striatum (DLS), which can drive habitual drug use.

craving, and drug seeking. The formation, storage, and maintenance of the associations between drugs of abuse and related stimuli are regulated by a series of processes known as consolidation and reconsolidation [111,112]. These processes have been extensively studied for other types of emotionally salient learning, particularly fear-associated memories [67,113,114]. Specifically, consolidation is a time-dependent process that occurs following initial learning and is necessary for the long-term storage of new information. Over time, consolidated memories can be retrieved, which reactivates and destabilizes the memory, causing it to enter a transiently labile state. After reactivation, the memory must then undergo reconsolidation to be restabilized in long-term memory [115,116]. During the labile period of destabilization, the now malleable memories are updated with new information, and potentially strengthened [117].

A multitude of cellular and molecular mechanisms underlie the consolidation, retrieval, and reconsolidation of drug-associated memories. However, due to many days of training required for animals to self-administer drugs of abuse (unlike single trial learning in fear conditioning), surprisingly few studies have directly investigated the molecular mechanisms required for initial drug-associated memory formation. However, it is generally assumed that consolidation of drug-associated memories requires the same molecular substrates that are required for the formation of any type of memory, including glutamate signaling, activation of intracellular signaling molecules including protein kinase A (PKA), Ca2+-calmodulin dependent kinase (CaMKII), cAMP response element binding protein (CREB), and the synthesis of new proteins [118–121].

Researchers have garnered more insight into the mechanisms regulating drug memory maintenance and reconsolidation, because it is easier to perform single manipulations once the memory is formed. These studies usually involve drug self-administration followed by brief presentation of the drug-associated cue or context to reactivate the memory. Memory reactivation can then be followed by manipulation of specific signaling molecules, and the integrity of the memory is then assessed by a subsequent cue-induced reinstatement test. If the memory was weakened by a manipulation, then the amount of cue-induced drug seeking is reduced. Early studies of reconsolidation of fear memories demonstrated that reconsolidation is a protein synthesis-dependent process [122]. This concept has been extended to drug-associated memories, as inhibition of protein synthesis in the BLA prevented the reconsolidation of cocaine-associated cue and contextual memories [112,123]. In addition, Everitt and colleagues determined that activity of the immediate early gene *zif268* as well as activation of NMDA receptors are required for the reconsolidation of drug-cue memories [112,124,125]. Intra-BLA infusions of the NMDA antagonist, D-APV, prior to reconsolidation caused a reduction in subsequent cue-dependent drug-seeking behavior, which correlated with a decrease in *zif268* expression. Furthermore, systemic injections of the NMDA antagonist MK-801 prior to reactivation also reduced cue-reinforced drug seeking [124]. Subsequent studies revealed that knockdown of *zif268* interfered with the reconsolidation of both remote and new memories [125a]. Furthermore, *zif268* knockdown in the BLA, but not the NAc core, impaired the reconsolidation of cocaine-cue memories, while knockdown in both the BLA and NAc core impaired cocaine-context memories [126].

Adrenergic signaling also regulates the reconsolidation of fear- and drug-associated memories [127–129]. Following reactivation of a fear-associated memory, treatment with the α2 agonist, clonidine, disrupts reconsolidation [130]. Meanwhile, systemic treatment with the β-adrenergic antagonist, propranolol, weakens the reinforcing capabilities of cocaine-paired cues [125], and blocks the reconsolidation of cocaine- and morphine-, but not alcohol-associated contextual memories [127,129,131]. One study indicates that propranolol given prior to memory reactivation may prevent the retrieval of memories [132], which is an important consideration for the timeframe of treatment, if propranolol is used as a therapeutic to prevent reconsolidation.

Several intracellular signaling pathways also mediate drug-memory consolidation and reconsolidation. Chronic cocaine exposure persistently increases PKA activity [133,134], and post-retrieval inhibition of PKA in the BLA reduces both cue-and context-induced, but not drug-primed reinstatement [73,74,135]. In addition, activation of a negative regulator of PKA, exchange protein activated by cAMP (Epac), also disrupts cocaine-cue memory reconsolidation [136]. The ERK signaling cascade has also been implicated in the reconsolidation of cocaine-associated contextual memories. Intra-NAc core inhibitors of the ERK kinase, MEK, decreases preference for a cocaine-conditioned context [137], and intra-BLA ERK inhibition following reactivation results in less context-induced reinstatement [138]. Finally, a recent study published by our group used phosphoproteomics to identify signaling cascades regulating reconsolidation of cocaine-cue memories in the BLA. Numerous signaling events were observed, including phosphorylation of ERK2 and protein synthesis regulatory proteins. CaMKIIα was also regulated by reconsolidation, and subsequent behavioral studies confirmed that inhibition of CaMKII in the BLA disrupts reconsolidation of cocaine memory (Fig. 4; [139,140]).

Epigenetic changes, including chromatin remodeling, also occur throughout the brain during the addiction cycle [141]. Modification of chromatin by histone acetylation/deacetylation and other mechanisms are important for the regulation of gene expression. Histone acetylation (HAT) is linked to the activation of genes, while histone deacetylation (HDAC) is associated with gene repression. HDAC inhibitors have been linked to beneficial learning-related neural changes, including increased synaptic plasticity and long-term memory [142]. Additionally, changes in HAT in the mesocorticolimbic system

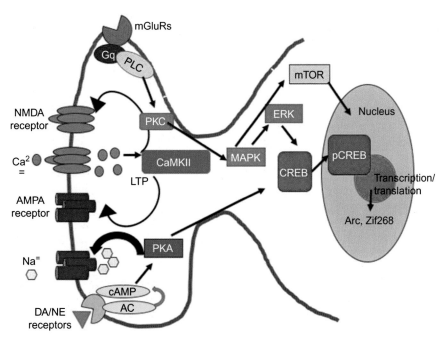

FIG. 4 Intracellular signaling mechanisms regulating the consolidation and reconsolidation of drug-associated memories. Glutamatergic and monoaminergic signaling via AMPA, NMDA, mGluR, and Gs-coupled dopamine and adrenergic receptors activate numerous signaling cascades that are necessary for the formation and maintenance of memory. Studies of the mechanisms underlying the reconsolidation of drug-associated memories have indicated that blocking many of these molecules can disrupt memory for drug-associated cues and contexts. These include NMDA receptors, β-adrenergic receptors, Ca2+-calmodulin-dependent kinase II (CaMKII), cAMP-dependent PKA, extracellular regulated kinases (ERK), protein kinase C (PKC), mammalian target of rapamycin (mTOR), cAMP response element binding protein (CREB), and immediate early genes, such as *zif268*.

(NAc, mPFC, and BLA) have been linked to drug-induced behavior and changes in neuroplasticity [143–145], suggesting that HAT might promote the formation of drug-associated memory. Indeed, HDAC inhibitors have been shown to enhance heroin- and morphine-associated contextual memories [146,147], and increase heroin-primed reinstatement [143]. In support of the hypothesis that HAT is important for the formation and maintenance of drug memories, another series of studies has found that a HAT inhibitor can prevent the reconsolidation of cocaine-associated memory to prevent reinstatement [148,149]. Thus, while there are some conflicting findings in the literature, including one study that found that HDAC inhibition can decrease incubated cocaine seeking [150], memory-associated HAT appears to be an important process for establishing drug-associated memories. In summary, studies of drug-associated memory reconsolidation generally confirm that the molecular mechanisms that regulate normal learning and memory are responsible for the formation of drug-associated memories. The degree to which drug-associated memories are stronger or more stable than other memories is less clearly established though there is evidence that exposure to drugs can enhance normal learning and memory processes that might explain why behaviors directed toward drug-associated stimuli are difficult to disrupt [98,151,152].

Molecular Mechanisms of Incubation of Craving

The observation of the incubation of craving phenomenon, where the degree of cue-induced drug seeking increases with increasing abstinence, suggests that after a drug associated memory is consolidated and stored, the memory can undergo continuous modification even in the absence of exposure to the cues or the drug itself. Incubation of craving has also been detected in human drug users, following withdrawal from cocaine [153], nicotine [154], methamphetamine [155], and alcohol [33,34], further supporting the importance of understanding this phenomenon in order to better treat addiction.

Glutamatergic signaling, specifically in the NAc appears to be heavily involved in the incubation of craving. Following extended withdrawal from cocaine or methamphetamine self-administration, there is an accumulation of GluA2-lacking calcium permeable (CP) AMPA receptors in the NAc [95,156]. Meanwhile, the surface expression of metabotropic glutamate receptors (mGluR1) progressively decreases throughout withdrawal. Inhibiting CP-AMPARs, or enhancing mGluR1 activity by application of a mGluR1 positive allosteric modulator (PAM), which prevents the accumulation of CP-AMPARs, inhibits time-dependent increases in drug seeking that occur with extended withdrawal [32,156]. Likewise, a recent study has found that administration of a novel PAM that targets the mGluR2 receptor can also reduce incubated cue-induced methamphetamine seeking [157].

Another emerging area of research into the neurobiology of relapse has come from Dong and colleagues who discovered that incubation is associated with the loss of silent synapses [immature synapses that express stable NMDA receptors but in which AMPA receptors are either labile or absent [96]] in projections to the NAc. Cocaine self-administration generates silent synapses at BLA, mPFC, and thalamic glutamatergic inputs onto MSNs [31,158,159]. However, with increasing abstinence from cocaine, the silent synapses become "unsilenced," a process involving the insertion of CP-AMPARs [95,96]. Critically, the re-silencing of these synapses, via optogenetic stimulation, downregulates CP-AMPARs, and decreases incubation of cue-induced cocaine seeking [31]. Therefore, the time-dependent "un-silencing" of silent synapses during abstinence is one of the mechanisms responsible for cue-driven drug seeking.

Interestingly, the role of PFC projections to the NAc in withdrawal-dependent silent synapse remodeling is more complex than the scenario described above. While silent synapses are generated in both the IL and PL projections to the NAc after cocaine self-administration, after 45 days of withdrawal, the IL-NAc silent synapses are unsilenced/matured by recruitment of CP-AMPARs, whereas the PL-NAc silent synapses mature through recruitment of non-CP-AMPARs. Moreover, optogenetic reversal of silent synapse-based remodeling in the IL-NAc and PL-NAc projections potentiated and inhibited incubation of cocaine craving, respectively [158]. Therefore, similar mechanisms (i.e., unsilencing of synapses) can both inhibit and promote drug-seeking depending on the neural circuit involved. Optogenetic stimulation of the IL-NAc shell pathway has also been shown to block conditioned place preference for morphine [160]. It would be interesting for future studies to examine if promoting the unsilencing of IL-NAc synapses alone can help prevent drug seeking.

In addition to the complexity of projection specific withdrawal effects, emerging research has identified drug-class-specific neuroplastic changes. For example, although cocaine and morphine both result in incubation of craving, the underlying mechanisms are different. Cocaine generates silent synapses in D1-MSNs via synaptogenesis; whereas, morphine generates silent synapses in D2-MSNs via internalization of AMPAR from preexisting synapses. During withdrawal, cocaine-generated silent synapses are "unsilenced" by recruiting AMPARs, which strengthens excitatory input to D1-MSNs. However, morphine-generated silent synapses are eliminated, thus weakening excitatory drive to D2-MSNs [160,161]. In both cases, the ratio of D1-D2 signaling is increased and incubation of craving is observed, but the results indicate that different treatment approaches may be needed based on the class of substance abused.

Neuronal Ensembles as an Emerging Model of Drug-Associated Memory Formation

Recently, it has been proposed that sparsely distributed patterns of neurons called neuronal ensembles, may be responsible for the encoding of learned associations, such as those seen in cue- and context-dependent drug-seeking behaviors [162,163]. Only 1–5% of neurons within a brain region make up an ensemble, but there may be several ensembles within a region, and individual neurons can contribute to more than one ensemble [164–166]. It is believed that the subset of neurons that receive the strongest, most persistent afferent input will be the ones recruited into any given ensemble. Neuronal ensembles are typically identified using correlational data between in vivo electrophysiology and Fos imaging in activated neurons, as strong, correlated activity generates action potentials and Fos expression [78]. Activity-dependent calcium influx influences plasticity at dendritic spines, but also activates the ERK/MAPK pathway, which stimulates c-fos promoter activity in the nucleus [167–170].

A novel approach known as the Daun02 inactivation procedure has been utilized to demonstrate a causal role for neuronal ensembles in mediating drug-seeking behaviors [75,91,164,171]. In these experiments, *c-fos-lacZ* transgenic rats are used, in which translation of β-galactosidase occurs in strongly activated, *c-fos*-expressing neurons. Daun02 is then injected and catalyzed to daunorubicin by β-galactosidase, which is believed to then induce apoptotic cell death of only the previously activated neurons [172]. Thus, the neural ensemble activated by a particular behavior, such as context-induced drug seeking, is then eliminated. Unlike studies where entire brain regions are inactivated, this approach targets a small number of neurons and allows the investigation of how neuronal ensembles regulate drug-associated behaviors. The elimination of a context-specific ensemble of neurons in the NAc interfered with the context-specific sensitization of cocaine-induced locomotion [171]. Furthermore, the activation of a NAc shell ensemble is necessary for context-induced reinstatement of cocaine seeking [75]. In addition, activation of an ensemble in the vmPFC is necessary for context-induced reinstatement of heroin seeking, while an ensemble in the OFC is necessary for context-specific cue-induced heroin seeking [91,164]. Finally, inactivation of a neuronal ensemble within the DLS can prevent context-induced seeking of methamphetamine, while inactivation of an ensemble within the DMS prevents incubation of methamphetamine craving [24,173]. More recent research is probing the molecular and physiological changes that might emerge in these behavior- or memory-specific ensembles. For example, exposure to a cocaine-conditioned context activates neurons in the NAc shell, core, and DS, but only neurons in the NAc core exhibit a change in excitability after cocaine conditioning, and this change was a suppression of non-ensemble neural activity, as opposed to a change in context-specific ensemble [174]. However, an additional study suggests that self-administration training for a food reward increases the excitability of ensemble neurons in the dmPFC [175]. Future research should begin to determine how circuits of neuronal ensembles across multiple brain regions work in unison to influence drug-associated behaviors.

EMERGING STRATEGIES FOR RELAPSE PREVENTION

Extinction as a Memory Process that Opposes Relapse

Extinction is a learning process that can occur if the cues, contexts, or responses associated with drug use are presented in sufficient number or over a long enough period without drug reinforcement that a reduction in drug seeking is observed [139,140,176,177]. Extinction involves the formation of a competing memory whereby the cue does not predict drug reinforcement, and this memory can then interfere with the expression of the original drug-paired association [178]. Thus, after sufficient extinction, subsequent encounters with the stimuli are less likely to trigger conditioned responses, including craving and relapse-like behavior [179]. The strength of the drug-associated memory is indicative of its potential to cause reinstatement; therefore, both interfering with reconsolidation and strengthening extinction have been proposed as therapeutic strategies to prevent relapse [180,181].

Molecular Mechanisms Underlying Extinction

Like reconsolidation, extinction was first characterized in studies of aversive learning, and has been shown to effectively attenuate maladaptive memories [182,183]. Recently, however, the idea of extinguishing reward-related memories has been proposed as a treatment for addiction, using the concepts derived from studies of conditioned fear. Extinction primarily involves activity in the BLA, NAc shell, and vmPFC [88,94,184]. As previously mentioned, long-term drug use induces glutamatergic alterations throughout the mesocorticolimbic system. For example, chronic cocaine exposure decreases GluA1 and GluA2/3 expression in the NAc shell. Interestingly, these cocaine-induced changes in plasticity are reversed by extinguishing an operant response for cocaine. Moreover, extinction is enhanced by viral overexpression of GluA1 and GluA2 in the NAc, demonstrating that extinction involves enhanced glutamatergic activity in the NAc shell [185,186], which is also consistent with the studies discussed above indicating that extinction learning is critical for the IL-NAc shell circuit to inhibit cocaine seeking [90]. Furthermore, glutamatergic activity in the BLA is involved in the extinction of drug-paired cue memories, as antagonism of NMDA receptors in this region results in disrupted extinction [81,187].

The involvement of glutamatergic activity during extinction has led to attempts to pharmacologically facilitate extinction via glutamate receptor agonists. D-Serine, an agonist at the glycine site of the NMDAR was shown to facilitate extinction, resulting in a decrease in cocaine-primed reinstatement [188]. Similarly, N-acetylcysteine, given in combination with instrumental extinction training, produces long-lasting reductions in cue- and heroin-primed drug seeking [189]. Finally, D-cycloserine (DCS), a partial agonist at the NMDA receptor, has also been shown to augment the extinction of drug-associated memories. For example, the extinction of responding for self-administered cocaine was enhanced by DCS in both rats and monkeys, thereby reducing cocaine reacquisition [190]. Additionally, DCS treatment following cue extinction in a novel context inhibits cue-induced renewal of drug seeking, but only when injected systemically or directly to the NAc core [179]. The effects of DCS on the extinction of drug memories are similar to results observed in studies of fear memory extinction in both animal models and human clinical studies [191,192]. However, in clinical studies of human drug using populations, the success of DCS has been limited. A randomized, placebo-controlled study showed that DCS was unable to facilitate extinction, and instead, enhanced craving in cocaine-dependent individuals [193]. Moreover, in preclinical studies, DCS treatment can increase reinstatement of drug seeking. Increased cue-induced drug seeking after DCS treatment appears to occur following a *brief* reexposure to the conditioned stimuli, suggesting that DCS may be strengthening the reconsolidation process as opposed to promoting extinction [194]. Indeed, in this study, DCS was administered after reexposure to 30 cue presentations, and studies from our lab indicate that 60 cue presentations only yield a mild reduction in reinstatement, indicating that 30 was unlikely to be sufficient to produce extinction [139,140]. Together, these studies highlight the complicated nature of reconsolidation and extinction following cue reactivation and demonstrate that a combined approach, in which a single pharmacological compound can simultaneously enhance extinction and inhibit reconsolidation and may have a better likelihood of weakening drug-associated memory to prevent relapse.

Bidirectional Targeting of Reconsolidation and Extinction

Both reconsolidation and extinction require the activity of various cellular signaling events, and many of these events are required for both extinction and reconsolidation to occur though others events may be regulated in an opposing manner [195,196]. Our lab recently conducted a study aimed at identifying specific phosphoproteins that are oppositely regulated by extinction and reconsolidation of a memory associated with self-administered cocaine [139,140]. Rats learned to associate cocaine infusions with an audiovisual cue and were later subjected to unreinforced presentations of the cue either three times to trigger reconsolidation or 120 times to trigger extinction. A phosphoproteomics analysis of the BLA identified a small number of signaling events that were oppositely regulated, including a novel phosphorylation event on calcium-calmodulin-dependent kinase II α (CaMKIIα) at serine 331. CaMKII is a well-characterized protein involved in various forms of memory and synaptic plasticity [120,197]. We then tested the effects of intra-BLA CaMKII inhibition on cocaine-cue memory reconsolidation and extinction, and discovered that in both conditions, this pharmacological manipulation resulted in a reduction in cue-induced reinstatement, relative to controls, suggesting the existence of molecular mechanisms capable of simultaneously augmenting extinction and disrupting reconsolidation [139,140]. Future studies should validate the long-term effects of CaMKII inhibition on synaptic plasticity as well as the long-term effects on drug-cue memory and determine if other bi-directionally identified proteins can be similarly targeted to prevent relapse-like behavior.

Translational Approaches for Enhancing Extinction/Inhibiting Reconsolidation

As described in section "Neuroanatomical Basis for Maladaptive Memory and Relapse", stimuli associated with drugs of abuse become sufficient to trigger conditioned responses that lead to craving and relapse and these responses can be weakened by extinction or by inhibiting reconsolidation. This behavioral strategy has long been established in the clinical setting and is known as cue exposure therapy (CET). Cues associated with drugs of abuse elicit increases in heart rate, skin temperature, and craving responses that can be extinguished with repetitive presentations of the cues [198]. CET has shown promise for maintaining abstinence among drug-dependent individuals. For example, extinction of cocaine-associated cues reduced craving, while prolonging abstinence among cocaine-addicted individuals [199]. Similarly, CET increased the latency to relapse and reduced alcohol intake in alcohol-dependent subjects [200]. Unfortunately, other studies have shown potential limitations of CET for long-term treatment, as extinction does not appear to generalize to other contexts, including the drug-taking environment [201,202]. In a group of opiate-dependent subjects, extinction decreased craving in the lab but was associated with higher relapse rates than controls [203]. It has been suggested that when the extinction and post-extinction environment are the same, renewal is less likely to occur [27,204]. As such, some recent studies have used a novel form of CET involving a virtual reality immersive environment in which drug-dependent individuals undergo extinction in a life-like simulation of their natural drug-using environment [205,206]. This form of CET is proposed to be more effective at eliciting conditioned responses during extinction, therefore decreasing subsequent cue-induced craving. The virtual reality approach seems promising but has not yet been implemented enough to know its long-term effectiveness.

In addition to context-dependency, other limitations such as spontaneous recovery of drug-seeking, suggest that CET may need to be supplemented with pharmacological interventions to strengthen the consolidation of extinction memory to overcome these treatment barriers. As described in section "Neuroanatomical Basis for Maladaptive Memory and Relapse", both reconsolidation and extinction are vulnerable to pharmacological manipulations [180,181], and reconsolidation and extinction manipulations have been successfully implemented clinically to treat anxiety disorders. For example, fMRI revealed that interfering with reconsolidation could abolish a fear-memory trace in the human BLA [207]. Similarly, Ressler and colleagues evaluated the efficacy of immersive CET in combination with DCS treatment at decreasing PTSD-like symptoms in a group of war veterans across six sessions. This combination reduced cortisol levels and startle-reactivity, and lowered PTSD symptoms, specifically among individuals that exhibited within-session learning [208]. However, as described above, clinical application of these approaches in substance-dependent individuals has had limited success at sustaining a long-term abstinent state [193,209].

Finally, a novel nonpharmacological strategy for the strengthening of extinction memory is to reactivate the original memory, and then subject individuals to an extinction session within the reconsolidation window [210–212]. Theoretically, the original memory will be made labile by reactivation, allowing it to be overwritten by extinction learning. LeDoux and colleagues reported that this procedure was less likely to produce the spontaneous recovery as well as renewal, resulting in a more persistent attenuation of fear-associated memories [210]. This strategy has also been effective at treating drug memories. In cocaine- and heroin-trained rats, retrieval-extinction attenuated drug-primed and context-induced reinstatement (renewal). Similarly, in heroin addicts, retrieval-extinction persistently diminished cue-induced craving for up to 180 days

[212]. Most recently, a novel form of retrieval-extinction in which noncontingent cocaine or methylphenidate injections 1 h prior to extinction resulted in a reduction in spontaneous recovery, cocaine-primed and context-induced reinstatement, as well as withdrawal-dependent incubated cocaine-seeking [213]. This strategy was also effective at reducing reinstatement to cocaine-paired cues that were not present during extinction. Interestingly, the inhibitory effect of UCS retrieval-extinction was mediated by the reversal of the insertion of CP-AMPARs described above. Future studies will have to determine if exposure to a small dose of the drug of abuse (UCS) prior to extinction training can produce similar results in humans.

Brain Stimulation

As described above, addiction is associated with dysfunction within the mesocorticolimbic circuit. Hypofunction of the PFC can result in the loss of inhibitory control over drug seeking [214,215]. Alterations in glutamatergic and dopaminergic transmission in regions, such as the NAc, also promote relapse [216,217]. Despite numerous efforts, systemic pharmacology-based treatments have yet to consistently reduce craving and drug-seeking behaviors. An alternative therapeutic approach is to stimulate affected brain areas to modulate neuronal firing, thus blocking drug-mediated behaviors. Two methods of brain stimulation that have recently arisen as prospective therapeutic candidates for substance use disorders are deep brain stimulation (DBS) and repetitive transcranial magnetic stimulation (rTMS).

DBS involves the surgical implantation of bipolar electrodes within specific subcortical brain regions in which a pulse generator delivers high-frequency stimulation that depolarizes or hyperpolarizes local neurons [218]. The mechanism of action remains unclear, but studies indicate that normal brain function can be reestablished by a synchronization or desynchronization of brain-wide circuits [219]. While DBS has been successful at improving outcomes for several neurological and psychiatric disorders, including Parkinson's disease and depression [220,221], relatively few studies have directly assessed the effects of DBS on addictive behaviors. In animal studies, high-frequency DBS has transiently suppressed cocaine-induced locomotor sensitization, reduced ethanol consumption, and attenuated cocaine-primed reinstatement [222–224]. However, most addiction-related DBS findings come from case reports or poorly controlled studies, but the results have yielded optimism. Studies have shown the effectiveness of DBS for alcohol, nicotine, and heroin. In one case report, an individual that previously used alcohol to relieve symptoms of anxiety, reported a complete cessation of drinking following DBS [225]. Similarly, in a small clinical study of five alcohol-dependent individuals, bilateral DBS in the NAc resulted in the remission of craving in all individuals, and long-term abstinence in two of the five [226]. Similar results were achieved in a study of smokers, where rates of smoking cessation 30 months post-DBS were higher than those of unaided smoking cessation in the general population (30% vs. 8.7%) [227]. Finally, two case reports of heroin-abusing individuals also yielded complete remission of drug use following DBS in the NAc [228,229]. Together, these studies demonstrate the potential relapse-preventing potential of DBS. However, the mechanism by which DBS is effective is currently unknown, making it unclear if altered plasticity of drug-associated memories mediates these effects.

rTMS is a noninvasive form of brain stimulation in which magnetic pulses are delivered on top of the skull to indirectly generate electrical currents [230,231]. The frequency of stimulation can be varied, whereby low-frequency stimulation (<1 Hz) reduces neuronal activity and high-frequency stimulation (between 5 and 20 Hz) enhances activity [232,233]. As with DBS, the mechanisms of action are not completely understood, but evidence suggests that TMS acts to locally alter blood flow, neuronal excitability and firing frequency, and neurotransmitter release [234]. Glutamatergic signaling appears to be involved, as NMDA antagonists block the long-term effects of high-frequency TMS [235]. Although the effects of rTMS are generally limited to cortical tissue, there are reports of possible downstream effects in deeper brain structures [236]. Most clinical TMS studies of addiction have utilized high-frequency stimulation in the dorsolateral PFC (dlPFC), which is involved in cognitive processing, working memory, and impulse control [237,238]. Studies of rTMS have produced mixed results in drug-dependent individuals; some have demonstrated acute reductions in craving and drug use [239–242], whereas others have shown no effect on craving [243]. Recently, Bonci and colleagues have suggested that rTMS rescues drug-induced hypoactivity of the PFC [244,245]. In a rodent model, long-term cocaine self-administration decreased excitability of PL pyramidal neurons, while in vivo optogenetic PL stimulation prevented compulsive drug seeking [244]. In a clinical study of cocaine abusers, rTMS-driven stimulation of the DLPFC resulted in a significantly higher number of cocaine-negative drug tests as well as significantly lower levels of craving compared to a control group [245].

Despite the mixed results of brain stimulation for the treatment of addiction, there is promise for the normalization of drug-induced deficits in brain function, which may ultimately prevent drug seeking. Future studies should develop placebo-controlled, randomized, double-blind studies with larger sample sizes to fully assess the efficacy of brain stimulation approaches at preventing relapse and to determine the mechanisms of action.

SUMMARY

In conclusion, one of the largest impediments to the successful treatment of addiction has been the inability to find methods to help individuals maintain abstinence. Repeated drug use induces long-lasting alterations within the brain's learning and memory circuitry, which can increase the vulnerability to relapse when an individual is reexposed to the abused drug or drug-associated cues. With the aid of human imaging studies and preclinical animal models, the last 10–20 years have provided a wealth of knowledge about the neuroanatomical, cellular, and molecular mechanisms responsible for promoting relapse. Coordinated synaptic activity across multiple brain regions initiates specific intracellular processes that are ultimately responsible for driving drug-seeking behavior. The recent explosion of technological advances in the broader neuroscience field has presented new opportunities to study the precise mechanisms underlying maladaptive memory formation and relapse. The development of novel cell-type and circuit-specific approaches for preventing relapse will undoubtedly lead to innovative strategies that may prove helpful in combatting addiction.

REFERENCES

[1] Sinha R, Shaham Y, Heilig M. Translational and reverse translational research on the role of stress in drug craving and relapse. Psychopharmacology 2011;218(1):69–82. https://doi.org/10.1007/s00213-011-2263-y.

[2] Gawin FH, Kleber HD. Abstinence symptomatology and psychiatric diagnosis in cocaine abusers. Clinical observations. Arch Gen Psychiatry 1986;43(2):107–13.

[3] Kassani A, Niazi M, Hassanzadeh J, Menati R. Survival analysis of drug abuse relapse in addiction treatment centers. Int J High Risk Behav Addict 2015;4(3). https://doi.org/10.5812/ijhrba.23402.

[4] Miller M, Chen ALC, Stokes SD, Silverman S, Bowirrat A, Manka M, Blum K. Early intervention of intravenous KB220IV-neuroadaptagen amino-acid therapy (NAAT) improves behavioral outcomes in a residential addiction treatment program: a pilot study. J Psychoactive Drugs 2012;44(5):398–409. https://doi.org/10.1080/02791072.2012.737727.

[5] Sinha R, Garcia M, Paliwal P, Kreek MJ, Rounsaville BJ. Stress-induced cocaine craving and hypothalamic-pituitary-adrenal responses are predictive of cocaine relapse outcomes. Arch Gen Psychiatry 2006;63(3):324. https://doi.org/10.1001/archpsyc.63.3.324.

[6] Witteman J, Post H, Tarvainen M, de Bruijn A, Perna EDSF, Ramaekers JG, Wiers RW. Cue reactivity and its relation to craving and relapse in alcohol dependence: a combined laboratory and field study. Psychopharmacology 2015;232(20):3685–96. https://doi.org/10.1007/s00213-015-4027-6.

[7] Back SE, Gros DF, McCauley JL, Flanagan JC, Cox E, Barth KS, Brady KT. Laboratory-induced cue reactivity among individuals with prescription opioid dependence. Addict Behav 2014;39(8):1217–23. https://doi.org/10.1016/j.addbeh.2014.04.007.

[8] Marchant NJ, Rabei R, Kaganovsky K, Caprioli D, Bossert JM, Bonci A, Shaham Y. A critical role of lateral hypothalamus in context-induced relapse to alcohol seeking after punishment-imposed abstinence. J Neurosci 2014;34(22):7447–57. https://doi.org/10.1523/JNEUROSCI.0256-14.2014.

[9] O'Brien CP, Childress AR, McLellan AT, Ehrman R. Classical conditioning in drug-dependent humans. Ann N Y Acad Sci 1992;654:400–15.

[10] Kalivas PW, Volkow ND. The neural basis of addiciton: a pathology of motivation and choice. Am J Psychiatry 2005;162(8):1403–13. https://doi.org/10.1176/appi.ajp.162.8.1403.

[11] Koob GF, Volkow ND. Neurocircuitry of Addiction. Neuropsychopharmacology 2010;35(1):217–38. https://doi.org/10.1038/npp.2009.110.

[12] Piazza PV, Deroche-Gamonet V. A multistep general theory of transition to addiction. Psychopharmacology 2013;229(3):387–413. https://doi.org/10.1007/s00213-013-3224-4.

[13] Engblom D, Bilbao A, Sanchis-Segura C, Dahan L, Perreau-Lenz S, Balland B, Spanagel R. Glutamate receptors on dopamine neurons control the persistence of cocaine seeking. Neuron 2008;59(3):497–508. https://doi.org/10.1016/j.neuron.2008.07.010.

[14] Mameli M, Bellone C, Brown MTC, Lüscher C. Cocaine inverts rules for synaptic plasticity of glutamate transmission in the ventral tegmental area. Nat Neurosci 2011;14(4):414–6. https://doi.org/10.1038/nn.2763.

[15] Nugent FS, Penick EC, Kauer JA. Opioids block long-term potentiation of inhibitory synapses. Nature 2007;446(7139):1086–90. https://doi.org/10.1038/nature05726.

[16] Grimm JW, Hope BT, Wise RA, Shaham Y. Neuroadaptation. Incubation of cocaine craving after withdrawal. Nature 2001;412(6843):141–2. https://doi.org/10.1038/35084134.

[17] Shaham Y, Shalev U, Lu L, De Wit H, Stewart J. The reinstatement model of drug relapse: history, methodology and major findings. Psychopharmacology 2003;168(1–2):3–20. https://doi.org/10.1007/s00213-002-1224-x.

[18] Bertholomey ML, Nagarajan V, Torregrossa MM. Sex differences in reinstatement of alcohol seeking in response to cues and yohimbine in rats with and without a history of adolescent corticosterone exposure. Psychopharmacology 2016;233(12):2277–87. https://doi.org/10.1007/s00213-016-4278-x.

[19] Feltenstein MW, Ghee SM, See RE. Nicotine self-administration and reinstatement of nicotine-seeking in male and female rats. Drug Alcohol Depend 2012;121(3):240–6. https://doi.org/10.1016/j.drugalcdep.2011.09.001.

[20] Shalev U, Grimm JW, Shaham Y. Neurobiology of relapse to heroin and cocaine seeking: a review. Pharmacol Rev 2002;54(1):1–42.

[21] Bossert JM, Poles GC, Wihbey KA, Koya E, Shaham Y. Differential effects of blockade of dopamine D1-family receptors in nucleus accumbens core or shell on reinstatement of heroin seeking induced by contextual and discrete cues. J Neurosci 2007;27(46):12655–63. https://doi.org/10.1523/JNEUROSCI.3926-07.2007.

[22] Leri F, Stewart J. Drug-induced reinstatement to heroin and cocaine seeking: a rodent model of relapse in polydrug use. Exp Clin Psychopharmacol 2001;9(3):297–306.

[23] Mantsch JR, Baker DA, Funk D, Lê AD, Shaham Y. Stress-induced reinstatement of drug seeking: 20 years of progress. Neuropsychopharmacology 2016;41(1):335–56. https://doi.org/10.1038/npp.2015.142.

[24] Rubio FJ, Liu Q-R, Li X, Cruz FC, Leao RM, Warren BL, Hope BT. Context-induced reinstatement of methamphetamine seeking is associated with unique molecular alterations in Fos-expressing dorsolateral striatum neurons. J Neurosci 2015;35(14):5625–39. https://doi.org/10.1523/JNEUROSCI.4997-14.2015.

[25] Stringfield SJ, Higginbotham JA, Wang R, Berger AL, McLaughlin RJ, Fuchs RA. Role of glucocorticoid receptor-mediated mechanisms in cocaine memory enhancement. Neuropharmacology 2017. https://doi.org/10.1016/j.neuropharm.2017.05.022.

[26] Koob GF. Neurobiology of addiction toward the development of new therapies. Ann New York Acad Sci 2000;909:170–85.

[27] Crombag HS, Shaham Y. Renewal of drug seeking by contextual cues after prolonged extinction in rats. Behav Neurosci 2002;116(1):169–73.

[28] Epstein DH, Preston KL. The reinstatement model and relapse prevention: a clinical perspective. Psychopharmacology 2003;168(1–2):31–41. https://doi.org/10.1007/s00213-003-1470-6.

[29] Katz JL, Higgins ST. The validity of the reinstatement model of craving and relapse to drug use. Psychopharmacology 2003;168(1–2):21–30. https://doi.org/10.1007/s00213-003-1441-y.

[30] Lu L, Grimm JW, Hope BT, Shaham Y. Incubation of cocaine craving after withdrawal: a review of preclinical data. Neuropharmacology 2004;47 (Suppl 1):214–26. https://doi.org/10.1016/j.neuropharm.2004.06.027.

[31] Lee BR, Ma Y-Y, Huang YH, Wang X, Otaka M, Ishikawa M, Dong Y. Maturation of silent synapses in amygdala-accumbens projection contributes to incubation of cocaine craving. Nat Neurosci 2013;16(11):1644–51. https://doi.org/10.1038/nn.3533.

[32] Loweth JA, Tseng KY, Wolf ME. Adaptations in AMPA receptor transmission in the nucleus accumbens contributing to incubation of cocaine craving. Neuropharmacology 2014;76:287–300. https://doi.org/10.1016/j.neuropharm.2013.04.061.

[33] Li P, Wu P, Xin X, Fan Y-L, Wang G-B, Wang F, Lu L. Incubation of alcohol craving during abstinence in patients with alcohol dependence. Addict Biol 2015;20(3):513–22. https://doi.org/10.1111/adb.12140.

[34] Li X, Rubio FJ, Zeric T, Bossert JM, Kambhampati S, Cates HM, Shaham Y. Incubation of methamphetamine craving is associated with selective increases in expression of Bdnf and Trkb, glutamate receptors, and epigenetic enzymes in cue-activated Fos-expressing dorsal striatal neurons. J Neurosci 2015;35(21):8232–44. https://doi.org/10.1523/JNEUROSCI.1022-15.2015.

[35] Theberge FR, Li X, Kambhampati S, Pickens CL, St. Laurent R, Bossert JM, Shaham Y. Effect of chronic delivery of the toll-like receptor 4 antagonist (+)-naltrexone on incubation of heroin craving. Biol Psychiatry 2013;73(8):729–37. https://doi.org/10.1016/j.biopsych.2012.12.019.

[36] Funk D, Coen K, Tamadon S, Hope BT, Shaham Y, Lê AD. Role of central amygdala neuronal ensembles in incubation of nicotine craving. J Neurosci 2016;36(33):8612–23. https://doi.org/10.1523/JNEUROSCI.1505-16.2016.

[37] Kirschmann EK, Pollock MW, Nagarajan V, Torregrossa MM. Effects of adolescent cannabinoid self-administration in rats on addiction-related behaviors and working memory. Neuropsychopharmacology 2017. https://doi.org/10.1038/npp.2016.178.

[38] Lüscher C, Malenka RC. Drug-evoked synaptic plasticity in addiction: from molecular changes to circuit remodeling. Neuron 2011;69(4):650–63. https://doi.org/10.1016/j.neuron.2011.01.017.

[39] Pierce RC, Kumaresan V. The mesolimbic dopamine system: the final common pathway for the reinforcing effect of drugs of abuse? Neurosci Biobehav Rev 2006;30(2):215–38. https://doi.org/10.1016/j.neubiorev.2005.04.016.

[40] Wise RA. Dopamine, learning and motivation. Nat Rev Neurosci 2004;5(6):483–94. https://doi.org/10.1038/nrn1406.

[41] Thomas MJ, Kalivas PW, Shaham Y. Neuroplasticity in the mesolimbic dopamine system and cocaine addiction. Br J Pharmacol 2008;154 (2):327–42. https://doi.org/10.1038/bjp.2008.77.

[42] Vlachou S, Guery S, Froestl W, Banerjee D, Benedict J, Finn MG, Markou A. Repeated administration of the GABAB receptor positive modulator BHF177 decreased nicotine self-administration, and acute administration decreased cue-induced reinstatement of nicotine seeking in rats. Psychopharmacology 2011;215(1):117–28. https://doi.org/10.1007/s00213-010-2119-x.

[43] Yamaguchi T, Wang H-L, Li X, Ng TH, Morales M. Mesocorticolimbic glutamatergic pathway. J Neurosci 2011;31(23):8476–90. https://doi.org/10.1523/JNEUROSCI.1598-11.2011.

[44] Tomasi D, Wang G-J, Wang R, Caparelli EC, Logan J, Volkow ND. Overlapping patterns of brain activation to food and cocaine cues in cocaine abusers: association to striatal D2/D3 receptors. Hum Brain Mapp 2015;36(1):120–36. https://doi.org/10.1002/hbm.22617.

[45] Volkow ND, Fowler JS, Wang GJ, Baler R, Telang F. Imaging dopamine's role in drug abuse and addiction. Neuropharmacology 2009;56(SUPPL. 1):3–8. https://doi.org/10.1016/j.neuropharm.2008.05.022.

[46] Robbins TW, Everitt BJ. Limbic-striatal memory systems and drug addiction. Neurobiol Learn Mem 2002;78(3):625–36.

[47] Schultz W, Tremblay L, Hollerman JR. Reward processing in primate orbitofrontal cortex and basal ganglia. Cereb Cortex 2000;10(3):272–84.

[48] Drevets WC, Gautier C, Price JC, Kupfer DJ, Kinahan PE, Grace AA, Mathis CA. Amphetamine-induced dopamine release in human ventral striatum correlates with euphoria. Biol Psychiatry 2001;49(2):81–96.

[49] Schultz W. Predictive reward signal of dopamine neurons. J Neurophysiol 1998;80(1):1–27.

[50] Phillips PEM, Stuber GD, Heien MLAV, Wightman RM, Carelli RM. Subsecond dopamine release promotes cocaine seeking. Nature 2003;422 (6932):614–8. https://doi.org/10.1038/nature01476.

[51] Stuber GD, Klanker M, de Ridder B, Bowers MS, Joosten RN, Feenstra MG, Bonci A. Reward-predictive cues enhance excitatory synaptic strength onto midbrain dopamine neurons. Science 2008;321(5896):1690–2. https://doi.org/10.1126/science.1160873.

[52] Chiu CQ, Puente N, Grandes P, Castillo PE. Dopaminergic modulation of endocannabinoid-mediated plasticity at GABAergic synapses in the prefrontal cortex. J Neurosci 2010;30(21):7236–48. https://doi.org/10.1523/JNEUROSCI.0736-10.2010.

[53] Wang W, Dever D, Lowe J, Storey GP, Bhansali A, Eck EK, Bamford NS. Regulation of prefrontal excitatory neurotransmission by dopamine in the nucleus accumbens core. J Physiol 2012;590(16):3743–69. https://doi.org/10.1113/jphysiol.2012.235200.

[54] Boudreau AC, Reimers JM, Milovanovic M, Wolf ME. Cell surface AMPA receptors in the rat nucleus accumbens increase during cocaine withdrawal but internalize after cocaine challenge in association with altered activation of mitogen-activated protein kinases. J Neurosci 2007;27(39):10621–35. https://doi.org/10.1523/JNEUROSCI.2163-07.2007.

[55] Hearing MC, Zink AN, Wickman K. Cocaine-induced adaptations in metabotropic inhibitory signaling in the mesocorticolimbic system. Rev Neurosci 2012;23(4):325–51. https://doi.org/10.1515/revneuro-2012-0045.

[56] McCracken CB, Grace AA. Persistent cocaine-induced reversal learning deficits are associated with altered limbic cortico-striatal local field potential synchronization. J Neurosci 2013;33(44):17469–82. https://doi.org/10.1523/JNEUROSCI.1440-13.2013.

[57] Saddoris MP, Sugam JA, Carelli RM. Prior cocaine experience impairs normal phasic dopamine signals of reward value in accumbens shell. Neuropsychopharmacology 2017;42(3):766–73. https://doi.org/10.1038/npp.2016.189.

[58] Ciccocioppo R, Sanna PP, Weiss F. Cocaine-predictive stimulus induces drug-seeking behavior and neural activation in limbic brain regions after multiple months of abstinence: reversal by D1 antagonists. Proc Natl Acad Sci U S A 2001;98(4):1976–81. https://doi.org/10.1073/pnas.98.4.1976.

[59] Hamlin AS, Newby J, McNally GP. The neural correlates and role of D1 dopamine receptors in renewal of extinguished alcohol-seeking. Neuroscience 2007;146(2):525–36. https://doi.org/10.1016/j.neuroscience.2007.01.063.

[60] Neisewander JL, Baker Da, Fuchs Ra, Tran-Nguyen LT, Palmer A, Marshall JF. Fos protein expression and cocaine-seeking behavior in rats after exposure to a cocaine self-administration environment. J Neurosci 2000;20(2):798–805. https://doi.org/10.1038/466194a.

[61] Nakazawa K, McHugh TJ, Wilson MA, Tonegawa S. NMDA receptors, place cells and hippocampal spatial memory. Nat Rev Neurosci 2004;5 (5):361–72. https://doi.org/10.1038/nrn1385.

[62] Chen C, Kim JJ, Thompson RF, Tonegawa S. Hippocampal lesions impair contextual fear conditioning in two strains of mice. Behav Neurosci 1996;110(5):1177–80.

[63] Maren S, Fanselow MS. Electrolytic lesions of the fimbria/fornix, dorsal hippocampus, or entorhinal cortex produce anterograde deficits in contextual fear conditioning in rats. Neurobiol Learn Mem 1997;67(2):142–9. https://doi.org/10.1006/nlme.1996.3752.

[64] Lasseter HC, Ramirez DR, Xie X, Fuchs RA. Involvement of the lateral orbitofrontal cortex in drug context-induced reinstatement of cocaine-seeking behavior in rats. Eur J Neurosci 2009;30(7):1370–81. https://doi.org/10.1111/j.1460-9568.2009.06906.x.

[65] Xie X, Ramirez DR, Lasseter HC, Fuchs RA. Effects of mGluR1 antagonism in the dorsal hippocampus on drug context-induced reinstatement of cocaine-seeking behavior in rats. Psychopharmacology 2010;208(1):1–11. https://doi.org/10.1007/s00213-009-1700-7.

[66] Fuchs RA, Evans KA, Ledford CC, Parker MP, Case JM, Mehta RH, See RE. The role of the dorsomedial prefrontal cortex, basolateral amygdala, and dorsal hippocampus in contextual reinstatement of cocaine seeking in rats. Neuropsychopharmacology 2005;30(2):296–309. https://doi.org/10.1038/sj.npp.1300579.

[67] Schafe GE, LeDoux JE. Memory consolidation of auditory pavlovian fear conditioning requires protein synthesis and protein kinase A in the amygdala. J Neurosci 2000;20(18):RC96.

[68] Lasseter HC, Wells AM, Xie X, Fuchs RA. Interaction of the basolateral amygdala and orbitofrontal cortex is critical for drug context-induced reinstatement of cocaine-seeking behavior in rats. Neuropsychopharmacology 2011;36(3):711–20. https://doi.org/10.1038/npp.2010.209.

[69] Lasseter HC, Xie X, Arguello AA, Wells AM, Hodges MA, Fuchs RA. Contribution of a mesocorticolimbic subcircuit to drug context-induced reinstatement of cocaine-seeking behavior in rats. Neuropsychopharmacology 2013; https://doi.org/10.1038/npp.2013.249.

[70] Stefanik MT, Kalivas PW. Optogenetic dissection of basolateral amygdala projections during cue-induced reinstatement of cocaine seeking. Front Behav Neurosci 2013;7: https://doi.org/10.3389/fnbeh.2013.00213.

[71] Keistler CR, Hammarlund E, Barker JM, Bond CW, DiLeone RJ, Pittenger C, Taylor JR. Regulation of alcohol extinction and cue-induced reinstatement by specific projections among medial prefrontal cortex, nucleus accumbens, and basolateral amygdala. J Neurosci 2017;37(17).

[72] Rich MT, Torregrossa MM. Molecular and synaptic mechanisms regulating drug-associated memories: towards a bidirectional treatment strategy. Brain Res Bull 2017; https://doi.org/10.1016/j.brainresbull.2017.09.003.

[73] Sanchez H, Quinn JJ, Torregrossa MM, Taylor JR. Reconsolidation of a cocaine-associated stimulus requires amygdalar protein kinase A. J Neurosci 2010;30(12):4401–7. https://doi.org/10.1523/JNEUROSCI.3149-09.2010.

[74] Sanchez H, Quinn JJ, Torregrossa MM, Taylor JR. Reconsolidation of a cocaine-associated stimulus requires amygdalar protein kinase A. J Neurosci 2010;30(12):4401–7. https://doi.org/10.1523/JNEUROSCI.3149-09.2010.

[75] Cruz FC, Babin KR, Leao RM, Goldart EM, Bossert JM, Shaham Y, Hope BT. Role of nucleus accumbens shell neuronal ensembles in context-induced reinstatement of cocaine-seeking. J Neurosci 2014;34(22):7437–46. https://doi.org/10.1523/JNEUROSCI.0238-14.2014.

[76] Hamlin AS, Blatchford KE, McNally GP. Renewal of an extinguished instrumental response: neural correlates and the role of D1 dopamine receptors. Neuroscience 2006;143(1):25–38. https://doi.org/10.1016/j.neuroscience.2006.07.035.

[77] Campioni MR, Xu M, McGehee DS. Stress-induced changes in nucleus accumbens glutamate synaptic plasticity. J Neurophysiol 2009;101 (6):3192–8. https://doi.org/10.1152/jn.91111.2008.

[78] Cruz FC, Javier Rubio F, Hope BT. Using c-fos to study neuronal ensembles in corticostriatal circuitry of addiction. Brain Res 2015;1628:157–73. https://doi.org/10.1016/j.brainres.2014.11.005.

[79] De Giovanni LN, Guzman AS, Virgolini MB, Cancela LM. NMDA antagonist MK 801 in nucleus accumbens core but not shell disrupts the restraint stress-induced reinstatement of extinguished cocaine-conditioned place preference in rats. Behav Brain Res 2016;315:150–9. https://doi.org/10.1016/j.bbr.2016.08.011.

[80] Fuchs RA, Evans KA, Parker MC, See RE. Differential involvement of the core and shell subregions of the nucleus accumbens in conditioned cue-induced reinstatement of cocaine seeking in rats. Psychopharmacology 2004;176(3–4):459–65. https://doi.org/10.1007/s00213-004-1895-6.

[81] Fuchs RA, Ramirez DR, Bell GH. Nucleus accumbens shell and core involvement in drug context-induced reinstatement of cocaine seeking in rats. Psychopharmacology 2008;200(4):545–56. https://doi.org/10.1007/s00213-008-1234-4.

[82] McFarland K, Davidge SB, Lapish CC, Kalivas PW. Limbic and motor circuitry underlying footshock-induced reinstatement of cocaine-seeking behavior. J Neurosci 2004;24(7):1551–60. https://doi.org/10.1523/JNEUROSCI.4177-03.2004.

[83] Peters J, Kalivas PW, Quirk GJ. Extinction circuits for fear and addiction overlap in prefrontal cortex. Learn Mem 2009;16(5):279–88. https://doi.org/10.1101/lm.1041309.

[84] Brog JS, Salyapongse A, Deutch AY, Zahm DS. The patterns of afferent innervation of the core and shell in the ?Accumbens? part of the rat ventral striatum: immunohistochemical detection of retrogradely transported fluoro-gold. J Comp Neurol 1993;338(2):255–78. https://doi.org/10.1002/cne.903380209.

[85] Sesack SR, Deutch AY, Roth RH, Bunney BS. Topographical organization of the efferent projections of the medial prefrontal cortex in the rat: an anterograde tract-tracing study with phaseolus vulgaris leucoagglutinin. J Comp Neurol 1989;290(2):213–42. https://doi.org/10.1002/cne.902900205.

[86] Bossert JM, Stern AL, Theberge FRM, Marchant NJ, Wang H-L, Morales M, Shaham Y. Role of projections from ventral medial prefrontal cortex to nucleus accumbens shell in context-induced reinstatement of heroin seeking. J Neurosci 2012;32(14):4982–91. https://doi.org/10.1523/JNEUROSCI.0005-12.2012.

[87] Rogers JL, Ghee S, See RE. The neural circuitry underlying reinstatement of heroin-seeking behavior in an animal model of relapse. Neuroscience 2008;151(2):579–88. https://doi.org/10.1016/j.neuroscience.2007.10.012.

[88] Peters J, LaLumiere RT, Kalivas PW. Infralimbic prefrontal cortex is responsible for inhibiting cocaine seeking in extinguished rats. J Neurosci 2008;28(23):6046–53. https://doi.org/10.1523/JNEUROSCI.1045-08.2008.

[89] Suto N, Laque A, De Ness GL, Wagner GE, Watry D, Kerr T, Weiss F. Distinct memory engrams in the infralimbic cortex of rats control opposing environmental actions on a learned behavior *eLife*, 2016;5. https://doi.org/10.7554/eLife.21920.

[90] Augur IF, Wyckoff AR, Aston-Jones G, Kalivas PW, Peters J. Chemogenetic activation of an extinction neural circuit reduces cue-induced reinstatement of cocaine seeking. J Neurosci 2016;36(39):10174–80. https://doi.org/10.1523/JNEUROSCI.0773-16.2016.

[91] Fanous S, Goldart EM, Theberge FRM, Bossert JM, Shaham Y, Hope BT. Role of orbitofrontal cortex neuronal ensembles in the expression of incubation of heroin craving. J Neurosci 2012;32(34):11600–9. https://doi.org/10.1523/JNEUROSCI.1914-12.2012.

[92] Madsen HB, Brown RM, Short JL, Lawrence AJ. Investigation of the neuroanatomical substrates of reward seeking following protracted abstinence in mice. J Physiol 2012;590(10):2427–42. https://doi.org/10.1113/jphysiol.2011.225219.

[93] Thiel KJ, Wenzel JM, Pentkowski NS, Hobbs RJ, Alleweireldt AT, Neisewander JL. Stimulation of dopamine D2/D3 but not D1 receptors in the central amygdala decreases cocaine-seeking behavior. Behav Brain Res 2010;214(2):386–94. https://doi.org/10.1016/j.bbr.2010.06.021.

[94] Koya E, Uejima JL, Wihbey KA, Bossert JM, Hope BT, Shaham Y. Role of ventral medial prefrontal cortex in incubation of cocaine craving. Neuropharmacology 2009;56:177–85. https://doi.org/10.1016/j.neuropharm.2008.04.022.

[95] Conrad KL, Tseng KY, Uejima JL, Reimers JM, Heng L-J, Shaham Y, Wolf ME. Formation of accumbens GluR2-lacking AMPA receptors mediates incubation of cocaine craving. Nature 2008;454(7200):118–21. https://doi.org/10.1038/nature06995.

[96] Huang YH, Lin Y, Mu P, Lee BR, Brown TE, Wayman G, Dong Y. In vivo cocaine experience generates silent synapses. Neuron 2009;63(1):40–7. https://doi.org/10.1016/j.neuron.2009.06.007.

[97] Belin D, Belin-Rauscent A, Murray JE, Everitt BJ. Addiction: failure of control over maladaptive incentive habits. Curr Opin Neurobiol 2013;23 (4):564–72. https://doi.org/10.1016/j.conb.2013.01.025.

[98] Torregrossa MM, Corlett PR, Taylor JR. Aberrant learning and memory in addiction. Neurobiol Learn Mem 2011. https://doi.org/10.1016/j.nlm.2011.02.014.

[99] Porrino LJ, Lyons D, Smith HR, Daunais JB, Nader MA. cocaine self-administration produces a progressive involvement of limbic, association, and sensorimotor striatal domains. J Neurosci 2004;24(14):3554–62. https://doi.org/10.1523/JNEUROSCI.5578-03.2004.

[100] Willuhn I, Burgeno LM, Everitt BJ, Phillips PEM. Hierarchical recruitment of phasic dopamine signaling in the striatum during the progression of cocaine use. Proc Natl Acad Sci U S A 2012;109(50):20703–8. https://doi.org/10.1073/pnas.1213460109.

[101] Corbit LH, Nie H, Janak PH. Habitual alcohol seeking: time course and the contribution of subregions of the dorsal striatum. Biol Psychiatry 2012;72(5):389–95. https://doi.org/10.1016/j.biopsych.2012.02.024.

[102] Yin HH, Knowlton BJ, Balleine BW. Lesions of dorsolateral striatum preserve outcome expectancy but disrupt habit formation in instrumental learning. Eur J Neurosci 2004;19(1):181–9.

[103] Barker JM, Torregrossa MM, Arnold AP, Taylor JR. Dissociation of genetic and hormonal influences on sex differences in alcoholism-related behaviors. J Neurosci 2010;30(27):9140–4. https://doi.org/10.1523/JNEUROSCI.0548-10.2010.

[104] Zapata A, Minney VL, Shippenberg TS. Shift from goal-directed to habitual cocaine seeking after prolonged experience in rats. J Neurosci 2010;30 (46):15457–63. https://doi.org/10.1523/JNEUROSCI.4072-10.2010.

[105] Pacchioni AM, Gabriele A, See RE. Dorsal striatum mediation of cocaine-seeking after withdrawal from short or long daily access cocaine self-administration in rats. Behav Brain Res 2011;218(2):296–300. https://doi.org/10.1016/j.bbr.2010.12.014.

[106] Murray JE, Belin D, Everitt BJ. Double dissociation of the dorsomedial and dorsolateral striatal control over the acquisition and performance of cocaine seeking. Neuropsychopharmacology 2012;37(11):2456–66. https://doi.org/10.1038/npp.2012.104.

[107] Murray JE, Belin-Rauscent A, Simon M, Giuliano C, Benoit-Marand M, Everitt BJ, Belin D. Basolateral and central amygdala differentially recruit and maintain dorsolateral striatum-dependent cocaine-seeking habits. Nat Commun 2015;6. https://doi.org/10.1038/ncomms10088.

[108] Hooks MS, Duffy P, Striplin C, Kalivas PW. Behavioral and neurochemical sensitization following cocaine self-administration. Psychopharmacology 1994;115(1–2):265–72.

[109] Koob GF, Ahmed SH, Boutrel B, Chen SA, Kenny PJ, Markou A, Sanna PP. Neurobiological mechanisms in the transition from drug use to drug dependence. Neurosci Biobehav Rev 2004;27(8):739–49. https://doi.org/10.1016/j.neubiorev.2003.11.007.

[110] See RE. Neural substrates of cocaine-cue associations that trigger relapse. Eur J Pharmacol 2005;526(1–3):140–6. https://doi.org/10.1016/j.ejphar.2005.09.034.

[111] Castellano C, Cestari V, Ciamei A. NMDA receptors and learning and memory processes. Curr Drug Targets 2001;2(3):273–83.

[112] Lee JLC, Di Ciano P, Thomas KL, Everitt BJ. Disrupting reconsolidation of drug memories reduces cocaine-seeking behavior. Neuron 2005;47(6):795–801. https://doi.org/10.1016/j.neuron.2005.08.007.

[113] Debiec J, LeDoux JE, Nader K. Cellular and systems reconsolidation in the hippocampus. Neuron 2002;36(3):527–38. https://doi.org/10.1016/S0896-6273(02)01001-2.

[114] Duvarci S, Nader K, LeDoux JE. De novo mRNA synthesis is required for both consolidation and reconsolidation of fear memories in the amygdala. Learn Mem 2008;15(10):747–55. https://doi.org/10.1101/lm.1027208.

[115] Pedreira ME, Pérez-Cuesta LM, Maldonado H. Reactivation and reconsolidation of long-term memory in the crab Chasmagnathus: protein synthesis requirement and mediation by NMDA-type glutamatergic receptors, J Neurosci 2002;22(18):8305–11. https://doi.org/10.1523/JNEUROSCI.22-18-08305.2002.

[116] Tronson NC, Wiseman SL, Olausson P, Taylor JR. Bidirectional behavioral plasticity of memory reconsolidation depends on amygdalar protein kinase A. Nat Neurosci 2006;9(2):167–9. https://doi.org/10.1038/nn1628.

[117] Tronson NC, Taylor JR. Molecular mechanisms of memory reconsolidation. Nat Rev Neurosci 2007;8(4):262–75. https://doi.org/10.1038/nrn2090.

[118] Abel T, Lattal KM. Molecular mechanisms of memory acquisition, consolidation and retrieval. Curr Opin Neurobiol 2001;11(2):180–7.

[119] Abel T, Nguyen PV. Regulation of hippocampus-dependent memory by cyclic AMP-dependent protein kinase. Prog Brain Res 2008;169:97–115. https://doi.org/10.1016/S0079-6123(07)00006-4.

[120] Sanhueza M, Lisman J. The CaMKII/NMDAR complex as a molecular memory. Mol Brain 2013;6(1):10. https://doi.org/10.1186/1756-6606-6-10.

[121] Yiu AP, Mercaldo V, Yan C, Richards B, Rashid AJ, Hsiang H-LL, Josselyn SA. Neurons are recruited to a memory trace based on relative neuronal excitability immediately before training. Neuron 2014;83(3):722–35. https://doi.org/10.1016/j.neuron.2014.07.017.

[122] Nader K, Schafe GE, Le Doux JE. Fear memories require protein synthesis in the amygdala for reconsolidation after retrieval. Nature 2000;406(6797):722–6. https://doi.org/10.1038/35021052.

[123] Fuchs RA, Bell GH, Ramirez DR, Eaddy JL, Su Z. Basolateral amygdala involvement in memory reconsolidation processes that facilitate drug context-induced cocaine seeking. Eur J Neurosci 2009;30(5):889–900. https://doi.org/10.1111/j.1460-9568.2009.06888.x.

[124] Milton AL, Lee JLC, Butler VJ, Gardner R, Everitt BJ. Intra-amygdala and systemic antagonism of NMDA receptors prevents the reconsolidation of drug-associated memory and impairs subsequently both novel and previously acquired drug-seeking behaviors. J Neurosci 2008;28(33):8230–7. https://doi.org/10.1523/JNEUROSCI.1723-08.2008.

[125] Milton AL, Lee JLC, Everitt BJ. Reconsolidation of appetitive memories for both natural and drug reinforcement is dependent on {beta}-adrenergic receptors. Learn Mem 2008;15(2):88–92. https://doi.org/10.1101/lm.825008.

[125a] Lee JL, Milton AL, Everitt BJ. Cue-induced cocaine seeking and relapse are reduced by disruption of drug memory reconsolidation. J Neurosci 2006;26(22):5881 7, https://doi.org/10.1523/JNEUROSCI.0323-06.2006.

[126] Theberge FRM, Milton AL, Belin D, Lee JLC, Everitt BJ. The basolateral amygdala and nucleus accumbens core mediate dissociable aspects of drug memory reconsolidation. Learn Mem 2010;17(9):444–53. https://doi.org/10.1101/lm.1757410.

[127] Bernardi RE, Lattal KM, Berger SP. Postretrieval propranolol disrupts a cocaine conditioned place preference. Neuroreport 2006;17(13):1443–7. https://doi.org/10.1097/01.wnr.0000233098.20655.26.

[128] Dębiec J, Bush DEA, LeDoux JE. Noradrenergic enhancement of reconsolidation in the amygdala impairs extinction of conditioned fear in rats: a possible mechanism for the persistence of traumatic memories in PTSD. Depress Anxiety 2011;28(3):186–93. https://doi.org/10.1002/da.20803.

[129] Robinson MJF, Franklin KBJ. Central but not peripheral beta-adrenergic antagonism blocks reconsolidation for a morphine place preference. Behav Brain Res 2007;182(1):129–34. https://doi.org/10.1016/j.bbr.2007.05.023.

[130] Gamache K, Pitman RK, Nader K. Preclinical evaluation of reconsolidation blockade by clonidine as a potential novel treatment for posttraumatic stress disorder. Neuropsychopharmacology 2012;37(13):2789–96. https://doi.org/10.1038/npp.2012.145.

[131] Milton AL, Schramm MJW, Wawrzynski JR, Gore F, Oikonomou-Mpegeti F, Wang NQ, Everitt BJ. Antagonism at NMDA receptors, but not β-adrenergic receptors, disrupts the reconsolidation of pavlovian conditioned approach and instrumental transfer for ethanol-associated conditioned stimuli. Psychopharmacology 2012;219(3):751–61. https://doi.org/10.1007/s00213-011-2399-9.

[132] Otis JM, Mueller D. Inhibition of β-adrenergic receptors induces a persistent deficit in retrieval of a cocaine-associated memory providing protection against reinstatement. Neuropsychopharmacology 2011;36(9):1912–20. https://doi.org/10.1038/npp.2011.77.

[133] Lynch WJ, Taylor JR. Persistent changes in motivation to self-administer cocaine following modulation of cyclic AMP-dependent protein kinase A (PKA) activity in the nucleus accumbens. Eur J Neurosci 2005;22(5):1214–20. https://doi.org/10.1111/j.1460-9568.2005.04305.x.

[134] Nestler EJ. Molecular mechanisms of drug addiction. Neuropharmacology 2004;47:24–32. https://doi.org/10.1016/j.neuropharm.2004.06.031.

[135] Arguello AA, Hodges MA, Wells AM, Lara H, Xie X, Fuchs RA. Involvement of amygdalar protein kinase A, but not calcium/calmodulin-dependent protein kinase II, in the reconsolidation of cocaine-related contextual memories in rats. Psychopharmacology 2014;231(1):55–65. https://doi.org/10.1007/s00213-013-3203-9.

[136] Wan X, Torregrossa MM, Sanchez H, Nairn AC, Taylor JR. Activation of exchange protein activated by cAMP in the rat basolateral amygdala impairs reconsolidation of a memory associated with self-administered cocaine. PLoS One 2014;9(9). https://doi.org/10.1371/journal.pone.0107359.

[137] Miller CA, Marshall JF. Molecular substrates for retrieval and reconsolidation of cocaine-associated contextual memory. Neuron 2005;47 (6):873–84. https://doi.org/10.1016/j.neuron.2005.08.006.

[138] Wells AM, Arguello AA, Xie X, Blanton MA, Lasseter HC, Reittinger AM, Fuchs RA. Extracellular signal-regulated kinase in the basolateral amygdala, but not the nucleus accumbens core, is critical for context-response-cocaine memory reconsolidation in rats. Neuropsychopharmacology 2013;38(5):753–62. https://doi.org/10.1038/npp.2012.238.

[139] Rich MT, Abbott TB, Chung L, Gulcicek EE, Stone KL, Colangelo CM, Torregrossa MM. Phosphoproteomic analysis reveals a novel mechanism of CaMKII regulation inversely induced by cocaine memory extinction versus reconsolidation. J Neurosci 2016;36(29):7613–27. https://doi.org/10.1523/JNEUROSCI.1108-16.2016.

[140] Rich MT, Abbott TB, Chung L, Gulcicek EE, Stone KL, Colangelo CM, Torregrossa MM. Phosphoproteomic analysis reveals a novel mechanism of CaMKIIα regulation inversely induced by cocaine memory extinction versus reconsolidation. J Neurosci 2016;36(29):7613–27. https://doi.org/10.1523/JNEUROSCI.1108-16.2016.

[141] Renthal W, Kumar A, Xiao G, Wilkinson M, Covington HE, Maze I, Nestler EJ. Genome-wide analysis of chromatin regulation by cocaine reveals a role for sirtuins. Neuron 2009;62(3):335–48. https://doi.org/10.1016/j.neuron.2009.03.026.

[142] Barrett RM, Wood MA. Beyond transcription factors: the role of chromatin modifying enzymes in regulating transcription required for memory, Learn Mem 2008;15(7):460–7. (Cold Spring Harbor, N.Y.), https://doi.org/10.1101/lm.917508.

[143] Chen W-S, Xu W-J, Zhu H-Q, Gao L, Lai M-J, Zhang F-Q, Liu H-F. Effects of histone deacetylase inhibitor sodium butyrate on heroin seeking behavior in the nucleus accumbens in rats. Brain Res 2016;1652:151–7. https://doi.org/10.1016/j.brainres.2016.10.007.

[144] Nestler EJ. Epigenetic mechanisms of drug addiction. Neuropharmacology 2014;76(Pt B):259–68. https://doi.org/10.1016/j.neuropharm.2013.04.004.

[145] Simon-O'Brien E, Alaux-Cantin S, Warnault V, Buttolo R, Naassila M, Vilpoux C. The histone deacetylase inhibitor sodium butyrate decreases excessive ethanol intake in dependent animals. Addict Biol 2015;20(4):676–89. https://doi.org/10.1111/adb.12161.

[146] Sheng J, Lv Zg, Wang L, Zhou Y, Hui B. Histone H3 phosphoacetylation is critical for heroin-induced place preference. Neuroreport 2011;22 (12):575–80. https://doi.org/10.1097/WNR.0b013e328348e6aa.

[147] Wang Y, Lai J, Cui H, Zhu Y, Zhao B, Wang W, Wei S. Inhibition of histone deacetylase in the basolateral amygdala facilitates morphine context-associated memory formation in rats. J Mol Neurosci 2015;55(1):269–78. https://doi.org/10.1007/s12031-014-0317-4.

[148] Dunbar AB, Taylor JR. Garcinol blocks the reconsolidation of multiple cocaine-paired cues after a single cocaine-reactivation session. Neuropsychopharmacology 2017;42(9):1884–92. https://doi.org/10.1038/npp.2017.27.

[149] Monsey MS, Sanchez H, Taylor JR. The naturally occurring compound Garcinia Indica selectively impairs the reconsolidation of a cocaine-associated memory. Neuropsychopharmacology 2017;42(3):587–97. https://doi.org/10.1038/npp.2016.117.

[150] Romieu P, Deschatrettes E, Host L, Gobaille S, Sandner G, Zwiller J. The inhibition of histone deacetylases reduces the reinstatement of cocaine-seeking behavior in rats. Curr Neuropharmacol 2011;9(1):21–5. https://doi.org/10.2174/157015911795017317.

[151] Olausson P, Jentsch JD, Taylor JR. Repeated nicotine exposure enhances reward-related learning in the rat. Neuropsychopharmacology 2003;28 (7):1264–71. https://doi.org/10.1038/sj.npp.1300173.

[152] Taylor JR, Jentsch JD. Repeated intermittent administration of psychomotor stimulant drugs alters the acquisition of Pavlovian approach behavior in rats: differential effects of cocaine, d-amphetamine and 3,4- methylenedioxymethamphetamine ("Ecstasy"). Biol Psychiatry 2001;50(2):137–43.

[153] Parvaz MA, Moeller SJ, Goldstein RZ. Incubation of cue-induced craving in adults addicted to cocaine measured by electroencephalography. JAMA Psychiat 2016;73(11):1127–34. https://doi.org/10.1001/jamapsychiatry.2016.2181.

[154] Bedi G, Preston KL, Epstein DH, Heishman SJ, Marrone GF, Shaham Y, de Wit H. Incubation of cue-induced cigarette craving during abstinence in human smokers. Biol Psychiatry 2011;69(7):708–11. https://doi.org/10.1016/j.biopsych.2010.07.014.

[155] Wang G, Shi J, Chen N, Xu L, Li J, Li P, Lu L. Effects of length of abstinence on decision-making and craving in methamphetamine abusers. PLoS One 2013;8(7). https://doi.org/10.1371/journal.pone.0068791.

[156] Scheyer AF, Loweth JA, Christian DT, Uejima J, Rabei R, Le T, Wolf ME. AMPA Receptor plasticity in accumbens core contributes to incubation of methamphetamine craving. Biol Psychiatry 2016;80(9):661–70. https://doi.org/10.1016/j.biopsych.2016.04.003.

[157] Caprioli D, Venniro M, Zeric T, Li X, Adhikary S, Madangopal R, Shaham Y. Effect of the novel positive allosteric modulator of metabotropic glutamate receptor 2 AZD8529 on incubation of methamphetamine craving after prolonged voluntary abstinence in a rat model. Biol Psychiatry 2015;78(7):463–73. https://doi.org/10.1016/j.biopsych.2015.02.018.

[158] Ma Y-Y, Lee BR, Wang X, Guo C, Liu L, Cui R, Dong Y. Bidirectional modulation of incubation of cocaine craving by silent synapse-based remodeling of prefrontal cortex to accumbens projections. Neuron 2014;83(6):1453–67. https://doi.org/10.1016/j.neuron.2014.08.023.

[159] Neumann PA, Wang Y, Yan Y, Wang Y, Ishikawa M, Cui R, Dong Y. Cocaine-induced synaptic alterations in thalamus to nucleus accumbens projection. Neuropsychopharmacology 2016;41(9):2399–410. https://doi.org/10.1038/npp.2016.52.

[160] Hearing MC, Jedynak J, Ebner SR, Ingebretson A, Asp AJ, Fischer RA, Thomas MJ. Reversal of morphine-induced cell-type-specific synaptic plasticity in the nucleus accumbens shell blocks reinstatement. Proc Natl Acad Sci 2016;113(3):757–62. https://doi.org/10.1073/pnas.1519248113.

[161] Graziane NM, Sun S, Wright WJ, Jang D, Liu Z, Huang YH, Dong Y. Opposing mechanisms mediate morphine- and cocaine-induced generation of silent synapses. Nat Neurosci 2016;19(7):915–25. https://doi.org/10.1038/nn.4313.

[162] Buzsáki G, Moser EI. Memory, navigation and theta rhythm in the hippocampal-entorhinal system. Nat Neurosci 2013;16(2):130–8. https://doi.org/10.1038/nn.3304.

[163] Pennartz CMA, Lee E, Verheul J, Lipa P, Barnes CA, McNaughton BL. The ventral striatum in off-line processing: ensemble reactivation during sleep and modulation by hippocampal ripples. J Neurosci 2004;24(29):6446–56. https://doi.org/10.1523/JNEUROSCI.0575-04.2004.

[164] Bossert JM, Stern AL, Theberge FRM, Cifani C, Koya E, Hope BT, Shaham Y. Ventral medial prefrontal cortex neuronal ensembles mediate context-induced relapse to heroin. Nat Neurosci 2011;14(4):420–2. https://doi.org/10.1038/nn.2758.

[165] Chawla MK, Guzowski JF, Ramirez-Amaya V, Lipa P, Hoffman KL, Marriott LK, Barnes CA. Sparse, environmentally selective expression of Arc RNA in the upper blade of the rodent fascia dentata by brief spatial experience. Hippocampus 2005;15(5):579–86. https://doi.org/10.1002/hipo.20091.

[166] Schwindel CD, McNaughton BL. Hippocampal–cortical interactions and the dynamics of memory trace reactivation. Prog Brain Res 2011;193:163–77. https://doi.org/10.1016/B978-0-444-53839-0.00011-9.

[167] Brami-Cherrier K, Valjent E, Hervé D, Darragh J, Corvol J-C, Pages C, Caboche J. Parsing molecular and behavioral effects of cocaine in mitogen- and stress-activated protein kinase-1-deficient mice. J Neurosci 2005;25(49):11444–54. https://doi.org/10.1523/JNEUROSCI.1711-05.2005.

[168] Cahill E, Salery M, Vanhoutte P, Caboche J. Convergence of dopamine and glutamate signaling onto striatal ERK activation in response to drugs of abuse. Front Pharmacol 2014;4. https://doi.org/10.3389/fphar.2013.00172.

[169] Deisseroth K, Tsien RW. Dynamic multiphosphorylation passwords for activity-dependent gene expression. Neuron 2002;34(2):179–82.

[170] Lüscher C, Nicoll Ra, Malenka RC, Muller D. Synaptic plasticity and dynamic modulation of the postsynaptic membrane. Nat Neurosci 2000;3 (6):545–50. https://doi.org/10.1038/75714.

[171] Koya E, Golden SA, Harvey BK, Guez-Barber DH, Berkow A, Simmons DE, Hope BT. Targeted disruption of cocaine-activated nucleus accumbens neurons prevents context-specific sensitization. Nat Neurosci 2009;12(8):1069–73. https://doi.org/10.1038/nn.2364.

[172] Farquhar D, Pan BF, Sakurai M, Ghosh A, Mullen CA, Nelson JA. Suicide gene therapy using E. coli beta-galactosidase. Cancer Chemother Pharmacol 2002;50(1):65–70. https://doi.org/10.1007/s00280-002-0438-2.

[173] Caprioli D, Venniro M, Zhang M, Bossert JM, Warren BL, Hope BT, Shaham Y. Role of dorsomedial striatum neuronal ensembles in incubation of methamphetamine craving after voluntary abstinence. J Neurosci 2017;37(4):1014–27. https://doi.org/10.1523/JNEUROSCI.3091-16.2016.

[174] Ziminski JJ, Hessler S, Margetts-Smith G, Sieburg MC, Crombag HS, Koya E. Changes in appetitive associative strength modulates nucleus accumbens, but not orbitofrontal cortex neuronal ensemble excitability. J Neurosci 2017;37(12):3160–70. https://doi.org/10.1523/JNEUROSCI.3766-16.2017.

[175] Whitaker LR, Warren BL, Venniro M, Harte TC, McPherson KB, Beidel J, Hope BT. Bidirectional modulation of intrinsic excitability in rat prelimbic cortex neuronal ensembles and non-ensembles after operant learning. J Neurosci 2017;37(36):8845–56. https://doi.org/10.1523/JNEUROSCI.3761-16.2017.

[176] Pedreira ME, Maldonado H. Protein synthesis subserves reconsolidation or extinction depending on reminder duration. Neuron 2003;38(6):863–9.

[177] Suzuki A. Memory reconsolidation and extinction have distinct temporal and biochemical signatures. J Neurosci 2004;24(20):4787–95. https://doi.org/10.1523/JNEUROSCI.5491-03.2004.

[178] Bouton ME, Moody EW. Memory processes in classical conditioning. Neurosci Biobehav Rev 2004;28(7):663–74. https://doi.org/10.1016/j.neubiorev.2004.09.001.

[179] Torregrossa MM, Sanchez H, Taylor JR. D-cycloserine reduces the context specificity of pavlovian extinction of cocaine cues through actions in the nucleus accumbens. J Neurosci 2010;30(31):10526–33. https://doi.org/10.1523/JNEUROSCI.2523-10.2010.

[180] Taylor JR, Olausson P, Quinn JJ, Torregrossa MM. Targeting extinction and reconsolidation mechanisms to combat the impact of drug cues on addiction. Neuropharmacology 2009;56(Suppl 1):186–95. https://doi.org/10.1016/j.neuropharm.2008.07.027.

[181] Torregrossa MM, Taylor JR. Learning to forget: manipulating extinction and reconsolidation processes to treat addiction. Psychopharmacology 2012. https://doi.org/10.1007/s00213-012-2750-9.

[182] Fucich EA, Paredes D, Morilak DA. Therapeutic effects of extinction learning as a model of exposure therapy in rats. Neuropsychopharmacology 2016;41(13):3092–102. https://doi.org/10.1038/npp.2016.127.

[183] Santini E. Consolidation of fear extinction requires protein synthesis in the medial prefrontal cortex. J Neurosci 2004;24(25):5704–10. https://doi.org/10.1523/JNEUROSCI.0786-04.2004.

[184] Lindgren JL, Gallagher M, Holland PC. Lesions of basolateral amygdala impair extinction of CS motivational value, but not of explicit conditioned responses, in Pavlovian appetitive second-order conditioning. Eur J Neurosci 2003;17(1):160–6.

[185] Self DW, Choi K-H. Extinction-induced neuroplasticity attenuates stress-induced cocaine seeking: a state-dependent learning hypothesis. Stress 2004;7(3):145–55. https://doi.org/10.1080/10253890400012677.

[186] Sutton MA, Schmidt EF, Choi K-H, Schad CA, Whisler K, Simmons D, Self DW. Extinction-induced upregulation in AMPA receptors reduces cocaine-seeking behaviour. Nature 2003;421(6918):70–5. https://doi.org/10.1038/nature01249.

[187] Feltenstein MW, See RE. NMDA receptor blockade in the basolateral amygdala disrupts consolidation of stimulus-reward memory and extinction learning during reinstatement of cocaine-seeking in an animal model of relapse. Neurobiol Learn Mem 2007;88(4):435–44. https://doi.org/10.1016/j.nlm.2007.05.006.

[188] Kelamangalath L, Seymour CM, Wagner JJ. d-Serine facilitates the effects of extinction to reduce cocaine-primed reinstatement of drug-seeking behavior. Neurobiol Learn Mem 2009;92(4):544–51. https://doi.org/10.1016/j.nlm.2009.07.004.

[189] Zhou W, Kalivas PW. N-acetylcysteine reduces extinction responding and induces enduring reductions in cue- and heroin-induced drug-seeking. Biol Psychiatry 2008;63(3):338–40. https://doi.org/10.1016/j.biopsych.2007.06.008.

[190] Nic Dhonnchadha BA, Szalay JJ, Achat-Mendes C, Platt DM, Otto MW, Spealman RD, Kantak KM. D-cycloserine deters reacquisition of cocaine self-administration by augmenting extinction learning. Neuropsychopharmacology 2010;35(2):357–67. https://doi.org/10.1038/npp.2009.139.

[191] Ledgerwood L, Richardson R, Cranney J. D-Cycloserine and the facilitation of extinction of conditioned fear: consequences for reinstatement. Behav Neurosci 2004;118(3):505–13. https://doi.org/10.1037/0735-7044.118.3.505.

[192] Ressler KJ, Rothbaum BO, Tannenbaum L, Anderson P, Graap K, Zimand E, Davis M. Cognitive enhancers as adjuncts to psychotherapy. Arch Gen Psychiatry 2004;61(11):1136. https://doi.org/10.1001/archpsyc.61.11.1136.

[193] Price KL, Baker NL, McRae-Clark AL, Saladin ME, Desantis SM, Santa Ana EJ, Brady KT. A randomized, placebo-controlled laboratory study of the effects of d-cycloserine on craving in cocaine-dependent individuals. Psychopharmacology 2013;226(4):739–46. https://doi.org/10.1007/s00213-011-2592-x.

[194] Lee JLC, Gardner RJ, Butler VJ, Everitt BJ. D-cycloserine potentiates the reconsolidation of cocaine-associated memories. Learn Mem 2009;16(1):82–5. https://doi.org/10.1101/lm.1186609.

[195] de la Fuente V, Freudenthal R, Romano A. Reconsolidation or extinction: transcription factor switch in the determination of memory course after retrieval. J Neurosci 2011;31(15):5562–73. https://doi.org/10.1523/JNEUROSCI.6066-10.2011.

[196] Merlo E, Milton AL, Goozee ZY, Theobald DE, Everitt BJ. Reconsolidation and extinction are dissociable and mutually exclusive processes: behavioral and molecular evidence. J Neurosci 2014;34(7):2422–31. https://doi.org/10.1523/JNEUROSCI.4001-13.2014.

[197] Coultrap SJ, Freund RK, O'Leary H, Sanderson JL, Roche KW, Dell'Acqua ML, Bayer KU. Autonomous CaMKII mediates both LTP and LTD using a mechanism for differential substrate site selection. Cell Rep 2014;6(3):431–7. https://doi.org/10.1016/j.celrep.2014.01.005.

[198] Foltin RW, Haney M. Conditioned effects of environmental stimuli paired with smoked cocaine in humans. Psychopharmacology 2000;149(1):24–33.

[199] O'Brien CP, Childress AR, McLellan T, Ehrman R. Integrating systemic cue exposure with standard treatment in recovering drug dependent patients. Addict Behav 1990;15(4):355–65.

[200] Drummond DC, Glautier S. A controlled trial of cue exposure treatment in alcohol dependence. J Consult Clin Psychol 1994;62(4):809–17.

[201] Bouton ME. Context, ambiguity, and unlearning: sources of relapse after behavioral extinction. Biol Psychiatry 2002;52(10):976–86.

[202] Peck JA, Ranaldi R. Drug abstinence: exploring animal models and behavioral treatment strategies. Psychopharmacology 2014;231(10):2045–58. https://doi.org/10.1007/s00213-014-3517-2.

[203] Marissen MAE, Franken IHA, Blanken P, van den Brink W, Hendriks VM. Cue exposure therapy for the treatment of opiate addiction: results of a randomized controlled clinical trial. Psychother Psychosom 2007;76(2):97–105. https://doi.org/10.1159/000097968.

[204] Thewissen R, Snijders SJBD, Havermans RC, van den Hout M, Jansen A. Renewal of cue-elicited urge to smoke: implications for cue exposure treatment. Behav Res Ther 2006;44(10):1441–9. https://doi.org/10.1016/j.brat.2005.10.010.

[205] Kuntze MF, Stoermer R, Mager R, Roessler A, Mueller-Spahn F, Bullinger AH. Immersive virtual environments in cue exposure. Cyberpsychol Behav 2001;4(4):497–501. https://doi.org/10.1089/109493101750527051.

[206] Lee J-H, Kwon H, Choi J, Yang B-H. Cue-exposure therapy to decrease alcohol craving in virtual environment. Cyberpsychol Behav 2007;10(5):617–23. https://doi.org/10.1089/cpb.2007.9978.

[207] Agren T, Engman J, Frick A, Bjorkstrand J, Larsson E-M, Furmark T, Fredrikson M. Disruption of Reconsolidation Erases a Fear Memory Trace in the Human Amygdala. Science 2012;337(6101):1550–2. https://doi.org/10.1126/science.1223006.

[208] Rothbaum BO, Price M, Jovanovic T, Norrholm SD, Gerardi M, Dunlop B, Ressler KJ. A randomized, double-blind evaluation of d-cycloserine or alprazolam combined with virtual reality exposure therapy for posttraumatic stress disorder in Iraq and Afghanistan war veterans. Am J Psychiatr 2014;171(6):640–8. https://doi.org/10.1176/appi.ajp.2014.13121625.

[209] Carroll KM, Onken LS. Behavioral therapies for drug abuse. Am J Psychiatry 2005;162(8):1452–60. https://doi.org/10.1176/appi.ajp.162.8.1452.

[210] Monfils M-H, Cowansage KK, Klann E, LeDoux JE. Extinction-reconsolidation boundaries: key to persistent attenuation of fear memories. Science 2009;324(5929):951–5. https://doi.org/10.1126/science.1167975.

[211] Schiller D, Raio CM, Phelps EA. Extinction training during the reconsolidation window prevents recovery of fear. J Vis Exp 2012;66. https://doi.org/10.3791/3893.

[212] Xue Y-X, Luo Y-X, Wu P, Shi H-S, Xue L-F, Chen C, Lu L. A memory retrieval-extinction procedure to prevent drug craving and relapse. Science 2012;336(6078):241–5. https://doi.org/10.1126/science.1215070.

[213] Luo Y, Xue Y, Liu J, Shi H, Jian M, Han Y, Lu L. A novel UCS memory retrieval-extinction procedure to inhibit relapse to drug seeking. Nat Commun 2015;6:7675. https://doi.org/10.1038/ncomms8675.

[214] Goldstein RZ, Volkow ND. Dysfunction of the prefrontal cortex in addiction: neuroimaging findings and clinical implications. Nat Rev Neurosci 2011;12(11):652–69. https://doi.org/10.1038/nrn3119.

[215] Jentsch JD, Taylor JR. Impulsivity resulting from frontostriatal dysfunction in drug abuse: implications for the control of behavior by reward-related stimuli. Psychopharmacology 1999;146(4):373–90.

[216] Pierce RC, Wolf ME. Psychostimulant-induced neuroadaptations in nucleus accumbens AMPA receptor transmission. Cold Spring Harb Perspect Med 2013;3(2). https://doi.org/10.1101/cshperspect.a012021.

[217] Shen H-w, Toda S, Moussawi K, Bouknight A, Zahm DS, Kalivas PW. Altered dendritic spine plasticity in cocaine-withdrawn rats. J Neurosci 2009;29(9):2876–84. https://doi.org/10.1523/JNEUROSCI.5638-08.2009.

[218] McIntyre CC, Grill WM, Sherman DL, Thakor NV. Cellular effects of deep brain stimulation: model-based analysis of activation and inhibition. J Neurophysiol 2004;91(4):1457–69. https://doi.org/10.1152/jn.00989.2003.

[219] Murrow RW. Penfield's prediction: a mechanism for deep brain stimulation. Front Neurol 2014;5. https://doi.org/10.3389/fneur.2014.00213.

[220] Riva-Posse P, Choi KS, Holtzheimer PE, McIntyre CC, Gross RE, Chaturvedi A, Mayberg HS. Defining critical white matter pathways mediating successful subcallosal cingulate deep brain stimulation for treatment-resistant depression. Biol Psychiatry 2014;76(12):963–9. https://doi.org/10.1016/j.biopsych.2014.03.029.

[221] Wagenbreth C, Zaehle T, Galazky I, Voges J, Guitart-Masip M, Heinze H-J, Düzel E. Deep brain stimulation of the subthalamic nucleus modulates reward processing and action selection in Parkinson patients. J Neurol 2015;262(6):1541–7. https://doi.org/10.1007/s00415-015-7749-9.

[222] Creed M, Pascoli VJ, Luscher C. Refining deep brain stimulation to emulate optogenetic treatment of synaptic pathology. Science 2015;347(6222):659–64. https://doi.org/10.1126/science.1260776.

[223] Knapp CM, Tozier L, Pak A, Ciraulo DA, Kornetsky C. Deep brain stimulation of the nucleus accumbens reduces ethanol consumption in rats. Pharmacol Biochem Behav 2009;92(3):474–9. https://doi.org/10.1016/j.pbb.2009.01.017.

[224] Vassoler FM, White SL, Hopkins TJ, Guercio LA, Espallergues J, Berton O, Pierce RC. Deep brain stimulation of the nucleus accumbens shell attenuates cocaine reinstatement through local and antidromic activation. J Neurosci 2013;33(36):14446–54. https://doi.org/10.1523/JNEUROSCI.4804-12.2013.

[225] Kuhn J, Lenartz D, Huff W, Lee S, Koulousakis A, Klosterkoetter J, Sturm V. Remission of alcohol dependency following deep brain stimulation of the nucleus accumbens: valuable therapeutic implications? J Neurol Neurosurg Psychiatry 2007;78(10):1152–3. https://doi.org/10.1136/jnnp.2006.113092.

[226] Müller U, Sturm V, Voges J, Heinze H-J, Galazky I, Büntjen L, Bogerts B. Nucleus accumbens deep brain stimulation for alcohol addiction—safety and clinical long-term results of a pilot trial. Pharmacopsychiatry 2016;49(4):170–3. https://doi.org/10.1055/s-0042-104507.

[227] Kuhn J, Bauer R, Pohl S, Lenartz D, Huff W, Kim EH, Sturm V. Observations on unaided smoking cessation after deep brain stimulation of the nucleus accumbens. Eur Addict Res 2009;15(4):196–201. https://doi.org/10.1159/000228930.

[228] Valencia-Alfonso C-E, Luigjes J, Smolders R, Cohen MX, Levar N, Mazaheri A, Denys D. Effective deep brain stimulation in heroin addiction: a case report with complementary intracranial electroencephalogram. Biol Psychiatry 2012;71(8):e35–7. https://doi.org/10.1016/j.biopsych.2011.12.013.

[229] Zhou H, Xu J, Jiang J. Deep brain stimulation of nucleus accumbens on heroin-seeking behaviors: a case report. Biol Psychiatry 2011;69(11):e41–2. https://doi.org/10.1016/j.biopsych.2011.02.012.

[230] Gorelick DA, Zangen A, George MS. Transcranial magnetic stimulation in the treatment of substance addiction. Ann N Y Acad Sci 2014;1327:79–93. https://doi.org/10.1111/nyas.12479.

[231] Rossini, P. M., Burke, D., Chen, R., Cohen, L. G., Daskalakis, Z., Di Iorio, R., Ziemann, U. (2015). Non-invasive electrical and magnetic stimulation of the brain, spinal cord, roots and peripheral nerves: basic principles and procedures for routine clinical and research application. An updated report from an I.F.C.N. Committee. Clin Neurophysiol, 126(6), 1071–1107. https://doi.org/10.1016/j.clinph.2015.02.001

[232] Salling MC, Martinez D. Brain stimulation in addiction. Neuropsychopharmacology 2016;41(12):2798–809. https://doi.org/10.1038/npp.2016.80.

[233] Speer AM, Kimbrell TA, Wassermann EM, D Repella J, Willis MW, Herscovitch P, Post RM. Opposite effects of high and low frequency rTMS on regional brain activity in depressed patients. Biol Psychiatry 2000;48(12):1133–41.

[234] Grall-Bronnec M, Sauvaget A. The use of repetitive transcranial magnetic stimulation for modulating craving and addictive behaviours: a critical literature review of efficacy, technical and methodological considerations. Neurosci Biobehav Rev 2014;47:592–613. https://doi.org/10.1016/j.neubiorev.2014.10.013.

[235] Huang Y-Z, Rothwell JC, Edwards MJ, Chen R-S. Effect of physiological activity on an NMDA-dependent form of cortical plasticity in human. Cereb Cortex 2008;18(3):563–70. https://doi.org/10.1093/cercor/bhm087.

[236] Fox P, Ingham R, George MS, Mayberg H, Ingham J, Roby J, Jerabek P. Imaging human intra-cerebral connectivity by PET during TMS. Neuroreport 1997;8(12):2787–91.

[237] Barbey AK, Koenigs M, Grafman J. Dorsolateral prefrontal contributions to human working memory. Cortex 2013;49(5):1195–205. https://doi.org/10.1016/j.cortex.2012.05.022.

[238] Steinbeis N, Bernhardt BC, Singer T. Impulse control and underlying functions of the left DLPFC mediate age-related and age-independent individual differences in strategic social behavior. Neuron 2012;73(5):1040–51. https://doi.org/10.1016/j.neuron.2011.12.027.

[239] Eichhammer P, Johann M, Kharraz A, Binder H, Pittrow D, Wodarz N, Hajak G. High-frequency repetitive transcranial magnetic stimulation decreases cigarette smoking. J Clin Psychiatry 2003;64(8):951–3.

[240] Johann M, Wiegand R, Kharraz A, Bobbe G, Sommer G, Hajak G, Eichhammer P. Transcranial magnetic stimulation for nicotine dependence. Psychiatr Prax 2003;30(Suppl. 2):S129–31.

[241] Mishra BR, Nizamie SH, Das B, Praharaj SK. Efficacy of repetitive transcranial magnetic stimulation in alcohol dependence: a sham-controlled study. Addiction 2010;105(1):49–55. https://doi.org/10.1111/j.1360-0443.2009.02777.x.

[242] Politi E, Fauci E, Santoro A, Smeraldi E. Daily sessions of transcranial magnetic stimulation to the left prefrontal cortex gradually reduce cocaine craving. Am J Addict 2008;17(4):345–6. https://doi.org/10.1080/10550490802139283.

[243] Herremans, S. C., Baeken, C., Vanderbruggen, N., Vanderhasselt, M. A., Zeeuws, D., Santermans, L., De Raedt, R. (2012). No influence of one right-sided prefrontal HF-rTMS session on alcohol craving in recently detoxified alcohol-dependent patients: results of a naturalistic study. Drug Alcohol Depend, 120(1–3), 209–213. https://doi.org/10.1016/j.drugalcdep.2011.07.021

[244] Chen BT, Yau H-J, Hatch C, Kusumoto-Yoshida I, Cho SL, Hopf FW, Bonci A. Rescuing cocaine-induced prefrontal cortex hypoactivity prevents compulsive cocaine seeking. Nature 2013;496(7445):359–62. https://doi.org/10.1038/nature12024.

[245] Terraneo A, Leggio L, Saladini M, Ermani M, Bonci A, Gallimberti L. Transcranial magnetic stimulation of dorsolateral prefrontal cortex reduces cocaine use: a pilot study. Eur Neuropsychopharmacol 2016;26(1):37–44. https://doi.org/10.1016/j.euroneuro.2015.11.011.

Chapter 9

Neural Morphology and Addiction

Natalie S. McGuier*, Joachim D. Uys[†], and Patrick J. Mulholland[‡]

*Department of Biological Sciences, Carnegie Mellon University, Pittsburgh, PA, United States, [†]Department of Cell and Molecular Pharmacology, Medical University of South Carolina, Charleston, SC, United States, [‡]Departments of Neuroscience and Psychiatry & Behavioral Science, Charleston Alcohol Research Center, Medical University of South Carolina, Charleston, SC, United States

Substance use disorders (SUDs) are characterized by the dysregulation of the mesolimbic dopaminergic pathway after prolonged drug use. This is expressed as an intense need to continue taking the drug despite physical and social ramifications. The list of addictive substances is long, but the most detrimental commonly abused drugs include cocaine, amphetamines, opiates/opioids, inhalants, and alcohol. Some of these, including cocaine and amphetamines, directly stimulate the dopaminergic pathway by blocking or reversing dopamine reuptake causing prolonged action of the neurotransmitter [1]. Because dopamine release during learning indicates the salience of an experience, the perceived importance of the drug of abuse is increased. Other drugs including opiates/opioids, inhalants, and alcohol act through a variety of molecular mechanisms and indirectly alter dopaminergic signaling by alternative molecular pathways or by producing euphoric or desirable emotional states. The brain-wide maladaptive alterations induced by addictive substances cause a crippling disease that requires enormous personal, social, and medicinal support to overcome. As a result, the cost of addiction is immense. The expense of addiction is highlighted through the lens of alcohol abuse. In the United States, there are an estimated 88,000 annual deaths from alcohol-related causes, making it the fourth leading preventable cause of death. Furthermore, an excessive alcohol consumption cost approximately $249 billion in expenses related to losses in workplace productivity, health care, criminal justice, and motor vehicle accidents in 1 year alone [2]. Considering these extreme societal and monetary costs, it is obvious that an effective and accessible treatment options for addiction are desperately needed.

Over the last 20 years, studies have provided a basis for experience-dependent morphological adaptations in the structure and function of neurons that underpin learning and memory. With our increasing knowledge that drugs of abuse highjack the processes underlying learning-induced morphological adaptations of brain cells, understanding these adaptations may identify novel targets for treating and preventing SUDs. In this chapter, we describe dendritic spines and mechanisms regulating their structure in the adult brain and briefly discuss approaches used to quantify morphological changes in brain cell structure. We then highlight recent findings revealing that chronic alcohol and drug exposure induces morphological plasticity of dendrites and dendritic spines in neurons within the corticostriatal circuitry. In doing so, we provide evidence that different classes of abused substances can produce similar albeit complex morphological adaptations supporting the hypothesis that there are shared molecular mechanisms underpinning drug-induced morphological plasticity [3]. We will highlight findings from recent studies using technological advances in imaging to explore the importance of dendritic spine morphology on biochemical compartmentalization and electrical signaling within neurons. Finally, we discuss additional aspects of morphological plasticity important for controlling function of neurons that are understudied in the field of addiction neuroscience.

DENDRITIC SPINES

Dendritic spines are the small, protruding, membranous organelles found on the dendritic processes of neurons where the majority of excitatory synaptic signaling occurs in brain. Spines are classified into three classes (i.e., stubby, mushroom, and thin) based on the morphological characteristics of the spine head, neck, and length (Fig. 1). In general, thin spines have long necks and a small head compared with mature, mushroom-shaped spines that have a wide head and thin neck. In contrast, stubby spines have no neck with the postsynaptic density (PSD) of their head in close proximity to the dendritic shaft. In addition to these types, there exist filopodia that resemble thin spines, but are longer and are devoid of bulbous heads and most cellular organelles. Thin and stubby spines likely represent the transient stages of spine formation and elimination whereas mushroom spines are thought to represent stable, mature synapses with presynapses opposite its PSD.

Neural Mechanisms of Addiction. https://doi.org/10.1016/B978-0-12-812202-0.00009-9
Copyright © 2019 Elsevier Inc. All rights reserved.

FIG. 1 Subclasses of dendritic spines observed on cortical pyramidal neurons and striatal medium spiny neurons. The four subclasses of dendritic spines are classified based on the morphological characteristics of the spine length and spine neck and head (L = length; H_D = diameter of spine head; N_D = diameter of spine neck).

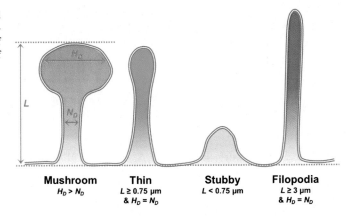

The shape of a spine links its physiological and biochemical properties. The head is a microenvironment isolated from the dendrite by a thin neck. A dense network of scaffolding and signaling proteins exists in the head of functional spines where it regulates the strength of synaptic transmission. Spine head volume is correlated with the overall composition of the PSD and the number of presynaptic vesicles [4]. Long-term potentiation and tetanic stimulation can further recruit AMPA and NMDA receptors to the PSD, which is often associated with head enlargement. In addition, shape can influence the local concentration of Ca^{2+} and expression of cAMP-regulated signaling proteins that affect stability and plasticity [5–8]. All of these factors are highly associated with learning and memory.

In young and maturing neurons, there is rapid synaptic formation and pruning as neuronal networks are formed and tuned to experience. This phenomenon, spine turnover, continues to a lesser extent throughout adulthood. This is perhaps unsurprising given that organisms are constantly learning and adapting to novel experiences. While we are only just beginning to elucidate the relationship between experience and morphological adaptations in the central nervous system, some fascinating points are becoming clear.

Spine growth, synapse formation, and synapse destabilization are activity-dependent processes, however, the directionality of these events is highly dependent on electrical input, age of the individual, brain region, and cell type [9]. For example, discrete sets of synapses in pyramidal neurons of rodent motor cortex are affected by different motor learning tasks. In fear conditioning and extinction learning tasks, the elimination and formation of spines in frontal association cortex occur in both a cue-dependent and location-specific manner with no net change in overall spine density [4]. In addition, environmental enrichment, which is associated with increased learning capacity in neurotypical and neurodivergent models, is also associated with an increase in synaptic remodeling and a net increase in spine density. Importantly, the stability and general thickness of a spine are associated with lifelong memories [10].

Regulation of spine plasticity likely takes place locally within a given section of dendrite. Spines adjacent to potentiated synapses tend to be co-activated than more distant neighbors. Indeed, new spines preferentially form near activated spines; providing a mechanism for how increased synaptic activity and LTP may strengthen connectivity between two neurons [11]. As described in detail below, a spine's actinous scaffold dictates its morphology, and structural stability and plasticity are largely regulated through actin-based mechanisms. Taken together, behavioral learning and experience-dependent adaptations in dendritic spine morphology can be considered a structural correlate of learning. Addiction can be defined as a disruption of normal learning behaviors where the salience of drugs of abuse is overrepresented in neural circuits leading to aberrant behaviors. With increasing knowledge of learning-induced morphological adaptations both in terms of structural change and the molecular mechanisms driving the change, we may be able to specifically target deviant memory formation associated with addiction at the synaptic level.

ACTIN CYCLING IN DENDRITIC SPINES

Actin and actin-binding proteins (ABPs) are essential regulators of morphological changes in dendritic spine formation that are, in turn, regulated by other proteins and molecules. While stabilization of mature spines is required for synaptic network maintenance, there are dynamic actin processes such as filamentous actin (F-actin) remodeling that are essential for synaptic plasticity. This results in spine turnover that leads to the formation of new spines and filopodia as well as pruning of existing spines [12,13]. Actin exists in either monomeric/globular (G-actin) or F-actin, the latter is formed by polymerizing adenosine-5′-triphosphate (ATP)-bound G-actin. Polymerization is bidirectional with the balance shifted toward the "barbed" end of the filament (named for the physical appearance of the molecule) [14]. Disassembly or depolymerization of F-actin occurs at the pointed end with the help of the actin depolymerizing factor (ADF)/cofilin family proteins. Actin filament depolymerization ensures the turnover of actin filaments within these structures and maintains a pool of actin monomers that permits the continual restructuring and growth of the actin cytoskeleton.

The activities of ABPs are tightly regulated by intracellular signaling events, which control their phosphorylation state and cellular localization. Changes in their activities can lead to rapid remodeling of the actin cytoskeleton and the subsequent dendritic spine morphology. ABPs regulate the balance between spine assembly and disassembly which is required for normal synaptic network maintenance [15]. ABPs such as Arp2/3, cortactin, and cofilin determine the rate of change of actin dynamics that ultimately regulates the morphology of dendritic spines and synaptic development. Protein (Arp) 2/3 complex consists of seven proteins that are structurally and functionally related to actin [16]. They create branch points by nucleating the assembly of filaments and capping the pointed ends. New monomers are added to the barbed end, thereby creating additional fast-growing barbed ends for further actin polymerization and elongation. In dendritic spines, Arp2/3 is localized in a ring-shaped region within the spinoplasm [17]. Such localization specificity suggests a restricted spine region dedicated to actin branching. Cortactin binds to Arp2/3 and promotes branching and stabilization of F-actin [18]. In the brain, cortactin is enriched in dendritic spines, where it co-localizes with F-actin. Different pools of cortactin can regulate spine morphology and synaptic function. For example, there is a large cortactin pool in the actin core within the dendritic spine, which regulates spine shape, and a smaller pool close to the PSD that may regulate synaptic function [19].

Resting cells have detectable levels of cofilin activity that contributes to both F-actin depolymerization and polymerization at the "pointed" ends and at the "barbed" ends, respectively, thereby causing a slow F-actin turnover rate [20]. Phosphorylation of cofilin by LIM kinase (LIMK) on serine 3, leads to dissociation from G-actin, inhibition of the depolymerizing activity of cofilin and stabilizing of the actin cytoskeleton [21]. The subcellular concentration of cofilin can differ between compartments leading to differential cofilin/actin ratios, thus affecting actin dynamics. High ratios of cofilin induce dissociation of the Arp2/3 complex and de-branching, resulting in formation of long, unbranched filaments [22]. Therefore, the actions of cofilin in the cell largely depend on its localization and the level of activity [22]. As described below, ABPs are key regulators of cocaine-induced plasticity of dendritic spines and are critical to understanding mechanisms associated with drug-induced remodeling of neurons within the corticostriatal circuitry.

APPROACHES TO ANALYZE CELL MORPHOLOGY

The majority of the morphological adaptations produced by abused substances that are discussed in this chapter were observed using one of two approaches (fluorescent dye labeling or the classical Golgi-Cox staining) to label neurons in fixed sections from rodent brain. Although the Golgi-Cox staining method has been utilized since the late 1800s to study brain morphology, this approach suffers from a number of limitations, most notably the underestimation of dendritic spine density and poor resolution of fine processes. In comparison, the use of fluorescent dyes (i.e., Lucifer Yellow, lipophilic dyes) offers improved resolution for labeling and analyzing dendrites and dendritic spines. For example, one study directly compared dendritic spine density in medium spiny neurons (MSNs) in the nucleus accumbens (NAc) in slices stained with the Golgi-Cox or diolistically labeled with the lipophilic dye DiI [23]. These authors reported more than a two-fold increase in dendritic spine density in the DiI-labeled MSNs in comparison with the Golgi-labeled MSNs. While the three-dimensional (3D) analysis of the DiI-labeled MSNs likely accounts for the enhanced spine density, DiI labels more spines with small head diameter suggesting that the lipophilic nature of the dye fills spines with thin necks better than the Golgi-Cox staining [23]. As described later in the chapter, fine morphological features of dendritic spine necks are crucial filters of spino-somatic voltage and biochemical signals that influence synaptic plasticity. Thus, fluorescent dyes provide superior ability to analyze micro-architectural characteristics of dendritic spines, especially when combined with recent advances in confocal microscopy (i.e., super-resolution imaging).

In addition to measuring total dendritic spine density, the density of the four morphological subclasses is also analyzed in addiction studies. Visual inspection of labeled neurons validates this approach for classification of spine subclass, as all three classes of spines and filopodia are present in adult neurons across multiple brain regions [24–26]. As described below, thin spines are highly prevalent in adult neurons across species and have been implicated in learning and addiction [25–27], demonstrating the usefulness of spine classification in behaviorally relevant paradigms. However, in contrast to this classification scheme, some studies suggest that dendritic spines do not exist in morphological subclasses. In support of this hypothesis, spine head diameter and other morphological characteristics of spines were reported on unimodal distributions and heterogeneous continuums along individual dendritic sections rather than a bimodal or multimodal distribution that would reflect multiple populations of spines [28–30]. Regardless of whether a categorization scheme exists for spines

FIG. 2 Structure and analysis of primary neurons in the corticostriatal circuitry. (A) The structure of typical cortical pyramidal neuron with basal and apical dendrites stemming from the soma. Shown are oblique apical dendrites proximal to the soma and apical tufts at the distal portion of the main apical dendrite. (B) An example of the centrifugal method of branch ordering of the apical and basal dendrites of a cortical pyramidal neuron. (C) The structure of a medium spiny neuron showing seven primary dendrites (range typically 4–8) stemming from a polygomial cell body that form a spherical dendritic field. (D) Superimposed concentric circles throughout the dendritic field of a medium spiny neuron with increasing radius used for Sholl analysis. The number of dendritic intersections at each equidistant concentric circle is counted.

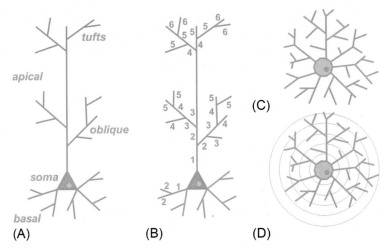

or not, it is clear that adaptations in dendritic spine density and morphology play crucial roles in the behavioral and functional impairments associated with detrimental neuropsychiatric disorders.

When combined with optical imaging, the Golgi-Cox staining, diolistic labeling with lipophilic dyes, transgenic reporter mice that express fluorescent proteins, and intracellular iontophoretic Lucifer yellow injections are standard methods for acquisition of full dendritic arbors of neurons. Once images of the arborization are acquired, they are typically collapsed prior to analysis in two dimensions. To avoid limitations associated with collapsing the 3D structures, recent technological advances in neuron reconstruction algorithms allow for analysis of noncollapsed dendritic arbors in 3D. Regardless of the number of dimensions, there are a number of different approaches for quantitative analysis of branching patterns including Scholl analysis, Strahler's centripetal ordering systems, and centrifugal ordering systems [31,32]. Fig. 2 shows examples of the cellular structure of two primary neurons (pyramidal neurons and MSNs) in the corticostriatal circuitry and two example approaches for analysis of the dendritic structure. The main objectives of dendritic arbor analysis are to determine the total length of the dendrites and dendritic substructures (apical vs. basal dendrites) and to order the segments into a definable structure. As one might expect, elucidating the 3D structure of a neuron is central to understanding its phenotype, function, and possible role in the neurocircuitry. In this chapter, we will review recent corticostriatal morphological adaptations in dendrites and dendritic spines in preclinical models of alcohol and drug exposure. A summary of the major and best characterized morphological adaptations induced by chronic alcohol and psychostimulant exposure is shown in Fig. 3.

ALCOHOL

Studies of alcohol dependence and chronic exposure in adult rodents report dendritic hypertrophy and enhanced dendritic spine density in pyramidal projection neurons in the prefrontal cortex (PFC) (Fig. 3). The increase in arborization by alcohol is most pronounced in proximal dendrites of layer II/III and V pyramidal neurons in prelimbic and infralimbic subregions of the medial PFC [33,34]. In contrast, alcohol-induced adaptations in dendritic arborization are absent in the orbitofrontal cortex (OFC) pyramidal neurons [35,36]. However, the effect of chronic alcohol exposure on dendritic spine density in all subregions of the PFC, including the OFC, is consistent with the majority of studies showing enhanced prevalence of spines [25,33,37]. Even though spine density is increased in the PFC subdivisions, alcohol differentially affects expression of spine subclasses. In the prelimbic medial PFC, chronic intermittent alcohol exposure increased mature, mushroom-shaped dendritic spines in layer V but not II/III [33,37] pyramidal neurons, while long, thin spines were enhanced in layer II/III of the lateral OFC pyramidal neurons [25] and layer V of the infralimbic cortex (P. Mulholland, *unpublished observation*). These studies show that alcohol-induced morphological adaptations in dendritic arborization and spine density can occur independently, which are supported by findings from previous reports where dendrite and spine reorganization do not always coincide [12,38]. Although there is consistency across studies in the PFC, alcohol-associated changes in dendrite and dendritic spine morphology are complex and depend on the cortical layer, subregion, and time since last alcohol exposure.

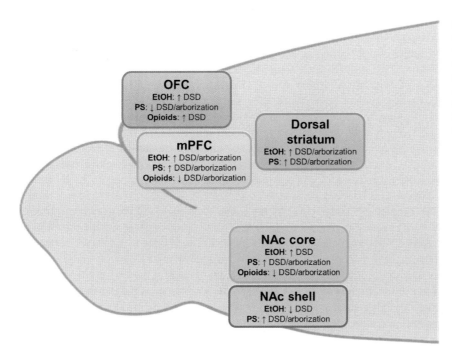

FIG. 3 Best characterized adaptations in dendritic spines and dendritic arborization in the corticostriatal circuitry in alcohol-, psychostimulant-, and opioid-treated animals. Coc, cocaine; DSD, dendritic spine density; EtOH, ethanol; NAc, nucleus accumbens; OFC, orbitofrontal cortex; mPFC, medial prefrontal cortex; PS, psychostimulants (including cocaine, amphetamine, and nicotine).

Similar to spine adaptations in the cortex, prolonged exposure to alcohol enhances the prevalence of dendritic spines and expands dendritic spine heads in NAc core MSNs [26,39,40] (Fig. 3). Withdrawal from chronic intermittent alcohol exposure produces dendritic hypertrophy in NAc core MSNs [39], and Zhou et al. [40] reported disoriented and "reactive" dendrites characterized by thickening and beading in NAc MSNs. In opposition to the NAc core, spine density in NAc shell MSNs was either unaltered or reduced by chronic alcohol exposure [39,41,42], and ethanol dependence did not alter dendritic arborization in NAc shell MSNs [39]. In the dorsomedial striatum (DMS), increased branching and dendritic length was observed only in dopamine D1R-expressing striatal MSNs in alcohol drinking mice [43]. These authors also reported that while drinking did not alter total spine density, there was a shift in the expression of mushroom and long spines. In a binge-like drinking model, total spine density was not altered in the dorsolateral striatum (DLS) of C57BL/6J mice [44]. However, dendritic spines were elevated in the putamen (a region analogous to the mouse DLS), but not caudate of nonhuman primates that had access to a long-term alcohol drinking and relapse model [45].

Comparable with adult alcohol models, emerging evidence shows that chronic and binge-like alcohol exposure during adolescence alters dendritic spine density in corticostriatal circuitry. Chronic intermittent ethanol exposure decreased total spine density while expanding head diameter of wide-headed spines in the infralimbic PFC of Thy-1 mice [46]. There are conflicting reports of adolescent CIE exposure on spines in the prelimbic subregion, with one study showing no change in total spine density and a study in rats reporting ~25% increase in total density and density of long "plasticity" spines [47]. The discrepancy could be explained by differences in the cortical layer or section of dendrites selected for analysis (layer II/III vs. layer V pyramidal neurons; apical vs. basal dendrites), the species, or time point since last alcohol exposure (3 d vs. 48 d). In a prenatal alcohol exposure model, complexity of layer II/III pyramidal neurons was reduced and there was a shift in spine subclass phenotype from immature to mature spines on basal dendrites in the adolescent brain [48]. Importantly, the spine adaptations that occur in response to developmental alcohol exposure are persistent, and they coincide with long-lasting behavioral and functional adolescent-typical phenotypes [49].

PSYCHOSTIMULANTS

Over the last two decades, numerous studies have shown that psychostimulants (cocaine, amphetamine, and nicotine) induce changes in dendrite and dendritic spine morphology in projection neurons within the corticostriatal pathway [3,50–52] (Fig. 3). Using Golgi staining and a variety of newer visualization methods, psychostimulants generally increase dendrite length and spine density in the medial PFC pyramidal neurons, while the reverse is true for pyramidal neurons in the OFC. In comparison, dendritic hypertrophy, enhanced spine density, and spine head expansion are reported in dorsal and ventral (NAc) striatum [3,52–54]. Studies using transgenic mice have investigated the effect of chronic cocaine treatment on dendritic spine density in distinct D1R- or D2R-expressing MSNs in the NAc core [55]. They found that after 28 days of cocaine treatment and 2 days of withdrawal, spine density increased in both Drd1-EGFP- and Drd2-EGFP-positive neurons but this increase was still evident 30 days after drug withdrawal in Drd1-EGFP-positive neurons. Other studies utilizing glutamate photo-uncaging and whole-cell recording to examine synaptic strength at individual spines found that average synaptic strength was reduced specifically at mushroom spines of D1R MSNs at 24 h of cocaine withdrawal [56]. Importantly, morphological adaptations produced by psychostimulants are similar to reported changes induced by alcohol exposure in the medial PFC, dorsal striatum, and NAc core. Opposite effects on dendritic spine density were reported in other brain regions in the corticostriatal pathway. For example, spine density was increased by alcohol and decreased by psychostimulants in the OFC.

In contrast to the unknown molecular mechanisms underlying alcohol-induced morphology changes, the molecular drivers of cocaine-induced spine morphology changes are known to occur at both the transcription and the protein level. In fact, expression of the transcription factor, ΔFosB, is associated with the formation and/or the maintenance of dendritic spines in D1R- and D2R-containing neurons in the NAc core in Drd1-EGFP or Drd2-EGFP mice following chronic cocaine treatment [55]. Further, elegant cell-type specific studies found that ΔFosB decreased silent synapses onto NAc shell, but not core, D2 dopamine receptor-expressing indirect pathway MSNs and that ΔFosB increased the density of immature spines in D1 but not D2 pathway MSNs [57]. Other studies have found that cocaine treatment induced changes in the nuclear subcellular location of the transcription factor, serum response factor (SRF). Furthermore, the same study also shows that regulation of SRF expression is critical for the effects of cocaine on dendritic spine formation [58]. Other transcription factors like MEF2 have been implicated in dendritic spine changes after cocaine treatment. In fact, reducing MEF2 activity in the NAc in vivo was required for the cocaine-induced increases in dendritic spine density. Interestingly, increasing MEF2 activity in the NAc, which blocked the cocaine-induced increase in dendritic spine density, enhanced sensitized behavioral responses to cocaine [59]. In addition, the transcription factor nuclear factor-kappa B (NF-κB) is increased after cocaine administration and regulates cocaine-induced dendritic spine changes on MSNs of the NAc [60]. Subsequent activation of NF-κB by inhibitor of κB kinase (IKKca and IKKdn) increased the number of dendritic spines on NAc neurons, while inhibition of NF-κB by IKKdn decreased basal dendritic spine number and blocked the increase in dendritic spines after chronic cocaine [60]. In addition to the changes on a morphological level, inhibition of NF-κB blocked the rewarding effects of cocaine and the expression of a previously formed conditioned place preference for cocaine. Together these data suggest that there are numerous transcription factors that act on the MSNs and regulate cocaine-induced changes in striatal dendritic spine morphology.

In addition to transcription factors, regulatory and signaling proteins regulate cocaine-induced striatal spine changes. In the NAc, cyclin-dependent kinase-5 (CDK5) is increased after cocaine administration, and intra-accumbens infusion of the CDK5 inhibitor roscovitine, attenuates cocaine-induced dendritic spine outgrowth in the NAc core and shell [61]. Other abundant protein kinases such as extracellular signal-regulated kinase (ERK) can regulate dendritic morphological changes induced by repeated cocaine administration [62]. It was shown that D1R signaling underpins cocaine-induced changes in both dendritic branching and spine density of the NAc shell MSNs [62]. Other cocaine studies have shown that chronic treatment decreased the active form of Rac1, a small GTPase. Furthermore, overexpression of a dominant negative mutant of Rac1 or local knockout of Rac1 is sufficient to increase the density of immature dendritic spines on NAc MSNs, whereas overexpression of a constitutively active Rac1 or light activation of a photoactivatable form of Rac1 blocked the chronic cocaine-induced changes [63].

While studies like these point to the involvement of specific regulatory proteins, the morphological changes in the spine are regulated by ABPs. In fact, cocaine administration and withdrawal induced changes in G- and F-actin together with actin cycling proteins that led to subsequent changes in dendritic spine plasticity in the NAc [23,53,64]. Interestingly, there was a decrease in G-actin, (p)-cofilin and LIMK after cocaine withdrawal, yet disrupting actin cycling with either an actin polymerization inhibitor, lantrunculin A, or accelerating actin depolymerization with a LIMK inhibitor, lead to augmentation of cocaine-induced reinstatement of drug seeking [64], suggesting potentially protective effects of morphological changes against cocaine-seeking behavior. Studies investigating the role of stress-induced motor cross-sensitization with cocaine have found a decrease in the phosphorylation of two ABPs, cofilin and cortactin, and an increase in the PSD size and the surface expression of GluA1 [65]. Furthermore, inhibiting actin cycling and polymerization with latrunculin A into the NAc inhibited the expression of cross-sensitization to cocaine and restored the expression of GluA1 to control levels. These data suggest that manipulating the actin cycling mechanism plays a crucial role in regulation of the addictive phenotype of cocaine, as well as alcohol adaptive plasticity of dendritic spines [66].

OPIOIDS

In contrast to psychostimulants, chronic opioid (i.e., morphine, heroin) exposure and withdrawal reduces the complexity of dendritic arbors and decreases dendritic spine density and head diameter in the PFC and NAc neurons [67–69]. Consistent with opposite effects of psychostimulants on morphological plasticity within subregions of the PFC, morphine exposure significantly increased dendritic spines in the OFC [70]. Like many of the other abused drugs, morphological adaptations induced by chronic heroin self-administration coincide with aberrant synaptic plasticity within the corticostriatal pathway [71].

INHALANTS

Although inhalant addiction is relatively rare, inhaling intoxicating amounts of volatile solvents found in paint, glues, and aerosol cans is extremely dangerous and has a high abuse potential. Interestingly, volatile solvent use is more prevalent in adolescents than most other abused drugs. Given that adolescence is period of significant brain development, it is critical to understand how abused inhalants affect neuron morphology and function. Acute inhalant exposure is known to reduce dendritic spine motility in cultured neurons [72]. Chronic toluene exposure in adults decreases the complexity of dendritic arbors in superficial layer cortical neurons [73], and increases basal dendritic spine density in deep-layer prelimbic PFC neurons of adult rodents exposed during adolescence [74].

LIMITATIONS AND FUTURE DIRECTIONS

A major limitation of postmortem approaches that measure dendrite and dendritic spine morphology is the missing link with functional adaptations in neuron firing. What is also missing is how morphological and functional changes in dendrites and spines influence drug-seeking behavior. Dendrites shape neuronal output by actively integrating synaptic information mediated by Ca^{2+}-, Na^{+}-, and NMDA receptor-dependent dendritic supralinear events [75]. Ion channels enriched in dendritic arbors enable dendritic excitability and limit the amplitude of back-propagating action potentials (bAP) in distal dendrites. Moreover, apical dendrite total length and branching structure influences firing patterns of cortical neurons [76], and dendritic diameter influences synaptic integration [36,77]. This is important as studies have shown that abused drugs can alter the diameter of dendrites, as discussed above, as well as ion channel function and intrinsic excitability [78–81]. There are approaches that can examine how morphological adaptations produced by drugs of abuse alter the relationship between synaptic and dendritic signaling and output of neurons. However, unraveling these complex mechanisms is an arduous task that has been largely unexplored in the addiction field. In an elegant study, Granato and colleagues showed that alcohol exposure early in development attenuated dendritic calcium spikes and reduced intrinsic excitability and dendritic arborization in cortical pyramidal neurons [82]. In an organotypic slice culture model, chronic alcohol reduced A-type currents and expression of Kv4.2 channel complexes [83]. These adaptations in K^{+} channel currents were associated with enhanced bAP-evoked Ca^{2+} transients in distal apical dendrites of CA1 pyramidal neurons. In the medial PFC pyramidal neurons, studies have demonstrated that chronic cocaine administration prolonged the duration of dendritic plateau potentials via increased voltage-sensitive Ca^{2+} currents [84], increased dendritic arborization [52], and enhanced intrinsic excitability [84,85]. Together, these findings indicate that the ability of drugs of abuse to modify dendritic arborization and integration is a possible critical mechanisms underpinning behavioral impairments driving drug-seeking and relapse-like behaviors that requires further study.

There are a number of recent developments in techniques and imaging technology that when combined with spine analysis can advance our understanding of morphological adaptations induced by abused substances. Because the approaches discussed in this chapter do not concurrently determine if there is an active presynaptic terminal opposing the spine's PSD, one can only deduce that labeled spines represent the postsynaptic site of *putative* glutamatergic synapses. Moreover, these approaches cannot determine where the spine is in its lifecycle. This is crucial because of the transient (1–4 days) subpopulation and experience-dependent turnover of spines in the adult cortex [7,86,87]. Importantly, not all new spines have opposing presynaptic terminals [88], suggesting that a proportion of spines analyzed in fixed tissue are not part of asymmetric synapses. Future studies should combine immunofluorescent staining of presynaptic proteins or transgenic mice [89] with diolistic labeling of neurons to determine if labeled spines in fixed tissue samples oppose an active presynaptic terminal. In addition, with advances in circuitry tract tracing techniques and whole brain imaging, it is now possible to determine drug-induced adaptations in the projectome (i.e., dendritic architecture and synaptic structure in projection-specific circuits).

Fluorescent light microscopy while allowing for immunodetection of spine proteins cannot resolve the nanoscale organization within spine subcompartments because of the inherent limitation of light diffraction. There are recent technological and computational advances in confocal microscopy that provide substantial improvements in resolution over traditional capabilities that can reach a lateral and axial resolutions of ∼200–250 and ∼500–800 nm, respectively. Termed super-resolution microscopy, these advance approaches (e.g., STORM, PALM, and STED) allow resolution beyond the diffraction limit into a biologically meaningful range (10–100 nm) where proteins in spine subcompartments can be resolved [90]. Indeed, super-resolution microscopy has already revealed molecular mechanisms underpinning the structural organization of dendritic spines [91]. For example, a recent study using STED nanoscopy demonstrated periodic ring-like F-actin structures in the neck of dendritic spines [92] (Fig. 4), similar to what was reported for axons. Much like the traditional Golgi-Cox staining was surpassed by diolistic labeling for more accurate identification of dendritic spine density [23], super-resolution technology offers an advanced method for the quantitation of dendritic spine density and head morphology [91]. As an example, when four matching dendritic sections were acquired with a GaAsP detector and an Airyscan super-resolution detector (Fig. 5A), there was better resolution of spine heads and necks (Fig. 5B) and significantly more dendritic spines identified in the dendritic sections that were acquired using the super-resolution detector (Fig. 5C). Thus, these technological advances provide exciting new opportunities to identify how abused drugs shape dendritic spines and how they modify molecular elements of spine subcompartments.

One area of morphological investigation that is overlooked in addiction studies is the plasticity of dendritic spine necks. Although optical and computational studies are inconsistent, spine necks are thought to act as diffusion barriers that electrically and biochemically compartmentalize the spine head from the parent dendrite. In an elegant study, Jayant and colleagues used quantam-dot-coated nanopipettes to record from spine heads while also simultaneously performing somatic patch-clamp recordings [93]. While bAPs fully invaded dendritic spines as expected, they reported an attenuation of the large excitatory postsynaptic potentials recorded at spine heads on proximal dendrites at the soma. This study provides the first direct evidence that spine necks isolate spino-somatic voltage signals. In addition, a recent publication using super-resolution imaging and two-photon glutamate uncaging demonstrated that spine neck morphology is a crucial regulator of biochemical diffusion from spines into parent dendritic shafts [29]. Thus, studies need to determine if drug-induced morphological changes influence spino-somatic voltage signaling and biochemical compartmentalization between the spine head and the dendritic shaft.

Interestingly, a recent study demonstrated that localization of βIII spectrin to the spine neck (Fig. 4) is important for controlling both its constricted characteristic shape and the amplitude of spontaneous glutamatergic events that reach the soma [94]. βIII spectrin is the first molecular target at the spine neck that acts as a barrier to isolate spino-somatic signaling. These results may be critical for understanding the plasticity of addiction because spine neck morphology is

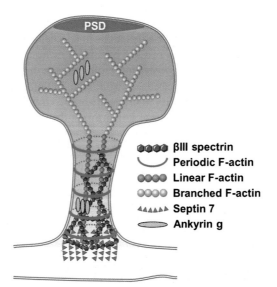

FIG. 4 Heterogeneous molecular organization of dendritic spines. Similar to actin structure in axons, F-actin is organized in periodic rings in spine necks. Branched F-actin is enriched in spine heads, and βIII spectrin and linear F-actin are present in spine necks. Septin 7 localizes in filaments to the base of dendritic spines where it functions as a diffusion barrier, and ankyrin g is expressed in nanoclusters within the spine neck and head.

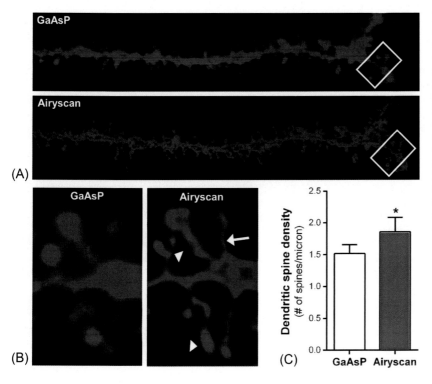

FIG. 5 Super-resolution imaging reveals enhanced structural details of dendritic spines and increased dendritic spine density. (A) Comparison of the same dendritic section acquired using GaAsP and Airyscan super-resolution detectors on a Zeiss LSM 880 confocal microscopy system with a 63 × oil objective (1.4 NA). (B) Magnified image from the white box shown in panel A. There is improved structural clarity of dendritic necks (arrow) and spine heads (arrowhead) in images acquired with the Airyscan detector that achieves a 1.7-fold increase in resolution. (C) Quantitation of four matching dendritic sections averaging $50.4 \pm 4.99\,\mu m$ in length revealed more dendritic spines in the images acquired with the super-resolution detector (paired t-test, $t(3) = 3.306$, $p < 0.05$).

regulated by activity-dependent processes [95] and spine necks become shorter and wider following induction of long-term potentiation [29]. In addition to βIII spectrin, dendritic spine neck morphology is also influenced by septin 7, a GTP-binding protein expressed in filaments at the neck of dendritic spines [96,97], and ankyrin g, a scaffolding protein that localizes to nanodomains in spine heads and necks [98]. To our knowledge, the role of βIII spectrin, septin 7, and ankyrin g in regulating morphological plasticity induced by drugs of abuse is unexplored, although transcriptome profiling of postmortem brain tissue from alcoholics and binge alcohol drinking rats revealed increases (*SPTBN2*) or decreases (*ANK3*, *SEPT7*) in transcript levels [99–102]. Future studies should examine the role of these spine neck proteins as possible molecular mechanisms that underpin aberrant dendritic spine morphology following drug intake and during abstinence.

In summary, we have outlined overwhelming evidence that abused drugs produce morphological plasticity of dendrites and dendritic spines in the corticostriatal pathway in preclinical rodent models of addiction. In general, psychostimulants and alcohol elicit similar morphological adaptations in the medial PFC, dorsal striatum, and NAc core, suggesting that there are possible shared mechanisms underpinning structural changes in dendrites and dendritic spines. It is also clear that psychostimulants and opiates produce opposite adaptations in the medial PFC and OFC subregions of the cortex. Finally, many of the changes that occur in response to abused drugs have been reported in the thin subclass of dendritic spines. Thus, morphological adaptations in the corticostriatal circuitry that occur during drug and alcohol intake and protracted withdrawal may represent neural correlates that drive drug-seeking and relapse behavior. However, a remaining limitation is the difficulty linking dendrite and dendritic spine morphological adaptations with behaviors associated with drug intake. With the advance in many techniques used to study neuronal morphology that can concurrently examine function, future studies will be better suited to identify novel therapeutic targets that underlie drug-induced dendritic adaptations and reduce the high rates of relapse that plague individuals with SUDs.

ACKNOWLEDGMENTS

These studies were supported by National Institutes of Health grants AA020930 (PJM), AA023288 (PJM), AA024603 (PJM), ODO21532 (PJM), and AA024426 (JDU). The authors declare no conflict of interest.

REFERENCES

[1] Zhu J, Reith ME. Role of the dopamine transporter in the action of psychostimulants, nicotine, and other drugs of abuse. CNS Neurol Disord Drug Targets 2008;7(5):393–409.

[2] Sacks JJ, Gonzales KR, Bouchery EE, Tomedi LE, Brewer RD. 2010 national and state costs of excessive alcohol consumption. Am J Prev Med 2015;49(5):e73–79. https://doi.org/10.1016/j.amepre.2015.05.031.

[3] Mulholland PJ, Chandler LJ, Kalivas PW. Signals from the fourth dimension regulate drug relapse. Trends Neurosci 2016;39(7):472–85. https://doi.org/10.1016/j.tins.2016.04.007.

[4] Rochefort NL, Konnerth A. Dendritic spines: from structure to in vivo function. EMBO Rep 2012;13(8):699–708. https://doi.org/10.1038/embor.2012.102.

[5] Arnsten AF, Wang MJ, Paspalas CD. Neuromodulation of thought: flexibilities and vulnerabilities in prefrontal cortical network synapses. Neuron 2012;76(1):223–39. https://doi.org/10.1016/j.neuron.2012.08.038.

[6] Bourne J, Harris KM. Do thin spines learn to be mushroom spines that remember? Curr Opin Neurobiol 2007;17(3):381–6. https://doi.org/10.1016/j.conb.2007.04.009.

[7] Holtmaat A, Wilbrecht L, Knott GW, Welker E, Svoboda K. Experience-dependent and cell-type-specific spine growth in the neocortex. Nature 2006;441(7096):979–83. https://doi.org/10.1038/nature04783.

[8] Paspalas CD, Wang M, Arnsten AF. Constellation of HCN channels and cAMP regulating proteins in dendritic spines of the primate prefrontal cortex: potential substrate for working memory deficits in schizophrenia. Cereb Cortex 2013;23(7):1643–54. https://doi.org/10.1093/cercor/bhs152.

[9] Sala C, Segal M. Dendritic spines: the locus of structural and functional plasticity. Physiol Rev 2014;94(1):141–88. https://doi.org/10.1152/physrev.00012.2013.

[10] Yang G, Pan F, Gan WB. Stably maintained dendritic spines are associated with lifelong memories. Nature 2009;462(7275):920–4. https://doi.org/10.1038/nature08577.

[11] Caroni P, Donato F, Muller D. Structural plasticity upon learning: regulation and functions. Nat Rev Neurosci 2012;13(7):478–90. https://doi.org/10.1038/nrn3258.

[12] Kolb B, Cioe J, Comeau W. Contrasting effects of motor and visual spatial learning tasks on dendritic arborization and spine density in rats. Neurobiol Learn Mem 2008;90(2):295–300. https://doi.org/10.1016/j.nlm.2008.04.012.

[13] Smart FM, Halpain S. Regulation of dendritic spine stability. Hippocampus 2000;10(5):542–54. https://doi.org/10.1002/1098-1063(2000)10:5<542::AID-HIPO4>3.0.CO;2-7.

[14] Woodrum DT, Rich SA, Pollard TD. Evidence for biased bidirectional polymerization of actin filaments using heavy meromyosin prepared by an improved method. J Cell Biol 1975;67(1):231–7.

[15] Sekino Y, Kojima N, Shirao T. Role of actin cytoskeleton in dendritic spine morphogenesis. Neurochem Int 2007;51(2–4):92–104. https://doi.org/10.1016/j.neuint.2007.04.029.

[16] Pollard TD. Regulation of actin filament assembly by Arp2/3 complex and formins. Annu Rev Biophys Biomol Struct 2007;36:451–77. https://doi.org/10.1146/annurev.biophys.35.040405.101936.

[17] Racz B, Weinberg RJ. Organization of the Arp2/3 complex in hippocampal spines. J Neurosci 2008;28(22):5654–9. https://doi.org/10.1523/JNEUROSCI.0756-08.2008.

[18] Ammer AG, Weed SA. Cortactin branches out: roles in regulating protrusive actin dynamics. Cell Motil Cytoskeleton 2008;65(9):687–707. https://doi.org/10.1002/cm.20296.

[19] Racz B, Weinberg RJ. The subcellular organization of cortactin in hippocampus. J Neurosci 2004;24(46):10310–7. https://doi.org/10.1523/JNEUROSCI.2080-04.2004.

[20] Bamburg JR, Wiggan OP. ADF/cofilin and actin dynamics in disease. Trends Cell Biol 2002;12(12):598–605.

[21] Ono S. Regulation of actin filament dynamics by actin depolymerizing factor/cofilin and actin-interacting protein 1: new blades for twisted filaments. Biochemistry 2003;42(46):13363–70. https://doi.org/10.1021/bi034600x.

[22] Andrianantoandro E, Pollard TD. Mechanism of actin filament turnover by severing and nucleation at different concentrations of ADF/cofilin. Mol Cell 2006;24(1):13–23. https://doi.org/10.1016/j.molcel.2006.08.006.

[23] Shen HW, Toda S, Moussawi K, Bouknight A, Zahm DS, Kalivas PW. Altered dendritic spine plasticity in cocaine-withdrawn rats. J Neurosci 2009;29(9):2876–84. https://doi.org/10.1523/JNEUROSCI.5638-08.2009.

[24] Harris KM, Jensen FE, Tsao B. Three-dimensional structure of dendritic spines and synapses in rat hippocampus (CA1) at postnatal day 15 and adult ages: implications for the maturation of synaptic physiology and long-term potentiation. J Neurosci 1992;12(7):2685–705.

[25] McGuier NS, Padula AE, Lopez MF, Woodward JJ, Mulholland PJ. Withdrawal from chronic intermittent alcohol exposure increases dendritic spine density in the lateral orbitofrontal cortex of mice. Alcohol 2015;49(1):21–7. https://doi.org/10.1016/j.alcohol.2014.07.017.

[26] Uys JD, McGuier NS, Gass JT, Griffin 3rd WC, Ball LE, Mulholland PJ. Chronic intermittent ethanol exposure and withdrawal leads to adaptations in nucleus accumbens core postsynaptic density proteome and dendritic spines. Addict Biol 2016;21(3):560–74. https://doi.org/10.1111/adb.12238.

[27] Dumitriu D, Hao J, Hara Y, Kaufmann J, Janssen WG, Lou W, Morrison JH. Selective changes in thin spine density and morphology in monkey prefrontal cortex correlate with aging-related cognitive impairment. J Neurosci 2010;30(22):7507–15. https://doi.org/10.1523/JNEUROSCI.6410-09.2010.

[28] Konur S, Rabinowitz D, Fenstermaker VL, Yuste R. Systematic regulation of spine sizes and densities in pyramidal neurons. J Neurobiol 2003;56(2):95–112. https://doi.org/10.1002/neu.10229.

[29] Tonnesen J, Katona G, Rozsa B, Nagerl UV. Spine neck plasticity regulates compartmentalization of synapses. Nat Neurosci 2014;17(5):678–85. https://doi.org/10.1038/nn.3682.

[30] Wallace W, Bear MF. A morphological correlate of synaptic scaling in visual cortex. J Neurosci 2004;24(31):6928–38. https://doi.org/10.1523/JNEUROSCI.1110-04.2004.

[31] Sholl DA. Dendritic organization in the neurons of the visual and motor cortices of the cat. J Anat 1953;87(4):387–406.

[32] Uylings HB, Smit GJ, Veltman WA. Ordering methods in quantitative analysis of branching structures of dendritic trees. Adv Neurol 1975;12:347–54.

[33] Kim A, Zamora-Martinez ER, Edwards S, Mandyam CD. Structural reorganization of pyramidal neurons in the medial prefrontal cortex of alcohol dependent rats is associated with altered glial plasticity. Brain Struct Funct 2015;220(3):1705–20. https://doi.org/10.1007/s00429-014-0755-3.

[34] Klenowski PM, Fogarty MJ, Shariff M, Belmer A, Bellingham MC, Bartlett SE. Increased synaptic excitation and abnormal dendritic structure of prefrontal cortex layer V pyramidal neurons following prolonged binge-like consumption of ethanol. eNeuro 2016;3(6). https://doi.org/10.1523/ENEURO.0248-16.2016.

[35] DePoy L, Daut R, Brigman JL, MacPherson K, Crowley N, Gunduz-Cinar O, Holmes A. Chronic alcohol produces neuroadaptations to prime dorsal striatal learning. Proc Natl Acad Sci U S A 2013;110(36):14783–8. https://doi.org/10.1073/pnas.1308198110.

[36] Holmes WR. The role of dendritic diameters in maximizing the effectiveness of synaptic inputs. Brain Res 1989;478(1):127–37.

[37] Kroener S, Mulholland PJ, New NN, Gass JT, Becker HC, Chandler LJ. Chronic alcohol exposure alters behavioral and synaptic plasticity of the rodent prefrontal cortex. PLoS One 2012;7(5). https://doi.org/10.1371/journal.pone.0037541.

[38] Drakew A, Frotscher M, Heimrich B. Blockade of neuronal activity alters spine maturation of dentate granule cells but not their dendritic arborization. Neuroscience 1999;94(3):767–74.

[39] Peterson VL, McCool BA, Hamilton DA. Effects of ethanol exposure and withdrawal on dendritic morphology and spine density in the nucleus accumbens core and shell. Brain Res 2015;1594:125–35. https://doi.org/10.1016/j.brainres.2014.10.036.

[40] Zhou FC, Anthony B, Dunn KW, Lindquist WB, Xu ZC, Deng P. Chronic alcohol drinking alters neuronal dendritic spines in the brain reward center nucleus accumbens. Brain Res 2007;1134(1):148–61. https://doi.org/10.1016/j.brainres.2006.11.046.

[41] Nona CN, Bermejo MK, Ramsey AJ, Nobrega JN. Changes in dendritic spine density in the nucleus accumbens do not underlie ethanol sensitization. Synapse 2015;69(12):607–10. https://doi.org/10.1002/syn.21862.

[42] Spiga S, Talani G, Mulas G, Licheri V, Fois GR, Muggironi G, Diana M. Hampered long-term depression and thin spine loss in the nucleus accumbens of ethanol-dependent rats. Proc Natl Acad Sci U S A 2014;111(35):E3745–3754. https://doi.org/10.1073/pnas.1406768111.

[43] Wang J, Cheng Y, Wang X, Roltsch Hellard E, Ma T, Gil H, Ron D. Alcohol elicits functional and structural plasticity selectively in dopamine D1 receptor-expressing neurons of the dorsomedial striatum. J Neurosci 2015;35(33):11634–43. https://doi.org/10.1523/JNEUROSCI.0003-15.2015.

[44] Wilcox MV, Cuzon Carlson VC, Sherazee N, Sprow GM, Bock R, Thiele TE, Alvarez VA. Repeated binge-like ethanol drinking alters ethanol drinking patterns and depresses striatal GABAergic transmission. Neuropsychopharmacology 2014;39(3):579–94. https://doi.org/10.1038/npp.2013.230.

[45] Cuzon Carlson VC, Seabold GK, Helms CM, Garg N, Odagiri M, Rau AR, Grant KA. Synaptic and morphological neuroadaptations in the putamen associated with long-term, relapsing alcohol drinking in primates. Neuropsychopharmacology 2011;36(12):2513–28. https://doi.org/10.1038/npp.2011.140.

[46] Jury NJ, Pollack GA, Ward MJ, Bezek JL, Ng AJ, Pinard CR, Holmes A. Chronic ethanol during adolescence impacts Corticolimbic dendritic spines and behavior. Alcohol Clin Exp Res 2017;41(7):1298–308. https://doi.org/10.1111/acer.13422.

[47] Trantham-Davidson H, Centanni SW, Garr SC, New NN, Mulholland PJ, Gass JT, Chandler LJ. Binge-like alcohol exposure during adolescence disrupts dopaminergic neurotransmission in the adult Prelimbic cortex. Neuropsychopharmacology 2017;42(5):1024–36. https://doi.org/10.1038/npp.2016.190.

[48] Hamilton GF, Whitcher LT, Klintsova AY. Postnatal binge-like alcohol exposure decreases dendritic complexity while increasing the density of mature spines in mPFC layer II/III pyramidal neurons. Synapse 2010;64(2):127–35. https://doi.org/10.1002/syn.20711.

[49] Spear LP, Swartzwelder HS. Adolescent alcohol exposure and persistence of adolescent-typical phenotypes into adulthood: a mini-review. Neurosci Biobehav Rev 2014;45:1–8. https://doi.org/10.1016/j.neubiorev.2014.04.012.

[50] DePoy LM, Gourley SL. Synaptic cytoskeletal plasticity in the prefrontal cortex following psychostimulant exposure. Traffic 2015;16(9):919–40. https://doi.org/10.1111/tra.12295.

[51] Gipson CD, Reissner KJ, Kupchik YM, Smith AC, Stankeviciute N, Hensley-Simon ME, Kalivas PW. Reinstatement of nicotine seeking is mediated by glutamatergic plasticity. Proc Natl Acad Sci U S A 2013;110(22):9124–9. https://doi.org/10.1073/pnas.1220591110.

[52] Robinson TE, Kolb B. Structural plasticity associated with exposure to drugs of abuse. Neuropharmacology 2004;47(Suppl. 1):33–46. https://doi.org/10.1016/j.neuropharm.2004.06.025.

[53] Dobi A, Seabold GK, Christensen CH, Bock R, Alvarez VA. Cocaine-induced plasticity in the nucleus accumbens is cell specific and develops without prolonged withdrawal. J Neurosci 2011;31(5):1895–904. https://doi.org/10.1523/JNEUROSCI.5375-10.2011.

[54] Li J, Zhang L, Chen Z, Xie M, Huang L, Xue J, Zhang L. Cocaine activates Rac1 to control structural and behavioral plasticity in caudate putamen. Neurobiol Dis 2015;75:159–76. https://doi.org/10.1016/j.nbd.2014.12.031.

[55] Lee KW, Kim Y, Kim AM, Helmin K, Nairn AC, Greengard P. Cocaine-induced dendritic spine formation in D1 and D2 dopamine receptor-containing medium spiny neurons in nucleus accumbens. Proc Natl Acad Sci U S A 2006;103(9):3399–404. https://doi.org/10.1073/pnas.0511244103.

[56] Khibnik LA, Beaumont M, Doyle M, Heshmati M, Slesinger PA, Nestler EJ, Russo SJ. Stress and cocaine trigger divergent and cell type-specific regulation of synaptic transmission at single spines in nucleus Accumbens. Biol Psychiatry 2016;79(11):898–905. https://doi.org/10.1016/j.biopsych.2015.05.022.

[57] Grueter BA, Robison AJ, Neve RL, Nestler EJ, Malenka RC. FosB differentially modulates nucleus accumbens direct and indirect pathway function. Proc Natl Acad Sci U S A 2013;110(5):1923–8. https://doi.org/10.1073/pnas.1221742110.

[58] Cahill ME, Walker DM, Gancarz AM, Wang ZJ, Lardner CK, Bagot RC, Nestler EJ. The dendritic spine morphogenic effects of repeated cocaine use occur through the regulation of serum response factor signaling. Mol Psychiatry 2017. https://doi.org/10.1038/mp.2017.116.

[59] Pulipparacharuvil S, Renthal W, Hale CF, Taniguchi M, Xiao G, Kumar A, Cowan CW. Cocaine regulates MEF2 to control synaptic and behavioral plasticity. Neuron 2008;59(4):621–33. https://doi.org/10.1016/j.neuron.2008.06.020.

[60] Russo SJ, Wilkinson MB, Mazei-Robison MS, Dietz DM, Maze I, Krishnan V, Nestler EJ. Nuclear factor kappa B signaling regulates neuronal morphology and cocaine reward. J Neurosci 2009;29(11):3529–37. https://doi.org/10.1523/JNEUROSCI.6173-08.2009.

[61] Norrholm SD, Bibb JA, Nestler EJ, Ouimet CC, Taylor JR, Greengard P. Cocaine-induced proliferation of dendritic spines in nucleus accumbens is dependent on the activity of cyclin-dependent kinase-5. Neuroscience 2003;116(1):19–22.

[62] Ren Z, Sun WL, Jiao H, Zhang D, Kong H, Wang X, Xu M. Dopamine D1 and N-methyl-D-aspartate receptors and extracellular signal-regulated kinase mediate neuronal morphological changes induced by repeated cocaine administration. Neuroscience 2010;168(1):48–60. https://doi.org/10.1016/j.neuroscience.2010.03.034.

[63] Dietz DM, Sun H, Lobo MK, Cahill ME, Chadwick B, Gao V, Nestler EJ. Rac1 is essential in cocaine-induced structural plasticity of nucleus accumbens neurons. Nat Neurosci 2012;15(6):891–6. https://doi.org/10.1038/nn.3094.

[64] Toda S, Shen HW, Peters J, Cagle S, Kalivas PW. Cocaine increases actin cycling: effects in the reinstatement model of drug seeking. J Neurosci 2006;26(5):1579–87. https://doi.org/10.1523/JNEUROSCI.4132-05.2006.

[65] Esparza MA, Bollati F, Garcia-Keller C, Virgolini MB, Lopez LM, Brusco A, Cancela LM. Stress-induced sensitization to cocaine: actin cytoskeleton remodeling within mesocorticolimbic nuclei. Eur J Neurosci 2012;36(8):3103–17. https://doi.org/10.1111/j.1460-9568.2012.08239.x.

[66] Mulholland PJ, Chandler LJ. The thorny side of addiction: adaptive plasticity and dendritic spines. ScientificWorldJournal 2007;7:9–21. https://doi.org/10.1100/tsw.2007.247.

[67] Robinson TE, Kolb B. Morphine alters the structure of neurons in the nucleus accumbens and neocortex of rats. Synapse 1999;33(2):160–2. https://doi.org/10.1002/(SICI)1098-2396(199908)33:2<160::AID-SYN6>3.0.CO;2-S.

[68] Shen H, Moussawi K, Zhou W, Toda S, Kalivas PW. Heroin relapse requires long-term potentiation-like plasticity mediated by NMDA2b-containing receptors. Proc Natl Acad Sci U S A 2011;108(48):19407–12. https://doi.org/10.1073/pnas.1112052108.

[69] Spiga S, Puddu MC, Pisano M, Diana M. Morphine withdrawal-induced morphological changes in the nucleus accumbens. Eur J Neurosci 2005;22(9):2332–40. https://doi.org/10.1111/j.1460-9568.2005.04416.x.

[70] Robinson TE, Gorny G, Savage VR, Kolb B. Widespread but regionally specific effects of experimenter- versus self-administered morphine on dendritic spines in the nucleus accumbens, hippocampus, and neocortex of adult rats. Synapse 2002;46(4):271–9. https://doi.org/10.1002/syn.10146.

[71] Shen H, Kalivas PW. Reduced LTP and LTD in prefrontal cortex synapses in the nucleus accumbens after heroin self-administration. Int J Neuropsychopharmacol 2013;16(5):1165–7. https://doi.org/10.1017/S1461145712001071.

[72] Kaech S, Brinkhaus H, Matus A. Volatile anesthetics block actin-based motility in dendritic spines. Proc Natl Acad Sci U S A 1999;96(18):10433–7.

[73] Pascual R, Bustamante C. Melatonin promotes distal dendritic ramifications in layer II/III cortical pyramidal cells of rats exposed to toluene vapors. Brain Res 2010;1355:214–20. https://doi.org/10.1016/j.brainres.2010.07.086.

[74] Braunscheidel KM, Gass JT, Mulholland PJ, Floresco SB, Woodward JJ. Persistent cognitive and morphological alterations induced by repeated exposure of adolescent rats to the abused inhalant toluene. Neurobiol Learn Mem 2017;144:136–46. https://doi.org/10.1016/j.nlm.2017.07.007.

[75] Palmer LM. Dendritic integration in pyramidal neurons during network activity and disease. Brain Res Bull 2014;103:2–10. https://doi.org/10.1016/j.brainresbull.2013.09.010.

[76] van Elburg RA, van Ooyen A. Impact of dendritic size and dendritic topology on burst firing in pyramidal cells. PLoS Comput Biol 2010;6(5). https://doi.org/10.1371/journal.pcbi.1000781.

[77] Branco T, Clark BA, Hausser M. Dendritic discrimination of temporal input sequences in cortical neurons. Science 2010;329(5999):1671–5. https://doi.org/10.1126/science.1189664.

[78] Kourrich S, Calu DJ, Bonci A. Intrinsic plasticity: an emerging player in addiction. Nat Rev Neurosci 2015;16(3):173–84. https://doi.org/10.1038/nrn3877.

[79] Nimitvilai S, Lopez MF, Mulholland PJ, Woodward JJ. Chronic intermittent ethanol exposure enhances the excitability and synaptic plasticity of lateral orbitofrontal cortex neurons and induces a tolerance to the acute inhibitory actions of ethanol. Neuropsychopharmacology 2016;41(4):1112–27. https://doi.org/10.1038/npp.2015.250.

[80] Nimitvilai S, Uys JD, Woodward JJ, Randall PK, Ball LE, Williams RW, Mulholland PJ. Orbitofrontal neuroadaptations and cross-species synaptic biomarkers in heavy-drinking macaques. J Neurosci 2017;37(13):3646–60. https://doi.org/10.1523/JNEUROSCI.0133-17.2017.

[81] Padula AE, Griffin 3rd WC, Lopez MF, Nimitvilai S, Cannady R, McGuier NS, Mulholland PJ. KCNN genes that encode small-conductance Ca2+-activated K+ channels influence alcohol and drug addiction. Neuropsychopharmacology 2015;40(8):1928–39. https://doi.org/10.1038/npp.2015.42.

[82] Granato A, Palmer LM, De Giorgio A, Tavian D, Larkum ME. Early exposure to alcohol leads to permanent impairment of dendritic excitability in neocortical pyramidal neurons. J Neurosci 2012;32(4):1377–82. https://doi.org/10.1523/JNEUROSCI.5520-11.2012.

[83] Mulholland PJ, Spencer KB, Hu W, Kroener S, Chandler LJ. Neuroplasticity of A-type potassium channel complexes induced by chronic alcohol exposure enhances dendritic calcium transients in hippocampus. Psychopharmacology (Berl) 2015;232(11):1995–2006. https://doi.org/10.1007/s00213-014-3835-4.

[84] Nasif FJ, Sidiropoulou K, Hu XT, White FJ. Repeated cocaine administration increases membrane excitability of pyramidal neurons in the rat medial prefrontal cortex. J Pharmacol Exp Ther 2005;312(3):1305–13. https://doi.org/10.1124/jpet.104.075184.

[85] Dong Y, Nasif FJ, Tsui JJ, Ju WY, Cooper DC, Hu XT, White FJ. Cocaine-induced plasticity of intrinsic membrane properties in prefrontal cortex pyramidal neurons: adaptations in potassium currents. J Neurosci 2005;25(4):936–40. https://doi.org/10.1523/JNEUROSCI.4715-04.2005.

[86] Holtmaat AJ, Trachtenberg JT, Wilbrecht L, Shepherd GM, Zhang X, Knott GW, Svoboda K. Transient and persistent dendritic spines in the neocortex in vivo. Neuron 2005;45(2):279–91. https://doi.org/10.1016/j.neuron.2005.01.003.

[87] Trachtenberg JT, Chen BE, Knott GW, Feng G, Sanes JR, Welker E, Svoboda K. Long-term in vivo imaging of experience-dependent synaptic plasticity in adult cortex. Nature 2002;420(6917):788–94. https://doi.org/10.1038/nature01273.

[88] Knott GW, Holtmaat A, Wilbrecht L, Welker E, Svoboda K. Spine growth precedes synapse formation in the adult neocortex in vivo. Nat Neurosci 2006;9(9):1117–24. https://doi.org/10.1038/nn1747.

[89] Heck N, Dos Santos M, Amairi B, Salery M, Besnard A, Herzog E, Caboche J. A new automated 3D detection of synaptic contacts reveals the formation of cortico-striatal synapses upon cocaine treatment in vivo. Brain Struct Funct 2015;220(5):2953–66. https://doi.org/10.1007/s00429-014-0837-2.

[90] Maglione M, Sigrist SJ. Seeing the forest tree by tree: super-resolution light microscopy meets the neurosciences. Nat Neurosci 2013;16(7):790–7. https://doi.org/10.1038/nn.3403.

[91] MacGillavry HD, Hoogenraad CC. The internal architecture of dendritic spines revealed by super-resolution imaging: what did we learn so far? Exp Cell Res 2015;335(2):180–6. https://doi.org/10.1016/j.yexcr.2015.02.024.

[92] Bar J, Kobler O, van Bommel B, Mikhaylova M. Periodic F-actin structures shape the neck of dendritic spines. Sci Rep 2016;6:37136. https://doi.org/10.1038/srep37136.

[93] Jayant K, Hirtz JJ, Plante IJ, Tsai DM, De Boer WD, Semonche A, Yuste R. Targeted intracellular voltage recordings from dendritic spines using quantum-dot-coated nanopipettes. Nat Nanotechnol 2017;12(4):335–42. https://doi.org/10.1038/nnano.2016.268.

[94] Efimova N, Korobova F, Stankewich MC, Moberly AH, Stolz DB, Wang J, Svitkina T. βIII Spectrin is necessary for formation of the constricted neck of dendritic spines and regulation of synaptic activity in neurons. J Neurosci 2017;37(27):6442–59. https://doi.org/10.1523/JNEUROSCI.3520-16.2017.

[95] Bloodgood BL, Sabatini BL. Neuronal activity regulates diffusion across the neck of dendritic spines. Science 2005;310(5749):866–9. https://doi.org/10.1126/science.1114816.

[96] Ewers H, Tada T, Petersen JD, Racz B, Sheng M, Choquet D. A septin-dependent diffusion barrier at dendritic spine necks. PLoS One 2014;9(12) https://doi.org/10.1371/journal.pone.0113916.

[97] Xie Y, Vessey JP, Konecna A, Dahm R, Macchi P, Kiebler MA. The GTP-binding protein Septin 7 is critical for dendrite branching and dendritic-spine morphology. Curr Biol 2007;17(20):1746–51. https://doi.org/10.1016/j.cub.2007.08.042.

[98] Smith KR, Kopeikina KJ, Fawcett-Patel JM, Leaderbrand K, Gao R, Schurmann B, Penzes P. Psychiatric risk factor ANK3/ankyrin-G nanodomains regulate the structure and function of glutamatergic synapses. Neuron 2014;84(2):399–415. https://doi.org/10.1016/j.neuron.2014.10.010.

[99] Lewohl JM, Wang L, Miles MF, Zhang L, Dodd PR, Harris RA. Gene expression in human alcoholism: microarray analysis of frontal cortex. Alcohol Clin Exp Res 2000;24(12):1873–82.

[100] McClintick JN, McBride WJ, Bell RL, Ding ZM, Liu Y, Xuei X, Edenberg HJ. Gene expression changes in glutamate and GABA-A receptors, neuropeptides, ion channels, and cholesterol synthesis in the periaqueductal gray following binge-like alcohol drinking by adolescent alcohol-preferring (P) rats. Alcohol Clin Exp Res 2016;40(5):955–68. https://doi.org/10.1111/acer.13056.

[101] McClintick JN, Xuei X, Tischfield JA, Goate A, Foroud T, Wetherill L, Edenberg HJ. Stress-response pathways are altered in the hippocampus of chronic alcoholics. Alcohol 2013;47(7):505–15. https://doi.org/10.1016/j.alcohol.2013.07.002.

[102] Ponomarev I, Wang S, Zhang L, Harris RA, Mayfield RD. Gene coexpression networks in human brain identify epigenetic modifications in alcohol dependence. J Neurosci 2012;32(5):1884–97. https://doi.org/10.1523/JNEUROSCI.3136-11.2012.

Chapter 10

Mechanisms Regulating Compulsive Drug Behaviors

Craig T. Werner*,a, Amy M. Gancarz†,a, and David M. Dietz*

*Department of Pharmacology and Toxicology, Research Institute on Addictions, Program in Neuroscience, The State University of New York at Buffalo, Buffalo, NY, United States, †Department of Psychology, California State University Bakersfield, Bakersfield, CA, United States

INTRODUCTION

Substance use disorder is a chronic brain disease that affects a subset of individuals who use addictive drugs [1]. It is a compulsive relapsing disorder characterized by (i) compulsion to seek and take the drug, (ii) loss of control in limiting intake, and (iii) emergence of a negative emotional state (e.g., dysphoria, anxiety, and irritability) reflecting a motivational withdrawal syndrome when access to the drug is prevented [2,3].

Substance use disorder goes well beyond the act of drug taking. Rather, individuals spend a great deal of time and effort to procure the drug. Drug-seeking behavior is often governed by exposure to drug-paired cues, which are environmental stimuli that gain incentive motivational properties due to association with the rewarding properties of a drug [4], and it is during long periods of drug seeking that the compulsive features of substance abuse are expressed. Compulsivity can be described as enduring, repetitive behaviors in the face of adverse consequences and in inappropriate situations, which closely resembles the symptoms of substance dependence outlined in the *Diagnostic and Statistical Manual of Mental Disorders*, 5th Edition (DSM-V): "continued substance use despite knowledge of having had a persistent or recurrent physical or psychological problem and a great deal of time spent in activities necessary to obtain the substance."

Whereas initial use of addictive drugs produces discrete molecular events in particular brain regions, chronic drug taking alters the brain in specific ways that may ultimately contribute to the development of compulsive drug use and relapse [5–8]. Preclinical models have been developed to study endophenotypes of compulsive drug use and compulsive relapse (see review [9] and Chapter 2). Animal models of compulsive drug behaviors are based on motivated abstinence. Rats are first trained to self-administer a drug, and then an adverse consequence is introduced (e.g., mild foot shocks) to self-impose abstinence. Compulsive drug use can be studied in rats that do not self-impose abstinence or by the readiness to resume drug responding during sessions in the face of potential punishment [10–13]. These models provide tools to understand the mechanisms that underlie the development and expression of compulsivity.

It has been firmly established that plasticity in the mesolimbic dopamine (DA) system circuitry, which mediates reward, motivation informational processing, and behavior, underlies substance use disorders. Compulsive drug use involves dysregulation within and between a number of regions of the mesolimbic DA system, including the prefrontal cortex, ventral striatum, dorsal striatum, and extended amygdala. These regions are involved in executive decision making, reward, motivation, and emotional states.

Understanding the neurobiological components of compulsive drug use is of great interest for improving treatments for substance use disorders. Here, we review mechanisms that underlie compulsive drug use and relapse in preclinical models, as well as studies in human substance abusers. We discuss compulsivity in relation to impulsivity, and review current theories of compulsive drug use. We then address the mechanisms of compulsive drug use, including circuitry, neurotransmitter systems, and epigenetics.

---

a. Authors contributed equally.

Neural Mechanisms of Addiction. https://doi.org/10.1016/B978-0-12-812202-0.00010-5
Copyright © 2019 Elsevier Inc. All rights reserved.

IMPULSIVITY AND COMPULSIVITY

Compulsive behaviors are inappropriate perseverative behaviors that persist in the face of known negative consequences. Individuals that engage in compulsive behaviors are cognizant that repetition of these behaviors is harmful, yet are compelled to continue to emit the behaviors. It has been hypothesized that compulsive behaviors continue because they relieve negative emotional states (e.g., stress, tension, and anxiety; [14,15]). The neurobiology of compulsivity is thought to result from loss of cognitive top-down control or loss of response inhibition [16].

In contrast, impulsivity is defined as an inability to wait or plan, an inability to inhibit behavior resulting in a pattern of socially inappropriate behavior, and insensitivity to negative or delayed consequences [17,18]. Impulsive behavior is believed to lead to feelings of gratification, but, like compulsivity, is also considered a loss of top-down control. Impulsivity and related behavioral traits also appear to be mediated by sensitivity to the value of positive reinforcers [19].

According to the opponent process theory (see *Theories of compulsivity*), throughout an individual's progressive drug use, there are alternations between impulsivity and compulsivity that leads to stages of (i) preoccupation/anticipation, (ii) binge/intoxication, and (iii) withdrawal/negative affect. It has been suggested that impulsivity dominates early stages of drug use, and compulsivity dominates more progressive intensive stages of drug use. As an individual's behavior moves from impulsivity to compulsivity, it is reflective of a shift from positive reinforcement to negative reinforcement that drives the motivated behavior [14]. The relationship between impulsivity and compulsivity is complex and is also interactive, as inhibitory control may also be recruited in the regulation of compulsive behavior (see *Theories of compulsivity*; see also [16]).

PRECLINICAL COMPULSIVE DRUG MODELS

There have been great strides over the last 20 years in the development of animal models to study the neurobiological basis of substance use disorder. Although it is very challenging to find an animal model of any psychiatric disorder that fully recapitulates the disease in its entirety, models have been established and validated to examine specific aspects of substance use disorder, including compulsive drug use. Studies utilizing these models have provided seminal contributions to the development and understanding of compulsivity. In this section, we describe preclinical models used to examine endophenotypes of compulsive drug-seeking behaviors.

Escalation of Drug Intake as a Model of Compulsive Drug Taking

One aspect of compulsive drug use is escalation of drug intake, as defined by a progressive increase in the frequency and/or intensity of drug use. Escalated drug use is a behavioral phenomenon that characterizes the development of the addicted state and meets the criteria of the DSM-V: "The substance is often taken in larger amounts and over a longer period than was intended." Early models of compulsive behavior shifted the focus of drug-self administration from limited access or short access (1–2-h daily sessions) to extended- or long-access conditions (6-h daily sessions). Notably, the behavior in the extended-access animals reflects a loss of control of drug intake, as an increase in drug intake occurred with extended self-administration experience [20]. Escalation is associated with an increase in hedonic set point [21–24], and enhanced efficacy of cocaine reward [25,26]. However, while escalated drug intake is related to compulsive drug behaviors, it is important to note that escalation of drug intake has not systematically been shown to facilitate compulsive drug seeking [22,27,28].

Three-Criteria Addiction-like Behavior

An animal model has been developed to study compulsive drug use by operationalizing the three main clinical criteria of addiction in the DSM-V: (i) difficulty limiting drug intake, (ii) high motivation to take drug, and (iii) continued use despite negative consequences. In this paradigm, rats are given prolonged exposure to drug self-administration (3 months, compared to more traditional 10–30 day durations). Subsequently, the three DSM criteria listed above are tested by (i) continued responding during signaled "no-drug" periods, (ii) performance during progressive-ratio schedules of reinforcement, and (iii) continued responding when drug is accompanied by a signaled footshock, respectively [22,29]. Propensity to relapse is also measured using traditional "reinstatement" procedures. This model reveals a subpopulation of rats that do not meet any of the three criteria, and are resistant to "addiction." Rats that meet all three criteria are considered "addicted" and represent 15%–20% of the population, which is similar to proportions of substance abusers observed in human populations. Interestingly, the subpopulations of rats do not differ significantly in initial cocaine self-administration rates, but those that are later identified as "addicted" develop higher motivation for the drug, inability to refrain from drug seeking, and resistance to punishment [29]. Subsequent studies have identified neurobiological and endophenotypical predispositions for "addicted" rats prior to any drug exposure [16,30].

Compulsive Drug Seeking in the Face of Probabilistic Punishment

Another model has been developed to explore compulsive drug-seeking behavior using probabilistic punishment of seeking responses [28]. Rats are first trained in a seeking-taking chained schedule [31,32] where rats must emit a variable number of responses on the drug-seeking lever (the first link in the chain), which then gives access to a separate lever that produces drug infusions (the taking link in the chain). The links alternate following time-out periods in which drug is unavailable. Thus, responding in the drug-seeking operandum does not produce drug infusions, and drug-seeking and drug-taking responses are separable [33]. Following this experience, rats are challenged with probabilistic punishment sessions in which completion of the seeking cycle results in the drug-taking lever that produces either an infusion of drug (no punishment trial) or presentation of a mild electric foot shock (punishment trial). While a majority of rats are not willing to take the risk of being punished after their seeking responses and stop cocaine seeking, compulsive rats are more likely to gamble under probabilistic punishment conditions and will eventually maintain seeking behavior [28].

Chronic Self-Administration and Incubated Drug Craving

Substance use disorder is characterized by compulsive drug-seeking and drug-taking behaviors with high rates of relapse. Cocaine self-administration in rats can produce a time-dependent intensification (or "incubation") or cue-induced drug seeking (craving) for a number of drugs of abuse [34] that is believed to contribute to persistent relapse vulnerability [35]. This incubation phenomenon has also been observed in human substance abusers for a number of drugs [36–38]. While incubated drug craving has not been systematically linked to compulsive drug use or drug craving, numerous associations between incubation and compulsion have been observed. Rats that show incubation of cocaine craving are less sensitive to punishment-induced suppression of cocaine taking during a binge [39], suggesting that cocaine taking has become compulsive. Furthermore, rats predisposed to risky decision making, which is associated with compulsivity [40], exhibit more robust incubation of cocaine craving [41], and compulsive methamphetamine taking under punishment is associated with greater cue-induced drug seeking [42].

THEORIES OF COMPULSIVITY

Opponent Process Theory

A main theory of compulsive drug-taking behavior suggests that two reinforcement processes contribute to the development of compulsive drug seeking: positive and negative reinforcement. This is referred to as opponent process theory, which was originally developed as a theory of emotion [43] prior to being applied to substance abuse [3,44]. Positive reinforcement is the process by which a specific behavior/response results in presentation of an appetitive stimulus, which increases the probability or frequency of the behavior that produced it. Negative reinforcement, on the other hand, is the termination or withdrawal of an aversive stimulus, which also increases the probability of the behavior/response in the future. Traditionally, negative reinforcement can be thought of as administration of an oral nonsteroidal anti-inflammatory drug (NSAID) in order to alleviate the negative symptoms of a headache. This logic can also be applied to drug users such that drug users continue to seek/take drugs in order to relieve the negative physiological/psychological withdrawal symptoms (e.g., dysphoria, anxiety, and irritability when access to the drug is prevented). Negative reinforcement has been considered the dominant motivational force that drives compulsivity [19], and Koob and Le Moal [44] coined such learning to seek drug to remove the negative emotional states in drug absence as the "Dark Side of Addiction" (see Fig. 1; [45]).

In the opponent process theory, there are two processes that are responsible for hedonic reactions (and changes in reactions) to stimuli. The a-process, or the primary process, is the process that immediately occurs after presentation of a stimulus. The a-process can have either positive or negative hedonic value, and continues for the duration of the stimulus, and terminates with the offset of the stimulus. The intensity of the a-process matches the saliency and quality of the stimulus. The b-process is considered the opponent process and moves counter to the a-process. Unlike the a-process, the onset of the b-process is delayed after the onset of the a-process, and is slow to reach maximum and does not decay until after the stimulus and the a-process have terminated. However, the b-process becomes strengthened with repeated exposure to the same stimulus, such that the onset occurs sooner and offset is more delayed with overall magnitude of the b-process increasing. Koob & Le Moal have suggested that this theory can be applied to understand drug addiction, namely drug tolerance and negative reinforcement processes motivate compulsive drug-taking behavior.

This homeostatic theory suggests that when exposed to a stimulus that initiates the a-process, the body attempts to return you back to your "baseline" (triggering the b-process). As this relates to drug addiction, the initial drug effects,

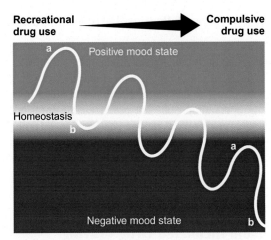

FIG. 1 Reduction in allostatic state associated with the development of compulsive drug behaviors. Diagram illustrating the opponent process model of motivation [43] in the framework of compulsive drug use [44]. The a-process is the activational process, and represents the positive hedonic state. The b-process is the counteradaptive opponent process, and represents the negative hedonic state. The allosteric state is a deviation of the regulatory system from homeostatic operating levels, and increases with frequent drug use due to an imbalance of the b-process and the a-process. This process is hypothesized to represent a transition to compulsive drug use and the development of substance abuse disorder.

which initiate the a-process, are opposed by the central nervous system, which involves the b-process. The first few exposures to drugs of abuse (e.g., opiates) triggers euphoria (via the a-process), after the drug effects subside, the b-process is now unopposed leaving the aversive negative emotional state as the counter reaction. Repeated exposure to the drug causes the b-process (the neural mechanisms opposing the drug) to become strengthened, leading to drug tolerance and dependence.

Evidence for the emergence of the b-process with extended drug exposure comes from seminal experiments using intracranial self-stimulation. Intracranial self-stimulation is a procedure where rats are implanted with electrodes directly aimed at the median forebrain bundle (MFB) and are allowed to emit an operant response in order to receive a low level of electrical stimulation. Stimulation of the MFB is highly rewarding, and researchers are able to reliably measure response rates across various intensities of stimulation in order to determine brain reward thresholds. Any manipulation that produces increases in brain reward thresholds is thought to reflect the establishment of an anhedonic state in the animal, as greater stimulation is required to achieve reward [46]. Withdrawal from cocaine self-administration produces marked increases in brain stimulation reward thresholds compared to drug naïve thresholds. Furthermore, the changes in brain stimulation reward threshold levels were correlated with the duration of self-administration sessions, which led the authors to conclude that the elevation in reward threshold was reflective of an altered anhedonic state [47]. A copious number of studies have since shown this effect occurs for various drugs of abuse ([19,45,48–51] for review). Moreover, subsequent studies demonstrated that reward thresholds are increased exclusively following long access, but not short access, to self-administration paradigms of cocaine [52], heroin [53], and methamphetamine [54], suggesting that the temporal duration and pattern of drug taking are essential components to the establishment of anhedonic processes.

These findings support the notion that negative reinforcement is a mechanism of compulsivity. Other support comes from studies illustrating escalation of drug intake with extended access to self-administration (see *Pre-clinical compulsive drug models*), illustrating an increase in hedonic set point with repeated drug exposure. Further evidence suggests animals with extended access to self-administration have greater motivation, and will work harder to gain access to drugs using progressive-ratio schedules of reinforcement [25,26,55–57]. Therefore, according to the opponent process theory, drug addiction is a function of homeostatic dysregulation, termed allostasis, produced by neural adaptations mediating hedonic functioning. Such allostasis has been suggested to occur in brain reward systems and by the recruitment of antireward/stress neural systems [44].

Aberrant Habitual Learning Theory of Compulsive Drug Behavior

Another theory of compulsive drug-taking behavior focuses on aberrant learning, and the motivational influences of conditioned stimuli that lead to habitual behaviors. While it is widely known that drugs' rewarding properties come from their ability to activate the mesolimbic DA circuitry [58], a great wealth of data has accumulated demonstrating that environmental stimuli, when paired with drug taking, gain motivational significance through Pavlovian (classical) conditioning [59–61]. Such drug-associated conditioned stimuli have profound effects on drug seeking and taking, and can induce drug craving long after drug cessation [62–65].

Researchers have argued that compulsive drug behaviors are a function of learning instrumental habitual stimulus-response (S-R) associations rather than declarative action-outcome (A-O) goal-directed behavior [66]. It is hypothesized that by learning S-R associations, instrumental behavior becomes a habit that is elicited by drug-associated conditioned stimuli, regardless of the value of the primary reinforcer. Therefore, devaluation of the primary reinforcer has little effect on seeking of the reinforcer. Aberrant habitual learning theory hypothesizes that drugs of abuse promote learning of strong stimulus response habits, and such habitual learning can become maladaptive and confer compulsivity to behavior [63,67–70].

Incentive Sensitization Theory of Drug Addiction

The incentive salience theory of addiction has also focused on classical conditioning as a mechanism to drive compulsive drug seeking [71]. In this theory, there is a sensitization to the incentive motivational effects of drugs and drug-paired cues. Consequently, incentive sensitization results in an increased focus toward drug cues and compulsive wanting and motivation for drugs [72]. This theory posits that more than simple S-R habitual learning is necessary to drive compulsive drug behaviors. Rather, additional motivational processes are needed.

Drug wanting and pathological motivation stem from sensitization in brain circuits that mediates Pavlovian conditioning that produce an increase in incentive salience and responsiveness to drug-related cues. This theory also states that repeated drug use sensitizes neural systems that mediate incentive salience, or "wanting," but not neural systems that produce the euphoric effects of drugs that are initially produced following drug consumption (e.g., "liking"), creating a dissociation between drug wanting and drug liking [72].

Neurobiological support for this theory stems from a classic experiment that recorded changes in midbrain DA activity following repeated exposure to a reward. Initially, these midbrain neurons fired in response to the novel reward. However, following repeated exposure, the neurons became activated in response to cues that were predictive of reward delivery. This experiment provides evidence that previously neutral stimuli gain incentive salience increasing motivation to seek a reward [73]. Subsequent work in human drug users also demonstrated hyperactivity in midbrain structures in response to drug-associated cues [64,74–80].

Inhibitory Control

Drug abusers show a variety of cognitive impairments, including impaired decision making, working memory, and response inhibition [81–83]. These deficits are described as dysfunctional inhibitory control and are hypothesized to contribute to compulsivity [84]. One example, is reversal learning, which involves the suppression of previously reinforced actions in parallel with learning a new action. The neural circuitry that mediates inhibitory control is not well characterized but appears to involve frontal cortical projections [85,86] to striatal regions [87–89]. Loss of executive inhibitory control involves reduced frontal cortical input and enhancement of striatum-mediated responding, which is believed to exacerbate compulsive behavior [69,90]. This hypothesis suggests that there is a shift in the balance of control from frontal cortex to striatum as drug seeking becomes compulsive, and is referred to as the "must do" (must seek and take drugs) of a compulsive habit rather than the "must have" of an increased motivational state [66].

NEURAL CIRCUITRY

Two approaches have been used when considering underlying neural mechanisms that lead to changes in reward thresholds and aversive motivational states that lead to compulsivity: within- and between-system neuroadaptations [91]. Within-system neuroadaptations are alterations in the cellular response to a drug with the intentions of ameliorating drug effects. These cellular adaptations oppose drug effects and continue following drug cessation, which lead to further adaptations. Alternatively, between-system neuroadaptations reflect a rewiring of drug circuitry in which a neural circuit is rerouted and/or recruited. Within- and between-system adaptations are not mutually exclusive, instead there are dynamic interactions between the two systems that alter responses and lead to dysregulation of reward circuits and recruitment of anti-reward circuits. This section will consider between-system neuroadaptations in circuits as they relate to compulsive drug behaviors with a focus on the prefrontal cortex, striatum, and extended amygdala (see Fig. 2).

Recreational drug behavior

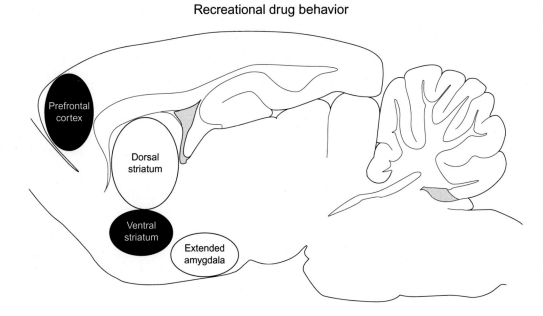

Compulsive drug behavior

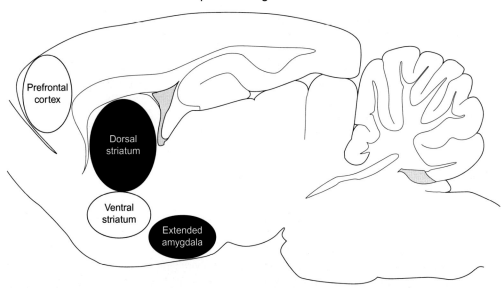

Prefrontal cortex	Ventral striatum	Dorsal striatum	Extended amygdala
Regions	Regions	Regions	Regions
Orbitofrontal cortex	Nucleus accumbens	Anterior dorsolateral striatum	Bed nucleus of stria terminalis
Prelimbic cortex	Olfactory tuburcle	Midlateral striatum	Central nucleus of the amygdala
			Nucleus accumbens shell transition zone
Function	Function	Function	
Executive function	Motivation	Responding rigidity	Function
Decision making	Reward	Instrumental responding	Emotion
			Negative reinforcement

FIG. 2   Brain regions involved in drug use and compulsive drug behaviors. There is a reduction in prefrontal cortex-mediated inhibitory control during compulsive drug behaviors. Activation of the ventral striatum enables drug-cue associations, and repeated drug exposure that leads to compulsive drug behaviors engages dorsal striatal-mediated control. Dysregulation of the extended amygdala is suggested to drive the increasing negative affective state as drug taking becomes compulsive like [19].

Prefrontal Cortex

Prefrontal cortical regions are involved in executive function and inhibitory control, and dysfunction in frontal cortices mediates compulsive drug-seeking behaviors. Impaired inhibitory control is linked to reduce but stable tonic DA levels in DA projections to the prefrontal cortex [92–94]. Reversal learning is impaired in both monkeys exposed to chronic cocaine administration [95] and human drug abusers [96], and appears to involve dysfunction in the orbitofrontal cortex [85,86]. Furthermore, functional connectivity studies have shown frontal cortical regions, including reduced orbitofrontal cortex activity, are correlated with measures of compulsivity in chronic stimulant-dependent individuals [97].

In an animal model that produces a subpopulation of compulsive cocaine-seeking rats using chronic cocaine exposure and intermittent punishment (see *Pre-clinical compulsive drug models*; [28]), frontal cortical regions have been implicated in compulsive stimulant drug seeking [98–101]. Specifically, pyramidal neurons in the prelimbic prefrontal cortex of punishment resistant, compulsive rats were hypoactive relative to punishment sensitive rats. Furthermore, optogenetic stimulation of the prelimbic cortex decreased compulsive cocaine seeking (i.e., cocaine seeking under punishment), and increased latencies to seek in compulsive rats. In contrast, inhibition of this brain area increased compulsive cocaine seeking [98] in the subpopulation of rats that previously suppressed cocaine seeking during punishment. Interestingly, lesions to the prelimbic cortical area prior to training did not result in compulsive cocaine seeking after limited cocaine history [101], suggesting impaired prefrontal cortical activity is a consequence of long-term cocaine use rather than a predisposition for compulsive cocaine use. In human neuroimaging studies, reductions in gray matter volume have been observed in frontal cortical regions in alcoholics [102], nicotine smokers [103,104], and stimulant abusers [105].

VENTRAL AND DORSAL STRIATUM

The basal ganglia circuit is a central component of the mesolimbic DA system that processes information related to reward and motivation. The basal ganglia features the striatum, which be divided into two subareas: the ventral striatum (nucleus accumbens and olfactory tuburcle) and the dorsal striatum (caudate nucleus and putamen). The striatum has an established role in the development and expression of compulsive responding for drugs. Chronic drug exposure is known to induce neuroadaptations in the basal ganglia that are triggered by phasic DA release. This phasic DA signaling enables drug-paired cues to increase DA levels and engages the ventral striatum. A shift in activation from the ventral to the dorsal striatum functioning following repeated exposure to drugs of abuse has been hypothesized to underlie the emergence of compulsive drug seeking [106]. Activity of the ventral striatum allows for recruitment of dorsal corticostriatal loops [107,108] and these "spiraling" connections between the nucleus accumbens to dorsal regions of the striatum have been reported in both primates and rats [109,110]. This circuitry has been shown to have functional significance in that disrupting the striato-midbrain-striatal serial connectivity by the combination of unilateral blockade of DA receptors in the dorsal striatum contralateral to a selective excitotoxic lesion of the nucleus accumbens, results in decreased drug-seeking behavior as measured by cocaine self-administration using a second-order schedule of reinforcement [29]. Within the dorsolateral striatum, inactivation of the anterior domain selectively disrupted punished, but not unpunished, drug seeking even after an extensive training. In contrast, inactivation of the midlateral striatum disrupted both punished and unpunished seeking [99]. These results indicate that the anterior dorsolateral striatum mediates responding rigidity while the midlateral striatal area is necessary for instrumental responding, which is likely occurring via motor cortical connectivity.

In contrast to the loss of gray matter volume in the prefrontal cortices reported across numerous substances of abuse (described in the previous section; [102–105]), studies have reported increases in gray matter in the basal ganglia and striatum specifically ([111]; but see [112]). Reduced gray matter in the prefrontal cortex and increased gray matter in the basal ganglia supports the S-R habit hypothesis [66] that loss of prefrontal-mediated control and strengthening of striatal-mediated control underlies compulsive behaviors.

EXTENDED AMYGDALA

The extended amygdala is involved in processing emotion and has been previously implicated in compulsivity as it relates to drug addiction [19,106,113,114]. In the opponent process theory, as drug taking becomes compulsive like, motivation to seek drug is hypothesized to be driven by both positive and negative reinforcement. The negative affective state that drives negative reinforcement is suggested to derive from dysregulation of the extended amygdala. The extended amygdala includes subregions that have similar cytoarchitectural and neuroanatomical connections, and is composed of the bed nucleus of the stria terminalis, the central nucleus of the amygdala, and a transition zone in the shell of the nucleus accumbens [115].

Neurobiological systems that are involved in stress are thought to be engaged to overcome the perturbing effects of chronic drug use to restore normal function in the domain of between-system neuroadaptations [113]. Both the hypothalamic-pituitary-adrenal (HPA) axis and extrahypothalamic brain stress system are dysregulated by chronic administration of all drugs of abuse [114,116–122] which includes amygdala activation during acute withdrawal [114,118,119,122]. The extended amygdala has been shown to mediate elevated reward threshold during nicotine withdrawal [123]. Furthermore, habit formation is believed to be mediated by plasticity in glutamatergic projects from the prefrontal cortex and amygdala to the ventral tegmental area and nucleus accumbens [124–127].

NEUROTRANSMISSION

The drug-induced changes in brain circuitry described in the previous section are accomplished through neuroadaptations in chemical messengers that are utilized to transmit signals between cells. Neurotransmitters are endogenous chemicals that mediate neurotransmission, and changes in neurotransmitter signaling underlie synaptic plasticity, including drug-induced alterations that rewire brain circuitry. In this section, we review some of the major neurotransmitter systems involved in compulsive drug seeking.

Dopamine

Dysfunction in DA transmission has long been understood to play a role in addiction-related behaviors. Expression of motor stereotypies, presumed to be a form of compulsive behavior, depends on enhanced DA function in the dorsal striatum [128]. In humans, long-term drug use is associated with a hypodopaminergic state, both in terms of reduced D2 DA receptors and DA release [129].

Recent studies have focused on the trait of impulsivity in rats, which is measured by excessive premature responses in a test of sustained attention, that is, the 5-choice serial reaction time task (5-CSRTT). High impulsivity has been shown to predict escalation of cocaine self-administration and compulsive cocaine intake [16], and has been associated with changes in DA signaling. Prior to any cocaine experience, high-impulsive rats display reductions in D2/3 DA receptor binding in the ventral, but not dorsal, striatum, changes in dopamine transporter (DAT) immunocytochemistry in the nucleus accumbens shell and reduction in gray matter in the nucleus accumbens core [130,131]. Similar D2/3 receptor binding properties have been reported in chronic stimulant abusers [129], and rhesus monkeys that exhibit high levels of stimulant self-administration [132]. Risky behavior in a decision-making task, which is related to lower striatal D2 receptor messenger RNA (mRNA) expression, led to enhanced intake of cocaine in adulthood [133], although compulsive cocaine self-administration has not yet been tested specifically.

Glutamate

A number of studies have established that cocaine exposure induces alterations in glutamate signaling and homeostasis [134]. N-acetyl cysteine (NAC), a cysteine substrate for the cysteine/glutamate antiporter, which increases basal levels of extracellular glutamate, was first shown to prevent relapse in both cue and drug reinstatement [135,136]. Subsequently, NAC was shown to significantly reduce habitual, anterior dorsolateral striatum dependent, cocaine seeking [137], and compulsive cocaine seeking in rats displaying addiction-like behavior after an extensive cocaine-taking history [138]. In addition, three-criteria rats (see *Theories of compulsivity*) show escalation of cocaine self-administration when given long-term access and a high vulnerability to relapse [139], which were attenuated by pretreatment with a metabotropic glutamate receptor (mGluR)2/3 agonist [140]. Habit formation that is hypothesized to underlie compulsive drug seeking is also associated with key synaptic changes in N-methyl-D-aspartate (NMDA) and α-amino-3-hydroxy-5-methyl-4-isoxazolepropionic acid (AMPA) receptors in glutamatergic inputs from the prefrontal cortex and amygdala to the ventral tegmental area and nucleus accumbens [124,125,127].

Serotonin

Substance abuse is associated with dysregulation of the serotonin (5-HT) neurotransmitter system [141], and Pelloux et al. [100] conducted compelling studies on 5-HT and compulsivity. In the compulsive subpopulation of rats that continued to seek cocaine despite punishment, but not non-compulsive rats that had very similar cocaine exposure (see *Pre-clinical compulsive drug models*), 5-HT was selectively reduced across prefrontal cortical areas (while at the same time DA utilization was decreased in the dorsal striatum; [100]). Systemic treatment with a 5-HT2C receptor antagonist or forebrain

5-HT depletion in rats that received a short cocaine history, in which none of the rats were compulsive, increased seeking under punishment [100]. These results demonstrate a causal role for low levels of forebrain 5-HT in compulsive drug seeking. Furthermore, treatment with citalopram, a 5-HT-selective reuptake inhibitor (SSRI), dose dependently reduced firmly established cocaine seeking under a second-order schedule of reinforcement and reduced compulsive seeking in rats [100]. This study suggests SSRIs may hold therapeutic potential to reduce compulsive drug seeking in human substance abusers, and the role of serotonin in treatment of substance abuse disorder is currently being investigated [142].

Pro- and Antistress Neurotransmitter Systems

There are a number of additional neural mechanisms that play important roles in emotional states that impact compulsive drug seeking. The extended amygdala-mediated stress response that occurs during chronic drug use involves multiple pro-dysphoria-inducing neurotransmitter systems that may contribute to negative emotional states associated with withdrawal. These pro-stress systems that converge on the amygdala include corticotropin-releasing factor (CRF), norepinephrine, vasopressin, substance P, hypocretin (orexin), and dynorphin. Blockade of these systems in the extended amygdala can block the development or expression of compulsive-like drug seeking or taking [19,55,143–151].

Neurotransmitter systems involved in antistress actions, including neuropeptide Y (NPY) and nociceptin, are hypothesized to oppose the actions of pro-stress systems, and are also activated in the extended amygdala with chronic drug use. Agonists for many of these systems reduce development and expression of dependence associated with chronic administration of alcohol [152–159]. Interestingly, stimulation of antistress systems or blockade of pro-stress neurotransmitters prevents alcohol-dependent gamma-Aminobutyric acid (GABA) release in the central nucleus of the amygdala, and can block the transition to excessive drinking [144,152]. The role of antistress systems in compulsive behaviors for other drugs of abuse is limited, but it is hypothesized that antistress systems may be a homeostatic response to between-system neuroadaptations [160,161].

EPIGENETIC MECHANISMS OF COMPULSIVE-RELATED BEHAVIORS

The complex behavioral changes that lead to drug addiction are thought to be a consequence of molecular and cellular modifications in specific brain regions [162], and these adaptations include alterations in gene expression following exposure to drugs of abuse [163]. Numerous studies have examined changes in gene expression after short-term exposure to drugs of abuse, but changes in gene expression using compulsive drug-seeking models is currently limited. Despite, it is important to consider what is known about gene expression in behaviors related to compulsive drug use.

Chronic exposure to various drugs of abuse alters gene expression profiles throughout key brain reward regions that are also involved in compulsive drug use [7,164–166]. A great deal of evidence has emerged that drugs of abuse cause activation of numerous immediate early genes that encode transcription factors in the central nervous system [167–171]. Such transcription factors are often considered "molecular switches" that coordinate alterations in gene expression that underlie neuronal plasticity and mediate behavioral changes.

The focus on gene expression in drug addiction has expanded beyond gene profiling to include epigenetic mechanisms [172], which regulate heritable alterations in gene expression that do not include changes to the DNA sequence [173–175]. Increasing evidence indicates that epigenetic mechanisms contribute to changes in morphological, synaptic, and behavioral plasticity following exposure to drugs of abuse by altering the expression of genes in reward-related brain regions [8,176]. There are numerous means by which epigenetic changes can occur within a neuron. Here, we discuss several mechanisms of epigenetic regulation following compulsive drug-taking behavior.

Chromatin Dynamics Following Chronic Drug Exposure

Chromatin structure regulates access of transcription factors and transcriptional machinery to DNA sequences, including promotor regions. Therefore, chromatin dynamics are essential for gene expression [177,178]. Chromatin exists in two basic states—heterochromatin and euchromatin—that are characterized by different levels of compression. Heterochromatin ("closed") is when chromatin is in a condensed state, winding histones tightly. This compact state is repressive to gene transcription, as the DNA is packaged tightly and prevents access for the gene transcription machinery. Conversely, in euchromatin ("open"), histones are in a more relaxed state, which allows access to DNA and is permissive to gene transcription [177]. The main factor influencing chromatin dynamics is the state of the histone tails, which constitute the location for various reversible posttranslational modifications that modulate gene transcription [179]. Histones act as spools around which DNA is wound, with each histone containing a tail. Within the histone tails are specific amino acid residues that are sites for posttranslational modifications such as acetylation and methylation [180]. By altering chromatin structure, gene transcription can be activated or repressed [181]. This is a very dynamic process, that varies in stability and duration, ranging from transient and easily reversible to more permanent, lifelong alterations [182].

Histone Acetylation

Histone acetylation, or the addition of an acetyl group to a histone tail, is associated with transcriptional activation, whereas histone deacetylation, removal of an acetyl group from the histone tail, is associated with transcriptional repression. Enzymes that facilitate this process are histone acetlytransferases (HATs) and histone deacetylases (HDACs). The function of HATs is to catalyze the addition of acetyl molecules to histone tails to create a more open chromatin configuration that is conducive to gene activation. In contrast, HDACs remove acetyl groups, fostering condensation of chromatin resulting in DNA coiling and inactivation of gene transcription and gene silencing [183–186]. The coordinated activity of HATs and HDACs regulate chromatin dynamics and gene transcription.

The epigenetic changes that have been observed in preclinical substance abuse disorder models are dependent on many variables, including (i) degree of drug exposure, (ii) duration of drug abstinence, (iii) brain region under investigation, (iv) histone and histone tail location, and (v) gene promotor. Chronic exposure to various psychomotor stimulants increases total cellular levels of acetylated histone H3 and H4 in the nucleus accumbens [187–189]. Chronic cocaine decreases HDAC5 activity in the nucleus accumbens enabling histone acetylation and increased expression of HDAC5 targeted genes, and disruption of HDAC5 activity results in increased response to chronic cocaine [190]. Increased acetylation of histone H3 proteins is associated with genes that regulate synaptic plasticity in the mesolimbic forebrain in mice following chronic methamphetamine exposure [191], and cocaine-induced alterations in histone H3 acetylation and corresponding changes in gene expression have been shown to persist during 100 days of abstinence [192]. These data suggest that long-lasting changes in chromatin dynamics occur following chronic drug exposure and may underlie compulsive drug use and relapse (for a review, see [193]).

Chronic cocaine is also associated with increased histone acetylation at distinct promoter regions [173,193]. Kumar et al. [187] showed increased H3 histone acetylation at cyclin-dependent kinase 5 (Cdk5) and brain-derived neurotrophic factor (BDNF) promotor regions in the striatum, accompanied by stable changes in mRNA, respectively. Cdk5 is a neuronal protein kinase that has previously been implicated in the cellular responses to drugs of abuse, and there is evidence that Cdk5 governs psychomotor stimulant drug-induced changes in DA neurotransmission and cytoskeletal structure [194]. BDNF has been implicated in memory formation and synaptic restructuring [195–200], and a growing body of evidence indicates time-dependent and regional-specific regulation of BDNF transcription occurs following chronic cocaine exposure. Abstinence from chronic cocaine resulted in increased BDNF mRNA expression accompanied by an increase in acetylated histone 3 (AcH3) at BDNF promoter regions in the nucleus accumbens, medial prefrontal cortex, and ventral tegmental area [201–203]. These data suggest that chromatin structure may be modified due to an increase in acetylation at BDNF promotor regions that results in an open chromatin state that leads to more rapid gene transcription of BDNF.

Chronic ethanol consumption results in an increased HDAC gene expression in rats and human, which does not occur following a single alcohol exposure [204]. Importantly, direct manipulation of histone acetylation and chromatin remodeling via pharmacological and viral-mediated transfer, produces functional drug-induced plasticity. The effects of HDAC inhibitors are inconsistent and depend on the specific drug of abuse, including history of drug exposure, withdrawal period, brain region specificity, and paradigm. For example, drug naive rats injected with systemic HDAC inhibitors administered before acquisition of cocaine self-administration dose dependently reduced number of cocaine infusions self-administered and motivation to work for infusions of cocaine, as measured by a progressive-ratio schedule of reinforcement, while not affecting responding for a natural reinforcer, sucrose [205]. HDAC inhibitors in the ventral tegmental area have also been shown to decrease ethanol self-administration [206], while increased cocaine intake was observed in cocaine-experienced animals pretreated with HDAC inhibitors [207]. In yet another study, an HDAC inhibitor infused directly into the nucleus accumbens increased, whereas viral-mediated overexpression of HDAC4 decreased, motivation for cocaine [208]. Cocaine-experienced rats exposed to HDAC inhibitors during an extended withdrawal period exhibited attenuated cocaine seeking as measured by drug and cue-induced reinstatement [209]. HDAC inhibitors also accelerated extinction and mitigated reinstatement in a cocaine conditioned place preference paradigm [210], and in rats previously trained to self-administer nicotine [211]. However, other studies have shown HDAC inhibitors increase heroin-primed seeking behavior [212]. Together, these data suggest that HDAC inhibitors differentially alter the reinforcing efficacy of drugs based on drug-taking experience and drug class (for a review, see [173]).

Histone Methylation

Histone methylation involves the addition of methyl groups to histone proteins, and generally has been considered a "gene silencer." Histone methylation has been suggested to be a complex, yet more durable and stable posttranslational modification relative to histone acetylation [213], and can occur in various states: mono-(me), di-(me2), or trimethylation (me3), with each methylation event having unique effects on gene transcription [213]. Furthermore, histone methylation is catalyzed by histone methyltransferases (HMTs) and histone demethylases (HDMs). A number of factors determine whether histone methylation at gene promoters either represses or promotes gene transcription, including (i) the target amino acid site, (ii) number of bound methyl groups, and (iii) specific enzymes present [213,214].

Chronic cocaine exposure during the adolescent period decreased histone H3 methylation in the medial prefrontal cortex, and led to drug-related behavioral adaptations during adulthood [215]. Specific histone methylation marks at numerous gene promoters were regulated by chronic cocaine administration in the mouse nucleus accumbens [216], and prefrontal cortex [217]. Furthermore, histone methylating enzymes, including G9a, have been shown to be downregulated in the nucleus accumbens by both experimenter- and self-administered cocaine in adult mice [218]. Consistent with molecular changes observed following cocaine exposure, alterations of G9a signaling resulted in drug-induced behavioral plasticity. Viral-mediated G9a overexpression significantly decreased, while pharmacological inhibition increased, preference for cocaine and increased H3K9me2 levels in the nucleus accumbens [218]. Together, these data suggest that alteration of enzymes involved in histone methylation results in changes in cocaine-induced behavioral plasticity that may contribute to compulsive aspects of drug addiction.

DNA METHYLATION AND OTHER HISTONE MODIFICATIONS

DNA methyltransferases (DNMTs) are enzymes that adjoin methyl groups directly to DNA cytosine-guanine dinucleotides (CpG) in the genome [219]. Methylation interferes with the ability of transcription factors to bind to target DNA sequences [181], are generally thought to be repressive to gene transcription (although, see [219]), and there is evidence that DNA methylation is altered by chronic exposure to drugs of abuse [193]. Cocaine self-administration produces increased gene and protein expression of MeCP2, a methyl-CpG-binding protein [220]. Furthermore, following chronic exposure to cocaine, protein-R-methyltransferase-6 (PRMT6), and its associated histone mark, were decreased in the nucleus accumbens in rodents, which was also found in human cocaine abusers [221]. Chronic exposure to drugs of abuse has also been shown to alter other histone modifications. For example, exposure to cocaine rapidly induces H3 phosphorylation in the striatum [162,187], and chronic cocaine exposure alters levels of poly-adenosinediphosphate (poly-ADP) ribosylation [222]. Further research is needed to understand the role of these drug-induced epigenetic alterations in the development of compulsive drug seeking.

MICRORNAs

Within noncoding RNA (98% of the human genome), which was traditionally thought to be "junk sequences" [223], important regulators of gene expression have been identified. A class of noncoding RNA, micro-RNAs (miRNAs) are short (19–24 nucleotides) nonprotein coding RNA molecules that have the ability to regulate gene expression at a posttranscriptional level, and have been shown to alter synaptic plasticity and behavior [224,225]. miRNAs are abundantly present in the brain, have an established role in synaptic plasticity, are rapidly transcribed, and are localized to dendrites [226–230]. Increasing evidence suggests that miRNAs have a variety of effects on gene expression, including mRNA degradation, increased mRNA translation, chromatin remodeling, and DNA methylation [225].

A growing body of evidence implicates miRNAs in compulsive drug behaviors. miRNA expression has been shown to be altered in the striatum in rats following long access to cocaine self-administration [231], which has been implicated in drug-related habit learning. Furthermore, rats with extended access to cocaine self-administration, had increased expression of miR-132 and miR-212 compared to limited-access self-administration, and yoked rats that received noncontingent injections of cocaine [231]. Viral-mediated overexpression of miR-212 in the dorsal striatum resulted in attenuated cocaine self-administration under unlimited access compared to controls, an effect that was not observed under restricted access conditions. Furthermore, viral overexpression of miR-212 produced a downward shift in dose response curve to cocaine, indicating a decreased motivation for drug, and decreased drug seeking during periods of drug unavailability [231]. Inhibition of miR-212 signaling produced increases in sensitivity to cocaine, as measured by increased cocaine self-administration, and increased drug seeking during periods of drug unavailability. There is evidence that MeCP2, a transcriptional factor, interacts with miR-212 to mediate BDNF expression [232], and other evidence indicates miR-212 interacts with other transcription factors to regulate gene expression and signaling [231]. In addition, researchers have found that miR-124 and let-7d are downregulated, whereas mir-181a is upregulated, in mesolimbic brain slices following chronic cocaine injections [233]. Viral overexpression of miR-124 and let-7d in the nucleus accumbens decreased conditioned place preference in rats, and viral overexpression of miR-181a produced the opposite effect [234]. Together, these data suggest that regulation of miR-212 and other miRNAs mediate cocaine-induced plasticity.

CONCLUSION

Compulsive drug seeking and use are hallmark symptoms of substance use disorders. These behaviors emerge as the brain undergoes changes in circuits and signaling pathways that lead to dysfunctional motivation and reward processing, which ultimately contribute to maintaining persistent relapse vulnerability. Preclinical models of compulsivity endophenotypes have expanded our understanding of the development of compulsion and provided evidence for theories that describe the pathophysiology and etiology of drug addiction.

REFERENCES

[1] Degenhardt L, Chiu WT, Sampson N, Kessler RC, Anthony JC, Angermeyer M, Wells JE. Toward a global view of alcohol, tobacco, cannabis, and cocaine use: findings from the WHO World Mental Health Surveys. PLoS Med 2008;5(7). https://doi.org/10.1371/journal.pmed.0050141.

[2] American Psychiatric Association. The diagnostic and statistical manual of mental disorders. 5th ed. Washington, DC: American Psychiatric Press; 2013.

[3] Koob GF, Le Moal M. Drug abuse: hedonic homeostatic dysregulation. Science 1997;278(5335):52–8.

[4] Kalivas PW, McFarland K. Brain circuitry and the reinstatement of cocaine-seeking behavior. Psychopharmacology (Berl) 2003;168(1-2):44–56. https://doi.org/10.1007/s00213-003-1393-2.

[5] Koob GF, Le Moal M. Plasticity of reward neurocircuitry and the 'dark side' of drug addiction. Nat Neurosci 2005;8(11):1442–4. https://doi.org/10.1038/nn1105-1442.

[6] Nestler EJ. Is there a common molecular pathway for addiction? Nat Neurosci 2005;8(11):1445–9. https://doi.org/10.1038/nn1578.

[7] Nestler EJ. Epigenetic mechanisms of drug addiction. Neuropharmacology 2014;76(Pt B):259–68. https://doi.org/10.1016/j.neuropharm.2013.04.004.

[8] Russo SJ, Dietz DM, Dumitriu D, Morrison JH, Malenka RC, Nestler EJ. The addicted synapse: mechanisms of synaptic and structural plasticity in nucleus accumbens. Trends Neurosci 2010;33(6):267–76. https://doi.org/10.1016/j.tins.2010.02.002.

[9] Belin-Rauscent A, Fouyssac M, Bonci A, Belin D. How preclinical models evolved to resemble the diagnostic criteria of drug addiction. Biol Psychiatry 2016;79(1):39–46. https://doi.org/10.1016/j.biopsych.2015.01.004.

[10] Economidou D, Pelloux Y, Robbins TW, Dalley JW, Everitt BJ. High impulsivity predicts relapse to cocaine-seeking after punishment-induced abstinence. Biol Psychiatry 2009;65(10):851–6. https://doi.org/10.1016/j.biopsych.2008.12.008.

[11] Marchant NJ, Khuc TN, Pickens CL, Bonci A, Shaham Y. Context-induced relapse to alcohol seeking after punishment in a rat model. Biol Psychiatry 2013;73(3):256–62. https://doi.org/10.1016/j.biopsych.2012.07.007.

[12] Marchant NJ, Li X, Shaham Y. Recent developments in animal models of drug relapse. Curr Opin Neurobiol 2013;23(4):675–83. https://doi.org/10.1016/j.conb.2013.01.003.

[13] Marchant NJ, Rabei R, Kaganovsky K, Caprioli D, Bossert JM, Bonci A, Shaham Y. A critical role of lateral hypothalamus in context-induced relapse to alcohol seeking after punishment-imposed abstinence. J Neurosci 2014;34(22):7447–57. https://doi.org/10.1523/JNEUROSCI.0256-14.2014.

[14] Berlin GS, Hollander E. Compulsivity, impulsivity, and the DSM-5 process. CNS Spectr 2014;19(1):62–8. https://doi.org/10.1017/S1092852913000722.

[15] Robbins TW, Gillan CM, Smith DG, de Wit S, Ersche KD. Neurocognitive endophenotypes of impulsivity and compulsivity: towards dimensional psychiatry. Trends Cogn Sci 2012;16(1):81–91. https://doi.org/10.1016/j.tics.2011.11.009.

[16] Dalley JW, Fryer TD, Brichard L, Robinson ES, Theobald DE, Lääne K, Robbins TW. Nucleus accumbens D2/3 receptors predict trait impulsivity and cocaine reinforcement. Science 2007;315(4816):1267–70.

[17] Ainslie G. Specious reward: a behavioral theory of impulsiveness and impulse control. Psychol Bull 1975;82(4):463–96.

[18] Eysenck HL. The nature of impulsivity: American Psychological Association. .

[19] Koob GF. Antireward, compulsivity, and addiction: seminal contributions of Dr. Athina Markou to motivational dysregulation in addiction. Psychopharmacology (Berl) 2017;234(9-10):1315–32. https://doi.org/10.1007/s00213-016-4484-6.

[20] Ahmed SH, Koob GF. Transition from moderate to excessive drug intake: change in hedonic set point. Science 1998;282(5387):298–300.

[21] Ahmed SH, Koob GF. Long-lasting increase in the set point for cocaine self-administration after escalation in rats. Psychopharmacology (Berl) 1999;146(3):303–12.

[22] Deroche-Gamonet V, Belin D, Piazza PV. Evidence for addiction-like behavior in the rat. Science 2004;305(5686):1014–7.

[23] Mantsch JR, Yuferov V, Mathieu-Kia AM, Ho A, Kreek MJ. Effects of extended access to high versus low cocaine doses on self-administration, cocaine-induced reinstatement and brain mRNA levels in rats. Psychopharmacology (Berl) 2004;175(1):26–36. https://doi.org/10.1007/s00213-004-1778-x.

[24] O'Dell LE, Roberts AJ, Smith RT, Koob GF. Enhanced alcohol self-administration after intermittent versus continuous alcohol vapor exposure. Alcoholism: Clinical & Experimental Research 2004;28(11):1676–82. https://doi.org/10.1097/01.alc.0000145781.11923.4e.

[25] Paterson NE, Markou A. Increased motivation for self-administered cocaine after escalated cocaine intake. Neuroreport 2003;14(17):2229–32. https://doi.org/10.1097/01.wnr.0000091685.94870.ba.

[26] Wee S, Mandyam CD, Lekic DM, Koob GF. Alpha 1-noradrenergic system role in increased motivation for cocaine intake in rats with prolonged access. Eur Neuropsychopharmacol 2008;18(4):303–11. https://doi.org/10.1016/j.euroneuro.2007.08.003.

[27] Ahmed SH, Lin D, Koob GF, Parsons LH. Escalation of cocaine self-administration does not depend on altered cocaine-induced nucleus accumbens dopamine levels. Journal of Neurochemistry 2004;86(1):102–13. https://doi.org/10.1046/j.1471-4159.2003.01833.x.

[28] Pelloux Y, Everitt BJ, Dickinson A. Compulsive drug seeking by rats under punishment: effects of drug taking history. Psychopharmacology (Berl) 2007;194(1):127–37. https://doi.org/10.1007/s00213-007-0805-0.

[29] Belin D, Mar AC, Dalley JW, Robbins TW, Everitt BJ. High impulsivity predicts the switch to compulsive cocaine-taking. Science 2008;320 (5881):1352–5. https://doi.org/10.1126/science.1158136.

[30] Dilleen R, Pelloux Y, Mar AC, Molander A, Robbins TW, Everitt BJ, Belin D. High anxiety is a predisposing endophenotype for loss of control over cocaine, but not heroin, self-administration in rats. Psychopharmacology (Berl) 2012;222(1):89–97. https://doi.org/10.1007/s00213-011-2626-4.

[31] Olmstead MC, Lafond MV, Everitt BJ, Dickinson A. Cocaine seeking by rats is a goal-directed action. Behav Neurosci 2001;115(2):394–402.

[32] Olmstead MC, Parkinson JA, Miles FJ, Everitt BJ, Dickinson A. Cocaine-seeking by rats: regulation, reinforcement and activation. Psychopharmacology (Berl) 2000;152(2):123–31.

[33] Vanderschuren LJ, Everitt BJ. Drug seeking becomes compulsive after prolonged cocaine self-administration. Science 2004;305(5686):1017–9. https://doi.org/10.1126/science.1098975.

[34] Pickens CL, Airavaara M, Theberge F, Fanous S, Hope BT, Shaham Y. Neurobiology of the incubation of drug craving. Trends Neurosci 2011;34 (8):411–20. https://doi.org/10.1016/j.tins.2011.06.001.

[35] Wolf ME. Synaptic mechanisms underlying persistent cocaine craving. Nat Rev Neurosci 2016;17(6):351–65. https://doi.org/10.1038/nrn.2016.39.

[36] Bedi G, Preston KL, Epstein DH, Heishman SJ, Marrone GF, Shaham Y, de Wit H. Incubation of cue-induced cigarette craving during abstinence in human smokers. Biol Psychiatry 2011;69(7):708–11. https://doi.org/10.1016/j.biopsych.2010.07.014.

[37] Parvaz MA, Moeller SJ, Goldstein RZ. Incubation of cue-induced craving in adults addicted to cocaine measured by electroencephalography. JAMA Psychiatry 2016;73(11):1127–34. https://doi.org/10.1001/jamapsychiatry.2016.2181.

[38] Wang G, Shi J, Chen N, Xu L, Li J, Li P, Lu L. Effects of length of abstinence on decision-making and craving in methamphetamine abusers. PLoS One 2013;8(7). https://doi.org/10.1371/journal.pone.0068791.

[39] Gancarz-Kausch AM, Adank DN, Dietz DM. Prolonged withdrawal following cocaine self-administration increases resistance to punishment in a cocaine binge. Sci Rep 2014;4. https://doi.org/10.1038/srep06876.

[40] El Massioui, N., Lamirault, C., Yague, S., Adjeroud, N., Garces, D., Maillard, A., Doyere, V. (2016). Impaired Decision Making and Loss of Inhibitory-Control in a Rat Model of Huntington Disease. Front Behav Neurosci, 10, 204. https://doi.org/10.3389/fnbeh.2016.00204

[41] Ferland JN, Winstanley CA. Risk-preferring rats make worse decisions and show increased incubation of craving after cocaine self-administration. Addict Biol 2017;22(4):991–1001. https://doi.org/10.1111/adb.12388.

[42] Torres OV, Jayanthi S, Ladenheim B, McCoy MT, Krasnova IN, Cadet JL. Compulsive methamphetamine taking under punishment is associated with greater cue-induced drug seeking in rats. Behav Brain Res 2017;326:265–71. https://doi.org/10.1016/j.bbr.2017.03.009.

[43] Solomon RL, Corbit JD. An opponent-process theory of motivation. I. Temporal dynamics of affect. Psychol Rev 1974;81(2):119–45.

[44] Koob GF, Le Moal M. Drug addiction, dysregulation of reward, and allostasis. Neuropsychopharmacology 2001;24(2):97–129. https://doi.org/10.1016/S0893-133X(00)00195-0.

[45] Schulteis G, Koob G. Neuropharmacology. Dark side of drug dependence. Nature 1994;371(6493):108–9. https://doi.org/10.1038/371108a0.

[46] Carlezon Jr. WA, Chartoff EH. Intracranial self-stimulation (ICSS) in rodents to study the neurobiology of motivation. Nat Protoc 2007;2 (11):2987–95. https://doi.org/10.1038/nprot.2007.441.

[47] Markou A, Koob GF. Postcocaine anhedonia. An animal model of cocaine withdrawal. Neuropsychopharmacology 1991;4(1):17–26.

[48] Epping-Jordan MP, Watkins SS, Koob GF, Markou A. Dramatic decreases in brain reward function during nicotine withdrawal. Nature 1998;393 (6680):76–9. https://doi.org/10.1038/30001.

[49] Gardner EL, Vorel SR. Cannabinoid transmission and reward-related events. Neurobiol Dis 1998;5(6 Pt B):502–33. https://doi.org/10.1006/nbdi.1998.0219.

[50] Paterson NE, Myers C, Markou A. Effects of repeated withdrawal from continuous amphetamine administration on brain reward function in rats. Psychopharmacology (Berl) 2000;152(4):440–6.

[51] Schulteis G, Markou A, Cole M, Koob GF. Decreased brain reward produced by ethanol withdrawal. Proc Natl Acad Sci U S A 1995;92 (13):5880–4.

[52] Ahmed SH, Kenny PJ, Koob GF, Markou A. Neurobiological evidence for hedonic allostasis associated with escalating cocaine use. Nat Neurosci 2002;5(7):625–6. https://doi.org/10.1038/nn872.

[53] Kenny PJ, Chen SA, Kitamura O, Markou A, Koob GF. Conditioned withdrawal drives heroin consumption and decreases reward sensitivity. J Neurosci 2006;26(22):5894–900. https://doi.org/10.1523/JNEUROSCI.0740-06.2006.

[54] Jang CG, Whitfield T, Schulteis G, Koob GF, Wee S. A dysphoric-like state during early withdrawal from extended access to methamphetamine self-administration in rats. Psychopharmacology (Berl) 2013;225(3):753–63. https://doi.org/10.1007/s00213-012-2864-0.

[55] Barbier E, Vendruscolo LF, Schlosburg JE, Edwards S, Juergens N, Park PE, Heilig M. The NK1 receptor antagonist L822429 reduces heroin reinforcement. Neuropsychopharmacology 2013;38(6):976–84. https://doi.org/10.1038/npp.2012.261.

[56] Cohen A, Koob GF, George O. Robust escalation of nicotine intake with extended access to nicotine self-administration and intermittent periods of abstinence. Neuropsychopharmacology 2012;37(9):2153–60. https://doi.org/10.1038/npp.2012.67.

[57] Walker BM, Koob GF. The gamma-aminobutyric acid-B receptor agonist baclofen attenuates responding for ethanol in ethanol-dependent rats. Alcohol Clin Exp Res 2007;31(1):11–8. https://doi.org/10.1111/j.1530-0277.2006.00259.x.

[58] Leshner AI, Koob GF. Drugs of abuse and the brain. Proc Assoc Am Physicians 1999;111(2):99–108.

[59] Gawin F, Kleber H. Pharmacologic treatments of cocaine abuse. Psychiatr Clin North Am 1986;9(3):573–83.

[60] Gawin FH, Kleber HD. Abstinence symptomatology and psychiatric diagnosis in cocaine abusers. Clinical observations. Arch Gen Psychiatry 1986;43(2):107–13.

[61] O'Brien CP, Childress AR, Ehrman R, Robbins SJ. Conditioning factors in drug abuse: can they explain compulsion? J Psychopharmacol 1998;12 (1):15–22. https://doi.org/10.1177/026988119801200103.

[62] Childress AR, Mozley PD, McElgin W, Fitzgerald J, Reivich M, O'Brien CP. Limbic activation during cue-induced cocaine craving. Am J Psychiatry 1999;156(1):11–8. https://doi.org/10.1176/ajp.156.1.11.

[63] Everitt BJ, Dickinson A, Robbins TW. The neuropsychological basis of addictive behaviour. Brain Res Brain Res Rev 2001;36(2-3):129–38.

[64] Garavan H, Pankiewicz J, Bloom A, Cho JK, Sperry L, Ross TJ, Stein EA. Cue-induced cocaine craving: neuroanatomical specificity for drug users and drug stimuli. Am J Psychiatry 2000;157(11):1789–98.

[65] Grant S, London ED, Newlin DB, Villemagne VL, Liu X, Contoreggi C, Margolin A. Activation of memory circuits during cue-elicited cocaine craving. Proc Natl Acad Sci U S A 1996;93(21):12040–5.

[66] Everitt BJ, Robbins TW. Neural systems of reinforcement for drug addiction: from actions to habits to compulsion. Nat Neurosci 2005;8 (11):1481–9. https://doi.org/10.1038/nn1579.

[67] Berke JD, Hyman SE. Addiction, dopamine, and the molecular mechanisms of memory. Neuron 2000;25(3):515–32.

[68] Hyman SE, Malenka RC, Nestler EJ. Neural mechanisms of addiction: the role of reward-related learning and memory. Annu Rev Neurosci 2006;29:565–98. https://doi.org/10.1146/annurev.neuro.29.051605.113009.

[69] Robbins TW, Everitt BJ. Drug addiction: bad habits add up. Nature 1999;398(6728):567–70. https://doi.org/10.1038/19208.

[70] Tiffany ST. A cognitive model of drug urges and drug-use behavior: role of automatic and nonautomatic processes. Psychol Rev 1990;97 (2):147–68.

[71] Robinson TE, Berridge KC. The neural basis of drug craving: an incentive-sensitization theory of addiction. Brain Res Brain Res Rev 1993;18 (3):247–91.

[72] Robinson TE, Berridge KC. Review. The incentive sensitization theory of addiction: some current issues. Philos Trans R Soc Lond B Biol Sci 2008;363(1507):3137–46. https://doi.org/10.1098/rstb.2008.0093.

[73] Schultz W, Dayan P, Montague PR. A neural substrate of prediction and reward. Science 1997;275(5306):1593–9.

[74] Due DL, Huettel SA, Hall WG, Rubin DC. Activation in mesolimbic and visuospatial neural circuits elicited by smoking cues: evidence from functional magnetic resonance imaging. Am J Psychiatry 2002;159(6):954–60. https://doi.org/10.1176/appi.ajp.159.6.954.

[75] Goldstein RZ, Tomasi D, Alia-Klein N, Honorio Carrillo J, Maloney T, Woicik PA, Volkow ND. Dopaminergic response to drug words in cocaine addiction. J Neurosci 2009;29(18):6001–6. https://doi.org/10.1523/JNEUROSCI.4247-08.2009.

[76] Goldstein RZ, Tomasi D, Rajaram S, Cottone LA, Zhang L, Maloney T, Volkow ND. Role of the anterior cingulate and medial orbitofrontal cortex in processing drug cues in cocaine addiction. Neuroscience 2007;144(4):1153–9. https://doi.org/10.1016/j.neuroscience.2006.11.024.

[77] Janes AC, Pizzagalli DA, Richardt S, de BFB, Chuzi S, Pachas G, Kaufman MJ. Brain reactivity to smoking cues prior to smoking cessation predicts ability to maintain tobacco abstinence. Biol Psychiatry 2010;67(8):722–9. https://doi.org/10.1016/j.biopsych.2009.12.034.

[78] Maas LC, Lukas SE, Kaufman MJ, Weiss RD, Daniels SL, Rogers VW, Renshaw PF. Functional magnetic resonance imaging of human brain activation during cue-induced cocaine craving. Am J Psychiatry 1998;155(1):124–6. https://doi.org/10.1176/ajp.155.1.124.

[79] Marhe R, Luijten M, van de Wetering BJ, Smits M, Franken IH. Individual differences in anterior cingulate activation associated with attentional bias predict cocaine use after treatment. Neuropsychopharmacology 2013;38(6):1085–93. https://doi.org/10.1038/npp.2013.7.

[80] Volkow ND, Wang GJ, Telang F, Fowler JS, Logan J, Childress AR, Wong C. Cocaine cues and dopamine in dorsal striatum: mechanism of craving in cocaine addiction. J Neurosci 2006;26(24):6583–8. https://doi.org/10.1523/jneurosci.1544-06.2006.

[81] Friedman NP, Miyake A, Corley RP, Young SE, Defries JC, Hewitt JK. Not all executive functions are related to intelligence. Psychol Sci 2006;17 (2):172–9. https://doi.org/10.1111/j.1467-9280.2006.01681.x.

[82] Ornstein TJ, Iddon JL, Baldacchino AM, Sahakian BJ, London M, Everitt BJ, Robbins TW. Profiles of cognitive dysfunction in chronic amphetamine and heroin abusers. Neuropsychopharmacology 2000;23(2):113–26. https://doi.org/10.1016/S0893-133X(00)00097-X.

[83] Rogers RD, Robbins TW. Investigating the neurocognitive deficits associated with chronic drug misuse. Curr Opin Neurobiol 2001;11 (2):250–7.

[84] Morein-Zamir S, Robbins TW. Fronto-striatal circuits in response-inhibition: Relevance to addiction. Brain Res 2015;1628(Pt A):117–29. https://doi.org/10.1016/j.brainres.2014.09.012.

[85] Chudasama Y, Robbins TW. Dissociable contributions of the orbitofrontal and infralimbic cortex to pavlovian autoshaping and discrimination reversal learning: further evidence for the functional heterogeneity of the rodent frontal cortex. J Neurosci 2003;23(25):8771–80.

[86] Dias R, Robbins TW, Roberts AC. Dissociation in prefrontal cortex of affective and attentional shifts. Nature 1996;380(6569):69–72. https://doi.org/10.1038/380069a0.

[87] Castane A, Theobald DE, Robbins TW. Selective lesions of the dorsomedial striatum impair serial spatial reversal learning in rats. Behav Brain Res 2010;210(1):74–83. https://doi.org/10.1016/j.bbr.2010.02.017.

[88] Clarke HF, Robbins TW, Roberts AC. Lesions of the medial striatum in monkeys produce perseverative impairments during reversal learning similar to those produced by lesions of the orbitofrontal cortex. J Neurosci 2008;28(43):10972–82. https://doi.org/10.1523/jneurosci.1521-08.2008.

[89] Groman SM, James AS, Seu E, Crawford MA, Harpster SN, Jentsch JD. Monoamine levels within the orbitofrontal cortex and putamen interact to predict reversal learning performance. Biol Psychiatry 2013;73(8):756–62. https://doi.org/10.1016/j.biopsych.2012.12.002.

[90] Jentsch JD, Taylor JR. Impulsivity resulting from frontostriatal dysfunction in drug abuse: implications for the control of behavior by reward-related stimuli. Psychopharmacology (Berl) 1999;146(4):373–90.

[91] Koob GF, Bloom FE. Cellular and molecular mechanisms of drug dependence. Science 1988;242(4879):715–23.

[92] Grace AA. Gating of information flow within the limbic system and the pathophysiology of schizophrenia. Brain Res Brain Res Rev 2000; 31(2-3):330–41.

[93] Grace AA. The tonic/phasic model of dopamine system regulation and its implications for understanding alcohol and psychostimulant craving. Addiction 2000;95(Suppl 2):S119–128.

[94] Volkow ND, Morales M. The Brain on Drugs: From Reward to Addiction. Cell 2015;162(4):712–25. https://doi.org/10.1016/j.cell.2015.07.046.

[95] Jentsch JD, Olausson P, De La Garza 2nd R, Taylor JR. Impairments of reversal learning and response perseveration after repeated, intermittent cocaine administrations to monkeys. Neuropsychopharmacology 2002;26(2):183–90. https://doi.org/10.1016/S0893-133X(01)00355-4.

[96] Ersche KD, Roiser JP, Robbins TW, Sahakian BJ. Chronic cocaine but not chronic amphetamine use is associated with perseverative responding in humans. Psychopharmacology (Berl) 2008;197(3):421–31. https://doi.org/10.1007/s00213-007-1051-1.

[97] Meunier D, Ersche KD, Craig KJ, Fornito A, Merlo-Pich E, Fineberg NA, Bullmore ET. Brain functional connectivity in stimulant drug dependence and obsessive-compulsive disorder. Neuroimage 2012;59(2):1461–8. https://doi.org/10.1016/j.neuroimage.2011.08.003.

[98] Chen BT, Yau HJ, Hatch C, Kusumoto-Yoshida I, Cho SL, Hopf FW, Bonci A. Rescuing cocaine-induced prefrontal cortex hypoactivity prevents compulsive cocaine seeking. Nature 2013;496(7445):359–62. https://doi.org/10.1038/nature12024.

[99] Jonkman S, Pelloux Y, Everitt BJ. Differential roles of the dorsolateral and midlateral striatum in punished cocaine seeking. J Neurosci 2012;32 (13):4645–50. https://doi.org/10.1523/jneurosci.0348-12.2012.

[100] Pelloux Y, Dilleen R, Economidou D, Theobald D, Everitt BJ. Reduced forebrain serotonin transmission is causally involved in the development of compulsive cocaine seeking in rats. Neuropsychopharmacology 2012;37(11):2505–14. https://doi.org/10.1038/npp.2012.111.

[101] Pelloux Y, Murray JE, Everitt BJ. Differential roles of the prefrontal cortical subregions and basolateral amygdala in compulsive cocaine seeking and relapse after voluntary abstinence in rats. Eur J Neurosci 2013;38(7):3018–26. https://doi.org/10.1111/ejn.12289.

[102] Fein G, Di Sclafani V, Cardenas VA, Goldmann H, Tolou-Shams M, Meyerhoff DJ. Cortical gray matter loss in treatment-naive alcohol dependent individuals. Alcohol Clin Exp Res 2002;26(4):558–64.

[103] Brody AL, Mandelkern MA, Jarvik ME, Lee GS, Smith EC, Huang JC, London ED. Differences between smokers and nonsmokers in regional gray matter volumes and densities. Biol Psychiatry 2004;55(1):77–84.

[104] Gallinat J, Meisenzahl E, Jacobsen LK, Kalus P, Bierbrauer J, Kienast T, Staedtgen M. Smoking and structural brain deficits: a volumetric MR investigation. Eur J Neurosci 2006;24(6):1744–50. https://doi.org/10.1111/j.1460-9568.2006.05050.x.

[105] Ersche KD, Williams GB, Robbins TW, Bullmore ET. Meta-analysis of structural brain abnormalities associated with stimulant drug dependence and neuroimaging of addiction vulnerability and resilience. Curr Opin Neurobiol 2013;23(4):615–24. https://doi.org/10.1016/j.conb.2013.02.017.

[106] Koob GF, Volkow ND. Neurobiology of addiction: a neurocircuitry analysis. Lancet Psychiatry 2016;3(8):760–73. https://doi.org/10.1016/S2215-0366(16)00104-8.

[107] Alexander GE, Crutcher MD, DeLong MR. Basal ganglia-thalamocortical circuits: parallel substrates for motor, oculomotor, "prefrontal" and "limbic" functions. Prog Brain Res 1990;85:119–46.

[108] Alexander GE, DeLong MR, Strick PL. Parallel organization of functionally segregated circuits linking basal ganglia and cortex. Annu Rev Neurosci 1986;9:357–81. https://doi.org/10.1146/annurev.ne.09.030186.002041.

[109] Haber SN, Fudge JL, McFarland NR. Striatonigrostriatal pathways in primates form an ascending spiral from the shell to the dorsolateral striatum. J Neurosci 2000;20(6):2369–82.

[110] Ikemoto S. Dopamine reward circuitry: two projection systems from the ventral midbrain to the nucleus accumbens-olfactory tubercle complex. Brain Res Rev 2007;56(1):27–78. https://doi.org/10.1016/j.brainresrev.2007.05.004.

[111] Robbins TW, Ersche KD, Everitt BJ. Drug addiction and the memory systems of the brain. Ann N Y Acad Sci 2008;1141:1–21. https://doi.org/10.1196/annals.1441.020.

[112] Sullivan EV, Deshmukh A, De Rosa E, Rosenbloom MJ, Pfefferbaum A. Striatal and forebrain nuclei volumes: contribution to motor function and working memory deficits in alcoholism. Biol Psychiatry 2005;57(7):768–76. https://doi.org/10.1016/j.biopsych.2004.12.012.

[113] Koob GF. A role for brain stress systems in addiction. Neuron 2008;59(1):11–34. https://doi.org/10.1016/j.neuron.2008.06.012.

[114] Koob GF. Brain stress systems in the amygdala and addiction. Brain Res 2009;1293:61–75. https://doi.org/10.1016/j.brainres.2009.03.038.

[115] Alheid GF, De Olmos JS, Beltramino CA. Amygdala and extended amygdala. In: Paxinos G, editor. The rat nervous system. 2nd ed. San Diego: Academic Press; 1995. p. 495–578.

[116] Delfs JM, Zhu Y, Druhan JP, Aston-Jones G. Noradrenaline in the ventral forebrain is critical for opiate withdrawal-induced aversion. Nature 2000;403(6768):430–4. https://doi.org/10.1038/35000212.

[117] Koob GF, Heinrichs SC, Menzaghi F, Pich EM, Britton KT. Corticotropin releasing factor, stress and behavior. Seminars in Neuroscience 1994; 6(4):221–9.

[118] Merlo Pich E, Lorang M, Yeganeh M, Rodriguez de Fonseca F, Raber J, Koob GF, Weiss F. Increase of extracellular corticotropin-releasing factor-like immunoreactivity levels in the amygdala of awake rats during restraint stress and ethanol withdrawal as measured by microdialysis. J Neurosci 1995;15(8):5439–47.

[119] Olive MF, Koenig HN, Nannini MA, Hodge CW. Elevated extracellular CRF levels in the bed nucleus of the stria terminalis during ethanol withdrawal and reduction by subsequent ethanol intake. Pharmacol Biochem Behav 2002;72(1-2):213–20.

[120] Rasmussen DD, Boldt BM, Bryant CA, Mitton DR, Larsen SA, Wilkinson CW. Chronic daily ethanol and withdrawal: 1. Long-term changes in the hypothalamo-pituitary-adrenal axis. Alcohol Clin Exp Res 2000;24(12):1836–49.

[121] Rivier C, Bruhn T, Vale W. Effect of ethanol on the hypothalamic-pituitary-adrenal axis in the rat: role of corticotropin-releasing factor (CRF). J Pharmacol Exp Ther 1984;229(1):127–31.

[122] Roberto M, Cruz MT, Gilpin NW, Sabino V, Schweitzer P, Bajo M, Parsons LH. Corticotropin releasing factor-induced amygdala gamma-aminobutyric Acid release plays a key role in alcohol dependence. Biol Psychiatry 2010;67(9):831–9. https://doi.org/10.1016/j.biopsych.2009.11.007.

[123] Marcinkiewcz CA, Prado MM, Isaac SK, Marshall A, Rylkova D, Bruijnzeel AW. Corticotropin-releasing factor within the central nucleus of the amygdala and the nucleus accumbens shell mediates the negative affective state of nicotine withdrawal in rats. Neuropsychopharmacology 2009;34(7):1743–52. https://doi.org/10.1038/npp.2008.231.

[124] Kalivas PW. The glutamate homeostasis hypothesis of addiction. Nat Rev Neurosci 2009;10(8):561–72. https://doi.org/10.1038/nrn2515.

[125] Luscher C, Malenka RC. Drug-evoked synaptic plasticity in addiction: from molecular changes to circuit remodeling. Neuron 2011;69(4):650–63. https://doi.org/10.1016/j.neuron.2011.01.017.

[126] Murray JE, Belin-Rauscent A, Simon M, Giuliano C, Benoit-Marand M, Everitt BJ, Belin D. Basolateral and central amygdala differentially recruit and maintain dorsolateral striatum-dependent cocaine-seeking habits. Nat Commun 2015;6. https://doi.org/10.1038/ncomms10088.

[127] Wolf ME, Ferrario CR. AMPA receptor plasticity in the nucleus accumbens after repeated exposure to cocaine. Neurosci Biobehav Rev 2010;35(2):185–211. https://doi.org/10.1016/j.neubiorev.2010.01.013.

[128] Kelly PH, Seviour PW, Iversen SD. Amphetamine and apomorphine responses in the rat following 6-OHDA lesions of the nucleus accumbens septi and corpus striatum. Brain Res 1975;94(3):507–22.

[129] Volkow ND, Fowler JS, Wang GJ, Swanson JM. Dopamine in drug abuse and addiction: results of imaging studies and treatment implications. Arch Neurol 2007;64(11):1575–9.

[130] Caprioli D, Sawiak SJ, Merlo E, Theobald DE, Spoelder M, Jupp B, Dalley JW. Gamma aminobutyric acidergic and neuronal structural markers in the nucleus accumbens core underlie trait-like impulsive behavior. Biol Psychiatry 2014;75(2):115–23. https://doi.org/10.1016/j.biopsych.2013.07.013.

[131] Dalley JW, Everitt BJ, Robbins TW. Impulsivity, compulsivity, and top-down cognitive control. Neuron 2011;69(4):680–94. https://doi.org/10.1016/j.neuron.2011.01.020.

[132] Nader M, Czoty PW, Gould RW, Riddick NV. Characterising organism X environment interactions in non-human primate models of addiction: PET imaging studies of dopamine D2 receptors. Oxford, UK: Oxford Univ. Press; 2010.

[133] Mitchell MR, Weiss VG, Beas BS, Morgan D, Bizon JL, Setlow B. Adolescent risk taking, cocaine self-administration, and striatal dopamine signaling. Neuropsychopharmacology 2014;39(4):955–62. https://doi.org/10.1038/npp.2013.295.

[134] Kalivas PW, Peters J, Knackstedt L. Animal models and brain circuits in drug addiction. Mol Interv 2006;6(6):339–44. https://doi.org/10.1124/mi.6.6.7.

[135] McClure EA, Gipson CD, Malcolm RJ, Kalivas PW, Gray KM. Potential role of N-acetylcysteine in the management of substance use disorders. CNS Drugs 2014;28(2):95–106. https://doi.org/10.1007/s40263-014-0142-x.

[136] Zhou W, Kalivas PW. N-acetylcysteine reduces extinction responding and induces enduring reductions in cue- and heroin-induced drug-seeking. Biol Psychiatry 2008;63(3):338–40. https://doi.org/10.1016/j.biopsych.2007.06.008.

[137] Murray JE, Everitt BJ, Belin D. N-Acetylcysteine reduces early- and late-stage cocaine seeking without affecting cocaine taking in rats. Addict Biol 2012;17(2):437–40. https://doi.org/10.1111/j.1369-1600.2011.00330.x.

[138] Ducret E, Puaud M, Lacoste J, Belin-Rauscent A, Fouyssac M, Dugast E, Belin D. N-acetylcysteine Facilitates Self-Imposed Abstinence After Escalation of Cocaine Intake. Biol Psychiatry 2016;80(3):226–34. https://doi.org/10.1016/j.biopsych.2015.09.019.

[139] Belin D, Jonkman S, Dickinson A, Robbins TW, Everitt BJ. Parallel and interactive learning processes within the basal ganglia: relevance for the understanding of addiction. Behav Brain Res 2009;199(1):89–102. https://doi.org/10.1016/j.bbr.2008.09.027.

[140] Cannella N, Halbout B, Uhrig S, Evrard L, Corsi M, Corti C, Spanagel R. The mGluR2/3 agonist LY379268 induced anti-reinstatement effects in rats exhibiting addiction-like behavior. Neuropsychopharmacology 2013;38(10):2048–56. https://doi.org/10.1038/npp.2013.106.

[141] Neisewander JL, Cheung TH, Pentkowski NS. Dopamine D3 and 5-HT1B receptor dysregulation as a result of psychostimulant intake and forced abstinence: Implications for medications development. Neuropharmacology 2014;76(Pt B):301–19. https://doi.org/10.1016/j.neuropharm.2013.08.014.

[142] Muller CP, Homberg JR. The role of serotonin in drug use and addiction. Behav Brain Res 2015;277:146–92. https://doi.org/10.1016/j.bbr.2014.04.007.

[143] Edwards S, Guerrero M, Ghoneim OM, Roberts E, Koob GF. Evidence that vasopressin V1b receptors mediate the transition to excessive drinking in ethanol-dependent rats. Addict Biol 2012;17(1):76–85. https://doi.org/10.1111/j.1369-1600.2010.00291.x.

[144] Gilpin NW, Koob GF. Effects of beta-adrenoceptor antagonists on alcohol drinking by alcohol-dependent rats. Psychopharmacology (Berl) 2010;212(3):431–9. https://doi.org/10.1007/s00213-010-1967-8.

[145] June HL, Liu J, Warnock KT, Bell KA, Balan I, Bollino D, Aurelian L. CRF-amplified neuronal TLR4/MCP-1 signaling regulates alcohol self-administration. Neuropsychopharmacology 2015;40(6):1549–59. https://doi.org/10.1038/npp.2015.4.

[146] Schlosburg, J. E., Whitfield, T. W., Jr., Park, P. E., Crawford, E. F., George, O., Vendruscolo, L. F., & Koob, G. F. (2013). Long-term antagonism of kappa opioid receptors prevents escalation of and increased motivation for heroin intake. J Neurosci, 33(49), 19384-19392. https://doi.org/10.1523/jneurosci.1979-13.2013

[147] Schmeichel BE, Barbier E, Misra KK, Contet C, Schlosburg JE, Grigoriadis D, Vendruscolo LF. Hypocretin receptor 2 antagonism dose-dependently reduces escalated heroin self-administration in rats. Neuropsychopharmacology 2015;40(5):1123–9. https://doi.org/10.1038/npp.2014.293.

[148] Vendruscolo LF, Barbier E, Schlosburg JE, Misra KK, Whitfield Jr. TW, Logrip ML, Koob GF. Corticosteroid-dependent plasticity mediates compulsive alcohol drinking in rats. J Neurosci 2012;32(22):7563–71. https://doi.org/10.1523/JNEUROSCI.0069-12.2012.

[149] Walker BM, Koob GF. Pharmacological evidence for a motivational role of kappa-opioid systems in ethanol dependence. Neuropsychopharmacology 2008;33(3):643–52. https://doi.org/10.1038/sj.npp.1301438.

[150] Walker BM, Rasmussen DD, Raskind MA, Koob GF. alpha1-noradrenergic receptor antagonism blocks dependence-induced increases in responding for ethanol. Alcohol 2008;42(2):91–7. https://doi.org/10.1016/j.alcohol.2007.12.002.

[151] Whitfield Jr. TW, Schlosburg JE, Wee S, Gould A, George O, Grant Y, Koob GF. kappa Opioid receptors in the nucleus accumbens shell mediate escalation of methamphetamine intake. J Neurosci 2015;35(10):4296–305. https://doi.org/10.1523/jneurosci.1978-13.2015.

[152] Economidou D, Hansson AC, Weiss F, Terasmaa A, Sommer WH, Cippitelli A, Heilig M. Dysregulation of nociceptin/orphanin FQ activity in the amygdala is linked to excessive alcohol drinking in the rat. Biol Psychiatry 2008;64(3):211–8. https://doi.org/10.1016/j.biopsych.2008.02.004.

[153] Gilpin NW, Stewart RB, Badia-Elder NE. Neuropeptide Y suppresses ethanol drinking in ethanol-abstinent, but not non-ethanol-abstinent, Wistar rats. Alcohol 2008;42(7):541–51. https://doi.org/10.1016/j.alcohol.2008.07.001.

[154] Gilpin NW, Stewart RB, Murphy JM, Li TK, Badia-Elder NE. Neuropeptide Y reduces oral ethanol intake in alcohol-preferring (P) rats following a period of imposed ethanol abstinence. Alcohol Clin Exp Res 2003;27(5):787–94. https://doi.org/10.1097/01.ALC.0000065723.93234.1D.

[155] Thorsell A. Neuropeptide Y (NPY) in alcohol intake and dependence. Peptides 2007;28(2):480–3. https://doi.org/10.1016/j.peptides.2006.11.017.

[156] Thorsell A, Repunte-Canonigo V, O'Dell LE, Chen SA, King AR, Lekic D, Sanna PP. Viral vector-induced amygdala NPY overexpression reverses increased alcohol intake caused by repeated deprivations in Wistar rats. Brain 2007;130(Pt 5):1330–7. https://doi.org/10.1093/brain/awm033.

[157] Thorsell A, Slawecki CJ, Ehlers CL. Effects of neuropeptide Y and corticotropin-releasing factor on ethanol intake in Wistar rats: interaction with chronic ethanol exposure. Behav Brain Res 2005;161(1):133–40. https://doi.org/10.1016/j.bbr.2005.01.016.

[158] Thorsell A, Slawecki CJ, Ehlers CL. Effects of neuropeptide Y on appetitive and consummatory behaviors associated with alcohol drinking in wistar rats with a history of ethanol exposure. Alcohol Clin Exp Res 2005;29(4):584–90.

[159] Thorsell A, Slawecki CJ, Khoury A, Mathe AA, Ehlers CL. Effect of social isolation on ethanol consumption and substance P/neurokinin expression in Wistar rats. Alcohol 2005;36(2):91–7. https://doi.org/10.1016/j.alcohol.2005.07.003.

[160] Heilig M, Koob GF. A key role for corticotropin-releasing factor in alcohol dependence. Trends Neurosci 2007;30(8):399–406. https://doi.org/10.1016/j.tins.2007.06.006.

[161] Valdez GR, Koob GF. Allostasis and dysregulation of corticotropin-releasing factor and neuropeptide Y systems: implications for the development of alcoholism. Pharmacol Biochem Behav 2004;79(4):671–89. https://doi.org/10.1016/j.pbb.2004.09.020.

[162] Brami-Cherrier K, Valjent E, Herve D, Darragh J, Corvol JC, Pages C, Caboche J. Parsing molecular and behavioral effects of cocaine in mitogen- and stress-activated protein kinase-1-deficient mice. J Neurosci 2005;25(49):11444–54. https://doi.org/10.1523/JNEUROSCI.1711-05.2005.

[163] Nestler EJ. Molecular neurobiology of addiction. Am J Addict 2001;10(3):201–17.

[164] Bibb JA, Chen J, Taylor JR, Svenningsson P, Nishi A, Snyder GL, Greengard P. Effects of chronic exposure to cocaine are regulated by the neuronal protein Cdk5. Nature 2001;410(6826):376–80. https://doi.org/10.1038/35066591.

[165] Bowers MS, McFarland K, Lake RW, Peterson YK, Lapish CC, Gregory ML, Kalivas PW. Activator of G protein signaling 3: a gatekeeper of cocaine sensitization and drug seeking. Neuron 2004;42(2):269–81.

[166] Nestler EJ. Genes and addiction. Nat Genet 2000;26(3):277–81. https://doi.org/10.1038/81570.

[167] Bhat RV, Baraban JM. Activation of transcription factor genes in striatum by cocaine: role of both serotonin and dopamine systems. J Pharmacol Exp Ther 1993;267(1):496–505.

[168] Hope B, Kosofsky B, Hyman SE, Nestler EJ. Regulation of immediate early gene expression and AP-1 binding in the rat nucleus accumbens by chronic cocaine. Proc Natl Acad Sci U S A 1992;89(13):5764–8.

[169] Humblot N, Thiriet N, Gobaille S, Aunis D, Zwiller J. The serotonergic system modulates the cocaine-induced expression of the immediate early genes egr-1 and c-fos in rat brain. Ann N Y Acad Sci 1998;844:7–20.

[170] Jouvert P, Dietrich JB, Aunis D, Zwiller J. Differential rat brain expression of EGR proteins and of the transcriptional corepressor NAB in response to acute or chronic cocaine administration. Neuromolecular Med 2002;1(2):137–51. https://doi.org/10.1385/NMM:1:2:137.

[171] Moratalla R, Robertson HA, Graybiel AM. Dynamic regulation of NGFI-A (zif268, egr1) gene expression in the striatum. J Neurosci 1992;12(7):2609–22.

[172] Colvis CM, Pollock JD, Goodman RH, Impey S, Dunn J, Mandel G, Nestler EJ. Epigenetic mechanisms and gene networks in the nervous system. J Neurosci 2005;25(45):10379–89. https://doi.org/10.1523/JNEUROSCI.4119-05.2005.

[173] Schmidt HD, McGinty JF, West AE, Sadri-Vakili G. Epigenetics and psychostimulant addiction. Cold Spring Harb Perspect Med 2013;3(3). https://doi.org/10.1101/cshperspect.a012047.

[174] Siegmund KD, Connor CM, Campan M, Long TI, Weisenberger DJ, Biniszkiewicz D, Akbarian S. DNA methylation in the human cerebral cortex is dynamically regulated throughout the life span and involves differentiated neurons. PLoS One 2007;2(9). https://doi.org/10.1371/journal. pone.0000895.

[175] Tsankova N, Renthal W, Kumar A, Nestler EJ. Epigenetic regulation in psychiatric disorders. Nat Rev Neurosci 2007;8(5):355–67. https://doi.org/ 10.1038/nrn2132.

[176] Renthal W, Nestler EJ. Epigenetic mechanisms in drug addiction. Trends Mol Med 2008;14(8):341–50. https://doi.org/10.1016/j.molmed. 2008.06.004.

[177] Berger SL. The complex language of chromatin regulation during transcription. Nature 2007;447(7143):407–12. https://doi.org/10.1038/nature 05915.

[178] Li B, Carey M, Workman JL. The role of chromatin during transcription. Cell 2007;128(4):707–19. https://doi.org/10.1016/j.cell.2007.01.015.

[179] Berger SL. Histone modifications in transcriptional regulation. Curr Opin Genet Dev 2002;12(2):142–8.

[180] Strahl BD, Allis CD. The language of covalent histone modifications. Nature 2000;403(6765):41–5. https://doi.org/10.1038/47412.

[181] Jaenisch R, Bird A. Epigenetic regulation of gene expression: how the genome integrates intrinsic and environmental signals. Nat Genet 2003;33 (Suppl):245–54. https://doi.org/10.1038/ng1089.

[182] Jaenisch R. DNA methylation and imprinting: why bother? Trends Genet 1997;13(8):323–9.

[183] Jenuwein T, Allis CD. Translating the histone code. Science 2001;293(5532):1074–80. https://doi.org/10.1126/science.1063127.

[184] Lopez-Rodas G, Brosch G, Georgieva EI, Sendra R, Franco L, Loidl P. Histone deacetylase. A key enzyme for the binding of regulatory proteins to chromatin. FEBS Lett 1993;317(3):175–80.

[185] Marks PA, Miller T, Richon VM. Histone deacetylases. Curr Opin Pharmacol 2003;3(4):344–51.

[186] Turner BM. Cellular memory and the histone code. Cell 2002;111(3):285–91.

[187] Kumar A, Choi KH, Renthal W, Tsankova NM, Theobald DE, Truong HT, Nestler EJ. Chromatin remodeling is a key mechanism underlying cocaine-induced plasticity in striatum. Neuron 2005;48(2):303–14. https://doi.org/10.1016/j.neuron.2005.09.023.

[188] Schroeder FA, Penta KL, Matevossian A, Jones SR, Konradi C, Tapper AR, Akbarian S. Drug-induced activation of dopamine D(1) receptor sig-naling and inhibition of class I/II histone deacetylase induce chromatin remodeling in reward circuitry and modulate cocaine-related behaviors. Neuropsychopharmacology 2008;33(12):2981–92. https://doi.org/10.1038/npp.2008.15.

[189] Shen HY, Kalda A, Yu L, Ferrara J, Zhu J, Chen JF. Additive effects of histone deacetylase inhibitors and amphetamine on histone H4 acetylation, cAMP responsive element binding protein phosphorylation and DeltaFosB expression in the striatum and locomotor sensitization in mice. Neuroscience 2008;157(3):644–55. https://doi.org/10.1016/j.neuroscience.2008.09.019.

[190] Renthal W, Maze I, Krishnan V, Covington 3rd HE, Xiao G, Kumar A, Nestler EJ. Histone deacetylase 5 epigenetically controls behavioral adap-tations to chronic emotional stimuli. Neuron 2007;56(3):517–29. https://doi.org/10.1016/j.neuron.2007.09.032.

[191] Shibasaki M, Mizuno K, Kurokawa K, Ohkuma S. L-type voltage-dependent calcium channels facilitate acetylation of histone H3 through PKCgamma phosphorylation in mice with methamphetamine-induced place preference. J Neurochem 2011;118(6):1056–66. https://doi.org/ 10.1111/j.1471-4159.2011.07387.x.

[192] Freeman WM, Patel KM, Brucklacher RM, Lull ME, Erwin M, Morgan D, Vrana KE. Persistent alterations in mesolimbic gene expression with abstinence from cocaine self-administration. Neuropsychopharmacology 2008;33(8):1807–17. https://doi.org/10.1038/sj.npp.1301577.

[193] Sadri-Vakili G. Cocaine triggers epigenetic alterations in the corticostriatal circuit. Brain Res 2015;1628(Pt A):50–9. https://doi.org/10.1016/j. brainres.2014.09.069.

[194] Benavides DR, Bibb JA. Role of Cdk5 in drug abuse and plasticity. Ann N Y Acad Sci 2004;1025:335–44. https://doi.org/10.1196/annals.1316.041.

[195] Lu B, Figurov A. Role of neurotrophins in synapse development and plasticity. Rev Neurosci 1997;8(1):1–12.

[196] Lu VB, Ballanyi K, Colmers WF, Smith PA. Neuron type-specific effects of brain-derived neurotrophic factor in rat superficial dorsal horn and their relevance to 'central sensitization'. J Physiol 2007;584(Pt 2):543–63. https://doi.org/10.1113/jphysiol.2007.141267.

[197] Lu Y, Christian K, Lu B. BDNF: a key regulator for protein synthesis-dependent LTP and long-term memory? Neurobiol Learn Mem 2008;89 (3):312–23. https://doi.org/10.1016/j.nlm.2007.08.018.

[198] Rex CS, Lin CY, Kramar EA, Chen LY, Gall CM, Lynch G. Brain-derived neurotrophic factor promotes long-term potentiation-related cytoskeletal changes in adult hippocampus. J Neurosci 2007;27(11):3017–29. https://doi.org/10.1523/JNEUROSCI.4037-06.2007.

[199] Schjetnan AG, Escobar ML. In vivo BDNF modulation of hippocampal mossy fiber plasticity induced by high frequency stimulation. Hippocampus 2012;22(1):1–8. https://doi.org/10.1002/hipo.20866.

[200] Yamada K, Mizuno M, Nabeshima T. Role for brain-derived neurotrophic factor in learning and memory. Life Sci 2002;70(7):735–44.

[201] Cleck JN, Ecke LE, Blendy JA. Endocrine and gene expression changes following forced swim stress exposure during cocaine abstinence in mice. Psychopharmacology (Berl) 2008;201(1):15–28. https://doi.org/10.1007/s00213-008-1243-3.

[202] Sadri-Vakili G, Kumaresan V, Schmidt HD, Famous KR, Chawla P, Vassoler FM, Cha JH. Cocaine-induced chromatin remodeling increases brain-derived neurotrophic factor transcription in the rat medial prefrontal cortex, which alters the reinforcing efficacy of cocaine. J Neurosci 2010;30 (35):11735–44. https://doi.org/10.1523/JNEUROSCI.2328-10.2010.

[203] Schmidt HD, Sangrey GR, Darnell SB, Schassburger RL, Cha JH, Pierce RC, Sadri-Vakili G. Increased brain-derived neurotrophic factor (BDNF) expression in the ventral tegmental area during cocaine abstinence is associated with increased histone acetylation at BDNF exon I-containing promoters. J Neurochem 2012;120(2):202–9. https://doi.org/10.1111/j.1471-4159.2011.07571.x.

[204] Lopez-Moreno JA, Marcos M, Calleja-Conde J, Echeverry-Alzate V, Buhler KM, Costa-Alba P, Gine E. Histone deacetylase gene expression following binge alcohol consumption in rats and humans. Alcohol Clin Exp Res 2015;39(10):1939–50. https://doi.org/10.1111/acer.12850.

[205] Romieu P, Host L, Gobaille S, Sandner G, Aunis D, Zwiller J. Histone deacetylase inhibitors decrease cocaine but not sucrose self-administration in rats. J Neurosci 2008;28(38):9342–8. https://doi.org/10.1523/JNEUROSCI.0379-08.2008.

[206] Ponomarev I, Stelly CE, Morikawa H, Blednov YA, Mayfield RD, Harris RA. Mechanistic insights into epigenetic modulation of ethanol consumption. Alcohol 2017;60:95–101. https://doi.org/10.1016/j.alcohol.2017.01.016.

[207] Sun J, Wang L, Jiang B, Hui B, Lv Z, Ma L. The effects of sodium butyrate, an inhibitor of histone deacetylase, on the cocaine- and sucrose-maintained self-administration in rats. Neurosci Lett 2008;441(1):72–6. https://doi.org/10.1016/j.neulet.2008.05.010.

[208] Wang L, Lv Z, Hu Z, Sheng J, Hui B, Sun J, Ma L. Chronic cocaine-induced H3 acetylation and transcriptional activation of CaMKIIalpha in the nucleus accumbens is critical for motivation for drug reinforcement. Neuropsychopharmacology 2010;35(4):913–28. https://doi.org/10.1038/npp.2009.193.

[209] Romieu P, Deschatrettes E, Host L, Gobaille S, Sandner G, Zwiller J. The inhibition of histone deacetylases reduces the reinstatement of cocaine-seeking behavior in rats. Curr Neuropharmacol 2011;9(1):21–5. https://doi.org/10.2174/157015911795017317.

[210] Malvaez M, Sanchis-Segura C, Vo D, Lattal KM, Wood MA. Modulation of chromatin modification facilitates extinction of cocaine-induced conditioned place preference. Biol Psychiatry 2010;67(1):36–43. https://doi.org/10.1016/j.biopsych.2009.07.032.

[211] Castino MR, Cornish JL, Clemens KJ. Inhibition of histone deacetylases facilitates extinction and attenuates reinstatement of nicotine self-administration in rats. PLoS One 2015;10(4). https://doi.org/10.1371/journal.pone.0124796.

[212] Chen WS, Xu WJ, Zhu HQ, Gao L, Lai MJ, Zhang FQ, Liu HF. Effects of histone deacetylase inhibitor sodium butyrate on heroin seeking behavior in the nucleus accumbens in rats. Brain Res 2016;1652:151–7. https://doi.org/10.1016/j.brainres.2016.10.007.

[213] Rice JC, Allis CD. Histone methylation versus histone acetylation: new insights into epigenetic regulation. Curr Opin Cell Biol 2001;13(3):263–73.

[214] Rice JC, Allis CD. Code of silence. Nature 2001;414(6861):258–61. https://doi.org/10.1038/35104721.

[215] Black, Y. D., Maclaren, F. R., Naydenov, A. V., Carlezon, W. A., Jr., Baxter, M. G., & Konradi, C. (2006). Altered attention and prefrontal cortex gene expression in rats after binge-like exposure to cocaine during adolescence. J Neurosci, 26(38), 9656-9665. https://doi.org/10.1523/JNEUROSCI.2391-06.2006

[216] Renthal W, Kumar A, Xiao G, Wilkinson M, Covington 3rd HE, Maze I, Nestler EJ. Genome-wide analysis of chromatin regulation by cocaine reveals a role for sirtuins. Neuron 2009;62(3):335–48. https://doi.org/10.1016/j.neuron.2009.03.026.

[217] Sadakierska-Chudy A, Frankowska M, Jastrzebska J, Wydra K, Miszkiel J, Sanak M, Filip M. Cocaine administration and its withdrawal enhance the expression of genes encoding histone-modifying enzymes and histone acetylation in the rat prefrontal cortex. Neurotox Res 2017;32(1):141–50. https://doi.org/10.1007/s12640-017-9728-7.

[218] Maze I, Covington 3rd HE, Dietz DM, LaPlant Q, Renthal W, Russo SJ, Nestler EJ. Essential role of the histone methyltransferase G9a in cocaine-induced plasticity. Science 2010;327(5962):213–6. https://doi.org/10.1126/science.1179438.

[219] Suzuki MM, Bird A. DNA methylation landscapes: provocative insights from epigenomics. Nat Rev Genet 2008;9(6):465–76. https://doi.org/10.1038/nrg2341.

[220] Host L, Dietrich JB, Carouge D, Aunis D, Zwiller J. Cocaine self-administration alters the expression of chromatin-remodelling proteins; modulation by histone deacetylase inhibition. J Psychopharmacol 2011;25(2):222–9. https://doi.org/10.1177/0269881109348173.

[221] Damez-Werno DM, Sun H, Scobie KN, Shao N, Rabkin J, Dias C, Nestler EJ. Histone arginine methylation in cocaine action in the nucleus accumbens. Proc Natl Acad Sci U S A 2016;113(34):9623–8. https://doi.org/10.1073/pnas.1605045113.

[222] Scobie KN, Damez-Werno D, Sun H, Shao N, Gancarz A, Panganiban CH, Nestler EJ. Essential role of poly(ADP-ribosyl)ation in cocaine action. Proc Natl Acad Sci U S A 2014;111(5):2005–10. https://doi.org/10.1073/pnas.1319703111.

[223] Mattick JS. Non-coding RNAs: the architects of eukaryotic complexity. EMBO Rep 2001;2(11):986–91. https://doi.org/10.1093/embo-reports/kve230.

[224] Ambros V. The functions of animal microRNAs. Nature 2004;431(7006):350–5. https://doi.org/10.1038/nature02871.

[225] Guarnieri DJ, DiLeone RJ. MicroRNAs: a new class of gene regulators. Ann Med 2008;40(3):197–208. https://doi.org/10.1080/07853890701771823.

[226] Didiano D, Hobert O. Molecular architecture of a miRNA-regulated 3' UTR. RNA 2008;14(7):1297–317. https://doi.org/10.1261/rna.1082708.

[227] Hobert O. Gene regulation by transcription factors and microRNAs. Science 2008;319(5871):1785–6. https://doi.org/10.1126/science.1151651.

[228] Lugli G, Torvik VI, Larson J, Smalheiser NR. Expression of microRNAs and their precursors in synaptic fractions of adult mouse forebrain. J Neurochem 2008;106(2):650–61. https://doi.org/10.1111/j.1471-4159.2008.05413.x.

[229] Sempere LF, Freemantle S, Pitha-Rowe I, Moss E, Dmitrovsky E, Ambros V. Expression profiling of mammalian microRNAs uncovers a subset of brain-expressed microRNAs with possible roles in murine and human neuronal differentiation. Genome Biol 2004;5(3):R13. https://doi.org/10.1186/gb-2004-5-3-r13.

[230] Siegel G, Obernosterer G, Fiore R, Oehmen M, Bicker S, Christensen M, Schratt GM. A functional screen implicates microRNA-138-dependent regulation of the depalmitoylation enzyme APT1 in dendritic spine morphogenesis. Nat Cell Biol 2009;11(6):705–16. https://doi.org/10.1038/ncb1876.

[231] Hollander JA, Im HI, Amelio AL, Kocerha J, Bali P, Lu Q, Kenny PJ. Striatal microRNA controls cocaine intake through CREB signalling. Nature 2010;466(7303):197–202. https://doi.org/10.1038/nature09202.

[232] Im HI, Hollander JA, Bali P, Kenny PJ. MeCP2 controls BDNF expression and cocaine intake through homeostatic interactions with microRNA-212. Nat Neurosci 2010;13(9):1120–7. https://doi.org/10.1038/nn.2615.

[233] Chandrasekar V, Dreyer JL. microRNAs miR-124, let-7d and miR-181a regulate cocaine-induced plasticity. Mol Cell Neurosci 2009;42(4):350–62. https://doi.org/10.1016/j.mcn.2009.08.009.

[234] Chandrasekar V, Dreyer JL. Regulation of MiR-124, Let-7d, and MiR-181a in the accumbens affects the expression, extinction, and reinstatement of cocaine-induced conditioned place preference. Neuropsychopharmacology 2011;36(6):1149–64. https://doi.org/10.1038/npp.2010.250.

Chapter 11

Role of Stress-Associated Signaling in Addiction

Elizabeth M. Doncheck and John R. Mantsch

Department of Biomedical Sciences, Marquette University, Milwaukee, WI, United States

INTRODUCTION

Stress and Addiction

Stress is essential for survival, as stress-related signaling mechanisms coordinate responses that are critical for adaptation when confronted with hazards [1]. Considering that survival under threatening circumstances requires active approach and avoidance behavioral responses, it is not surprising that stress-responsive systems interface with those that mediate motivation and reward. Dysregulation of these interconnected systems can promote maladaptive behaviors such as those that define addiction. In this chapter, we will provide an overview of how stress engages reward- and motivation-related neurocircuitry to promote drug seeking and addiction. Moreover, we will review evidence that excessive drug use can recruit and/or alter stress-associated signaling to establish or exaggerate its role in drug addiction. The result of the bidirectional relationship between stress and addiction is a vicious cycle within which stress is both a causative factor for and consequence of drug use. Understanding the signaling mechanisms that mediate this relationship is critical for establishing new and more effective intervention approaches for the management of substance use disorders.

Stress and the Progression and Cycle of Drug Use in Addiction

Drug addiction is progressive and cyclic. Drug use is initially recreational and driven by rewarding effects but progresses to compulsive patterns fueled by negative reinforcement and reward deficits/withdrawal-related symptoms. Once addiction develops, periods of drug "bingeing" and intoxication are punctuated with withdrawal followed by an emergent and persistent preoccupation with drug use/drug craving that often leads to relapse, thus reinitiating the cycle. However, not everyone who engages in drug use is susceptible to developing addiction. Moreover, the addiction cycle of drug use, abstinence/withdrawal, and relapse can differ across individuals. These individual differences are determined by a variety of factors, including genetics and environmental features—most notably stress. The contribution of stress to addiction is particularly problematic as it is both unpredictable and unavoidable in everyday life. Importantly, the degree to which stress contributes to drug addiction may be a determinant of effective treatment. For this reason, understanding the signaling mechanisms through which stress regulates various phases of the addiction cycle may reveal opportunities for new interventions aimed at managing addiction specifically in subpopulations of addicts whose drug use is stress related.

Addiction-Related Neurocircuitry

Although addiction involves widespread neuroadaptations in multiple neural systems, key features appear to be related to dysregulation of the mesocorticolimbic system [2,3], which includes a glutamatergic projection from the prelimbic prefrontal cortex (PrL-PFC) to the nucleus accumbens core (NAc-core) that is, regulated, in part, by midbrain dopaminergic projections from the ventral tegmental area (VTA). The mesocorticolimbic system is highly regulated by stress through the actions of hormones such as glucocorticoids and innervation from stress-response regions. As a result, stress can influence drug use on both rapid and protracted timescales, thereby acutely promoting drug seeking and producing long-term plasticity that leads to addiction (see Fig. 1 for reference).

Neural Mechanisms of Addiction. https://doi.org/10.1016/B978-0-12-812202-0.00011-7
Copyright © 2019 Elsevier Inc. All rights reserved.

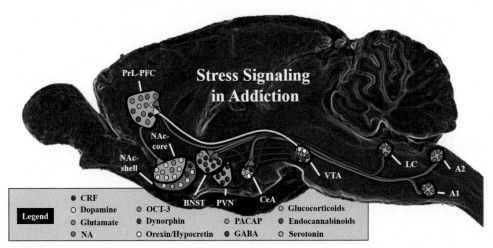

FIG. 1   Stress Signaling & Related Neurocircuitry in Addiction. CRF (red): CRF inputs to the BNST arise from the CeA, which also sends CRF efferents to the PVN, VTA, and LC. The VTA receives CRF inputs from the PVN, BNST, and CeA. Dopamine (yellow) and glutamate (green): The VTA sends dopaminergic projections to the PrL-PFC, which in turn sends glutamatergic projections to dopamine receptor-expressing medium spiny neurons in the NAc. Stress effects on NAc dopamine signaling critically regulate drug seeking. Local glutamatergic regulation in the VTA is also critical for stress-triggered drug seeking. NA (dark blue): The BNST receives NA input primarily from the A1/A2 medullary cell groups, while the CeA receives NA input from both the LC and A1/A2 medullary groups. The PrL-PFC primarily receives NA input from the LC. Dynorphin (or κOR activity; purple), endocannabinoids (brown), and serotonin (light gray): κOR activation in the VTA attenuates inhibitory neurotransmission to promote drug seeking, and dynorphin signaling has also been implicated in both the BNST and NAc-shell. In the NAc-shell, dynorphin stimulates serotonin reuptake to produce aversion and potentiate reward. Orexin/hypocretin (white): Orexin/hypocretin promotes drug seeking through either endocannabinoid-disinhibition of dopamine neurons or enhanced excitation in the VTA and has also been implicated in coordinating CeA activity. PACAP (pink): PACAP activity in the BNST is both necessary and sufficient for stress-triggered reinstatement. GABA (dark gray): GABA regulation has been reported to be particularly important for stress effects on drug seeking in the VTA and PrL-PFC. Glucocorticoids (orange), endocannabinoids (brown), and OCT-3 (light blue): Activation of the glucocorticoid receptor in the CeA has been implicated in stress-triggered drug seeking, glucocorticoids act acutely through endocannabinoid mobilization in the PrL-PFC and through inhibition of OCT-3 in the NAc to potentiate reinstatement of drug seeking.

Chapter Goals

This chapter will provide an overview of the neurobiological signaling mechanisms that underlie the influence of stress on substance use disorders. The work summarized will highlight how stress can act as a catalyst in addiction at every stage and every age, regardless of drug class or biological sex. The specific stress-induced signaling processes and adaptations underlying this relationship will be discussed, with a focus on mechanisms occurring within brain regions wherein the stress and motivation circuits intersect. We will primarily review studies investigating the role of stress signaling in cocaine addiction, as that has been the research focus of many laboratories, including our own. However, as many (although not all) of the contributions of stress to addiction appear to be similar across drug classes, major similarities and differences will be discussed. Additionally, as there are sex differences in both stress responsivity and drug addiction, with female cocaine addicts generally reported to experience greater stress reactivity [4,5], consideration will be given to sex differences in signaling mechanisms. Finally, given the pervasive and pathogenic influence of stress in the life of addicts, and the reciprocal relationship between stress and drug use, there is interest in developing pharmacotherapeutic interventions aimed at minimizing stress contributions to addiction. Therefore, this chapter will also consider potential treatment implications.

OVERVIEW OF STRESS INFLUENCE ON ADDICTION

Overview of the Stress Response

Our survival requires maintaining homeostasis, despite constant challenges by intrinsic and extrinsic factors. Stress can be defined as a state of challenged homeostasis, during which collective neuroendocrine and behavioral reactions coordinate homeostatic restoration. The stress response is coordinated by peripheral hormones, such as glucocorticoids and (nor)adrenaline [aka, (nor)epinephrine], and involves parallel activation of brain systems. While moderate and short-term activation of the stress response is healthy and adaptive, more severe and protracted stress can lead to disease states [6]. This results in an inverted U-shaped relationship between stress response activation and optimal health outcomes, with too much or too little activation producing maladaptive consequences [7], including depression, anxiety, and addiction [8].

Modes of Stress Regulation of Addiction

Primary Stress Response Signaling and Circuitry

Key components of the stress response include the hypothalamic-pituitary-adrenal (HPA) axis and the autonomic sympathetic nervous system. The initiator in the HPA axis is corticotropin-releasing factor (CRF), a neuropeptide which is released from the parvocellular neurons of the hypothalamic paraventricular nucleus (PVN). CRF stimulates adrenocorticotropic hormone (ACTH) release from anterior pituitary corticotrophs through activation of its G_s-protein-coupled CRF-R1 receptor, which is also widely expressed in the brain. Following release into circulation, ACTH in turn targets the adrenal cortex to promote glucocorticoid secretion. Glucocorticoids (cortisol in humans and corticosterone in rodents) are the final effectors of the HPA axis and readily access the brain. Glucocorticoids exert many of their effects through regulation of gene transcription by binding their ubiquitously distributed intracellular glucocorticoid receptors (GRs), although several nongenomic GR-independent effects have been characterized. The stress response also involves the sympathetic branch of the autonomic nervous system, which provides widespread adrenergic and noradrenergic regulation of effectors throughout the body via activation of α and β-adrenergic receptors (ARs). Notably, as peripherally released adrenaline does not penetrate the blood-brain barrier, peripheral adrenergic signaling is paralleled by activation of central noradrenergic systems. The coordinated activation of glucocorticoid and noradrenergic signaling allows for adaptive responding to perturbations in homeostasis, and it is through dysregulation and maladaptation of these processes that drug addiction is promoted (see Fig. 2 for reference).

Secondary Stress Circuitry

Critical structures involved in the integration of stress-related signaling include a complex of brain regions termed the extended amygdala, which includes the bed nucleus of the stria terminalis (BNST), central amygdala (CeA), and NAc-shell. Components of the extended amygdala receive noradrenergic innervation from the locus coeruleus (LC) and medullary cell groups [9] and interface closely with neurocircuitry that subserves reward and motivation, most notably, the mesocorticolimbic system. For this reason, the extended amygdala and the regions therein have been a primary focus of research investigating the contribution of stress to drug addiction.

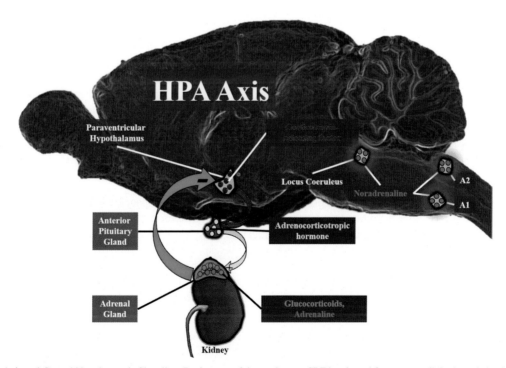

FIG. 2 HPA Axis and Central Noradrenergic Signaling. During stressful experiences, CRF is released from parvocellular hypothalamic PVN neurons and stimulates corticotrophs within the anterior pituitary to release ACTH into circulation. Following subsequent stimulation of the adrenal gland, ACTH induces the release of glucocorticoids (i.e., cortisol in humans, corticosterone in rodents) from the adrenal cortex and adrenaline from the adrenal medulla. While peripheral adrenaline cannot cross the blood-brain barrier, glucocorticoids can, which allows for negative feedback effects on the neurons of the PVN to shut down further HPA axis activity. Central noradrenaline is derived from the locus coeruleus and medullary A1/A2 cell groups.

Neuropeptide Mediators of Stress Responses

Among the many brain signaling molecules that contribute to the stress response, several neuropeptides appear to play important roles through their influence on addiction-related neurocircuitry. Coreleased with GABA and glutamate, neuropeptides can coordinate regional stress-related signaling through complex actions on multiple synapses and cell types via G-protein coupled receptor activation. Important neuropeptides involved in the stress response include CRF, dynorphin, orexin, substance P, and pituitary adenylate cyclase-activating peptide (PACAP). Most notably, the neuropeptide CRF has been identified as a critical mediator of stress-triggered drug seeking in preclinical rodent models, as CRF receptor antagonism prevents stress-triggered reinstatement across a range of drug classes while delivery of CRF into the brain is sufficient to induce drug seeking [10].

Influence of Stress on Addiction Process

Initiation/Predisposition

There is high comorbidity between stress-related conditions and substance use disorders [11]. Moreover, both the occurrence and severity of substance use disorders have been associated with prior trauma and high-stress environments [11,12], pointing to stress as a likely predisposing factor for addiction. Preclinical animal models have also provided evidence that stress can predispose individuals to substance use. Such observations provide opportunities to unravel the mechanisms through which stress increases susceptibility to drug addiction and therefore may provide insight into more effective prevention strategies.

Early Life

Childhood and adolescence are periods of high vulnerability to stress. The brain is still developing during this time, and is therefore malleable [13]. It has been reported that early life stress produces profound effects on reward processing and drug use in adulthood [14–17], suggesting that, during development, reward-related circuits are vulnerable to stress-induced maladaptation [18]. However, there is some disagreement on the effects stress exerts on later drug-seeking behavior.

Prenatal/Perinatal Stress In rodents, stress exposure prior to birth can affect illicit drug responses later in life. Prenatal stress, experienced by the fetus through maternal restraint, enhances adult drug seeking in rats [19]. Similar effects can be observed with postnatal stress through maternal separation [14,15,20–22]. For example, neonatal maternal isolation stress can enhance cocaine self-administration [23,24] and elevate cocaine-seeking responding during both extinction [20] and testing for cue-induced reinstatement [20]. However, some studies have failed to show enhanced cocaine seeking during adulthood following neonatal isolation [23,24]. Despite these discrepancies, most evidence suggests that early life stress enhances the propensity for drug seeking later in adulthood, consistent with reports in human addicts that early life trauma is associated with the onset and severity of substance use disorders [25].

Juvenile/Adolescent Stress Juvenile and adolescent stress also subsequently impact cocaine-seeking behavior in rodents later in adulthood—observations that are relevant considering the link between stress and substance use in teenage populations. For example, isolation stress and social stress in juvenile mice have been reported to increase drug seeking in adulthood [24]. Adolescent rodents have also been reported to be especially vulnerable to reinstatement precipitated by stress [26], and adolescent social defeat stress has also been shown to enhance cocaine-seeking behavior later in adulthood [29]. Moreover, stress-evoked relapse vulnerability is heightened in adolescent-onset versus adult-onset cocaine users [27,28]. These studies suggest that the age of exposure and stressor subtype may factor into predisposing individuals to different components of the addiction process.

Adulthood

Acquisition In rodents, the stress-induced facilitation of the initiation of drug use can be studied by assessing the acquisition of drug self-administration. Using this approach, it has been demonstrated that exposure to a number of stressors, including uncontrollable electric footshock [30], social defeat [31], witnessing shock in a conspecific [32], and social isolation [33], promote the onset of cocaine self-administration in adult rats.

Escalation

Following the acquisition of cocaine self-administration, loss of control over and escalation of intake is an indicator of a developing addiction. Escalated self-administration corresponds to increased thresholds for intracranial self-stimulation and sensitized or altered anxiety-related behaviors [34], and is symptomatic, in part, of an imbalance between reward and antireward systems. We and others have demonstrated that escalating patterns of cocaine intake can also be induced by exposing adult rats to stress (footshock, social defeat stress) at the time of, or prior to, self-administration [35–37], indicating that stress can accelerate the transition from initial use to addiction.

Relapse

The sudden and often unpredictable relapse to drug seeking that persists even after protracted periods of drug abstinence represents one of the most difficult challenges when managing addiction. For this reason, identifying situations in which drug users are particularly vulnerable to relapse and the underlying neurobiological processes are particularly critical to the development of new and more effective intervention strategies.

Stress-Triggered Relapse

Reports by drug users that drug craving is intensified during periods of stress are paralleled by reports that stress imagery can increase subjective measures of craving in a laboratory setting [38]. In rodents, the contribution of stress to relapse can be studied using reinstatement designs in which drug seeking is reestablished by exposure to various stressors following drug self-administration and extinction [10]. Using this approach, a number of stressors have been demonstrated to reinstate extinguished drug seeking following drug self-administration, including electric footshock [39], food restriction [40], forced swim [41], yohimbine administration [5], and exposure to social defeat-predictive cues [42]. We have found that the ability of stress to trigger drug seeking following cocaine self-administration is more readily observed in rats with a history of daily long-access, high-intake cocaine self-administration ($14 \times 6\,h/day$) and is not observed in rats with short daily drug access ($14 \times 2\,h/day$) [43], suggesting that susceptibility to stress-triggered relapse is established in an intake-dependent manner as a result of drug-induced neuroadaptations. This observation is consistent with clinical reports that stress-induced craving and related physiological responses are heightened in drug users with a history of high-frequency vs. low-frequency use [44].

Stress-Potentiated Relapse

Recent clinical reports suggest that the contribution of stress to drug seeking is complex such that rather than directly evoking drug use, stressors may interact with other triggers for use, such as drug-associated cues, to "set the stage" for relapse [45]. We have established a preclinical rodent model for these stage-setting effects wherein, under conditions where stress alone fails to reinstate extinguished cocaine seeking, it promotes reinstatement to an otherwise subthreshold priming dose of cocaine [46,47]. Notably, the neurobiological mechanisms that underlie the stage-setting effects of stress appear to be distinct from those that mediate stress-triggered cocaine seeking. Better understanding of the signaling mechanisms that are involved in the stage-setting effects of stress on drug seeking may reveal new opportunities for medication development aimed at relapse prevention.

Withdrawal

The observation that susceptibility to stress-triggered relapse depends on the amount/pattern of prior drug use suggests that stress-responsive systems may be sensitized or recruited with repeated drug exposure, thereby increasing the contribution of stress to the addiction cycle. Indeed, heightened activity of stress-responsive processes may contribute to the negative subjective symptoms that emerge with repeated use. Such symptoms are often manifested during drug withdrawal as anxiety and dysphoria, and likely contribute to relapse and escalating use patterns. The development of this aversive emotional state that drives the negative reinforcement of addiction has been referred to as the "dark side" of addiction [48].

Summary of the Contribution of Stress to Cocaine Addiction

The evidence that stress contributes to drug addiction is overwhelming and comes from both clinical observations and preclinical animal research. Episodic stress can promote drug use and relapse, while chronic stress can establish or facilitate neuroadaptations that accelerate the onset and increase the severity of addiction. The ability of stress to determine addiction susceptibility begins very early in life. Considering its powerful influence on the addiction cycle, identifying and/or developing interventions that target the contribution of stress to drug use and addiction should improve patient care. However, the development of such interventions relies on our understanding of the underlying neurobiology. With this in mind, the remainder of this chapter will focus on the signaling mechanisms through which stress contributes to drug addiction.

STRESS SIGNALING MECHANISMS IN ADDICTION

Considering the nature of the stress response, it is not surprising that stress can influence addiction through a diverse range of signaling mechanisms. As a result, the contributions of stress to different aspects of addiction are often mediated by distinct neurobiological processes. For this reason, we will separately address relevant signaling mechanisms for each of the phases of addiction (see Fig. 1 for reference).

Initiation/Predisposition

Early Life

Stress during early life is a significant risk factor for life-long substance use disorder through a combination of hyper-reactive stress systems [49], lack of inhibitory control by an immature PFC, and striatal adaptations regulating hedonics [50,51]. Early life stress experiences activate and alter the molecular organization of stress response systems, disrupting negative feedback on the HPA axis to enhance subsequent stress responses and bias coping strategies [52,53]. Early life stress also negatively affects motivation circuitry maturation through NAc dopamine alterations, thereby disrupting reward processing [19,54–56]. Concurrently, the protracted development of the PFC during adolescence opens a susceptibility window for addiction during which reductions in volume and cortical thickness, as well as reduced functional activity, predict emergent substance use disorder [57]. Fittingly, the high impulsivity in children and adolescents that predicts the switch to compulsive drug use [58] corresponds to both PFC [59] and striatal dysfunction [60,61]. Altogether, early life stress dysregulates development of both stress and corticolimbic systems in a manner which subsequently hastens transitions from abuse to dependence.

Accordingly, stress during prenatal, early life, juvenile, and adolescent periods produces persistent modifications of the stress response and corticolimbic systems at the molecular level. These modifications include prenatal maternal restraint stress-increased PFC dopamine levels, exaggerated NAc glutamate and dopamine release, and increased locomotor activity following cocaine challenge during adulthood [19], as well as abnormal glutamatergic synaptic Homer protein expression in both PFC and NAc [62]. Modifications to stress systems are also elicited by lack of maternal care during early life, and include elevated CRF mRNA levels in amygdala and hypothalamus, decreased α2-AR expression in LC [52,63], and altered GR methylation in hippocampus [53], all of which disrupt negative feedback on the HPA axis. Early life stress can likewise dysregulate NAc activity by enhancing baseline dopamine sensitivity [56]. Juvenile social stress also enhances reward sensitivity, stress responsivity, and decreases top-down control, effects which coincide with significant NAc synaptic gene expression changes which have been established in the modulation of cocaine use disorder [24,64,65]. Adolescent stress effects may also arise through alterations in glutamate signaling in the medial-PFC. For example, in stress-exposed adolescent rats, enhanced presynaptic vesicular glutamate release, reduced glutamate clearance, and increased postsynaptic GluN1 NMDA responsiveness all coincide with adolescent cocaine exposure, and may sensitize medial-PFC glutamatergic synapses to stress and contribute to increased stress sensitivity and stress-provoked reinstatement of cocaine seeking [28]. In sum, stress-induced alterations to both the stress response and corticolimbic systems during critical developmental periods may underlie the persistent behavioral phenotype displayed following early life stress.

Adulthood

Acquisition

The effects of stress on the onset the of drug use likely involve interactions between stress hormones and the mesolimbic dopamine system [66]. Social defeat-induced enhancement of the acquisition of cocaine self-administration in rats is associated with increases in NAc dopamine [67], likely attributable in part to CRF actions in the VTA [68]. Moreover, glucocorticoid hormones are both necessary and sufficient to promote the acquisition of cocaine-seeking behavior, as elimination or suppression of the corticosterone response prevents [69,70], while elevation of corticosterone accelerates [70] acquisition. While the precise signaling mechanisms through which glucocorticoids influence the acquisition of drug use are unclear, actions at GRs are likely involved, as GR deficiency has been reported to interfere with cocaine self-administration [71].

Escalation

Adrenal hormones and CRF actions in the VTA appear to contribute to escalating patterns of drug use resulting from chronic stress. We have reported that the ability of daily electric footshock at the time of cocaine self-administration to escalate drug intake is adrenal-dependent, with a likely role for glucocorticoids [35]. Interestingly, while an adrenal response is necessary for the escalation of cocaine seeking elicited by footshock, stress-level corticosterone alone is insufficient to produce these effects, suggestive of a likely contribution of other stressor-responsive signals, potentially adrenaline.

Others have reported that extrahypothalamic CRF signaling is critical for both the establishment and expression of escalating patterns of self-administration following repeated social defeat stress. Specifically, CRF signaling in the VTA is both necessary and sufficient for the ability of repeated social defeat to escalate cocaine self-administration [72] with apparent contributions of both CRFR1 and R2 receptors [73]. Increased CRF signaling in the VTA and enhanced regulation of NAc dopamine also appears to represent an important neuroadaptation that leads to the expression of escalating self-administration patterns [37,68] and escalated cocaine self-administration in socially defeated animals can be reduced by intra-VTA CRFR1 antagonist administration [74]. Notably a similar recruitment of VTA CRF signaling appears to contribute to the escalation of alcohol self-administration by social defeat stress [75].

Relapse

While a number of brain circuits contribute to relapse, a particularly well-studied pathway originates in the PrL-PFC and projects to the NAc-core [2]. This pathway is comprised of glutamatergic pyramidal neurons and is highly regulated by dopaminergic projections from the VTA, as well as inputs from other brain regions that contribute to relapse such as the ventral hippocampus and basolateral amygdala (BLA). Notably parallel pathways, including an infralimbic (IL)-PFC to NAc-shell projection implicated in exerting inhibitory constraint on goal-directed behavior, and cortical inputs to the dorsal striatum implicated in mediating compulsive drug taking, are also important determinants of drug seeking. However, if or how stress regulates these pathways to promote drug use has not been established. Furthermore, there are individual differences in the ability of stress to promote relapse, as in some stress alone is sufficient to directly induce relapse, while in others its effects are more nuanced. We will contrast the mechanisms for *stress-triggered relapse* with the "stage-setting" effects involved in *stress-potentiated relapse*.

Stress-Triggered Relapse

While the contributions of several pathways to stress-triggered relapse have been studied [10], this chapter will primarily focus on noradrenergic innervation of the extended amygdala and its regulation of VTA dopaminergic inputs to the PrL-PFC, which in turn can promote excitability of pyramidal neuron outputs to the NAc core.

The Extended Amygdala

The extended amygdala, comprised of the BNST, CeA, and NAc-shell, plays a critical role in processing/integrating stress-related information and closely regulates mesocorticolimbic neurocircuitry. A key node of connectivity between the extended amygdala and the mesocorticolimbic system is the VTA, which receives afferents from each of these regions that are active during stress-triggered drug seeking [76]. Inactivation studies have demonstrated that each of these regions is required for stress-triggered drug seeking [77]. Specifically, research has focused on the contributions of signaling in the BNST and CeA to stress-triggered relapse.

Central Amygdala The CeA is a likely convergence point for stress-related stimuli that influence drug seeking. The CeA is highly stress-reactive and plays a critical role in processing stress-related information and regulating behavioral responses. Notably, stressor regulation of the CeA is heightened following repeated exposure to drugs of abuse [78]. Consistent with a role in relapse, inhibition of the CeA prevents shock-induced cocaine seeking in rats [77]. Numerous studies have implicated noradrenergic signaling in stress-triggered drug seeking [79,80]. The CeA receives noradrenergic innervation from both the LC and medullary cell groups [81], and drug seeking induced by intracranial NA delivery is associated with a CeA Fos response [82]. It has, furthermore, been reported that noradrenergic signaling via CeA β-ARs is required for stress-triggered drug seeking [83]. Additionally, the CeA receives top-down regulation from cortical regions such as the insular cortex, which appears critical for regulating drug seeking [84]. During stress, the CeA undergoes both hormonal and peptidergic regulation, and both influence drug seeking. CeA activity can be coordinated by several peptides, including hypocretins/orexins [85], neuropeptide Y [86], and nociceptin [87]. Moreover, stress-triggered drug seeking may involve glucocorticoid signaling via GR in the CeA [88]. CeA outputs include direct projections to the VTA, as well as other regions of the extended amygdala, which in some cases appear to corelease CRF [89,90]. Notably, the pathway from the CeA to the BNST is required for stress-triggered drug seeking [91].

Bed Nucleus of the Stria Terminalis The BNST is a highly complex structure that is reciprocally connected with the mesocorticolimbic system. Although the contributions of BNST projections to the PFC and NAc are not well-studied, we have found that a projection from the ventrolateral BNST to the VTA is required for stress-triggered relapse [92]. Like the CeA, the BNST receives dense ascending noradrenergic innervation. However, in contrast to the CeA, noradrenergic innervation of the BNST that is pertinent to drug seeking appears to arise exclusively from the medullary cell groups via the ventral noradrenergic bundle [93]. The noradrenergic regulation of the BNST is complex, and involves both α- and β-ARs located at both pre- and post-synaptic sites [94]. We and others have demonstrated that β-AR activation, specifically β2 receptor activation, in the ventral BNST is both necessary and sufficient for stress-triggered drug seeking [83,92]. In the BNST, noradrenergic signaling coordinates with converging inputs to regulate drug seeking. These inputs include ascending monoaminergic projections, top-down cortical innervation, microcircuits within the BNST complex, and afferents from other extended amygdala structures, most notably the CeA. In the latter case, a CRF-releasing CeA to BNST pathway has been implicated in stress-triggered drug seeking [91] although some evidence suggests that intrinsic BNST cell populations may also release CRF [95]. Regardless, CRF signaling in the BNST, via CRF-R1 receptors, is also required for stress-triggered drug seeking [96,97]. CRF release into the BNST appears to be regulated by local β adrenergic signaling, as cocaine-seeking following intra-BNST β2-AR activation is prevented by local CRF-R1 antagonism [92], and systemic β2-AR antagonism prevents both forced swim-induced cocaine seeking in mice and swim-induced increases in BNST CRF mRNA [98]. Once released, CRF facilitates excitatory regulation of efferent projections, including those that innervate the VTA, likely via activation of CRF-R1 receptors on glutamatergic terminals [95]. Furthermore, a subpopulation of BNST VTA-projecting neurons that express CRF [89,92,95] are activated through a CRFR1-dependent mechanism [95], and pharmacological disconnection of this β2-AR regulated, CRF-releasing pathway to the VTA prevents stressor-induced drug seeking [92]. Although the actions of CRF in the BNST have been extensively studied, several other neuropeptides in the BNST likely contribute to stress-triggered relapse. For example, it has been reported that activation of the receptor for PACAP, pituitary adenylate-cyclase receptor-1, is both necessary and sufficient for stress-triggered cocaine seeking [99]. Likewise, dynorphin actions at BNST kappa opioid receptors (κORs) have been implicated in stress-triggered alcohol seeking [100].

The Mesocorticolimbic System

Although inputs that relay stress-related information feed into the mesocorticolimbic system to regulate drug seeking at multiple sites, this section will primarily focus on inputs into the VTA and their regulation of VTA projections to the PFC and NAc.

Ventral Tegmental Area The VTA is a midbrain structure that disperses both dopaminergic and non-dopaminergic projections throughout the mesocorticolimbic system to exert neuromodulatory actions critical for a range of reward and motivation processes, including hedonics, reward-related learning, and behavioral adaptation. Stressful/aversive stimuli regulate the VTA through multiple excitatory and inhibitory pathways, which have effects on projection neurons as well as intrinsic interneuron populations in the VTA. It has become clear that the VTA is a heterogeneous and highly complex structure comprised of distinct output pathways that can be differentially regulated depending on afferent, local, and retrograde signaling. As a result, it has been demonstrated that there are subpopulations of VTA projecting neurons that respond differently to stressful/aversive stimuli [101]—an observation that has important implications for understanding the mechanisms that mediate stress-triggered drug seeking. Considering the complexity of the VTA, it should not be surprising that the signaling mechanisms therein through which stress promotes drug seeking are equally complicated.

Much of the research investigating the contribution of the VTA in stress-triggered drug seeking has focused on CRF. Indeed, it has been demonstrated that stress-triggered drug seeking is associated with elevated CRF levels in the VTA [102] and that intra-VTA CRF delivery is sufficient to reinstate extinguished cocaine seeking [102,103]. However, the mechanisms through which CRF acts in the VTA to produce drug seeking are not entirely clear. While we and others have demonstrated that CRF-R1 but not CRF-R2 receptor activation is both necessary and sufficient for stress-triggered drug seeking [92,103–105], other laboratories have reported a role for the CRF-R2 receptor, activated through a mechanism that appears to require the CRF binding protein [102,106].

Although CRF-expressing afferents to the VTA from the PVN and the CeA have been identified [89] and CRF release from VTA dopamine neurons has been reported [107], a key source of CRF-releasing afferents into the VTA appears to be the BNST [89,95]. We have demonstrated that disrupting a CRF-releasing pathway from the BNST to the VTA prevents stress-triggered cocaine seeking [92]. While the mechanisms through which CRF regulates VTA function are complex and involve both pre- and post-synaptic regulation of multiple cell types, much evidence suggests that CRF promotes drug seeking via effects on glutamatergic neurotransmission [102,108,109]. However, it is likely that CRF is also coreleased with GABA and/or has effects on adjacent inhibitory synapses through which it can regulate drug seeking. Consistent with this possibility, we have reported that both stress and intra-VTA CRF-induced drug seeking involves signaling through GABA$_B$ receptors in the VTA [110]. In addition to the BNST, several other stress-responsive structures send afferents to the VTA, including the amygdala, hypothalamus, NAc-shell, septum, and lateral habenula (directly and via the rostro-medial tegmental nucleus). However, the degree to which projections from these regions to the VTA contribute to stress-triggered relapse remains to be characterized.

CRF is not the only stress-regulated VTA neuropeptide that can regulate drug seeking. Evidence suggests that the opioid peptide dynorphin can promote drug seeking via actions at VTA κORs. Prior work has demonstrated that κOR activation contributes to stress-triggered drug seeking [111,112] and, in the VTA, dynorphin actions at κORs attenuate GABAergic neurotransmission to produce drug seeking [113]. Another neuropeptide that can act in the VTA to contribute to stress-triggered drug seeking is orexin/hypocretin, which induces drug seeking when microinjected into the VTA [114], likely through disinhibition of dopaminergic outputs [115] and/or regulation of glutamatergic neurotransmission [116]. In addition, several other neuropeptides, including neurotensin and tachykinins, regulate the VTA and may also contribute to stress-triggered drug seeking.

Although the VTA sends projections to a number of brain regions, including the BNST and amygdala, the contributions of VTA projections to the PFC and NAc to stress-triggered relapse have been most intensely studied. Notably, evidence suggests that PFC-projecting VTA neurons are more robustly activated by stressful stimuli relative to those that project to the NAc [117], the latter of which display increased or decreased activity depending on the type of stressor, timeframe of measurement, and to which subregion of the NAc they project (see e.g., [105,118–122]).

Prefrontal Cortex Craving in human addicts is associated with altered activity of the prefrontal cortex, as measured using imaging-based approaches. Several prefrontal cortical regions have been implicated in craving—including the dorsolateral-PFC which, in humans, corresponds to the PrL-PFC in rodents [123]. Inhibition of the PrL-PFC using either TTX [124] or baclofen/muscimol [77] prevents stress-triggered reinstatement of extinguished drug seeking. We have found that stress-triggered cocaine seeking is associated with a robust Fos response in the PrL-PFC that is attributable to CRF-dependent activation of PrL-projecting VTA neurons, and that inactivation of VTA projections to the PrL-PFC using intersectional chemogenetics prevents stress-triggered cocaine seeking (unpublished findings). The stress-regulated PrL-projecting VTA neurons that promote cocaine seeking are likely dopaminergic. Several groups have reported that D1 receptor antagonism in the PrL-PFC prevents stress-triggered reinstatement [77,124], while stress-triggered cocaine seeking is associated with elevated PrL-PFC dopamine [77]. The effects of D1 receptor activation in the PrL-PFC are complex and involve action on both pyramidal and nonpyramidal neurons, but the overall result appears to be facilitated excitation of pyramidal neurons which renders some output pathways more susceptible to glutamatergic stimulation [125,126]. As a result, inputs that relay drug-related information are more likely to activate output pathways that mediate drug seeking in the presence of cortical D1 receptor activation. Although untested, it is thought that one key affected pathway is comprised of pyramidal neuron projections to the NAc-core. Accordingly, inhibition of the PrL-PFC using baclofen/muscimol prevents both shock-induced drug-seeking behavior and increases in NAc-core glutamate levels [127].

Notably, the PrL-PFC is not the only cortical region implicated in stress-triggered drug seeking. For example, consistent with reports of altered orbitofrontal cortical (OFC) activity during drug craving in humans, Capriles et al. [124] demonstrated that either TTX or D1 receptor antagonist micro-infusions into the OFC attenuated shock-induced cocaine seeking in rats. As the OFC plays an important role in risk-based decision making [128], stress-induced dysregulation of OFC function could favor behavioral patterns that include relapse to drug use.

Dopamine is not the only regulator of PrL-PFC function. The PFC is highly regulated by other monoamines, and the noradrenergic innervation of the PrL-PFC from the LC plays a critical role in attention and arousal. Indeed, it has been reported that α-AR signaling in the PFC promotes drug-seeking behavior [129]. Moreover, although not well studied in the context of drug seeking, PFC activity is also coordinated by neuropeptides, including CRF [130].

Nucleus Accumbens The NAc can be segregated into core and shell subregions, which have distinct anatomy and function. While the NAc-shell appears to encode unconditioned reward, the NAc-core is purported to be important for goal-directed behavior guided by predictive cues and context. Based on these roles, it is not surprising that both subregions have been implicated in drug seeking.

NAc-core. Consistent with its role in goal-directed behavior, the NAc-core, and elevated glutamate therein, plays a critical role in drug seeking and relapse. Shock-induced cocaine seeking is both prevented by inhibition of the NAc-core and associated with NAc-core glutamate elevations, with the primary implicated source of stress-regulated glutamate being the PrL-PFC [77]. Stress-induced increases in NAc glutamate likely regulate cocaine seeking via AMPA receptor signaling on medium spiny neurons that project to the ventral pallidum [131]. The contribution of NAc-core dopamine to stress-triggered drug seeking is less clear. While elevated dopamine in the NAc-core does not seem necessary for cocaine-primed reinstatement [132], the requirement for NAc-core dopamine in stress-triggered reinstatement has not been well studied.

NAc-shell. The role of the NAc-shell in stress-triggered cocaine seeking is more complicated. Like the NAc-core, inactivation of the NAc-shell prevents stress-triggered cocaine seeking [77]. However, consistent with its classification as a component of the extended amygdala, the NAc-shell appears to function as an upstream regulator of the VTA-PrL-NAc-core circuit [76,77]. In line with its role in reward and aversion, the NAc-shell may provide hedonic context that guides drug seeking via direct or indirect inputs into the VTA. Decades of research has revealed that dopamine in the NAc-shell is critical for encoding hedonics. However, surprisingly, the effects of stressors and aversive stimuli on the NAc-shell are controversial with evidence that, depending on the parameters, timing, and context, stress can produce either increases or decreases in NAc-shell dopamine. Thus, stress-triggered drug seeking has been associated with both increases [133] and decreases [105] in NAc-shell dopamine signaling. Moreover, both increases and decreases in NAc-shell dopamine have been attributed to stress-related signaling in the VTA through CRF [105,134,135]. Better understanding the relationship between stress and NAc dopamine, and how acute stress-related alterations in hedonic state contribute to drug seeking, represents an important goal for future research.

In addition to dopamine, several other stress-related signaling processes regulate the NAc-shell, most notably serotonin [136] and a wide range of neuropeptides including substance P [137], dynorphin [137,138], CRF [139], and cholecystokinin [140]. How the actions of these signaling molecules in the NAc-shell contribute to stress-triggered drug seeking requires further characterization.

Stress-Potentiated Reinstatement

Although stress does not directly trigger relapse in all, it does not mean that it goes without consequence. We have found that under conditions where it does not directly trigger drug seeking, a stressor can potentiate the reinstatement of drug seeking in response to an otherwise subthreshold low dose of cocaine, thereby setting the stage for relapse. The signaling mechanisms that underlie these *stage-setting* effects of stress are distinct from those that mediate stress-triggered drug seeking. Most notably, stress-potentiated drug seeking requires elevated glucocorticoids [46], while stress-triggered and cocaine-primed reinstatement do not [47,141,142]. The potentiating effects of stress on drug seeking are reproduced by corticosterone administration at a dose that establishes stress levels of the hormone in blood [46]. Consistent with the rapid time course of corticosterone, its effects on cocaine seeking appear to be produced independently of canonical GR activation, as the GR antagonist RU-486 fails to prevent corticosterone-potentiated reinstatement. Corticosterone produces it effects via multiple brain regions including the NAc [46] and PrL-PFC [143], through at least two distinct mechanisms of action.

In the NAc, corticosterone's effects on drug seeking are dopamine-dependent and involve reduced dopamine clearance via the organic cation transporter 3 (OCT3) [46,144]. OCT3 is a low affinity, high-capacity monoamine transporter that contributes to DAT-independent dopamine clearance, previously referred to as uptake 2. Corticosterone binds to and inhibits OCT3 [145] in the NAc, thereby augmenting dopamine signaling and corresponding behavioral effects (e.g., drug seeking). The effects of corticosterone on cocaine seeking are reproduced by the non-glucocorticoid OCT3 inhibitor, normetanephrine [46] and are not observed in OCT3-deficient mice [144].

In the PrL-PFC, corticosterone potentiates cocaine seeking via the mobilization of endocannabinoids and cannabinoid receptor-1 (CB1)-dependent reductions in inhibitory synaptic transmission [143] (see Fig. 3 for reference). Elevation

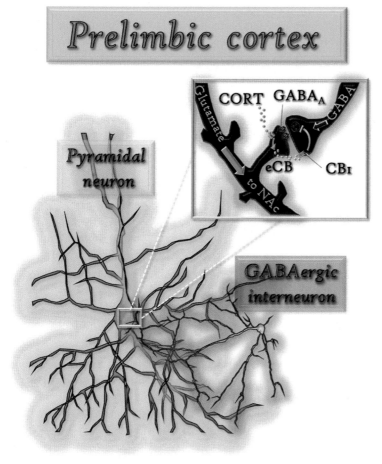

FIG. 3 Stress-Potentiated Reinstatement: Endocannabinoid-Mobilization in the Prelimbic Cortex. (1) Corticosterone suppresses inhibition of layer V prelimbic prefrontal cortical pyramidal neurons through mobilization of the endocannabinoid, 2-arachidonoylglycerol, which travels retrogradely to bind presynaptic $G_{i/o}$-coupled cannabinoid type-1 receptors on GABAergic interneurons. (2) Drug-related stimuli activate afferents from brain regions, such as the ventral tegmental area and amygdala, which release dopamine and glutamate onto layer V pyramidal neurons. (3) Endocannabinoid-disinhibited layer V pyramidal neurons are more readily excited by incoming drug-stimulated afferents, resulting in facilitated activation of glutamatergic projections to the nucleus accumbens and drug seeking.

of the endocannabinoid 2-arachidonoylglycerol (2-AG) in the PrL-PFC is both necessary and sufficient for corticosterone-potentiated cocaine seeking. Via an unknown mechanism that appears to involve $G_{q\alpha}$ signaling, corticosterone promotes 2-AG production in PrL-PFC pyramidal neurons, including those that project to the NAc-core (unpublished data). 2-AG diffuses retrogradely and binds to CB1 receptors on GABA terminals, thereby reducing GABA release. Through this disinhibition, stress and corticosterone can facilitate excitation of the PrL-PFC projection pathway to the NAc-core, rendering it more susceptible to glutamatergic inputs that promote drug seeking.

DRUG-INDUCED ADAPTATIONS IN STRESS-SIGNALING MECHANISMS

Key aspects of drug addiction are attributable to disrupted homeostasis fueled by neuroadaptations that counter the rewarding effects of abused drugs. Many of these adaptations involve the recruitment of stress-responsive signaling systems in the brain. The heightened activity of "antireward" systems is manifest as escalating use patterns driven in part by negative reinforcement, the onset of stress-related responses during withdrawal, and heightened stressor-reactivity that can further increase relapse susceptibility. Collectively, the contribution of these allostatic processes has been termed the "dark side of addiction." [48].

Escalating Use Patterns

Drug addiction is a progressive condition that includes a transition to escalating use patterns. In rodents, such patterns of self-administration can be established by prolonging daily drug access. For example, rats provided longer access to cocaine

(6–10 h/day) for self-administration progressively escalate their cocaine intake, while rats provide shorter daily cocaine access 1–2 h/day) do not [146,147]. Escalated drug self-administration has been attributed to emergent deficits in reward function and negative reinforcement [148], resulting in part from the recruitment of a number of stress-related systems [48]. Prominent among these are the CRF, noradrenergic, dynorphin/κOR, substance P/neurokinin receptor, and orexin/hypocretin systems. The contribution of each of these systems to drug intake is established or exaggerated in rats displaying escalated patterns of self-administration (CRF [149,150], dynorphin [151–153], NA [154–156], substance P [157], and orexin [85]). The adaptations in each of these systems that emerge with repeated drug exposure are numerous and providing a detailed overview of them is beyond the scope of this chapter. For example, repeated cocaine exposure produces widespread increases in the activity of the CRF system [158–161], and augments CRF sensitivity [109,162–164] in brain regions that regulate drug self-administration.

Withdrawal

In drug-dependent individuals, detoxification following bouts of drug use is associated with the onset of withdrawal symptoms that include hedonic deficits, dysphoria, and often intense anxiety. Many of these withdrawal symptoms are mediated by overactivity of some of the same signaling systems that contribute to escalated self-administration. For example, irritability- and anxiety-related symptoms resulting from alcohol [165,166], cocaine [167], and heroin [168] withdrawal are all CRF-dependent. The persistently heightened activity of these stress signaling systems during periods of abstinence and the resulting symptomology likely contribute to relapse susceptibility.

Heightened Stressor Reactivity

Additionally, there is evidence that stressor reactivity is heightened with repeated drug use. For example, consistent with clinical reports that individuals with a higher frequency of recent cocaine and alcohol use display heightened anxiety and associated cardiovascular and HPA responses to stress imagery [44], we have found that rats with a history of long-access cocaine self-administration show a heightened corticosterone response to restraint stress that is associated with impaired GR-mediated negative feedback regulation of the HPA axis [169]. Importantly, this intake-dependent increase in stressor reactivity translates to heightened risk for stress-triggered relapse. Fox et al. [44] found that self-reports of drug craving in response to stress imagery are also heightened in high vs. low frequency users. We have found that footshock-induced drug seeking is more readily observed in rats with a history of long-access self-administration [141]. Heightened susceptibility to stress-triggered drug seeking has also been reported following long-access heroin self-administration in rats [170].

The observation that susceptibility to stress-triggered drug seeking emerges in an intake-dependent manner with repeated use suggests that, in addicted individuals, drug-induced neuroadaptations reinforce communication between stress-responsive and reward/motivational circuitry in the brain. While such adaptations are likely widespread, we and others have focused on changes in drug-induced alterations in CRF signaling in the VTA. Similar to our observation with electric footshock [141], we have found that the ability of intra-VTA CRF micro-infusions to reinstate cocaine seeking in rats is only observed in rats with a history of long-access self-administration [103], suggesting that intake-dependent alterations in CRF responsive processes in, or downstream from, the VTA are critical for the emergent ability of stress to promote drug seeking. More recently, we have found that long-access cocaine self-administration establishes CRF control over dopaminergic projections from the VTA to the PrL-PFC by increasing the expression of VTA CRF-R1 receptors (unpublished findings), consistent with earlier reports using autoradiography that repeated cocaine can increase CRF binding in the VTA [171]. Notably, these findings are also consistent with prior observations that stressor and CRF regulation of somatodendritic dopamine release in the VTA is only observed in rats with a history of cocaine exposure [102].

Glucocorticoid Regulation

Research from our lab has found that elevated glucocorticoids at the time of cocaine self-administration are critical for the development of later susceptibility to stress-triggered relapse. Cocaine self-administration activates the HPA axis, elevating cortisol in humans and corticosterone in rats. With excessive/prolonged patterns of cocaine use, such as those observed under long-access testing conditions in rats, increases in glucocorticoids are augmented and prolonged, providing greater opportunity for maladaptive effects on brain function and neuropathology [172]. When the glucocorticoid response during long-access self-administration is eliminated by surgical adrenalectomy, the effects of extended-access self-administration on later stress-triggered reinstatement are prevented [142]. By contrast, when rats are adrenalectomized after self-administration, but prior to reinstatement testing, footshock-induced reinstatement is unaffected, suggesting that elevated glucocorticoids are critical for the establishment of neuroplasticity that determines later susceptibility to stress-triggered relapse but are not necessary for its expression. Glucocorticoid-dependent neuroadaptations that promote stressor-induced drug seeking involve the CRF system. Similar to shock-induced drug seeking, the ability of centrally administered CRF to reinstate extinguished cocaine seeking requires an intact adrenal response during prior self-administration [142].

MECHANISTIC CONTRASTS IN STRESS-ASSOCIATED SIGNALING BETWEEN DRUG CLASSES

While there are certainly unique interactions between stress and different illicit drugs, there are also commonalities. For example, early-life stress appears to enhance amphetamine, methamphetamine, and ethanol seeking through modulation of NAc dopamine signaling [36,173], and GR activation is required for ethanol intake escalation [174]. Both CRF-R1 and kOR activation are required for stress-triggered reinstatement of morphine [175] and ethanol seeking [100,176] as well as a nicotine-associated conditioned-place preference [177–179], and hypocretin/orexin and AR signaling are also required for stress-triggered reinstatement of ethanol [176,180] and morphine [97,181] seeking. Finally, stress-induced signaling adaptations that promote drug seeking in other drug classes have been localized to similar brain regions, including the extended amygdala [97,100,180,182] and mesocorticolimbic system [178,183]. Therefore, there are common mechanisms whereby stress facilitates drug seeking, regardless of drug type.

However, there may also be distinct stress signaling mechanisms which promote seeking different drugs of abuse. For example, CRF-R1 activation is required for heroin intake escalation [150] and, unlike cocaine seeking, GR activation appears necessary for stress-triggered reinstatement of morphine seeking [184], as is CB1 activation for nicotine [185]. Moreover, unique signaling mechanisms have been characterized, such as the necessity for melanocortin-4 receptor activation for nicotine seeking, dorsal raphe GABA-A activity for morphine seeking [186], and oxytocin signaling for both methamphetamine [183] and ethanol seeking [187,188]. Several other novel mechanisms have been characterized specifically in ethanol seeking [86,176,189–192] within the extended amygdala, and supplementary brain regions have been implicated for several drugs of abuse [176,186,192,193]. In sum, there are stress-signaling distinctions in the reinstatement of seeking other drugs of abuse.

CONSIDERATION OF SEX DIFFERENCES IN STRESS-ASSOCIATED SIGNALING IN ADDICTION

Although rates of substance use disorder are generally higher in males [194], females progress through the stages of addiction at an accelerated rate [195,196]. This "telescoping" phenomenon exhibited by female addicts is further compounded by the effects of stress. Females are more predisposed to drug dependence following prenatal stress [197], early life trauma [198], and chronic adolescent corticosterone exposure [199]. Females also acquire cocaine and heroin self-administration faster than males [200], are more reactive to stress following the development of dependence [4] and exhibit greater PrL-PFC activity during stress-induced craving [201]. Moreover, females are more prone to stress-induced escalation of cocaine self-administration [202] as well as stress-triggered reinstatement [5]. Understanding the basis for these sex differences likely has relevance for treatment. Indeed, some medications that target the stress response [203] appear to be more effective in managing cocaine craving in women.

An important consideration for the greater stress-provoked susceptibility to reinstatement in females is the interaction between stress effectors and ovarian hormones. Greater reinstatement responding in females not only corresponds to heightened corticosterone [199], CRF, and NA [204], but also heightened 17β-estradiol levels [5,199,205]. Conversely, lower reinstatement responding occurs when progesterone [5] and its 5α-reductase metabolite allopregnanolone are elevated [206], indicating interactive anti-stress effects of progesterone and pro-stress effects of 17β-estradiol. Indeed, elevated estrogens left unopposed by progesterone correspond to increased drug cue-elicited craving [207,208], impulsive reward seeking [209], and positive subjective effects of cocaine [210,211]. We and others have found that higher reinstatement is generally observed during proestrus and estrus, when 17β-estradiol (E2) levels peak in the rodent estrous cycle [5,205,212], and 17β-estradiol-promoted reinstatement has been localized to the mesocorticolimbic system [205,213,214]. Importantly, the elevations in corticosterone levels observed in proestrus females following stress exposure [215] are likely affected by 17β-estradiol regulation of the HPA axis [216–218], which may produce interactions that engender a more permissive state for reinstatement.

There may be sex differences in the signaling mechanisms underlying stress-promoted cocaine seeking. The effects of corticosterone on potentiated reinstatement in females may uniquely interact with ovarian hormones, as we have found that females display higher corticosterone-potentiated reinstatement responding when 17β-estradiol levels are high while progesterone is low, and suppressed reinstatement when progesterone levels are high (unpublished findings). During stress-potentiated reinstatement, males are more sensitively dependent on hypocretin signaling than females, a sex difference also potentially resulting from gonadal hormone influence [219]. Furthermore, females may be more sensitive to CRF-induced LC firing rates, due to both enhanced CRF sensitivity at concentrations that are ineffective in males [220] and a lack of CRF-R1 internalization following CRF-binding [221], indicating that the LC-noradrenergic system may be particularly affected by stress in females. These effects of CRF on LC firing may be promoted through 17β-estradiol excitation of CRF projections from the PVN [222]. Furthermore, stress-promotion of drug seeking through the dynorphin-κOR system may differ in females due to a higher prevalence of pDYN gene single-nucleotide polymorphisms associated with opiate dependence, as well as estrogen-regulated κOR receptor heteromers and dynorphin expression [223]. These differences highlight developmental and acute stress response elements in females that may differ from males and provide greater opportunity for stress to catalyze the addiction process in females.

TREATMENT IMPLICATIONS

Understanding the signaling mechanisms through which stress influences drug seeking should guide the development of new and more effective interventions. However, thus far, the results of clinical studies based on preclinical findings have been mixed. Medications targeting noradrenergic signaling have been the most promising. Alpha-2 AR agonist drugs, such as guanfacine, which functionally inhibit noradrenergic signaling, attenuate stress-induced drug craving in laboratory settings and have shown some effectiveness in out-patient studies [203]. Alpha-1 AR antagonists [224] and dopamine beta hydroxylase inhibitors (which inhibit NA synthesis) [225] have also shown some promise. Unfortunately, the results of clinical studies using medications targeting the CRF system have been negative. Despite likely brain receptor occupancy, the CRF receptor antagonist pexacerfont failed to alter stress-induced craving in alcohol-dependent individuals [226]. Considering the demonstrated effectiveness of CRF antagonists in rodent models, these results are disappointing and may, in fact, raise questions about the translational value of current preclinical models. The hope is that ongoing clinical evaluation of other medications targeting stress-responsive systems, including kOR [227], neurokinin receptor [228], and GR [229] antagonists, will yield more encouraging results.

SUMMARY

Stress signaling fuels the cycle of addiction, from early life stress effects on the acquisition of drug-seeking behavior to relapse vulnerability in adults. Additionally, abused drugs themselves can influence stress responsivity, suggesting that the stress signaling mechanisms activated during periods of drug self-administration and/or drug withdrawal may also contribute to drug-induced neuroplasticity that leads to subsequent use. Understanding how stress signaling regulates drug-induced neuroadaptations should provide important insights into the neurobiological events that underlie the pervasive nature of substance use disorder and may lead to new and more effective treatment approaches.

REFERENCES

[1] McEwen BS. Central effects of stress hormones in health and disease: understanding the protective and damaging effects of stress and stress mediators. Eur J Pharmacol 2008;583:174–85.

[2] Kalivas PW, McFarland K. Brain circuitry and the reinstatement of cocaine-seeking behavior. Psychopharmacology 2003;168:44–56.

[3] Koob GF, Volkow ND. Neurocircuitry of addiction. Neuropsychopharmacology 2010;35:217–38.

[4] Back SE, Brady KT, Jackson JL, Salstrom S, Zinzow H. Gender differences in stress reactivity among cocaine-dependent individuals. Psychopharmacology 2005;180:159–76.

[5] Feltenstein MW, Henderson AR, See RE. Enhancement of cue-induced reinstatement of cocaine-seeking in rats by yohimbine: sex differences and the role of the estrous cycle. Psychopharmacology 2011;216:53–62.

[6] Selye H. Stress and the general adaptation syndrome. Br Med J 1950;1:1383–92.

[7] Herman JP. Neural control of chronic stress adaptation. Front Behav Neurosci 2013;7:61.

[8] Chrousos GP, Gold PW. The concepts of stress and stress system disorders. JAMA 1992;267:1244–52.

[9] Smith RJ, Aston-Jones G. Noradrenergic transmission in the extended amygdala: role of increased drug-seeking and relapse during protracted drug abstinence. Brain Struct Funct 2008;213:43–61.

[10] Mantsch JR, Baker DA, Funk D, Lê AD, Shaham Y. Stress-induced reinstatement of drug seeking: 20 years of progress. Neuropsychopharmacology 2016;41:335–56.

[11] Brady KT, Sinha R. Co-occurring mental and substance use disorders: the neurobiological effects of chronic stress. Am J Psychiatry 2005;162:1483–93.

[12] Jacobsen LK, Southwick SM, Kosten TR. Substance use disorders in patients with posttraumatic stress disorder: a review of the literature. Am J Psychiatry 2001;158:1184–90.

[13] Gogtay N, Greenstein D, Lenane M, Clasen L, Sharp W, Gochman P, Butler P, Evans A, Rapoport J. Cortical brain development in nonpsychotic siblings of patients with childhood-onset schizophrenia. Arch Gen Psychiatry 2007;64:772–80.

[14] Flagel SB, Vazquez DM, Robinson TE. Manipulations during the second, but not the first, week of life increase susceptibility to cocaine self-administration in female rats. Neuropsychopharmacology 2003;28:1741–51.

[15] Kosten TA, Miserendino MJ, Kehoe P. Enhanced acquisition of cocaine self-administration in adult rats with neonatal isolation stress experience. Brain Res 2000;875:44–50.

[16] Kosten TA, Zhang XY, Kehoe P. Neurochemical and behavioral responses to cocaine in adult male rats with neonatal isolation experience. J Pharmacol Exp Ther 2005;314:661–7.

[17] Kosten TA, Zhang XY, Kehoe P. Heightened cocaine and food self-administration in female rats with neonatal isolation experience. Neuropsychopharmacology 2006;31:70–6.

[18] De Bellis MD. Developmental traumatology: a contributory mechanism for alcohol and substance use disorders. Psychoneuroendocrinology 2002;27:155–70.

[19] Kippin TE, Szumlinski KK, Kapasova Z, Rezner B, See RE. Prenatal stress enhances responsiveness to cocaine. Neuropsychopharmacology 2008;33:769–82.

[20] Lynch WJ, Mangini LD, Taylor JR. Neonatal isolation stress potentiates cocaine seeking behavior in adult male and female rats. Neuropsychopharmacology 2005;30:322–9.

[21] Martini M, Valverde O. A single episode of maternal deprivation impairs the motivation for cocaine in adolescent mice. Psychopharmacology 2012;219:149–58.

[22] O'Connor RM, Moloney RD, Glennon J, Vlachou S, Cryan JF. Enhancing glutamatergic transmission during adolescence reverses early-life stress-induced deficits in the rewarding effects of cocaine in rats. Neuropharmacology 2015;99:168–76.

[23] Zhang XY, Sanchez H, Kehoe P, Kosten TA. Neonatal isolation enhances maintenance but not reinstatement of cocaine self-administration in adult male rats. Psychopharmacology 2005;177:391–9.

[24] Lo Iacono L, Valzania A, Visco-Comandini F, Viscomi MT, Felsani A, Puglisi-Allegra S, Carola V. Regulation of nucleus accumbens transcript levels in mice by early-life social stress and cocaine. Neuropharmacology 2016;103:183–94.

[25] Khoury L, Tang YL, Bradley B, Cubells JF, Ressler KJ. Substance use, childhood traumatic experience, and posttraumatic stress disorder in an urban civilian population. Depress Anxiety 2010;27:1077–86.

[26] Anker JJ, Carroll ME. Reinstatement of cocaine seeking induced by drugs, cues, and stress in adolescent and adult rats. Psychopharmacology 2010;208:211–22.

[27] Wong WC, Marinelli M. Adolescent-onset of cocaine use is associated with heightened stress-induced reinstatement of cocaine seeking. Addict Biol 2016;21:634–45.

[28] Caffino L, Calabrese F, Gianotti G, Barbon A, Verheij MM, Racagni G, Fumagalli F. Stress rapidly dysregulates the glutamatergic synapse in the prefrontal cortex of cocaine-withdrawn adolescent rats. Addict Biol 2015;20:158–69.

[29] Burke AR, Miczek KA. Escalation of cocaine self-administration in adulthood after social defeat of adolescent rats: role of social experience and adaptive coping behavior. Psychopharmacology 2015;232:3067–79.

[30] Goeders NE, Guerin GF. Non-contingent electric footshock facilitates the acquisition of intravenous cocaine self-administration in rats. Psychopharmacology 1994;114:63–70.

[31] Haney M, Maccari S, Le Moal M, Simon H, Piazza PV. Social stress increases the acquisition of cocaine self-administration in male and female rats. Brain Res 1995;698:46–52.

[32] Ramsey NF, Van Ree JM. Emotional but not physical stress enhances intravenous cocaine self-administration in drug-naïve rats. Brain Res 1993;608:216–22.

[33] Schenk S, Lacelle G, Gorman K, Amit Z. Cocaine self-administration in rats influenced by environmental conditions: implications for the etiology of drug abuse. Neurosci Lett 1987;81:227–31.

[34] Koob GF. Neurobiological substrates for the dark side of compulsivity in addiction. Neuropharmacology 2009;56(Suppl. 1):18–31.

[35] Mantsch JR, Katz ES. Elevation of glucocorticoids is necessary but not sufficient for the escalation of cocaine self-administration by chronic electric footshock stress in rats. Neuropsychopharmacology 2007;32:367–76.

[36] Cruz FC, Quadros IM, Hogenelst K, Planeta CS, Miczek KA. Social defeat stress in rats: escalation of cocaine and "speedball" bing self-administration, but not heroin. Psychopharmacology 2011;215:165–75.

[37] Boyson CO, Holly EN, Shimamoto A, Albrechet-Souza L, Weiner LA, DeBold JF, Miczek KA. Social stress and CRF-dopamine interactions in the VTA: role of long-term escalation of cocaine self-administration. J Neurosci 2014;34:6659–67.

[38] Sinha R. Modeling stress and drug craving in the laboratory: implications for addiction treatment development. Addiction Biology 2009;14:84–98.

[39] Erb S, Shaham Y, Stewart J. Stress reinstates cocaine-seeking behavior after prolonged extinction and a drug-free period. Psychopharmacology 1996;128:408–12.

[40] Shalev U, Marinelli M, Baumann MH, Piazza PV, Shaham Y. The role of corticosterone in food deprivation-induced reinstatement of cocaine seeking in the rat. Psychopharmacology 2003;168:170–6.

[41] Conrad KL, McCutcheon JE, Cotterly LM, Ford KA, Beales M, Marinelli M. Persistent increases in cocaine-seeking behavior after acute exposure to cold swim stress. Biol Psychiatry 2010;68:303–5.

[42] Manvich DF, Stowe TA, Godfrey JR, Weinshenker D. A method for psychosocial stress-induced reinstatement of cocaine seeking in rats. Biol Psychiatry 2016;79:940–6.

[43] Mantsch JR, Baker DA, Serge JP, Hoks MA, Francis DM, Katz ES. Surgical adrenalectomy with diurnal corticosterone replacement slows escalation and prevents the augmentation of cocaine-induced reinstatement in rats self-administering cocaine under long-access conditions. Neuropsychopharmacology 2008;33:814–26.

[44] Fox HC, Talih M, Malison R, Anderson GM, Kreek MJ, Sinha R. Frequency of recent cocaine and alcohol use affects drug craving and associated responses to stress and drug-related cues. Psychoneuroendocrinology 2005;30:880–91.

[45] Preston KL, Kowalczyk WJ, Phillips KA, Jobes ML, Vahabzadeh M, Lin JL, Mezghanni M, Epstein DH. Exacerbated craving in the presence of stress and drug cues in drug-dependent patients. Neuropsychopharmacology 2017; https://doi.org/10.1038/npp.2017.275. Epub ahead of print.

[46] Graf EN, Wheeler RA, Baker DA, Ebben AL, Hill JE, McReynolds JR, Robble MA, Vranjkovic O, Wheeler DS, Mantsch JR, Gasser PJ. Corticosterone acts in the nucleus accumbens to enhance dopamine signaling and potentiate reinstatement of cocaine seeking. J Neurosci 2013;33:11800–10.

[47] McReynolds JR, Doncheck EM, Vranjkovic O, Ganzman GS, Baker DA, Hillard CJ, Mantsch JR. CB1 receptor antagonism blocks stress-potentiated reinstatement of cocaine seeking in rats. Psychopharmacology 2016;233:99–109.

[48] Koob GF, Le Moal M. Plasticity of reward neurocircuitry and the 'dark side' of drug addiction. Nat Neurosci 2005;8:1442–4.

[49] Hyman SM, Garcia M, Sinha R. Gender specific associations between types of childhood maltreatment and the onset, escalation and severity of substance use in cocaine dependent adults. Am J Drug Alcohol Abuse 2006;32:655–64.

[50] Andersen SL, Teicher MH. Desperately driven and no brakes: developmental stress exposure and subsequent risk for substance abuse. Neurosci Biobehav Rev 2009;33:516–24.

[51] Jordan CJ, Andersen SL. Sensitive periods of substance abuse: early risk for the transition to dependence. Dev Cogn Neurosci 2017;25:29–44.

[52] Caldji C, Tannenbaum B, Sharma S, Francis D, Plotsky PM, Meaney MJ. Maternal care during infancy regulates the development of neural systems mediating the expression of fearfulness in the rat. Proc Natl Acad Sci USA 1998;95:5335–40.

[53] Weaver IC, Cervoni N, Champagne FA, D'Alessio AC, Sharma S, Seckl JR, Dymov S, Szyf M, Meaney MJ. Epigenetic programming by maternal behavior. Nat Neurosci 2004;7:847–54.

[54] Matthews K, Robbins TW. Early experience as a determinant of adult behavioral responses to reward: the effects of repeated maternal separation in the rat. Neurosci Biobehav Rev 2003;27:45–55.

[55] Nestler EJ, Carlezon Jr. WA. The mesolimbic dopamine reward circuit in depression. Biol Psychiatry 2006;59:1151–9.

[56] Andersen SL, Lyss PJ, Dumont NL, Teicher MH. Enduring neurochemical effects of early maternal separation on limbic structures. Ann N Y Acad Sci 1999;877:756–9.

[57] Jones SA, Morales AM, Lavine JB, Nagel BJ. Convergent neurobiological predictors of emergent psychopathology during adolescence. Birth Defects Res 2017;109:1613–22.

[58] Nigg JT, Stavro G, Ettenhofer M, Hambrick DZ, Miller T, Henderson JM. Executive functions and ADHD in adults: evidence for selective effects on ADHA symptom domains. J Abnorm Psychol 2005;114:706–17.

[59] Jentsch JD, Taylor JR. Impulsivity resulting from frontostriatal dysfunction in drug abuse: implications for the control of behavior by reward-related stimuli. Psychopharmacology 1999;146:373–90.

[60] Belin D, Jonkman S, Dickinson A, Robbins TW, Everitt BJ. Parallel and interactive learning processes within the basal ganglia: relevance for the understanding of addiction. Behav Brain Res 2009;199:89–102.

[61] Chambers RA, Taylor JR, Potenza MN. Developmental neurocircuitry of motivation in adolescence: a critical period of addiction vulnerability. Am J Psychiatry 2003;160:1041–52.

[62] Ary AW, Auilar VR, Szumlinski KK, Kippin TE. Prenatal stress alters limbo-corticostriatal Homer protein expression. Synapse 2007;61:938–41.

[63] Liu D, Diorio J, Tannenbaum B, Caldji C, Francis D, Freedman A, Sharma S, Pearson D, Plotsky PM, Meaney MJ. Maternal care, hippocampal glucocorticoid receptors, and hypothalamic-pituitary-adrenal responses to stress. Science 1997;277:1659–62.

[64] Grimm JW, Lu L, Hayashi T, Hope BT, Su TP, Shaham Y. Time-dependent increases in brain-derived neurotrophic factor protein levels within the mesolimbic dopamine system after withdrawal from cocaine: implications for incubation of cocaine craving. J Neurosci 2003;23:742–7.

[65] Eipper-Mains JE, Kiraly DD, Duff MO, Horowitz MJ, McManus CJ, Eipper BA, Graveley BR, Mains RE. Effects of cocaine and withdrawal on the mouse nucleus accumbens transcriptome. Genes Brain Behav 2013;12:21–33.

[66] Marinelli M, Piazza PV. Interaction between glucocorticoid hormones, stress and psychostimulant drugs. Eur J Neurosci 2002;16:387–94.

[67] Tidey JW, Miczek KA. Acquisition of cocaine self-administration after social stress: role of accumbens dopamine. Psychopharmacology 1997;130:203–12.

[68] Holly EN, Boyson CO, Montagud-Romero S, Stein DJ, Gobrogge KL, DeBold JF, Miczek KA. Episodic social stress-escalated cocaine self-administration: role of phasic and tonic corticotropin releasing factor in the anterior and posterior ventral tegmental area. J Neurosci 2016;36:4093–105.

[69] Deroche V, Marinelli M, Le Moal M, Piazza PV (1997). Glucocorticoids and behavioral effects of psychostimulants. II: cocaine intravenous self-administration and reinstatement depend on glucocorticoid levels. J Pharmacol Exp Ther, 281: 1401–1407.

[70] Mantsch JR, Saphier D, Goeders NE. Corticosterone facilitates the acquisition of cocaine self-administration in rats: opposite effects of the type II glucocorticoid receptor agonist dexamethasone. J Pharmacol Exp Ther 1998;287:72–80.

[71] Ambroggi F, Turiault M, Milet A, Deroche-Gamonet V, Parnaudeau S, Balado E, Barik J, van der Veen R, Maroteaux G, Lemberger T, Schutz G, Lazar M, Marinelli M, Piazza PV, Tronche F. Stress and addiction: glucocorticoid receptor in dopaminoceptive neurons facilitates cocaine seeking. Nat Neurosci 2009;12:247–9.

[72] Leonard MZ, DeBold JF, Miczek KA. Escalated cocaine "binges" in rats: enduring effects of social defeat stress or intra-VTA CRF. Psychopharmcology 2017;234:2823–36.

[73] Holly EN, DeBold JF, Miczek KA. Increased mesocorticolimbic dopamine during acute and repeated social defeat stress: modulation by corticotropin releasing factor receptors in the ventral tegmental area. Psychopharmacology 2015;232:4469–79.

[74] Han X, DeBold JF, Miczek KA. Prevention and reversal of social stress-escalated cocaine self-administration in mice by intra-VTA CRFR1 antagonism. Psychopharmacology 2017;234:2813–21.

[75] Hwa LS, Holly EN, DeBold JF, Miczek KA. Social stress-escalated intermittent alcohol drinking: modulation by CRF-R1 in the ventral tegmental area and accumbal dopamine in mice. Psychopharmacology 2016;233:681–90.

[76] Briand LA, Vassoler FM, Pierce RC, Valentino RJ, Blendy JA. Ventral tegmental afferents in stress-induced reinstatement: the role of cAMP response element-binding protein. J Neurosci 2010;30:16149–59.

[77] McFarland K, Davidge SB, Lapish CC, Kalivas PW. Limbic and motor circuitry underlying footshock-induced reinstatement of cocaine-seeking behavior. J Neurosci 2004;24:1551–60.

[78] Erb S, Lopak V, Smith C. Cocaine pre-exposure produces a sensitized and context-specific c-fos mRNA response to footshock stress in the central nucleus of the amygdala. Neuroscience 2004;129:719–25.

[79] Erb S, Hitchcott PK, Rejabi H, Mueller D, Shaham Y, Stewart J. Alpha-2 adrenergic receptor agonists block stress-induced reinstatement of cocaine seeking. Neuropsychopharmacology 2000;23:138–50.

[80] Mantsch JR, Weyer A, Vranjkovic O, Beyer CE, Baker DA, Caretta H. Involvement of noradrenergic neurotransmission in the stress- but not cocaine-induced reinstatement of extinguished cocaine-induced conditioned place preference in mice: role for β-2 adrenergic receptors. Neuropsychopharmacology 2010;35:2165–78.

[81] Fallon JH, Koziell DA, Moore RY. Catecholamine innervation of the basal forebrain. II. Amygdala, suprarhinal cortex and entorhinal cortex. J Comp Neurol 1978;180:509–32.

[82] Brown ZJ, Nobrega JN, Erb S. Central injections of noradrenaline induce reinstatement of cocaine seeking and increase c-fos mRNA expression in the extended amygdala. Behav Brain Res 2011;217:472–6.

[83] Leri F, Flores J, Rodaros D, Stewart J. Blockade of stress-induced but not cocaine-induced reinstatement by infusion of noradrenergic antagonists into the bed nucleus of the stria terminalis or the central nucleus of the amygdala. J Neurosci 2002;22:5713–8.

[84] Venniro M, Caprioli D, Zhang M, Whitaker LR, Zhang S, Warren BL, Cifani C, Marchant NJ, Yizhar O, Bossert JM, Chiamulera C, Morales M, Shaham Y. The anterior insular cortex→central amygdala glutamatergic pathway is critical to relapse after contingency management. Neuron 2017;96:414–27.

[85] Schmeichel BE, Herman MA, Roberto M, Koob GF. Hypocretin neurotransmission within the central amygdala mediates escalated cocaine self-administration and stress-induced reinstatement in rats. Biol Psychiatry 2017;81:606–15.

[86] Cippitelli A, Damadzic R, Hansson AC, Singley E, Sommer WH, Eskay R, Thorsell A, Heilig M. Neuropeptide Y (NPY) suppresses yohimbine-induced reinstatement of alcohol seeking. Psychopharmacology 2010;208:417–26.

[87] Ciccocioppo R, de Glulielmo G, Hansson AC, Ubaldi M, Kallupi M, Cruz MT, Oleata CS, Heilig M, Roberto M. Restraint stress alters nociception/orphanin FQ and CRF systems in the rat central amygdala: significance for anxiety-like behaviors. J Neurosci 2014;34:363–72.

[88] Simms JA, Haass-Koffler CL, Bito-Onon J, Li R, Bartlett SE. Mifepristone in the central nucleus of the amygdala reduces yohimbine stress-induced reinstatement of ethanol seeking. Neuropsychopharmacology 2012;37:906–18.

[89] Rodaros D, Caruana DA, Amir S, Stewart J. Corticotropin-releasing factor projections from limbic forebrain and paraventricular nucleus of the hypothalamus to the region of the ventral tegmental area. Neuroscience 2007;150:8–13.

[90] Beckerman MA, Van Kempen TA, Justice NJ, Milner TA, Glass MJ. Corticotropin-releasing factor in the mouse central nucleus of the amygdala: ultrastructural distribution in the NMDA-NR1 receptor subunit expressing neurons as well as projection neurons to the bed nucleus of the stria terminalis. Exp Neurol 2013;239:120–32.

[91] Erb S, Salmaso N, Rodaros D, Stewart J. A role for the CRF-containing pathway from central nucleus of the amygdala to bed nucleus of the stria terminalis in the stress-induced reinstatement of cocaine seeking in rats. Psychopharmacology 2001;158:360–5.

[92] Vranjkovic O, Gasser PJ, Gerndt CH, Baker DA, Mantsch JR. Stress-induced cocaine seeking requires a beta-2 adrenergic receptor-regulated pathway from the ventral bed nucleus of the stria terminalis that regulates CRF actions in the ventral tegmental area. J Neurosci 2014;34:12504–14.

[93] Shaham Y, Erb S, Stewart J. Stress-induced relapse to heroin and cocaine seeking in rats: a review. Brain Res Brain Res Rev 2000;33:13–33.

[94] Vranjkovic O, Pina M, Kash TL, Winder DG. The bed nucleus of the stria terminalis in drug-associated behavior and affect: a circuit-based perspective. Neuropharmacology 2017;122:100–6.

[95] Silberman Y, Matthews TR, Winder DG. A corticotropin releasing factor pathway for ethanol regulation of the ventral tegmental area in the bed nucleus of the stria terminalis. J Neurosci 2013;33:950–60.

[96] Erb S, Stewart J. A role for the bed nucleus of the stria terminalis, but not the amygdala, in the effects of corticotropin-releasing factor on stress-induced reinstatement of cocaine seeking. J Neurosci 1999;19:RC35.

[97] Wang X, Cen X, Lu L. Noradrenaline in the bed nucleus of the stria terminalis is critical for stress-induced reactivation of morphine-conditioned place preference in rats. Eur J Pharmacol 2001;432:153–61.

[98] McReynolds JR, Vranjkovic O, Thao M, Baker DA, Makky K, Lim Y, Mantsch JR. Beta-2 adrenergic receptors mediate stress-evoked reinstatement of cocaine-induced conditioned place preference and increases in CRF mRNA in the bed nucleus of the stria terminalis in mice. Psychopharmacology 2014;231:3953–63.

[99] Miles OW, Thrailkil EA, Linden AK, May V, Bouton ME, Hammack SE. Pituitary adenylate cyclase-activating peptide in the bed nucleus of the stria terminalis mediates stress-induced reinstatement of cocaine seeking in rats. Neuropsychopharmacology 2017; https://doi.org/10.1038/npp.2017.135. epub ahead of print.

[100] Lê AD, Funk D, Coen K, Tamadon S, Shaham Y. Role of κ-opioid receptors in the bed nucleus of stria terminalis in reinstatement of alcohol seeking. Neuropsychopharmacology 2017; https://doi.org/10.1038/npp.2017.120. epub ahead of print.

[101] Lammel S, Lim BK, Ran C, Huang KW, Betley MJ, Tye KM, Deisseroth K, Malenka RC. Input-specific control of reward and aversion in the ventral tegmental area. Nature 2012;491:212–7.

[102] Wang B, Shaham Y, Zitzman D, Azari S, Wise RA, You ZB. Cocaine experience establishes control of midbrain glutamate and dopamine by corticotropin-releasing factor: a role in stress-induced relapse to drug seeking. J Neurosci 2005;25:5380–96.

[103] Blacktop JM, Seubert C, Baker DA, Ferda N, Lee G, Graf EN, Mantsch JR. Augmented cocaine seeking in response to stress or CRF delivered into the ventral tegmental area following long-access self-administration is mediated by CRF receptor type 1 but not CRF receptor type 2. J Neurosci 2011;31:11396–403.

[104] Chen NA, Jupp B, Sztainberg Y, Lebow m BRM, Kim JH, Chen A, Lawrence AJ. Knockdown of CRF1 receptors in the ventral tegmental area attenuates cue- and acute food deprivation stress-induced cocaine seeking in mice. J Neurosci 2014;34:11560–70.

[105] Twining RC, Wheeler DS, Ebben AL, Jacobsen AJ, Robble MA, Mantsch JR, Wheeler RC. Aversive stimuli drive drug seeking in a state of low dopamine tone. Biol Psychiatry 2015;77:895–902.

[106] Wang B, You ZB, Rice KC, Wise RA. Stress-induced relapse to cocaine seeking: roles for the CRF(2) receptor and CRF-binding protein in the ventral tegmental area of the rat. Psychopharmacology 2007;193:283–94.

[107] Grieder TE, Herman MA, Contet C, Tan LA, Vargas-Perez H, Cohen A, Chwalek M, Maal-Bared G, Freiling J, Scholsburg JE, Clarke L, Crawford E, Koebel P, Repunte-Canonigo V, Sanna PP, Tapper AR, Roberto M, Kieffer BL, Sawchenko PE, Koob GF, van der Kooy D, George O. VTA CRF neurons mediate the aversive effects of nicotine withdrawal and promote intake escalation. Nat Neurosci 2014;17:1751–8.

[108] Tagliaferro P, Morales M. Synapses between corticotropin-releasing factor-containing axon terminals and dopaminergic neurons in the ventral tegmental area are predominantly glutamatergic. J Comp Neurol 2008;506:616–26.

[109] Hahn J, Hopf FW, Bonci A. Chronic cocaine enhances corticotropin-releasing factor-dependent potentiation of excitatory transmission in ventral tegmental area dopamine neurons. J Neurosci 2009;29:6535–44.

[110] Blacktop JM, Vranjkovic O, Mayer M, Van Hoof M, Baker DA, Mantsch JR. Antagonism of GABA-B but not GABA-A receptors in the VTA prevents stress- and intra-VTA CRF-induced reinstatement of extinguished cocaine seeking in rats. Neuropharmacology 2016;102:197–206.

[111] Redila VA, Chavkin C. Stress-induced reinstatement of cocaine seeking is mediated by the kappa opioid system. Psychopharmacology 2008;200:59–70.

[112] Bruchas MR, Land BB, Lemos JC, Chavkin C. CRF1-R activation of the dynorphin/kappa opioid system in the mouse basolateral amygdala mediates anxiety-like behavior. PLoS One 2009;4.

[113] Polter AM, Bishop RA, Briand LA, Graziane NM, Pierce RC, Kauer JA. Poststress block of kappa opiod receptors rescues long-term potentiation of inhibitory synapses and prevents reinstatement of cocaine seeking. Biol Psychiatry 2014;76:785–93.

[114] Wang B, You ZB, Wise RA. Reinstatement of cocaine seeking by hypocretin (orexin) in the ventral tegmental area: independence from the local corticotropin-releasing factor network. Biol Psychiatry 2009;65:857–62.

[115] Tung LW, Lu GL, Lee YH, Lu L, Lee JH, Leishman E, Bradshaw H, Hwang LL, Hung MS, Mackie K, Zimmer A, Chiou LC. Orexins contribute to restraint stress-induced cocaine relapse by endocannabinoid-mediated disinhibition of dopaminergic neurons. Nat Commun 2016;7:12199.

[116] Mahler SV, Smith RJ, Aston-Jones G. Interactions between VTA orexin and glutamate in cue-induced reinstatement of cocaine seeking in rats. Psychopharmacology 2013;226:687–98.

[117] Deutch AY, Lee MC, Gillham MH, Cameron DA, Goldstein M, Iadarola MJ. Stress selectively increases fos protein in dopamine neurons innervating the prefrontal cortex. Cereb Cortex 1991;1:273–92.

[118] Holly EN, Miczek KA. Ventral tegmental area dopamine revisited: effects of acute and repeated stress. Psychopharmacology 2016;233:163–86.

[119] Lammel S, Lim BK, Malenka RC. Reward and aversion in a heterogeneous midbrain dopamine system. Neuropharmacology 2014;76(Pt B):351–9.

[120] Puglisi-Allegra S, Imperato A, Angelucci L, Cabib S. Acute stress induces time-dependent responses in dopamine mesolimbic system. Brain Res 1991;554:217–22.

[121] Badrinarayan A, Wescott SA, Vander Weele CM, Saunders BT, Couturier BE, Maren S, Aragona BJ. Aversive stimuli differentially modulate real-time dopamine transmission dynamics within the nucleus accumbens core and shell. J Neurosci 2012;32:15779–90.

[122] Roitman MF, Wheeler RA, Wightman RM, Carelli RM. Real-time chemical responses in the nucleus accumbens differentiate rewarding and aversive stimuli. Nature Neurosci 2008;11:1376–7.

[123] Vogt BA, Paxinos G. Cytoarchitecture of mouse and rat cingulate cortex with human homologies. Brain Struct Funct 2014;219:185–92.

[124] Capriles N, Rodaros D, Sorge RE, Stewart J. A role of the prefrontal cortex in stress- and cocaine-induced reinstatement of cocaine seeking in rats. Psychopharmacology 2003;168:66–74.

[125] Tseng KY, O'Donnell P. Dopamine-glutamate interactions controlling prefrontal cortical pyramidal cell excitability involve multiple signaling mechanisms. J Neurosci 2004;24:5131–9.

[126] Thurley K, Senn W, Luscher HR. Dopamine increases the gain of the input-output response of rat prefrontal pyramidal neurons. J Neurophysiol 2008;99:2985–97.

[127] McFarland K, Lapish CC, Kalivas PW. Prefrontal glutamate release into the core of the nucleus accumbens mediates cocaine-induced reinstatement of drug-seeking behavior. J Neurosci 2003;23:3531–7.

[128] O'Neill M, Schultz W. Economic risk coding by single neurons in the orbitofrontal cortex. J Physiol 2015;109:70–7.

[129] Schmidt KT, Schroeder JP, Foster SL, Squires K, Smith BM, Pitts EG, Epstein MP, Weinshenker D. Norepinephrine regulates cocaine-primed reinstatement via α1-adrenergic receptors in the medial prefrontal cortex. Neuropharmacology 2017;119:134–40.

[130] Uribe-Mariño A, Gassen NC, Wiesbeck MF, Balsevich G, Santarelli S, Solfrank B, Dournes C, Fries GR, Masana M, Labermeier C, Wang XD, Hafner K, Schmid B, Rein T, Chen A, Deussing JM, Schmidt MV. Prefrontal cortex corticotropin-releasing factor receptor 1 conveys acute stress-induced executive dysfunction. Biol Psychiatry 2016;80:743–53.

[131] Stefanik MT, Moussawi K, Kupchik YM, Smith KC, Miller RL, Huff ML, Deisseroth K, Kalivas PW, LaLumiere RT. Addict Biol 2013;18:50–3.

[132] Cornish JL, Kalivas PW. Glutamate transmission in the nucleus accumbens mediates relapse in cocaine addiction. J Neurosci 2000;20:RC89.

[133] Xi ZX, Gilbert J, Campos AC, Kline N, Ashby Jr. CR, Hagan JJ, Heidbreder CA, Gardner EL. Blockade of mesolimbic dopamine D3 receptors inhibits stress-induced reinstatement of cocaine-seeking in rats. Psychopharmacology 2004;176:57–65.

[134] Lodge DJ, Grace AA. Acute and chronic corticotropin-releasing factor 1 receptor blockade inhibits cocaine-induced dopamine release: correlation with dopamine neuron activity. J Pharmacol Exp Ther 2005;314:201–6.

[135] Wanat MJ, Bonci A, Phillips PE. CRF acts in the midbrain to attenuate accumbens dopamine release to rewards but not their predictors. Nat Neurosci 2013;16:383–5.

[136] Schindler AG, Messinger DI, Smith JS, Shankar H, Gustin RM, Schattauer SS, Lemos JC, Chavkin NW, Hagan CE, Neumaier JF, Chavkin C. Stress produces aversion and potentiates cocaine reward by releasing endogenous dynorphins in the ventral striatum to locally stimulate serotonin reuptake. J Neurosci 2012;32:17582–96.

[137] Land BB, Bruchas MR, Lemos JC, Xu M, Melief EJ, Chavkin C. The dysphoric component of stress is encoded by activation of the dynorphin kappa-opioid system. J Neurosci 2008;28:407–14.

[138] Al-Hasani R, McCall JG, Shin G, Gomez AM, Schmitz GP, Bernardi JM, Pyo CO, Park SI, Marcinkiewcz CM, Crowley NA, Krashes MJ, Lowell BB, Kash TL, Rogers JA, Bruchas MR. Distinct subpopulations of nucleus accumbens dynorphin neurons drive aversion and reward. Neuron 2015;87:1063–77.

[139] Lemos JC, Wanat MJ, Smith JS, Reyes BA, Hollon NG, Van Bockstaele EJ, Chavkin C, Phillips PE. Severe stress switches CRF action in the nucleus accumbens from appetitive to aversive. Nature 2012;490:402–6.

[140] Beinfeld MC, Connolly JK, Pierce RC. Cocaine treatment increases extracellular cholecystokinin (CCK) in the nucleus accumbens shell of awake, freely moving rats, an effect that is enhanced in rats that are behaviorally sensitized to cocaine. J Neurochem 2002;81:1021–7.

[141] Mantsch JR, Baker DA, Francis DM, Katz ES, Hoks MA, Serge JP. Stressor- and corticotropin releasing factor-induced reinstatement and active stress-related behavioral responses are augmented following long-access cocaine self-administration by rats. Psychopharmacology 2008;195:591–603.

[142] Graf EN, Hoks MA, Baumgardner J, Sierra J, Vranjkovic O, Bohr C, Baker DA, Mantsch JR. Adrenal activity during repeated long-access cocaine self-administration is required for later CRF-induced and CRF-dependent stressor-induced reinstatement in rats. Neuropsychopharmacology 2011;36:1444–54.

[143] McReynolds JR, Doncheck EM, Li Y, Vranjkovic O, Graf EN, Ogasawara D, Cravatt BF, Baker DA, Liu QS, Hillard CJ, Mantsch JR. Stress promotes drug seeking through glucocorticoid-dependent endocannabinoid mobilization in the prelimbic cortex, Biol Psychiatry 2017; pii: S0006-3223 (17)32046-2, https://doi.org/10.1016/j.biopsych.2017.09.024. epub ahead of print.

[144] McReynolds JR, Taylor A, Vranjkovic O, Ambrosius T, Derricks O, Nino B, Kurtoglu B, Wheeler RA, Baker DA, Gasser PJ, Mantsch JR. Corticosterone potentiation of cocaine-induced reinstatement of conditioned place preference in mice is mediated by blockade of the organic cation transporter 3. Neuropsychopharmacology 2017;42:757–65.

[145] Gasser PJ, Lowry CA, Orchinik M. Corticosterone-sensitive monoamine transport in the rat dorsomedial hypothalamus: potential role for organic cation transporter 3 in stress-induced modulation of monoaminergic neurotransmission. J Neurosci 2006;26:8758–66.

[146] Ahmed SH, Koob GF. Cocaine- but not food-seeking behavior is reinstated by stress after extinction. Psychopharmacology 1997;132:289–95.

[147] Mantsch JR, Yuferov V, Mathieu-Kia AM, Ho A, Kreek MJ. Effects of extended access to high versus low cocaine doses on self-administration, cocaine-induced reinstatement and brain mRNA levels in rats. Psychopharmacology 2004;175:26–36.

[148] Ahmed SH, Koob GF. Transition to drug addiction: a negative reinforcement model based on an allostatic decrease in reward function. Psychopharmacology 2005;180:473–90.

[149] Specio SE, Wee S, O'Dell LE, Boutrel B, Zorrilla EP, Koob GF. CRF(1) receptor antagonists attenuate escalated cocaine self-administration in rats. Psychopharmacology 2008;196:473–82.

[150] Park PE, Scholsburg JE, Vendruscolo LF, Schulteis G, Edwards S, Koob GF. Chronic CRF1 receptor blockade reduces heroin intake escalation and dependence-induced hyperalgesia. Addict Biol 2015;20:275–84.

[151] Wee S, Koob GF. The role of the dynorphin-kappa opioid system in the reinforcing effects of drugs of abuse. Psychopharmacology 2010;210:121–35.

[152] Walker BM, Zorrilla EP, Koob GF. Systemic κ-opioid receptor antagonism by nor-binaltorphimine reduces dependence-induced excessive alcohol self-administration in rats. Addict Biol 2011;16:116–9.

[153] Schlosburg JE, Whitfield Jr. TW, Park PE, Crawford EF, George O, Vendruscolo LF, Koob GF. Long-term antagonism of κ-opioid receptors prevents escalation of and increased motivation for heroin intake. J Neurosci 2013;33:19384–92.

[154] Wee S, Orio L, Ghirmai S, Cashman JR, Koob GF. Inhibition of kappy opioid receptors attenuated increased cocaine intake in rats with extended access to cocaine. Psychopharmacology 2009;205:565–75.

[155] Walker BM, Rasmussen DD, Raskind MA, Koob GF. Alpha1-noradrenergic receptor antagonism blocks dependence-induced increases in responding for ethanol. Alcohol 2008;42:91–7.

[156] Greenwell TN, Funk CK, Cottone P, Richardson HN, Chen SA, Rice KC, Zorrilla EP, Koob GF. Corticotropin-releasing factor-1 receptor antagonists decrease heroin self-administration in long- but not short-access rats. Addict Biol 2009;14:130–43.

[157] Barbier E Vendruscolo LF, Schlosburg JE, Edwards S, Juergens N, Park PE, Misra KK, Cheng K, Rice KC, Schank J, Schulteis G, Koob GF, Heilig M. The NK1 receptor antagonist L822429 reduces heroin reinforcement. Neuropsychopharmacology 2013;38:976–84.

[158] Zhou Y, Spangler R, LaForge KS, Maggos CE, Ho A, Kreek MJ. Corticotropin-releasing factor and type 1 corticotropin-releasing factor receptor messenger RNAs in rat brain and pituitary during "binge"-pattern cocaine administration and chronic withdrawal. J Pharmacol Exp Ther 1996;279 (1):351–8.

[159] Richter RM, Weiss F. In vivo CRF release in rat amygdala is increased during cocaine withdrawal in self-administering rats. Synapse 1999;32:254–61.

[160] Zorrilla EP, Valdez GR, Weiss F. Changes in levels of regional CRF-like immunoreactivity and plasma corticosterone during protracted drug withdrawal in dependent rats. Psychopharmacology 2001;158:374–81.

[161] Zorrilla EP, Wee S, Zhao Y, Specio S, Boutrel B, Koob GF, Weiss F. Extended access cocaine self-administration differentially activates dorsal raphe and amygdala corticotropin-releasing factor systems in rats. Addict Biol 2012;17:300–8.

[162] Erb S, Fuk D, Lê AD. Cocaine pre-exposure enhances CRF-induced expression of c-fos mRNA in the central nucleus of the amygdala: an effect that parallels the effects of cocaine pre-exposure on CRF-induced locomotor activity. Neurosci Lett 2005;383:209–14.

[163] Orozco-Cabal L, Pollandt S, Liu J, Shinnick-Gallagher P, Gallagher JP. Regulation of synaptic transmission by CRF receptors. Rev Neuosci 2006;17:279–307.

[164] Nobis WP, Kash TL, Silberman Y, Winder DG. β-Adrenergic receptors enhance excitatory transmission in the bed nucleus of the stria terminalis through a corticotrophin-releasing factor receptor-dependent and cocaine-regulated mechanism. Biol Psychiatry 2011;69:1083–90.

[165] Valdez GR, Roberts AJ, Chan K, Davis H, Brennan M, Zorrilla EP, Koob GF. Increased ethanol self-administration and anxiety-like behavior during acute ethanol withdrawal and protracted abstinence: regulation by corticotropin-releasing factor. Alcohol Clin Exp Res 2002;26:1494–501.

[166] Kimbrough A, de Guglielmo G, Kononoff J, Kallupi M, Zorrilla EP, George O. CRF$_1$ receptor-dependent increases in irritability-like behavior during abstinence from chronic intermittent ethanol vapor exposure. Alcohol Clin Exp Res 2017;41:1886–95.

[167] Sarnyai Z, Biro E, Gardi J, Vecsernyes M, Julesz J, Telegdy G. Brain corticotropin-releasing factor mediates 'anxiety-like' behavior induced by cocaine withdrawal in rats. Brain Res 1995;675:89–97.

[168] Park PE, Vendurscolo LF, Schlosburg JE, Edwards S, Schulteis G, Koob GF. Corticotropin-releasing factor (CRF) and α 2 adrenergic receptors mediate heroin withdrawal-potentiated startle in rats. Int J Neuropsychopharmacol 2013;16:1867–75.

[169] Mantsch JR, Cullinan WE, Tang LC, Baker DA, Katz ES, Hoks MA, Ziegler DR. Daily cocaine self-administration under long-access conditions augments restraint-induced increases in plasma corticosterone and impairs glucocorticoid receptor-mediated negative feedback in rats. Brain Res 2007;1167:101–11.

[170] Ahmed SH, Walker JR, Koob GF. Persistent increase in the motivation to take heroin in rats with a history of drug escalation. Neuropsychopharmacology 2000;22:413–21.

[171] Goeders NE, Bienvenu OJ, De Souza EB. Chronic cocaine administration alters corticotropin-releasing factor receptors in the rat brain. Brain Res 1990;531:322–8.

[172] Mantsch JR, Yuferov V, Mathieu-Kia AM, Ho A, Kreek MJ. Neuroendocrine alterations in a high-dose, extended-access rat self-administration model of escalating cocaine use. Psychoneuroendocrinology 2003;28:836–62.

[173] Karkhanis AN, Rose JH, Weiner JL, Jones SR. Early-life social isolation stress increases kappa opioid receptor responsiveness and downregulates the dopamine system. Neuropsychopharmacology 2016;41:2263–74.

[174] Vendruscolo LF, Barbier E, Scholsburg JE, Misra KK, Whitfield TW Jr, Logrip ML, Rivier C, Repunte-Canonigo V, Zorrilla EP, Sanna PP, Heilig M, Koob GF (2012). Corticosteroid-dependent plasticity mediates compulsive alcohol drinking in rats. J Neurosci, 32: 7563–7571.

[175] Sedki F, Eigenmann K, Gelinas J, Schouela N, Courchesne S, Shalev U. A role for kappa-, but not mu-opioid, receptor activation in acute food deprivation-induced reinstatement of heroin seeking in rats. Addict Biol 2015;20:423–32.

[176] Walker LC, Kastman HE, Koeleman JA, Smith CM, Perry CJ, Krstew EV, Gundlach AL, Lawrence AJ. Nucleus incertus corticotrophin-releasing factor 1 receptor signaling regulates alcohol seeking in rats. Addict Biol 2017;22:1641–54.

[177] Jackson KJ, McLaughlin JP, Carroll FI, Damaj MI. Effects of the kappa opioid receptor antagonist, norbinaltorphimine, on stress and drug-induced reinstatement of nicotine-conditioned place preference in mice. Psychopharmacology 2013;226:763–8.

[178] Funk D, Coen K, Lê AD. The role of kappa opioid receptors in stress-induced reinstatement of alcohol seeking in rats. Brain Behav 2014;4:356–67.

[179] Nygard SK, Hourguettes NJ, Sobczak GG, Carlezon WA, Bruchas MR. Stress-induced reinstatement of nicotine preference requires dynorphin/ kappa opioid activity in the basolateral amygdala. J Neurosci 2016;36:9937–48.

[180] Funk D, Coen K, Tamadon S, Li Z, Loughlin A, Lê AD. Effects of prazosin and doxazosin on yohimbine-induced reinstatement of alcohol seeking in rats. Psychopharmacology 2016;233:2197–207.

[181] Qi K, Wei C, Li Y, Sui N. Orexin receptors within the nucleus accumbens shell mediate the stress but not drug priming-induced reinstatement of morphine conditioned place preference. Front Behav Neurosci 2013;7:144.

[182] Qi J, Zhang S, Wang HL, Wang H, de Jesus Aceves Buendia J, Hoffman AF, Lupica CR, Seal RP, Morales M (2014). A glutamatergic reward input from the dorsal raphe to ventral tegmental area dopamine neurons. Nat Commun, 5: 5390.

[183] Han X, Albrechet-Souza L, Doyle MR, Shimamoto A, DeBold JF, Miczek KA. Social stress and escalated drug self-administration in mice II. Cocaine and dopamine in the nucleus accumbens. Psychopharmacology 2015;232:1003–10.

[184] Karimi S, Attarzadeh-Yazdi G, Yazdi-Ravandi S, Hesam S, Azizi P, Razavi Y, Haghparast A. Forced swim stress but not exogenous corticosterone could induce the reinstatement of extinguished morphine conditioned place preference in rats: involvement of glucocorticoid receptors in the baso-lateral amygdala. Behave Brain Res 2014;264:43–50.

[185] Gueye AB, Pryslawsky Y, Trigo JM, Poulia N, Delis F, Antoniou K, Loureiro M, Laviolette SR, Vemuri K, Makriyannis A, Le Foll B. The CB1 neutral antagonist AM4113 retains the therapeutic efficacy of the inverse agonist rimonabant for nicotine dependence and weight loss with better psychiatric tolerability. Int J Neuropsychopharmacol 2016;19:1–11.

[186] Li C, Staub DR, Kirby LG. Role of GABAA receptors in dorsal raphe nucleus in stress-induced reinstatement of morphine-conditioned place preference in rats. Psychopharmacology 2013;230:537–45.

[187] Zanos P, Georgiou P, Wright SR, Hourani SM, Kitchen I, Winsky-Sommerer R, Bailey A. The oxytocin analogue carbetocin prevents emotional impairment and stress-induced reinstatement of opioid-seeking in morphine-abstinent mice. Neuropsychopharmacology 2014;39:855–65.

[188] Georgiou P, Zanos P, Garcia-Carmona JA, Hourani S, Kitchen I, Kieffer BL, Laorden ML, Bailey A. The oxytocin analogue carbetocin prevents priming-induced reinstatement of morphine-seeking: involvement of dopaminergic, noradrenergic and MOPr systems. Eur Neuropsychopharmacol 2015;25:2459–64.

[189] Bhutada P, Mundhada Y, Ghodki Y, Dixit P, Umathe S, Jain K. Acquisition, expression, and reinstatement of ethanol-induced conditioned place preference in mice: effects of exposure to stress and modulation by mecamylamine. J Psychopharmacol 2012;26:315–23.

[190] Lê AD, Poulos CX, Harding S, Watchus J, Juzytsch W, Shaham Y. Effects of naltrexone and fluoxetine on alcohol self-administration and reinstatement of alcohol seeking induced by priming injections of alcohol and exposure to stress. Neuropsychopharmacology 1999;21:435–44.

[191] Heilig M, Egli M, Crabbe JC, Becker HC. Acute withdrawal, protracted abstinence and negative affect in alcoholism: are they linked? Addict Biol 2010;15:169–84.

[192] Augier E, Dulman RS, Damadzic R, Pilling A, Hamilton JP, Heilig M. The GABAB positive modulator ADX71441 attenuates alcohol self-administration and relapse to alcohol seeking in rats. Neuropsychopharmacology 2017;42:1789–99.

[193] Sobieraj JC, Kim A, Fannon MJ, Mandyam CD. Chronic wheel running-induced reduction of extinction and reinstatement of methamphetamine seeking in methamphetamine dependent rats is associated with reduced number of periaqueductal gray dopamine neurons. Brain Struct Funct 2016;221:261–76.

[194] Substance Abuse and Mental Health Services Administration. Behavioral health treatments and services, Retrieved from, http://www.samsha.gov/treatment; 2015.

[195] Kosten TA, Gawin FH, Kosten TR, Rounsaville BJ. Gender differences in cocaine use and treatment response. J Subst Abuse Treat 1993;10:63–6.

[196] Brady KT, Randall CL. Gender differences in substance use disorders. Psychiatr Clin North Am 1999;22:241–52.

[197] Thomas MB, Becker JB. Sex differences in prenatal stress effects on cocaine pursuit in rats. Physiol Behav 2017; https://doi.org/10.1016/j.physbeh.2017.10.019; pii: S0031-9384(17)30358-X.

[198] Hyman SM, Paliwal P, Chaplin TM, Mazure CM, Rounsaville BJ, Sinha R. Severity of childhood trauma is predictive of cocaine relapse outcomes in women but not men. Drug Alcohol Depend 2008;92:208–16.

[199] Bertholomey ML, Nagarajan V, Torregrossa MM. Sex differences in reinstatement of alcohol seeking in response to cues and yohimbine in rats with and without a history of adolescent corticosterone exposure. Psychopharmacology 2016;233:2277–87.

[200] Lynch WJ, Carroll ME. Sex differences in the acquisition of intravenously self-administered cocaine and heroin in rats. Psychopharmacology 1999;144:77–82.

[201] Li CS, Kosten TR, Sinha R. Sex differences in brain activation during stress imagery in abstinent cocaine users: a functional magnetic resonance imaging study. Biol Psychiatry 2005;57:487–94.

[202] Holly EN, Shimamoto A, Debold JF, Miczek KA. Sex differences in behavioral and neural cross-sensitization and escalated cocaine taking as a result of episodic social defeat stress in rats. Psychopharmacology 2012;224:179–88.

[203] Fox H, Sinha R. The role of guanfacine as a therapeutic agent to address stress-related pathophysiology in cocaine-dependent individuals. Adv Pharmacol 2014;69:217–65.

[204] McRae-Clark AL, Cason AM, Kohtz AS, Moran Santa-Maria M, Aston-Jones G, Brady KT. Impact of gender on corticotropin-releasing factor and noradrenergic sensitivity in cocaine use disorder. J Neurosci Res 2017;95:320–7.

[205] Doncheck EM, Urbanik LA, DeBaker MC, Barron LM, Liddiard GT, Tuscher JJ, Frick KM, Hillard CJ, Mantsch JR. 17β-estradiol potentiates the reinstatement of cocaine seeking in female rats: role of the prelimbic prefrontal cortex and cannabinoid type-1 receptors. Neuropsychopharmacology 2018;43:781–90. https://doi.org/10.1038/npp.2017.170. epub ahead of print.

[206] Regier PS, Claxton AB, Zlebnik NE, Carroll ME. Cocaine-, caffeine-, and stress-evoked cocaine reinstatement in high vs. low impulsive rats: treatment with allopregnanolone. Drug Alcohol Depend 2014;143:58–64.

[207] Sinha R, Fox H, Hong KI, Sofuoglu M, Morgan PT, Bergquist KT. Sex steroid hormones, stress response, and drug craving in cocaine-dependent women: implications for relapse susceptibility. Exp Clin Psychopharmacol 2007;15:445–52.

[208] Anker JJ, Carroll ME. Females are more vulnerable to drug abuse than males: evidence from preclinical studies and the role of ovarian hormones. Curr Top Behav Neurosci 2011;8:73–96.

[209] Reimers L, Buchel C, Diekhof EK. How to be patient. The ability to wait for a reward depends on menstrual cycle phase and feedback-related activity. Front Neurosci 2014;8:401.

[210] Evans SM, Foltin RW. Exogenous progesterone attenuates the subjective effects of smoked cocaine in women, but not in men. Neuropsychopharmacology 2006;31:659–74.

[211] Terner JM, de Wit H (2006). Menstrual cycle phase and responses to drugs of abuse in humans. Drug Alcohol Depend, 84: 1–13.

[212] Bertholomey ML, Torregrossa MM. Gonadal hormones affect alcohol drinking, but not cue+yohimbine-induced alcohol seeking, in male and female rats, Physiol Behav 2017; pii: S0031-9384(17)30376-1, https://doi.org/10.1016/j.physbeh.2017.10.025. epub ahead of print.

[213] Cummings JA, Jagannathan L, Jackson LR, Becker JB. Sex differences in the effects of estradiol in the nucleus accumbens and striatum on the response to cocaine: neurochemistry and behavior. Drug Alcohol Depend 2014;135:22–8.

[214] Calipari ES, Juarez B, Morel C, Walker DM, Cahill ME, Ribeiro E, Roman-Ortiz C, Ramakrishnan C, Deisseroth K, Han MH, Nestler EJ. Dopaminergic dynamics underlying sex-specific cocaine reward. Nat Commun 2017;8:13877.

[215] Iwasaki-Sekino A, Mano-Otagiri A, Ohata H, Yamauchi N, Shibasaki T. Gender differences in corticotropin and corticosterone secretion and corticotropin-releasing factor mRNA expression in the paraventricular nucleus of the hypothalamus and the central nucleus of the amygdala in response to footshock stress or psychological stress in rats. Psychoneuroendocrinology 2009;34:226–37.

[216] Lo MJ, Chang LL, Wang PS. Effects of estradiol on corticosterone secretion in ovariectomized rats. J Cell Biochem 2000;77:560–8.

[217] Niyomchai T, Russo SJ, Festa ED, Akhavan A, Jenab S, Quiñones-Jenab V. Progesterone inhibits behavioral responses and estrogen increases corticosterone levels after acute cocaine administration. Pharmacol Biochem Behav 2005;80:603–10.

[218] Weiser MJ, Handa RJ. Estrogen impairs glucocorticoid dependent negative feedback on the hypothalamic-pituitary-adrenal axis via estrogen receptor alpha within the hypothalamus. Neuroscience 2009;159:883–95.

[219] Zhou L, Ghee SM, Chan C, Lin L, Cameron MD, Kenny PJ, See RE. Orexin-1 receptor mediation of cocaine seeking in male and female rats. J Pharmacol Exp Ther 2012;340:801–9.

[220] Curtis AL, Bethea T, Valentino RJ. Sexually dimorphic responses of the brain norepinephrine system to stress and corticotropin-releasing factor. Neuropsychopharmacology 2006;31:544–54.

[221] Bangasser DA, Reyes BA, Piel D, Garacch V, Zhang XY, Plona ZM, Van Bockstaele EJ, Beck SG, Valentino RJ. Increased vulnerability of the brain norepinephrine system of females to corticotropin-releasing factor overexpression. Mol Psychiatry 2013;18:166–73.

[222] Hu P, Liu J, Yasrebi A, Gotthardt JD, Bello NT, Pang ZP, Roepke TA. Gq protein-coupled membrane-initiated estrogen signaling rapidly excites corticotropin-releasing hormone neurons in the hypothalamic paraventricular nucleus in female mice. Endocrinology 2016;157:3604–20.

[223] Chartoff EH, Mavrikaki M. Sex differences in kappa opioid receptor function and their potential impact on addiction. Front Neurosci 2015;9:466.

[224] Shorter D, Lindsay JA, Kosten TR. The alpha-1 adrenergic antagonist doxazosin for treatment of cocaine dependence: a pilot study. Drug Alcohol Depend 2013;131:66–70.

[225] De La Garza 2nd R, Bubar MJ, Carbone CL, Moeller FG, Newton TF, Anastasio NC, Harper TA, Ware DL, Fuller MA, Holstein GJ, Jayroe JB, Bandak SI, Reiman KZ, Neale AC, Pickford LB, Cunningham KA. Evaluation of the dopamine β-hydroxylase (DβH) inhibitor nepicastat in particpants who meet criteria for cocaine use disorder. Prog Neuropsychopharmacol Biol Psychiatry 2015;59:40–8.

[226] Kwako LE, Spagnolo PA, Schwandt ML, Thorsell A, George DT, Momenan R, Dio DE, Huestis M, Anizan S, Concheiro M, Sinha R, Heilig M. The corticotropin releasing hormone-1 (CRH1) receptor antagonist pexacerfont in alcohol dependence: a randomized controlled experimental medicine study. Neuropsychopharmacology 2015;40:1053–63.

[227] Buda JJ, Carroll FI, Kosten TR, Swearingen D, Walters BB. A double-blind, placebo-controlled trial to evaluate the safety, tolerability, and pharmacokinetics of single, escalating oral doses of JDTic. Neuropsychopharmacology 2015;40:2059–65.

[228] Kwako LE, George DT, Schwandt ML, Spagnolo PA, Momenan R, Hommer DW, Diamond CA, Sinha R, Shaham Y, Heilig M. The neurokinin-1 receptor antagonist aprepitant in co-morbid alcohol dependence and posttraumatic stress disorder: a human experimental study. Psychopharmacology 2015;232:295–304.

[229] Vendruscolo LF, Estey D, Goodell V, Macshane LG, Logrip ML, Schlosburg JE, McGinn MA, Zamora-Martinez ER, Belanoff JK, Hunt HJ, Sanna PP, George O, Koob GF, Edwards S, Mason BJ. Glucocorticoid receptor antagonism decreases alcohol seeking in alcohol-dependent individuals. J Clin Invest 2015;125:193–7.

Chapter 12

Negative Reinforcement Mechanisms in Addiction

Olivier George

Department of Neuroscience, The Scripps Research Institute, La Jolla, CA, United States

PSYCHOPATHOLOGICAL FRAMEWORK

Stages of the Addiction Cycle

Substance use disorder or drug addiction is a chronically relapsing disorder that is characterized by compulsion to seek and take the drug, loss of control in limiting drug intake, and emergence of a negative emotional state, reflecting a motivational withdrawal syndrome, when access to the drug is prevented [1]. Drug addiction includes three stages: *preoccupation/anticipation*, *binge/intoxication*, and *withdrawal/negative affect*. These three stages feed into each other to produce an addiction cycle. Each stage becomes more intense after each cycle, leading to the pathological state of addiction. These three stages reflect incentive reward deficits/stress surfeit, salience/pathological habits, and executive function deficits, respectively, providing a powerful impetus for compulsive drug-seeking behavior that is associated with drug addiction. These domains of dysfunction correspond to neuroadaptations that reflect allostatic changes in three key neurocircuits that mediate compulsive drug seeking: basal ganglia, extended amygdala, and prefrontal cortex, respectively [2] (Fig. 1). A key aspect that has been often neglected is that the relative contribution of each of these three stages to drug use and addiction varies both between and within individuals across time and is dependent on the specific drug of abuse.

From Positive to Negative Reinforcement

A second level of complexity is the fact that substance use disorder includes a transition from impulsive to compulsive behaviors and from positive to negative reinforcement [5] (Fig. 2). Impulsivity is characterized by "a predisposition toward rapid, unplanned reactions to internal and external stimuli without regard for the negative consequences of these reactions to themselves or others" [7]. Impulsivity is believed to be produced by positive reinforcement mechanisms. Positive reinforcement is defined as the process by which the presentation of a stimulus increases the probability of a response and is believed to be mainly driven by intense arousal and excitement before the stimulus and euphoria, pleasure, and reward immediately following the behavior, although this state is often followed by guilt and regret. Compulsivity is characterized as "perseverative, repetitive actions that are excessive and inappropriate" [8]. Negative reinforcement is defined as the process by which the removal of an aversive stimulus (or aversive state in the case of addiction) increases the probability of a response [9] and is believed to be preceded by anxiety, stress, and/or pain before performing the action, followed by relief from the negative emotional state following the action. Impulsivity dominates at initial stages of drug addiction. Individuals seek and take the drug for its initial pleasurable and reinforcing effects, without regard for the potential future negative consequences of using drugs. Compulsivity dominates at later stages of drug addiction. Such compulsivity leads to the escalation of drug intake and perseverative drug use despite adverse consequences. The transition from positive to negative reinforcement reflects a change in the underlying psychological and neurobiological mechanisms of motivation. Motivation can be defined as a "tendency of the whole animal to produce organized activity" [10]. The neural substrates for the two sources of reinforcement that play a key role in allostatic neuroadaptations derived from two key motivational systems that are required for survival: the brain reward system and the brain stress system.

The concept of motivation can be linked to hedonic, affective, and emotional states in the context of temporal dynamics that are elaborated by Solomon's opponent-process theory of motivation [11]. Hedonic, affective, or emotional states, once initiated, are modulated by the central nervous system through mechanisms that reduce the intensity of hedonic feelings.

Neural Mechanisms of Addiction. https://doi.org/10.1016/B978-0-12-812202-0.00012-9
Copyright © 2019 Elsevier Inc. All rights reserved.

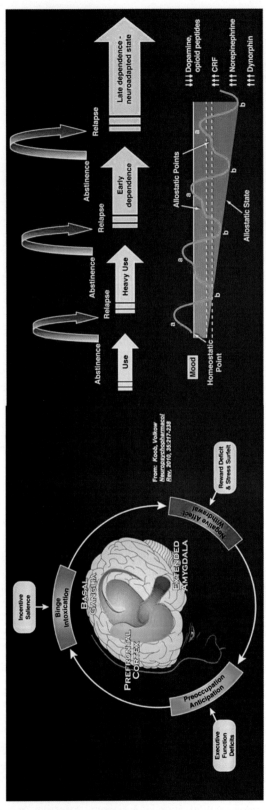

FIG. 1 (A) Three stages of the addiction cycle: *binge/intoxication, withdrawal/negative affect,* and *preoccupation/anticipation.* These three stages reflect incentive salience/pathological habits, reward deficits/stress surfeit, and executive function deficits, respectively, to provide a powerful impetus for compulsive drug-seeking behavior associated with drug addiction. These domains of dysfunction correspond to neuroadaptations that reflect allostatic changes in three key neurocircuits to mediate compulsive drug seeking: basal ganglia, extended amygdala, and prefrontal cortex, respectively. (B) The *a-process* represents a positive hedonic or positive mood state, and the *b-process* represents a negative hedonic or negative mood state. The affective stimulus (state) has been argued to be the sum of both the *a-process* and *b-process.* An individual who experiences a positive hedonic mood state from a drug of abuse with sufficient time between readministering the drug is hypothesized to retain the *a-process.* An appropriate counteradaptive opponent process (*b-process*) that balances the activational process (*a-process*) does not lead to an allostatic state. Changes in the affective stimulus (state) in an individual with repeated frequent drug use may represent a transition to an allostatic state in the brain reward systems and, by extrapolation, a transition to addiction. Notice that the apparent *b-process* never returns to the original homeostatic level before drug taking begins again, thus creating a progressively greater allostatic state in the brain reward system. The counteradaptive opponent-process (*b-process*) does not balance the activational process (*a-process*) but in fact shows residual hysteresis. Although these changes that are illustrated in the figure are exaggerated and condensed overtime, the hypothesis is that even during postdetoxification (a period of protracted abstinence), the reward system still bears allostatic changes. The following definitions apply: *allostasis,* the process of achieving stability through change; *allostatic state,* a state of chronic deviation of the regulatory system from its normal (homeostatic) operating level; *allostatic load,* the cost to the brain and body of the deviation, accumulating over time, and reflecting in many cases pathological states and the accumulation of damage. *[Top right from Heilig and Koob [3]. Bottom right from Koob and Le Moal [4]. Left from Koob and Volkow Neuropsychopharmacol Rev, 2010, 35:217–238].*

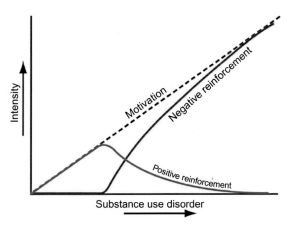

FIG. 2 Schematic of the progression of drug addiction over time, illustrating the shift in underlying motivational mechanisms. From initial, positive reinforcing, pleasurable drug effects, the addictive process progresses over time to being maintained by negative-reinforcing relief from a negative emotional state. *[From Koob [6]].*

This theory postulates that any motivational stimulus activates two opposing motivational processes. The *a-process* consists of positive or negative hedonic responses, has a fast onset and offset, correlates with the intensity, quality, and duration of the stimulus, and shows tolerance. The *b-process* appears after the *a-process* has terminated, is opposite in direction, is sluggish in onset, is slow to build up and decay, and gets larger with repeated exposure. The initial acute effect of a drug of abuse (i.e., the *a-process* or positive hedonic response) was hypothesized to be opposed or counteracted by the *b-process* as homeostatic changes in brain systems. With repeated exposure to drugs, the *b-process* sensitizes, appears earlier after the unconditioned stimulus, lasts longer, and masks the *a-process*, leading to apparent tolerance [12]. Two types of biological processes have been proposed to describe the mechanisms that underlie the neuroadaptations that are associated with generation of the opponent process in drug addiction: within-system neuroadaptations and between-system neuroadaptations [13]. In the within-system process, the drug elicits an opposing, neutralizing reaction within the same system in which the drug elicits its primary and unconditioned reinforcing actions. In the between-system process, neurobiological systems are recruited that are different from those that were initially activated by the drug.

NEUROBIOLOGICAL MECHANISMS OF NEGATIVE REINFORCEMENT

Within-System Neuroadaptations

Our understanding of the neurobiology of the *withdrawal/negative affect* stage has dramatically improved over the past decade (Fig. 3). This stage includes different sources of motivation to take drugs, including chronic irritability, emotional pain, malaise, dysphoria, alexithymia, states of stress, and the loss of motivation for natural rewards. For example, chronic administration of all major drugs of abuse leads to stress and anxiety-like responses during acute and protracted abstinence [14].

One explanation for the blunted function of the reward system during abstinence involves within-system neuroadaptations, in which the primary target of the drug rapidly adapts to neutralize the effect of the drug. Long-lasting within-system neuroadaptations can then lead to a decrease in brain reward function when the drug is removed [13]. For example, cocaine acutely produces dopamine and serotonin release, but decreases in dopaminergic and serotonergic transmission have been observed in the ventral striatum during cocaine withdrawal in rats [15]. Even more compelling are studies in humans that reported lower self-reported rewarding effects of drugs and a lower striatal dopamine response after amphetamine/methylphenidate challenges in active and detoxified abusers compared with controls [16–19]. Similar neuroadaptations are hypothesized to occur for other classes of drugs, including increases in μ-opioid receptor responsivity [20,21], decreases in γ-aminobutyric acid (GABA)ergic transmission in the ventral striatum, and increases in N-methyl-D-aspartate glutamatergic transmission in the ventral striatum [22,23] during opioid withdrawal. Complex regional changes in nicotinic acetylcholine receptor (nAChR) function in key brain regions, including the ventral tegmental area, ventral striatum, interpeduncular nucleus (IPN), amygdala, and habenula, have been reported for nicotine and alcohol, among other addictions [24,25].

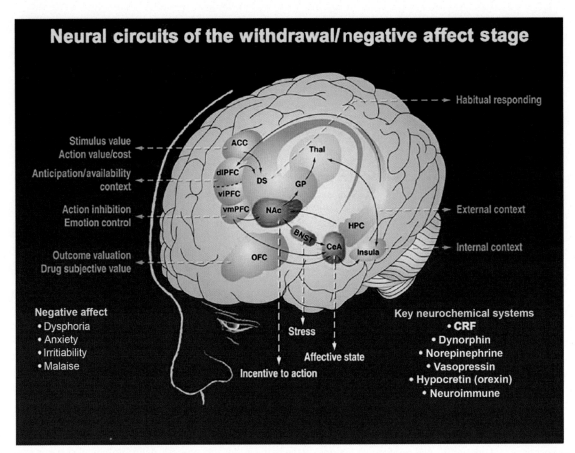

FIG. 3 Neural circuitry associated with the *withdrawal/negative affect* stage. The negative emotional state of withdrawal may engage activation of the extended amygdala. The extended amygdala is composed of several basal forebrain structures, including the bed nucleus of the stria terminalis (BNST), central nucleus of the amygdala (CeA), and possibly a transition area in the medial portion (or shell) of the nucleus accumbens. Major neurotransmitters in the extended amygdala that are hypothesized to play a role in negative reinforcement are corticotropin-releasing factor, norepinephrine, and dynorphin. The extended amygdala has major projections to the hypothalamus and brainstem. ACC, anterior cingulate cortex; dlPFC, dorsolateral prefrontal cortex; vlPFC, ventrolateral prefrontal cortex; vmPFC, ventromedial prefrontal cortex; OFC, orbitofrontal cortex; DS, dorsal striatum; NAc, nucleus accumbens; GP, globus pallidus; Thal, thalamus; BNST, bed nucleus of the stria terminalis; HPC, hippocampus; CeA, central nucleus of the amygdala. *[Modified with permission from George O, Koob GF. Proc Natl Acad Sci USA, 2013, 110:4165–4166].*

Between-System Neuroadaptations

Another explanation for the lower function of the reward system during abstinence involves between-system neuroadaptations, in which systems other than those that are involved in the positive rewarding effects of drugs are recruited or dysregulated by chronic drug use to oppose the rewarding effects of drugs of abuse [13]. A central component of this between-system neuroadaptation is activation of the stress pathways, including the hypothalamic-pituitary-adrenal (HPA) axis and extrahypothalamic brain stress systems that are mediated by corticotropin-releasing factor (CRF), norepinephrine, and dynorphin [26–28]. Withdrawal from drugs of abuse acutely increases adrenocorticotropic hormone, corticosterone, and extended amygdala CRF and dynorphin during withdrawal [29–36].

Two main brain circuits are likely to contribute to these opponent-like processes that lower brain reward function and increase brain stress system function. Both of these circuits are heavily influenced by CRF. One circuit involves the extended amygdala, which encompasses the central nucleus of the amygdala (CeA), bed nucleus of the stria terminalis (BNST), and part of the nucleus accumbens. The extended amygdala integrates brain arousal-stress and reward system information [37] to produce the between-system opponent process that is elaborated above. The CRF system in the extended amygdala is activated during acute withdrawal from cocaine, alcohol, opioids, Δ^9-tetrahydrocannabinol, and nicotine [33,38–40]. Similar effects have been observed with alcohol in the lateral BNST [34]. Cocaine withdrawal produces anxiety-like responses that can be reversed by a CRF receptor antagonist [41,42]. Similar results have been observed with nicotine [38,43,44], alcohol [32,45], and opioids [39,41,46]. Moreover, the ability of CRF receptor antagonists to block the anxiogenic- and aversive-like motivational effects of drug withdrawal predicts their efficacy in reducing the compulsive-like self-administration of cocaine [47], nicotine [38], and heroin [48] in rats.

The excessive release of dopamine and opioid peptides produces subsequent activation of the dynorphin system in the basal ganglia and extended amygdala, which has been hypothesized to feed back to decrease dopamine release and contribute to the dysphoric syndrome that is associated with cocaine dependence [49]. Dynorphins produce aversive dysphoric-like effects in animals and humans and have been hypothesized to mediate negative emotional states [50,51], depressive-like, aversive responses to stress, and dysphoric-like responses during withdrawal from drugs of abuse. Recent evidence suggests that the dynorphin-κ opioid receptor system in the extended amygdala also mediates compulsive-like responding for methamphetamine, heroin, and alcohol with extended access and dependence [50].

Another between-system neuroadaptation involves the habenula-to-IPN pathway. The lateral habenula plays a key role in encoding aversive states [52,53], in part by decreasing dopamine neuron firing in the ventral tegmental area after failure to receive an expected reward [52,53]. This hypothesis is consistent with the finding that nAChRs in the habenula-IPN pathway appear to modulate aversive responses to nicotine [54] and nicotine withdrawal [55,56]. We recently reported that the habenula-IPN pathway is also influenced by the CRF system [44,56]. Activation of this pathway during nicotine withdrawal was potentiated by CRF-producing neurons in the ventral tegmental area that project to the IPN. The down-regulation of CRF mRNA in the ventral tegmental area and CRF_1 receptor blockade in the IPN prevented the emergence of negative emotional states associated with withdrawal and reduced excessive nicotine intake after abstinence [44,56].

In addition to these subcortical circuits that involve the brain reward and stress systems, the insular cortex is an important cortical region for emotional aspects of the *withdrawal/negative affect* stage. Cravings for food, cocaine, and nicotine have been shown to activate the insular cortex [57–59], and tobacco smokers with damage to the insular cortex were able to stop smoking easily with little, if any, withdrawal symptoms, craving, or relapse [60]. The insula is hypothesized to integrate autonomic, visceral, and emotional information [61] during withdrawal and abstinence to produce the motivation to obtain the drug within a negative reinforcement framework (i.e., obtain relief from negative emotional states associated with withdrawal). Supporting this hypothesis, imaging studies have reported differential activation of the insula during craving, possibly reflecting interoceptive cues, and these effects were hypothesized to involve the CRF activation [62,63]. Moreover, reactivity of the insular cortex has been suggested to serve as a biomarker to help predict relapse [64].

INDIVIDUAL DIFFERENCES

Negative Emotional States and the Extended Amygdala

Converging evidence suggests that there are major individual differences in the response of the extended amygdala to emotional stimuli [65–70]. Individual differences in a personality trait that is linked to the drive to gain reward (increased reward drive) are positively correlated with activation of the amygdala and negatively correlated with activity of the ventromedial prefrontal cortex [65]. A heightened predisposition to aggression is also associated with an increased amygdala response to aggressive facial expressions [71,72]. The CeA (dorsal amygdala in humans) was also recruited during the conscious processing of fearful faces in healthy volunteers, and individual differences in trait anxiety predicted the response of a key input to the CeA, the BLA, to unconsciously process fearful faces [73]. The amygdala is also activated during affective judgment and during the expression of emotional responses when viewing affective pictures [74,75]. Moreover, the amygdala is activated during drug craving [76–80]. Smaller amygdala volumes [81] and abnormalities in the fiber tracts that connect the orbitofrontal cortex and anterior cingulate cortex to the amygdala have been observed in some cocaine-dependent individuals [82]. Interestingly, the changes in amygdala volume in cocaine-dependent individuals were observed even in subjects who were recently exposed to cocaine (<1–2 years), suggesting that a decrease in amygdala volume represents either early drug-induced impairment or an individual developmental predisposition [82].

Individual differences have also been observed in rodents. Differences in anxiety-like behavior have been related to differential levels of vasopressin and androgen receptors [83], glucocorticoid receptors, and cholecystokinin-B receptors [84] in the extended amygdala. Individuals with increased anxiety-like responding exhibited a downregulation of amygdala cholecystokinin-B receptor binding, possibly reflecting compensation for increased cholecystokinin activity [84]. Individual differences in the sensitivity of the CeA to inactivation by a $GABA_A$ receptor agonist have also been observed in rats that self-administer amphetamine. Inactivation of the CeA only decreased amphetamine self-administration in rats with excessive drug intake, suggesting that individual differences in the recruitment of the CeA might predispose an individual to excessive drug intake [85].

Pain and the Opioid Spinomesothalamocortical System

Individual differences have been well described in studies of pain. Pain thresholds greatly vary by gender, age, and the subject's behavioral state [86–88]. For example, under states of high emotion or intense concentration, people often report few signs of pain despite severe injuries [89]. Imaging studies have shown that activity of the midbrain, medial thalamus, and cortical nociceptive-receiving areas (e.g., insula and anterior cingulate cortex) is correlated with pain intensity [90]. Individuals with low pain thresholds exhibit higher activation of the primary somatosensory cortex, anterior cingulate cortex, and prefrontal cortex than individuals with high pain thresholds [91]. Interestingly, individual differences did not correlate with differential activation of the thalamus despite its key involvement in the encoding of nociceptive information [91]. Moreover, the impact of perceived controllability on pain perception varies highly between individuals and is correlated with differential activation of the anterior cingulate cortex, insula, and ventrolateral prefrontal cortex [92]. Activity of the anterior cingulate cortex, dorsolateral prefrontal cortex, insular cortex, nucleus accumbens, thalamus, and periaqueductal gray correlated with the magnitude of placebo analgesia in humans [93,94] and was mediated in part by activation of the endogenous opioid system, specifically the activation of μ-opioid receptors [94]. Abnormal pain perceptions have been reported in opiate addiction during the development, maintenance, and withdrawal periods [95–99], and preexisting pain problems are a key factor in the transition to opiate abuse and addiction [100]. Altogether, these results suggest that dysregulation of the "affective" pain system, particularly the insula, anterior cingulate cortex, and CeA, because of the crosstalk between pain pathways and negative emotional state pathways, might confer vulnerability to drug addiction in some individuals.

Altogether, these results demonstrate that individual differences in activity of the extended amygdala and spinomesothalamocortical system may predispose individuals to heightened anxiety, stress, pain, and general negative emotional states and represent a key factor in the transition from positive to negative reinforcement in the transition to addiction.

IMPLICATIONS FOR PERSONALIZED MEDICINE

There are considerable individual differences in the patterns of drug use and the psychological mechanisms that drive drug use. Drug use may be driven by the *binge/intoxication* stage for some individuals and by the *withdrawal/negative affect* stage for others. With psychostimulants and even alcohol, binge-like patterns can predominate in some individuals. Such individuals may escalate their drug intake in a binge-like pattern for various reasons, including peer pressure, sensation seeking, externalizing disorders, and drug-induced cognitive impairment (e.g., decision making, monitoring, renegade attention, and transcendence failure) with little, if any, initial negative emotional symptoms. Other individuals may quickly develop a pattern of chronic and heavy use that is caused by either conscious or unconscious attempts to self-medicate existing negative emotional states. Such individuals often have preexisting conditions that generate powerful negative emotional states, such as posttraumatic stress disorder, sexual abuse, major depressive disorder, or anxiety disorder, and will use drugs to obtain relief from these negative emotional states. However, chronic high-dose binge-like patterns of drug intake can cause the development of negative emotional states and ultimately drive self-medication of a state that is created by the drug itself. Ultimately, both the *binge/intoxication* stage and *withdrawal/negative affect* stage will contribute to a pathological state of compulsive drug seeking and taking. One intriguing area of research is the identification of genetic, biological, and psychological subpopulations of humans with substance use disorder within the framework of the three stages of the addiction cycle to better understand the drug addiction process and potentially predict treatment efficacy.

There are individual differences in executive function, prefrontal cortex function, and dopamine signaling, and the genetics of negative emotional states may help identify subgroups of patients with substance use disorder that may help predict treatment outcome. Executive function deficits in addiction and gray matter loss have been linked to deficits in the ability of behavioral treatments to effect recovery [101]. In addition, methamphetamine-dependent subjects with high striatal dopamine D_2 receptor levels had better treatment outcomes than individuals with low striatal D_2 receptor levels [102].

Attempts are being made to identify genetic markers, including single-nucleotide polymorphisms (SNPs), that may predict the vulnerability to substance use disorders and responsiveness to treatment. Several SNPs that are associated with the CRF system have been associated with excessive alcohol use. An association was found between SNPs that are related to the CRF_1 receptor gene (*Crhr1*) and binge drinking in adolescents and alcohol-dependent adults [103–105]. Another important genetic association has been found between alcohol dependence and SNPs that are related to the gene that encodes neuropeptide Y (NPY). NPY is an anxiolytic peptide that is involved in emotional regulation and stress coping and is known to antagonize the effects of CRF on addiction-like behaviors. Studies have linked SNPs of the NPY Y_2 receptor gene (*NPY2R*) and alcohol dependence, alcohol withdrawal symptoms, comorbid alcohol and cocaine dependence, and cocaine dependence [106]. The G1258A polymorphism of the NPY gene has been linked to alcohol dependence [107]. The rs16147 SNP of the NPY promoter gene was linked to tobacco addiction [108]. Should a medication become available that modulates CRF or NPY, such genetic analysis may reveal that subpopulations of subjects who carry specific SNPs might be more responsive than others.

CONCLUSIONS

Drug addiction is a chronically relapsing disorder that is associated with compulsive drug seeking and taking that progress through the *binge/intoxication*, *withdrawal/negative affect*, and *preoccupation/anticipation* stages. Upregulation of the CRF, norepinephrine, and dynorphin systems in the extended amygdala and habenular-IPN pathways appears to be critical for the development of negative reinforcement and the transition to compulsive drug use. Characterization of the neurobiological mechanisms of negative reinforcement in drug addiction is a key to understanding the causes of addiction and development of personalized medicine.

REFERENCES

[1] Koob GF, Le Moal M. Drug abuse: hedonic homeostatic dysregulation. Science 1997;278:52–8.

[2] Koob GF, Volkow ND. Neurobiology of addiction: a neurocircuitry analysis. Lancet Psychiatry 2016;3:760–73.

[3] Heilig M, Koob GF. A key role for corticotropin-releasing factor in alcohol dependence. Trends Neurosci 2007;30:399–406.

[4] Koob GF, Le Moal M. Drug addiction, dysregulation of reward, and allostasis. Neuropsychopharmacology 2001;24:97–129.

[5] Koob GF, Le Moal M. Addiction and the brain antireward system. Annu Rev Psychol 2008;59:29–53.

[6] Koob GF. Theoretical frameworks and mechanistic aspects of alcohol addiction: alcohol addiction: alcohol addiction as a reward deficit disorder. In: Sommer WH, Spanagel R, editors. Behavioral neurobiology of alcohol addiction. Current topics in behavioral neuroscience, vol. 13. Berlin: Springer-Verlag; 2013. p. 3–30.

[7] Moeller FG, Barratt ES, Dougherty DM, Schmitz JM, Swann AC. Psychiatric aspects of impulsivity. Am J Psychiatry 2001;158:1783–93.

[8] Berlin GS, Hollander E. Compulsivity, impulsivity, and the DSM-5 process. CNS Spectr 2014;19:62–8.

[9] Skinner BF. The behavior of organisms: an experimental analysis. New York: Apple-Century-Crofts; 1938.

[10] Hebb DO. Textbook of psychology. 3rd ed. Philadelphia: W.B. Saunders; 1972.

[11] Solomon RL, Corbit JD. An opponent-process theory of motivation: 1. Temporal dynamics of affect. Psychol Rev 1974;81:119–45.

[12] Laulin JP, Celerier E, Larcher A, Le Moal M, Simonnet G. Opiate tolerance to daily heroin administration: an apparent phenomenon associated with enhanced pain sensitivity. Neuroscience 1999;89:631–6.

[13] Koob GF, Bloom FE. Cellular and molecular mechanisms of drug dependence. Science 1988;242:715–23.

[14] Koob GF, Le Moal M. Plasticity of reward neurocircuitry and the "dark side" of drug addiction. Nat Neurosci 2005;8:1442–4.

[15] Weiss F, Markou A, Lorang MT, Koob GF. Basal extracellular dopamine levels in the nucleus accumbens are decreased during cocaine withdrawal after unlimited-access self-administration. Brain Res 1992;593:314–8.

[16] Martinez D, Narendran R, Foltin RW, Slifstein M, Hwang DR, Broft A, Huang Y, Cooper TB, Fischman MW, Kleber HD, Laruelle M. Amphetamine-induced dopamine release: markedly blunted in cocaine dependence and predictive of the choice to self-administer cocaine. Am J Psychiatry 2007;164:622–9.

[17] Volkow ND, Tomasi D, Wang GJ, Logan J, Alexoff DL, Jayne M, Fowler JS, Wong C, Yin P, Du C. Stimulant-induced dopamine increases are markedly blunted in active cocaine abusers. Mol Psychiatry 2014;19:1037–43.

[18] Volkow ND, Wang GJ, Fowler JS, Logan J, Gatley SJ, Hitzemann R, Chen AD, Dewey SL, Pappas N. Decreased striatal dopaminergic responsiveness in detoxified cocaine-dependent subjects. Nature 1997;386:830–3.

[19] Volkow ND, Wang GJ, Telang F, Fowler JS, Logan J, Jayne M, Ma Y, Pradhan K, Wong C. Profound decreases in dopamine release in striatum in detoxified alcoholics: possible orbitofrontal involvement. J Neurosci 2007;27:12700–6.

[20] Minkowski CP, Epstein D, Frost JJ, Gorelick DA. Differential response to IV carfentanil in chronic cocaine users and healthy controls. Addict Biol 2012;17:149–55.

[21] Stinus L, Le Moal M, Koob GF. Nucleus accumbens and amygdala are possible substrates for the aversive stimulus effects of opiate withdrawal. Neuroscience 1990;37:767–73.

[22] Dahchour A, de Witte P, Bolo N, Nedelec JF, Muzet M, Durbin P, Macher JP. Central effects of acamprosate: part 1. Acamprosate blocks the glutamate increase in the nucleus accumbens microdialysate in ethanol withdrawn rats. Psychiatry Res 1998;82:107–14.

[23] Davidson M, Shanley B, Wilce P. Increased NMDA-induced excitability during ethanol withdrawal: a behavioural and histological study. Brain Res 1995;674:91–6.

[24] Dani JA, Heinemann S. Molecular and cellular aspects of nicotine abuse. Neuron 1996;16:905–8.

[25] Tolu S, Eddine R, Marti F, David V, Graupner M, Pons S, Baudonnat M, Husson M, Besson M, Reperant C, Zemdegs J, Pagès C, Hay YA, Lambolez B, Caboche J, Gutkin B, Gardier AM, Changeux JP, Faure P, Maskos U. Co-activation of VTA DA and GABA neurons mediates nicotine reinforcement. Mol Psychiatry 2013;18:382–93.

[26] Schlosburg JE, Whitfield Jr. TW, Park PE, Crawford EF, George O, Vendruscolo LF, Koob GF. Long-term antagonism of κ opioid receptors prevents escalation of and increased motivation for heroin intake. J Neurosci 2013;33:19384–92.

[27] Walker BM, Koob GF. Pharmacological evidence for a motivational role of κ-opioid systems in ethanol dependence. Neuropsychopharmacology 2008;33:643–52.

[28] Whitfield TW Jr, Schlosburg J, Wee S, Gould A, George O, Grant Y, Zamora-Martinez ER, Edwards S, Crawford E, Vendruscolo LF, Koob GF κ Opioid receptors in the nucleus accumbens shell mediate escalation of methamphetamine intake. J Neurosci 2015;35:4296–4305.

[29] Delfs JM, Zhu Y, Druhan JP, Aston-Jones G. Noradrenaline in the ventral forebrain is critical for opiate withdrawal-induced aversion. Nature 2000;403:430–4.

[30] Koob GF. Brain stress systems in the amygdala and addiction. Brain Res 2009;1293:61–75.

[31] Koob GF, Buck CL, Cohen A, Edwards S, Park PE, Schlosburg JE, Schmeichel B, Vendruscolo LF, Wade CL, Whitfield Jr. TW, George O. Addiction as a stress surfeit disorder. Neuropharmacology 2014;76:370–82.

[32] Koob GF, Heinrichs SC, Menzaghi F, Pich EM, Britton KT. Corticotropin releasing factor, stress and behavior. Semin Neurosci 1994;6:221–9.

[33] Merlo-Pich E, Lorang M, Yeganeh M, Rodriguez de Fonseca F, Raber J, Koob GF, Weiss F. Increase of extracellular corticotropin-releasing factor-like immunoreactivity levels in the amygdala of awake rats during restraint stress and ethanol withdrawal as measured by microdialysis. J Neurosci 1995;15:5439–47.

[34] Olive MF, Koenig HN, Nannini MA, Hodge CW. Elevated extracellular CRF levels in the bed nucleus of the stria terminalis during ethanol withdrawal and reduction by subsequent ethanol intake. Pharmacol Biochem Behav 2002;72:213–20.

[35] Rasmussen DD, Boldt BM, Bryant CA, Mitton DR, Larsen SA, Wilkinson CW. Chronic daily ethanol and withdrawal: 1. Long-term changes in the hypothalamo-pituitary-adrenal axis. Alcohol Clin Exp Res 2000;24:1836–49.

[36] Rivier C, Bruhn T, Vale W. Effect of ethanol on the hypothalamic-pituitary-adrenal axis in the rat: role of corticotropin-releasing factor (CRF). J Pharmacol Exp Ther 1984;229:127–31.

[37] Heimer L, Alheid G. Piecing together the puzzle of basal forebrain anatomy. In: Napier TC, Kalivas PW, Hanin I, editors. The basal forebrain: anatomy to function. Advances in experimental medicine and biology, vol. 295. New York: Plenum Press; 1991. p. 1–42.

[38] George O, Ghozland S, Azar MR, Cottone P, Zorrilla EP, Parsons LH, O'Dell LE, Richardson HN, Koob GF. CRF-CRF$_1$ system activation mediates withdrawal-induced increases in nicotine self-administration in nicotine-dependent rats. Proc Natl Acad Sci U S A 2007;104:17198–203.

[39] Heinrichs SC, Menzaghi F, Schulteis G, Koob GF, Stinus L. Suppression of corticotropin-releasing factor in the amygdala attenuates the aversive consequences of morphine withdrawal. Behav Pharmacol 1995;6:.

[40] Richter RM, Weiss F. In vivo CRF release in rat amygdala is increased during cocaine withdrawal in self-administering rats. Synapse 1999;32:254–61.

[41] Basso AM, Spina M, Rivier J, Vale W, Koob GF. Corticotropin-releasing factor antagonist attenuates the "anxiogenic-like" effect in the defensive burying paradigm but not in the elevated plus-maze following chronic cocaine in rats. Psychopharmacology 1999;145:21–30.

[42] Sarnyai Z, Biro E, Gardi J, Vecsernyes M, Julesz J, Telegdy G. Brain corticotropin-releasing factor mediates "anxiety-like" behavior induced by cocaine withdrawal in rats. Brain Res 1995;675:89–97.

[43] Cohen A, Treweek J, Edwards S, Leão RM, Schulteis G, Koob GF, George O. Extended access to nicotine leads to a CRF$_1$ receptor dependent increase in anxiety-like behavior and hyperalgesia in rats. Addict Biol 2015;20:56–68.

[44] Grieder TE, Herman MA, Contet C, Tan LA, Vargas-Perez H, Cohen A, Chwalek M, Maal-Bared G, Freiling J, Schlosburg JE, Clarke L, Crawford E, Koebel P, Repunte-Cononigo V, Sanna PP, Tapper AR, Roberto M, Kieffer BL, Sawchenko PE, Koob GF, van der Kooy D, George O. VTA CRF neurons mediate the aversive effects of nicotine withdrawal and promote intake escalation. Nat Neurosci 2014;17:1751–8.

[45] Rassnick S, Heinrichs SC, Britton KT, Koob GF. Microinjection of a corticotropin-releasing factor antagonist into the central nucleus of the amygdala reverses anxiogenic-like effects of ethanol withdrawal. Brain Res 1993;605:25–32.

[46] Schulteis G, Markou A, Gold LH, Stinus L, Koob GF. Relative sensitivity to naloxone of multiple indices of opiate withdrawal: a quantitative dose-response analysis. J Pharmacol Exp Ther 1994;271:1391–8.

[47] Goeders NE, Guerin GF. Effects of the CRH receptor antagonist CP-154,526 on intravenous cocaine self-administration in rats. Neuropsychopharmacology 2000;23:577–86.

[48] Greenwell TN, Funk CK, Cottone P, Richardson HN, Chen SA, Rice K, Zorrilla EP, Koob GF. Corticotropin-releasing factor-1 receptor antagonists decrease heroin self-administration in long-, but not short-access rats. Addict Biol 2009;14:130–43.

[49] Carlezon Jr. WA, Nestler EJ, Neve RL. Herpes simplex virus-mediated gene transfer as a tool for neuropsychiatric research. Crit Rev Neurobiol 2000;14:47–67.

[50] Chavkin C, Koob GF. Dynorphin, dysphoria and dependence: the stress of addiction. Neuropsychopharmacology 2016;41:373–4.

[51] Koob GF. The dark side of emotion: the addiction perspective. Eur J Pharmacol 2015;753:73–87.

[52] Hikosaka O. The habenula: from stress evasion to value-based decision-making. Nat Rev Neurosci 2010;11:503–13.

[53] Matsumoto M, Hikosaka O. Lateral habenula as a source of negative reward signals in dopamine neurons. Nature 2007;447:1111–5.

[54] Fowler CD, Lu Q, Johnson PM, Marks MJ, Kenny PJ. Habenular α5 nicotinic receptor subunit signalling controls nicotine intake. Nature 2011;471:597–601.

[55] Salas R, Sturm R, Boulter J, De Biasi M. Nicotinic receptors in the habenulo-interpeduncular system are necessary for nicotine withdrawal in mice. J Neurosci 2009;29:3014–8.

[56] Zhao-Shea R, DeGroot SR, Liu L, Vallaster M, Pang X, Su Q, Gao G, Rando OJ, Martin GE, George O, Gardner PD, Tapper AR. Increased CRF signalling in a ventral tegmental area-interpeduncular nucleus-medial habenula circuit induces anxiety during nicotine withdrawal. Nat Commun 2015;6:6770.

[57] Bonson KR, Grant SJ, Contoreggi CS, Links JM, Metcalfe J, Weyl HL, Kurian V, Ernst M, London ED. Neural systems and cue-induced cocaine craving. Neuropsychopharmacology 2002;26:376–86.

[58] Pelchat ML, Johnson A, Chan R, Valdez J, Ragland JD. Images of desire: food-craving activation during fMRI. Neuroimage 2004;23:1486–93.

[59] Wang Z, Faith M, Patterson F, Tang K, Kerrin K, Wileyto EP, Detre JA, Lerman C. Neural substrates of abstinence-induced cigarette cravings in chronic smokers. J Neurosci 2007;27:14035–40.

[60] Naqvi NH, Rudrauf D, Damasio H, Bechara A. Damage to the insula disrupts addiction to cigarette smoking. Science 2007;315:531–4.

[61] Naqvi NH, Bechara A. The hidden island of addiction: the insula. Trends Neurosci 2009;32:56–67.

[62] Goudriaan AE, De Ruiter MB, van den Brink W, Oosterlaan J, Veltman DJ. Brain activation patterns associated with cue reactivity and craving in abstinent problem gamblers, heavy smokers and healthy controls: an fMRI study. Addict Biol 2010;15:491–503.

[63] Sanchez MM, Young LJ, Plotsky PM, Insel TR. Autoradiographic and in situ hybridization localization of corticotropin-releasing factor 1 and 2 receptors in nonhuman primate brain. J Comp Neurol 1999;408:365–77.

[64] Janes AC, Pizzagalli DA, Richardt S, deB Frederick B, Chuzi S, Pachas G, Culhane MA, Holmes AJ, Fava M, Evins AE, Kaufman MJ. Brain reactivity to smoking cues prior to smoking cessation predicts ability to maintain tobacco abstinence. Biol Psychiatry 2010;67:722–9.

[65] Beaver JD, Lawrence AD, Passamonti L, Calder AJ. Appetitive motivation predicts the neural response to facial signals of aggression. J Neurosci 2008;28:2719–25.

[66] Bishop SJ, Duncan J, Lawrence AD. State anxiety modulation of the amygdala response to unattended threat-related stimuli. J Neurosci 2004;24:10364–8.

[67] Canli T. Functional brain mapping of extraversion and neuroticism: learning from individual differences in emotion processing. J Pers 2004;72:1105–32.

[68] Canli T, Gabrieli JD. Imaging gender differences in sexual arousal. Nat Neurosci 2004;7:325–6.

[69] Hamann S, Canli T. Individual differences in emotion processing. Curr Opin Neurobiol 2004;14:233–8.

[70] Mather M, Canli T, English T, Whitfield S, Wais P, Ochsner K, Gabrieli JD, Carstensen LL. Amygdala responses to emotionally valenced stimuli in older and younger adults. Psychol Sci 2004;15:259–63.

[71] Coccaro EF, McCloskey MS, Fitzgerald DA, Phan KL. Amygdala and orbitofrontal reactivity to social threat in individuals with impulsive aggression. Biol Psychiatry 2007;62:168–78.

[72] Passamonti L, Rowe JB, Ewbank M, Hampshire A, Keane J, Calder AJ. Connectivity from the ventral anterior cingulate to the amygdala is modulated by appetitive motivation in response to facial signals of aggression. Neuroimage 2008;43:562–70.

[73] Etkin A, Klemenhagen KC, Dudman JT, Rogan MT, Hen R, Kandel ER, Hirsch J. Individual differences in trait anxiety predict the response of the basolateral amygdala to unconsciously processed fearful faces. Neuron 2004;44:1043–55.

[74] Phan KL, Taylor SF, Welsh RC, Decker LR, Noll DC, Nichols TE, Britton JC, Liberzon I. Activation of the medial prefrontal cortex and extended amygdala by individual ratings of emotional arousal: a fMRI study. Biol Psychiatry 2003;53:211–5.

[75] Taylor SF, Phan KL, Decker LR, Liberzon I. Subjective rating of emotionally salient stimuli modulates neural activity. Neuroimage 2003;18:650–9.

[76] Childress AR, Mozley PD, McElgin W, Fitzgerald J, Reivich M, O'Brien CP. Limbic activation during cue-induced cocaine craving. Am J Psychiatry 1999;156:11–8.

[77] Kilts CD, Schweitzer JB, Quinn CK, Gross RE, Faber TL, Muhammad F, Ely TD, Hoffman JM, Drexler KP. Neural activity related to drug craving in cocaine addiction. Arch Gen Psychiatry 2001;58:334–41.

[78] Volkow ND, Wang GJ, Fowler JS, Hitzemann R, Angrist B, Gatley SJ, Logan J, Ding YS, Pappas N. Association of methylphenidate-induced craving with changes in right striato-orbitofrontal metabolism in cocaine abusers: implications in addiction. Am J Psychiatry 1999;156:19–26.

[79] Wang GJ, Volkow ND, Fowler JS, Cervany P, Hitzemann RJ, Pappas NR, Wong CT, Felder C. Regional brain metabolic activation during craving elicited by recall of previous drug experiences. Life Sci 1999;64:775–84.

[80] Wexler BE, Gottschalk CH, Fulbright RK, Prohovnik I, Lacadie CM, Rounsaville BJ, Gore JC. Functional magnetic resonance imaging of cocaine craving. Am J Psychiatry 2001;158:86–95.

[81] Makris N, Gasic GP, Seidman LJ, Goldstein JM, Gastfriend DR, Elman I, Albaugh MD, Hodge SM, Ziegler DA, Sheahan FS, et al. Decreased absolute amygdala volume in cocaine addicts. Neuron 2004;44:729–40.

[82] Lim KO, Choi SJ, Pomara N, Wolkin A, Rotrosen JP. Reduced frontal white matter integrity in cocaine dependence: a controlled diffusion tensor imaging study. Biol Psychiatry 2002;51:890–5.

[83] Linfoot I, Gray M, Bingham B, Williamson M, Pinel JP, Viau V. Naturally occurring variations in defensive burying behavior are associated with differences in vasopressin, oxytocin, and androgen receptors in the male rat. Prog Neuropsychopharmacol Biol Psychiatry 2009;33:1129–40.

[84] Wunderlich GR, Raymond R, DeSousa NJ, Nobrega JN, Vaccarino FJ. Decreased CCK(B) receptor binding in rat amygdala in animals demonstrating greater anxiety-like behavior. Psychopharmacology (Berl) 2002;164:193–9.

[85] Cain ME, Denehy ED, Bardo MT. Individual differences in amphetamine self-administration: the role of the central nucleus of the amygdala. Neuropsychopharmacology 2008;33:1149–61.

[86] Fillingim RB, King CD, Ribeiro-Dasilva MC, Rahim-Williams B, Riley 3rd JL. Sex, gender, and pain: a review of recent clinical and experimental findings. J Pain 2009;10:447–85.

[87] Gibson SJ, Helme RD. Age-related differences in pain perception and report. Clin Geriatr Med 2001;17:433–56 v–vi.

[88] Nielsen CS, Staud R, Price DD. Individual differences in pain sensitivity: measurement, causation, and consequences. J Pain 2009;10:231–7.

[89] Fields H. State-dependent opioid control of pain. Nat Rev Neurosci 2004;5:565–75.

[90] Casey KL. Concepts of pain mechanisms: the contribution of functional imaging of the human brain. Prog Brain Res 2000;129:277–87.

[91] Coghill RC, McHaffie JG, Yen YF. Neural correlates of interindividual differences in the subjective experience of pain. Proc Natl Acad Sci U S A 2003;100:8538–42.

[92] Salomons TV, Johnstone T, Backonja MM, Shackman AJ, Davidson RJ. Individual differences in the effects of perceived controllability on pain perception: critical role of the prefrontal cortex. J Cogn Neurosci 2007;19:993–1003.

[93] Wager TD, Rilling JK, Smith EE, Sokolik A, Casey KL, Davidson RJ, Kosslyn SM, Rose RM, Cohen JD. Placebo-induced changes in FMRI in the anticipation and experience of pain. Science 2004;303:1162–7.

[94] Zubieta JK, Bueller JA, Jackson LR, Scott DJ, Xu Y, Koeppe RA, Nichols TE, Stohler CS. Placebo effects mediated by endogenous opioid activity on mu-opioid receptors. J Neurosci 2005;25:7754–62.

[95] Compton MA. Cold-pressor pain tolerance in opiate and cocaine abusers: correlates of drug type and use status. J Pain Symptom Manage 1994;9:462–73.

[96] Compton P, Charuvastra VC, Kintaudi K, Ling W. Pain responses in methadone-maintained opioid abusers. J Pain Symptom Manage 2000;20:237–45.

[97] Compton P, Estepa CA. Addiction in patients with chronic pain. Lippincotts Prim Care Pract 2000;4:254–72.

[98] Doverty M, Somogyi AA, White JM, Bochner F, Beare CH, Menelaou A, Ling W. Methadone maintenance patients are cross-tolerant to the anti-nociceptive effects of morphine. Pain 2001;93:155–63.

[99] Doverty M, White JM, Somogyi AA, Bochner F, Ali R, Ling W. Hyperalgesic responses in methadone maintenance patients. Pain 2001;90:91–6.

[100] Brands B, Blake J, Sproule B, Gourlay D, Busto U. Prescription opioid abuse in patients presenting for methadone maintenance treatment. Drug Alcohol Depend 2004;73:199–207.

[101] Rando K, Hong KI, Bhagwagar Z, Li CS, Bergquist K, Guarnaccia J, Sinha R. Association of frontal and posterior cortical gray matter volume with time to alcohol relapse: a prospective study. Am J Psychiatry 2011;168:183–92.

[102] Wang GJ, Smith L, Volkow ND, Telang F, Logan J, Tomasi D, Wong CT, Hoffman W, Jayne M, Alia-Klein N, Thanos P, Fowler JS. Decreased dopamine activity predicts relapse in methamphetamine abusers. Mol Psychiatry 2012;17:918–25.

[103] Chen AC, Manz N, Tang Y, Rangaswamy M, Almasy L, Kuperman S, Nurnberger Jr. J, O'Connor SJ, Edenberg HJ, Schuckit MA, Tischfield J, Foroud T, Bierut LJ, Rohrbaugh J, Rice JP, Goate A, Hesselbrock V, Porjesz B. Single-nucleotide polymorphisms in corticotropin releasing hormone receptor 1 gene (*CRHR1*) are associated with quantitative trait of event-related potential and alcohol dependence. Alcohol Clin Exp Res 2010;34:988–96.

[104] Schmid B, Blomeyer D, Treutlein J, Zimmermann US, Buchmann AF, Schmidt MH, Esser G, Rietschel M, Banaschewski T, Schumann G, Laucht M. Interacting effects of CRHR1 gene and stressful life events on drinking initiation and progression among 19-year-olds. Int J Neuropsychopharmacol 2010;13:703–14.

[105] Treutlein J, Kissling C, Frank J, Wiemann S, Dong L, Depner M, Saam C, Lascorz J, Soyka M, Preuss UW, Rujescu D, Skowronek MH, Rietschel M, Spanagel R, Heinz A, Laucht M, Mann K, Schumann G. Genetic association of the human corticotropin releasing hormone receptor 1 (*CRHR1*) with binge drinking and alcohol intake patterns in two independent samples. Mol Psychiatry 2006;11:594–602.

[106] Wetherill L, Schuckit MA, Hesselbrock V, Xuei X, Liang T, Dick DM, Kramer J, Nurnberger Jr. JI, Tischfield JA, Porjesz B, Edenberg HJ, Foroud T. Neuropeptide Y receptor genes are associated with alcohol dependence, alcohol withdrawal phenotypes, and cocaine dependence. Alcohol Clin Exp Res 2008;32:2031–40.

[107] Bhaskar LV, Thangaraj K, Kumar KP, Pardhasaradhi G, Singh L, Rao VR. Association between neuropeptide Y gene polymorphisms and alcohol dependence: a case-control study in two independent populations. Eur Addict Res 2013;19:307–13.

[108] Mutschler J, Abbruzzese E, von der Goltz C, Dinter C, Mobascher A, Thiele H, Diaz-Lacava A, Dahmen N, Gallinat J, Majic T, Petrovsky N, Kornhuber J, Thuerauf N, Gründer G, Brinkmeyer J, Wienker T, Wagner M, Winterer G, Kiefer F. Genetic variation in the neuropeptide Y gene promoter is associated with increased risk of tobacco smoking. Eur Addict Res 2012;18:246–52.

Chapter 13

Circadian Rhythms and Addiction

Kelly Barko*, Micah A. Shelton*, Joseph A. Seggio†, and Ryan W. Logan*,‡

*Translational Neuroscience Program, Department of Psychiatry, University of Pittsburgh School of Medicine, Pittsburgh, PA, United States

†Department of Biological Sciences, Bridgewater State University, Bridgewater, MA, United States, ‡The Jackson Laboratory, Bar Harbor, ME, United States

INTRODUCTION

There are complex and bidirectional relationships between circadian rhythms and substance use and abuse disorders (SUDs), whereby genetic and/or environmental perturbations to the circadian system contribute to addiction vulnerability, while chronic drug administration disrupts rhythms. Evening preference (e.g., "night owls") and social jet lag (environmental schedule is "misaligned" with internal timing) are putatively linked to mood and addiction disorders [1]. Shift work is a risk factor for a host of health-related problems and diseases, including greater substance use and dependence. Moreover, substance use may be an attempt to ameliorate mood problems and sleep issues often accompanying rhythm disruptions, in turn further exacerbating health problems.

We are beginning to appreciate the complexity of the impact of circadian rhythms on brain function and behaviors. The circadian system is capable of regulating mood and reward neural circuitry at the cellular and molecular levels. For example, the molecular clock regulates dopaminergic transmission through midbrain and striatal pathways. On the other hand, substance use also impacts the circadian system directly, and these rhythm disruptions can persist during abstinence and may contribute to relapse. Stabilization of rhythms and/or sleep may therefore be an effective target for addiction treatments.

Continued investigation of the mechanisms underlying these relationships may prove useful for further understanding of addiction vulnerability and also may reveal novel, effective targets for therapeutic interventions to treat addiction. Our chapter discusses the organization of the circadian system and how rhythms regulate mood and reward-related neural circuitry and behaviors. We also highlight key findings demonstrating the impact of rhythm disruptions on substance use, and discuss circadian disruptions as a risk factor for addiction vulnerability.

THE CIRCADIAN TIMING SYSTEM

Circadian rhythms are endogenously generated oscillations occurring with an approximate period of 24 h, persisting in the absence of environmental cues (e.g., light or food). Rhythms can be entrained, shifted, or modified by light, food, and drugs of abuse. Regularly occurring stimuli, such as light, entrain our rhythms to remain synchronized to the external environment. Changes in light schedules accompanied by jet lag or shift work may lead to desynchronization of internal clocks with the environment, requiring an individual to re-entrain to the new schedule. Individuals who experience repeated phase differences (e.g., chronic shift workers) are at a higher risk for cancer, diabetes, obesity, and addiction [2,3].

Hierarchical Organization of the Circadian System

The mammalian circadian pacemaker of the suprachiasmatic nucleus (SCN) located within the anterior hypothalamus is responsible for orchestrating and maintaining rhythms in the brain and other tissues. Lesions of the SCN eliminate rhythms of hormone release, locomotor activity, and sleep-wake cycles [4,5]. Photic (i.e., light) information is relayed from retinal ganglion cells expressing the photopigment melanopsin to the SCN and other brain areas ([6–8]). The SCN also receives projections from serotonergic neurons of the midbrain raphe nuclei and the intergeniculate leaflet (IGL) of the thalamus, which are primarily responsible for relaying nonphotic information (e.g., arousal) to the SCN [9] (Fig. 1).

Neural Mechanisms of Addiction. https://doi.org/10.1016/B978-0-12-812202-0.00013-0
Copyright © 2019 Elsevier Inc. All rights reserved.

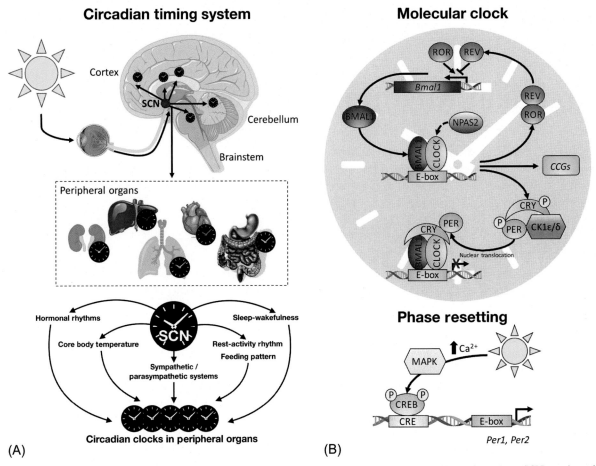

FIG. 1 Hierarchical organization of the circadian timing system and the molecular clock. (A) The suprachiasmatic nucleus (SCN) receives photic (i.e., light) information through the retinohypothalamic tract (RHT) via specialized melanopsin containing retinal ganglion cells. The SCN relays timing information to the rest of the brain and body via multiple pathways, including hormonal and neural pathways. These pathways maintain coordination among tissues and synchronization to the external environment. Extra-SCN brain regions and peripheral tissues contain functional molecular clocks to maintain tissue and cell-type homeostasis and metabolic programing. (B) Almost every cell in the body expressed the molecular clockwork. The circadian transcription factors CLOCK and BMAL1 dimerize to promote the transcription of various genes. Several of these genes called *Per* and *Cry*, once translated, dimerize and translocate to the nucleus to interact directly with CLOCK-BMAL1 complexes, effectively inhibiting their own transcription. Other genes controlled by the clock including nuclear orphan receptors (*Ror* and *Rev-erb*), which modulate the robustness of the molecular clock by directly modulating the expression of *Bmal1*. Casein kinases (CK1ε/δ) regulate the degradation of PER/CRY complexes via phosphorylation to modulate the timing and phase of the molecular clock. Photic input to the SCN activates cAMP response element-binding protein (CREB) signaling to induce the expression of *Per1* and *Per2*. Light pulses presented during specific times of the dark phase reset the phase of the molecular clock through CREB-mediated signaling.

Photic and nonphotic inputs converge on distinct neuronal ensembles within the SCN. The dorsomedial "shell" and ventrolateral "core" of the SCN contains vasopressin (VP) containing cells and vasoactive intestinal peptide (VIP) containing cells, respectively. VIP neurons receive direct glutamatergic retinal inputs, whereas the VP neurons receive GABAergic and VIP input from the core. Coupling the oscillatory network between core and shell generates self-sustaining pacemaker rhythms [10]. Thus, VP neurons of the SCN are rhythm generations, while VIP neurons relay light information to synchronize the SCN to the light-dark schedule. Astrocytes are also critical for generating and maintaining rhythms by the SCN [11–13]. Individual SCN neurons are autonomous oscillators fully capable of generating independent rhythms of gene expression and activity [14–16]. Cellular and molecular rhythms are tightly coupled to maintain coordination of rhythms across the SCN and perturbations within activity and molecular pathways control behavioral rhythms [17–19] (Fig. 2).

Timing information is relayed to other brain regions and tissues by the SCN. The SCN outputs to the peripheral tissues are necessary to maintain phase synchronization between central and peripheral oscillators [20]. The SCN coordinates rhythms across the brain via direct and indirect projections and across the body via temporal control of hormone release and sympathetic and parasympathetic terminals [21,22].

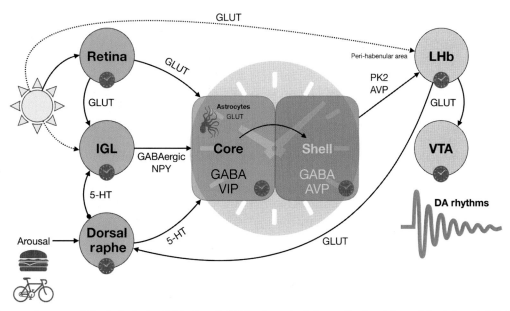

FIG. 2 SCN pacemaker network. Both photic (sunlight), through GLU and nonphotic inputs (food, activity, and drug) through DR 5-HT pathways converge upon the SCN. In addition, these signals project to the IGL, which is necessary for entrainment, and the SCN and IGL are interconnected by NPY and GABAergic signaling. The SCN core contains a group of VIP-containing neurons, which express circadian clock genes in response to photic input. The SCN astrocytes act in antiphase with the SCN neurons and modulate extracellular glutamate and act to inhibit the SCN-core neurons. The SCN shell contains AVP-containing cells, receives no retinal photic input, but receives GABAergic and VIP signaling from the core and is responsible for coupling the oscillatory network by generating a self-sustaining pacemaker rhythm. Output signals from the core (PK2, AVP) reach the LHb (with photic signaling) which then modulates rhythmic 5-HT and DA release from the DR and VTA, respectively. *AVP*, arginine vasopressin; *DA*, dopamine; *DR*, dorsal raphe nucleus; *GLU*, glutamatergic; *IGL*, intergeniculate leaflet of thalamus; *LHb*, lateral habenula; *NPY*, neuropeptide Y; *PK2*, prokineticin 2; *VIP*, vasoactive intestinal peptide; *VTA*, ventral tegmental area.

The Molecular Clock

Almost every cell in the body expresses the necessary machinery for molecular clock function. Molecular rhythms are generated by a series of interlocking autoregulatory transcriptional and transitional feedback loops with near 24-h oscillations [23–25]. The positive loop is driven by the circadian transcription factors CLOCK and BMAL1, which dimerize and bind to E-box elements within gene promoters. CLOCK-BMAL1 complexes transcribe hundreds to thousands of clock-controlled genes (CCGs) depending on the tissue and celltype [26]. Several of these CCGs are the circadian genes *Period (Per1,2,3)* and *Cryptochrome (Cry1,2)*. period (PER) and cryptochrome (CRY) proteins build within the cytoplasm and form heterodimers to translocate to the nucleus, directly interacting with CLOCK-BMAL1 complexes to inhibit their own transcription. Timing of nuclear entry is controlled by several kinases which phosphorylate PERs and CRYs for degradation [27,28]. The nuclear orphan receptor REV-ERBα is also a CCG which inhibits CLOCK-BMAL1 transcriptional activity [29]. The CLOCK homolog NPAS2 is capable of dimerizing with BMAL1 to control circadian-dependent transcription. NPAS2 is preferentially expressed in the mammalian forebrain and peripheral tissues [30–34]. Following loss of CLOCK, NPAS2 is able to compensate to restore cellular and behavioral rhythms [31,33]. Functional redundancy suggests rhythms are important for tissue and cell functions. Estimated 30%–50% of genes are rhythmically expressed in the body [35–37], and many of these genes have shown to be important for homeostasis, basic functions, and organismal health.

CIRCADIAN GENES AND DRUG REWARD

Mutations of circadian genes are linked to substance use and dependence in humans [38–40], suggesting genetic disruption of the molecular clock contributes to addiction vulnerability. Most of these findings reveal variants within *CLOCK* or *PER* genes as major players in addiction phenotypes. Preclinical studies have supported this notion by showing disruption of these genes alters reward circuitry and behavior.

Clock

Alcohol use and dependence, depression, and other comorbid disorders are associated with variants within the *CLOCK* gene [41,42]. The C allele (3111 T/C SNP; rs1801260) is linked to schizophrenia, bipolar disorder, metabolic disorders, sleep and circadian phenotypes, and alcohol use [43–49]. The 3111 T/C SNP is within the 3′ UTR, which likely affects the stability and transcriptional activation of *CLOCK* [50]. Mice with a defective CLOCK (*Clock*Δ19) protein voluntarily consume more alcohol at higher concentrations than wild-type mice [51]. These mutant mice also self-administer higher amounts of cocaine with greater motivation and enhanced sensitivity to cocaine-conditioned reward [52,53]. Through a series of studies, McClung and colleagues discovered these mutant behavioral phenotypes are primarily driven by hyper-dopaminergic ventral tegmental area (VTA) neurons and secondary postsynaptic plasticity of glutamatergic excitatory synapses in the nucleus accumbens (NAc) medium spiny neurons (MSNs) ([52,54–62]). The Δ19 mutation renders CLOCK as a dominant negative protein with impaired transcriptional activation and a loss of CCG rhythms [63–65]. Thus, these behavioral phenotypes are likely a consequence of impaired molecular clock function and direct transcriptional targets of the CLOCK-BMAL1 complex. Recent efforts are elucidating relevant targets and their potential impact of reward-related behaviors [59,61].

Period

Genetics studies have also found associations between *PER* gene variants and substance use. Cocaine dependence, for example, was significantly associated with a polymorphic repeat *PER2* variant [66]. Cocaine abusers were more likely to harbor the shorter allele, which was correlated with lower dopamine (DA) receptor 2 (D2) availability [66]. Lower D2 availability may be characteristic of addicted individuals [67]. Another *PER2* variant (rs56013859) predicts drinking amount and frequency in alcohol-dependent individuals, with no association to amount of alcohol drinking or alcohol misuse in social drinkers [40,41]. In a larger, longitudinal study, *PER1* and *PER2* variants were related to compulsive drinking during adolescence and heavy drinking during adulthood [39]. These variants were associated with stress-induced drinking [68] and increased frequency to report disrupted sleep [38]. Sample cohort selection is important for detecting these associations, since other studies have found no relationships between circadian gene variants and substance when excluding those with comorbid psychiatric disorders [69–75]. Genetic disruption of the molecular clock may be a common mechanism underlying comorbid mood and addiction disorders.

While the functional consequences of circadian gene variants on function and pathways to disease are largely unknown, studies with *Per* mutant mice have provided some foundation for mechanistic links between circadian genes and substance use. For example, *Per1* mutant mice voluntarily consumed more alcohol depending on genetic background [39,76,77]. *Per2* mutant mice also consumed more alcohol at higher preferences under two-bottle free-choice conditions [76,78–80]. In *Drosophila melanogaster*, deletion of *Per* (or *Timeless; Per* binding partner) abolished tolerance to repeated alcohol vapor exposure [81]. *Per* mutant flies with longer periods had prolonged recovery times and increased mortality after alcohol exposure due to impaired alcohol metabolism [82]. Interestingly, *Per2* mutant mice displayed increased locomotor sensitization to repeated cocaine injections, while *Per1* mutant mice failed to have any locomotor sensitization to cocaine [83], which was similar to *Per* mutant flies [84], suggesting differential roles of *Per* genes in the regulation of drug behaviors. There was a positive correlation between striatal PER1 rhythms and cocaine-induced locomotor response [85], which may be driven by melatonin rhythms [86]. Cocaine-conditioned reward was also attenuated in *Per1* mutant mice, with no effects on cocaine self-administration or extinction [83,87]. These studies suggest molecular rhythms are connected to the hedonic value of drugs of abuse and drug-induced plasticity leading to sensitization.

Sleep-wake disturbances are often reported by opiate-dependent patients [88] and morphine withdrawal disrupts circadian rhythms in rodents [89,90]. In limbic forebrain and frontal cortex, the amplitude of *Per2* rhythms was enhanced during withdrawal from chronic opiates [91–93]. *Per2* mutant mice had reduced tolerance to the analgesic and withdrawal effects of morphine [94]. Additional studies are necessary to further dissect the role of molecular rhythms or circadian genes on the tolerance and withdrawal effects of opiates, and whether these are involved in drug-induced plasticity underlying dependence.

The transcriptional and posttranslational activity of PERIOD is regulated by BMAL1 and several kinases, respectively. Nominal associations were found between *BMAL1* and alcohol dependence [41,95,96]. Furthermore, casein-kinase 1 epsilon(CK1ε) and delta (CK1δ) were tied to alcohol dependence and relapse [97] and the behavioral responses to other drugs of abuse [98,99]. These kinases sustain the pace and robustness of the molecular clock by phosphorylating PERs and other circadian proteins [100,101]. Importantly, pharmacological inhibition of CK1ε/δ, which also phase shifts the circadian pacemaker, prevented increased alcohol drinking following alcohol deprivation when administered at specific circadian phases [97], suggesting stabilizing high-amplitude rhythms by CK1ε/δ inhibition may exert therapeutic effects on addiction-related behaviors.

CIRCADIAN REGULATION OF MOOD AND REWARD NEURAL CIRCUITRY

Almost every cell in the body and brain expresses the necessary machinery for molecular clock function, yet we are just beginning to appreciate the tissue and cell-type specific role of circadian genes and transcription factors. In humans and animals, circadian-dependent neural activity and reward responses seem to modulate variation of mood and motivation across the day. For example, diurnal variation of dopaminergic transmission in striatal circuits corresponds to time-dependent changes in anxiety and depressive behaviors in rodents. The molecular clock controls region and cell-type specific mechanisms that regulate multiple neurotransmitter systems within the VTA and striatum. Clock dysfunction can impact these systems with consequences on drug reward and motivated behaviors.

The SCN

DA and glutamatergic rhythms in the mesolimbic circuitry may also be modulated via indirect projections from the SCN. The SCN sends multisynaptic projections to the VTA through the medial preoptic area [102–104]. The paraventricular thalamus (PVT) is also involved in reward-related behavior [105–107] and has reciprocal connections to the SCN [108]. The PVT projects to the NAc and other regions involved in motivated behaviors, and PVT neurons display rhythmicity of neural activity. The SCN lesions are known to produce an antidepressant effect, likely through these connections to relevant circuits [109]. Even the antidepressant effects of agomelatine, a melatonin receptor agonist and serotonin (5-HT$_{2C}$) antagonist, requires an intact SCN [110], further supporting the role of the SCN to modulate the behavioral effects of stress and mood behaviors. However, there is no direct evidence SCN lesions also impact reward and drug-taking behavior.

The discovery of novel wide-spike VTA neurons, which have robust firing activity rhythms, has provided hints of evidence the VTA may be entrained by the SCN [104,111]. Exvivo slices of VTA fail to display rhythms of circadian gene expression [112] or *Per2:luciferase* reporter rhythms [113], suggesting removal from timing input from the SCN leads to loss of rhythms in the VTA. Interestingly, the SCN lesions alter the rhythms of dopamine transporter (DAT) and tyrosine hydroxylase (TH) in the NAc [114], although extra-SCN oscillators may be involved [115].

The SCN projections release humoral factors and peptides in a circadian-dependent manner. For example, prokineticin 2 (PK2) is generated in the SCN and known to be critical for circadian locomotor rhythms [116]. Primary targets of the SCN express the PK2 receptor, including those regions projecting directly to the VTA, such as the lateral habenula and medial preoptic area [117]. Mood-related behaviors are reduced in PK2 knockout mice and intracerebroventricular (ICV) infusions of PK2 rescue these behaviors [118]. Factors released by the SCN influence mood and affect through mesolimbic action, but additional studies are necessary to determine whether these are relevant for reward and motivated behaviors.

Molecular Clocks and VTA Pathways

The VTA is primarily composed of dopaminergic and GABAergic cells, which are differentially distributed across the rostral-caudal axis. Several genes of the DA signaling pathways are expressed rhythmically. For example, diurnal rhythms of the rate limiting enzyme, TH, along with DA receptors 1 and 2, and monoamine oxidase (MAOA), enzyme responsible for DA degradation, have been observed in the VTA [52,119]. Each of these genes contain an E-box promoter sequence, and CLOCK/BMAL1 heterodimers bind directly to these sequences for *TH* and *Maoa* [114,120]. The circadian transcription factor CLOCK is highly expressed in the VTA and less so in the NAc [52,86]. Characterization of the alterations of mesolimbic function due to impaired CLOCK function has provided tremendous insight into molecular clock regulation of mood and reward neural circuitry. Mice harboring a mutation causing a dominant negative CLOCK protein (*Clock*Δ19 mutant mice) express higher TH and synthesize more DA in the VTA. These VTA DA neurons also have higher burst firing activity and an overall loss of diurnal DA rhythms [52,56]. The VTA-specific knockdown of CLOCK recapitulated many of these phenotypes, further suggesting CLOCK exerts local modulation of DA neurotransmission [58,61]. CLOCK also regulates the transcription of *cholecystokinin* (*Cck*). CCK is a neuropeptide co-released with DA to inhibit further DA release [120a,121]. *Cck* is a CCG and is significantly reduced in the VTA of *Clock*Δ19 mutant mice [55]. The VTA-specific knockdown of *Cck* resembles mania-like behaviors of *Clock*Δ19 mutant mice. The net effect of CLOCK disruption in the VTA is increased extracellular DA in striatal regions and hypersensitivity to the rewarding properties of several drugs of abuse including alcohol and cocaine.

Other circadian mechanisms regulate midbrain DA transmission. For example, diurnal variation of DAT activity modulates DA rhythms in the striatum [122]. Genes of the stabilizing loops of the molecular clock are also involved. The circadian nuclear receptor REV-ERBα, which represses the transcription of *Bmal1*, competes with the DA-enriched nuclear receptor NURR1 to also repress *TH* expression and contributes to the diurnal variation of DAergic transmission and mood-related behaviors. Similar to the effects of the *Clock* mutation, *Reverbα* mutant mice exhibit reduced anxiety and depressive behaviors and elevated midbrain DAergic activity [123]. *Reverbα* is also among the most affected genes in the striatum following methamphetamine and heroin administration [124]. Hyperdopaminergic activity induces secondary changes in GABAergic signaling in the VTA [52,55,58], and potentially postsynaptic adaptations mediating excitatory transmission in the NAc [59], all of which likely have a role in the regulation of mood and reward-related behaviors.

Molecular Clocks and NAc Pathways

A majority of the NAc (~95%–98%) is composed of GABAergic MSNs predominantly expressing dopamine 1 or 2 receptors (D1-MSNs or D2-MSNs). Activity of MSNs is modulated by glutamatergic input from the prefrontal cortex and amygdala, and dopaminergic input from the VTA. Diurnal DA release from the VTA projections to the NAc may modulate variation of motivated behaviors [121]. Loss of circadian modulation of DAergic signaling from the VTA reduces excitatory signaling and reduces the D1 to D2 receptor ratio by upregulating D2 receptor expression in the NAc [62]. A shift toward D2 enhances the locomotor response to D2 agonists, despite normal locomotor sensitization to cocaine. In addition, *Per2* mutant mice have reduced D1 to D2 ratios and elevated DA in the NAc, likely due to reduced *Maoa* expression which impairs DA turnover [125]. Functional *Per* genes may be protective against drug-related behaviors under baseline and stress conditions. Both *Per1* and *Per2* transcription can be induced by glucocorticoid release and direct transcriptional regulation at glucocorticoid receptor response promoter elements [126–128]. *PER* variants identified in high drinking stressed adolescents reduces glucocorticoid-induced transcription of *Per1* [39]. These impairments of *PER* function may contribute to elevated alcohol drinking and dependence [39,129,130].

Little is known about the mechanisms underlying the impact of *Per* on cellular and molecular functions relevant for reward behavior, however, a few studies have provided insights. *Per2* mutant mice voluntarily consume copious amounts of alcohol [78–80], which can be attenuated by acamprosate treatment, which reduces hyper-glutamatergic signaling often induced by repeated binge and withdrawal episodes [131,132]. Acamprosate reduces extracellular glutamate levels in the brain and reduces alcohol drinking during free choice and relapse paradigms in *Per2* mutant mice [78,79,132]. These effects seem to be independent of any direct modulation of the circadian pacemaker, however, acamprosate directly infused into the SCN also reduces alcohol intake in *Per* mutant mice during a relapse paradigm [78,79]. Elevated DA and glutamatergic transmission may be responsible for alcohol reward sensitivity. These mice have lowered expression of the glutamate transporter *Eaat1*, decreased efficiency of synaptic glutamate clearing by astrocytes, and increased extracellular glutamate in the striatum [40]. Direct infusion of acamprosate into the NAc reduces alcohol intake similar to the effects of the SCN [78,79]. Treatments targeting the circadian system may prove effective for treating alcohol-use disorders.

Molecular clock machinery is expressed in almost every cell of the body, and while the circadian transcription factor CLOCK is ubiquitously expressed in the brain, its homolog NPAS2 is highly enriched within D1 MSNs of the NAc [133]. NPAS2 also seems to be particularly responsive to stress and drugs of abuse relative to other circadian transcription factors, as repeated cocaine administration significantly increases NPAS2 expression in the NAc with no change in CLOCK or BMAL1 [134]. Furthermore, NPAS2-deficient mice have attenuated cocaine conditioned place preference (CPP) at lower cocaine doses, and the NAc-specific knockdown of NPAS2 recapitulates these behavioral effects. Thus, NPAS2 may be a key circadian modulator of drug reward in the NAc, potentially through cell-type-specific activation in D1-MSNs.

DRUGS OF ABUSE IMPACT CIRCADIAN GENES AND BEHAVIORAL RHYTHMS

Circadian dysregulation impacts cellular and molecular pathways contributing to mood and reward-related behaviors. Drugs of abuse impact circadian genes and rhythm function, which could also contribute to disease vulnerability and progression. Disrupted sleep-wake and other rhythms, such as hormonal variations, persist during acute, and protracted abstinence and may be involved in withdrawal syndrome. These persisting disturbances in sleep and rhythms may then contribute to relapse as dependent individuals may seek to alleviate these issues.

Drugs of Abuse Alter Circadian Gene Expression

Acute and chronic drug administration alters the rhythms of molecular, physiological, and behavioral factors. Few studies have examined the impact of drugs of abuse on rhythms in tissues from actively using, dependent, or abstinent individuals. Compared with healthy controls, human alcoholics have reduced amplitude of circadian gene expression in peripheral blood mononuclear cells (PBMCs), which persisted at least 1 week into withdrawal [96]. Although amplitude tended to increase, overall amplitude never reached that of healthy controls, although rhythm parameters had no correlation with withdrawal severity [96]. *Per2:luciferase* reporter skin fibroblast rhythms were similar between healthy controls and alcohol-dependent patients [135]. Notably, fibroblast rhythm period was inversely correlated with illness severity—shorter period predicted severity of alcohol dependence [135]. Isolation of peripheral cells may be useful for characterizing rhythms in dependent individuals to further determine whether rhythm changes are important for disease-related symptoms or ultimately the chronicity of the disease.

Alcohol administration changes the expression of circadian genes in almost every rodent tissue, depending on alcohol dose and duration. For example, chronic alcohol administration provided as liquid diet shifted the expression of *Per* rhythms in the SCN and arcuate nucleus of the hypothalamus [136], while minimally affecting other circadian genes in the SCN [137]. Alcohol drinking for a more extended period (e.g., 6–12 months) reduced the expression of arginine vasopressin (AVP), vasoactive-intestinal peptide (VIP), and somatostatin (SST) in the SCN [138–140], all of which are important for intracellular communication and time keeping [141,142]. These changes persisted well into alcohol withdrawal, suggesting a plausible mechanism by which alcohol drinking and withdrawal impacts circadian period and photic phase shifting [143–146].

Psychostimulants, opiates, and other drugs of abuse change the expression of circadian genes in the SCN and reward-related brain regions. Methamphetamine-altered rhythms of *Per1/2* in the SCN and *Per1/2, Bmal1,* and *Rev-erbα* in the striatum [119,147–151]. Altered circadian gene rhythms persisted during acute and protracted withdrawal in the dorsal striatum, prefrontal cortex, and NAc [92,148]. In addition, repeated methamphetamine administration lengthened circadian period and restored behavioral rhythmicity in arrhythmic animals [152,153]. Inhibiting *Per1* activation in the frontal cortex, hippocampus, and striatum prevented morphine CPP and activation of extracellular signal-regulated kinases (ERK) and cAMP response element-binding protein (CREB) signaling [148]. Drugs of abuse may activate immediate early expression of *Per1* to elicit downstream activation of CREB-mediated gene transcription important for drug-induced cellular and behavioral plasticity.

Cocaine differentially affects circadian gene expression across brain regions. Repeated cocaine administration enhanced the amplitude of *Per1/2* and reduced the amplitude of *Bmal1* and *Cry1* in the hippocampus, accompanied by selective increases of *Per1* and *Clock* in the caudate putamen [85]. Similar changes were observed in the dorsal striatum [154]. In the NAc, repeated cocaine administration, but not acute, altered the rhythms of *Npas2, Per1/3*[134]. NPAS2 binding to the promoters of *Per1* and *Per3* was also enhanced following repeated cocaine [134], suggesting cocaine impacts the positive transcriptional loop of the molecular clock.

Drug-induced changes of circadian gene rhythms are likely mediated by DA receptor signaling pathways. Extensive crosstalk between the molecular clock and dopaminergic pathways has been established. For example, the locomotor response to D2/D3 receptor agonists is circadiandependent [155]. The activation of these receptors in the striatum inhibits the expression of *Clock* and *Per1*, whereas the activation of the D1 receptor drives their expression including *Npas2* and *Bmal1* [156]. Altered circadian transcription factor expression likely impacts their activation and function, which may have downstream effects on signaling pathways involved in the response to drugs of abuse and long-term plasticity important for dependence and addiction.

Drugs of Abuse Impact the Master Circadian Pacemaker

The SCN is vulnerable to the impact of drugs of abuse. The SCN receives afferent projections from several other brain regions to integrate photic (i.e., light) and nonphotic information (e.g., arousal). Light is relayed through retinohypothalamic tract (RHT) projections activating glutamatergic synapses onto the SCN neurons. Glutamatergic activation recapitulates several of the phase shifting effects of light at certain circadian phases. Alcohol has been shown to prevent glutamate-induced phase delays of the SCN neural firing and co-treatment with $5HT_{1A}R$ agonist (8-OH-DPAT) enhanced phase advances [157,158]. In addition, alcohol modifies the free-running period and disrupts phase-resetting of the master circadian pacemaker [146,159,160]. Other drugs of abuse, including methamphetamine, cannabinoids, nicotine, and cocaine, impacted the pattern and rate of the SCN neural firing [161–163]. For example, activation of M1 and M4 acetylcholine receptors by nicotine phase advances the SCN neural firing rhythms [163,164]. Sensitivity to nicotine seemingly varies across the day and may involve the SCN-dependent pathways [165,166]. Acute cocaine infused directly into the SCN produced large phase advances and could be prevented by coadministration of a serotonin receptor antagonist [160]. These effects may be mediated by presynaptic receptors on projections from the raphe nucleus or postsynaptic SCN neurons (Fig. 3).

Many of these studies investigated the acute drug effects on the SCN function. A wide-open question is whether these effects persist following acute administration or whether repeated administration of drugs of abuse have long-term consequences on the SCN plasticity. It has been put forth that the SCN may develop rapid tolerance to the acute effects of

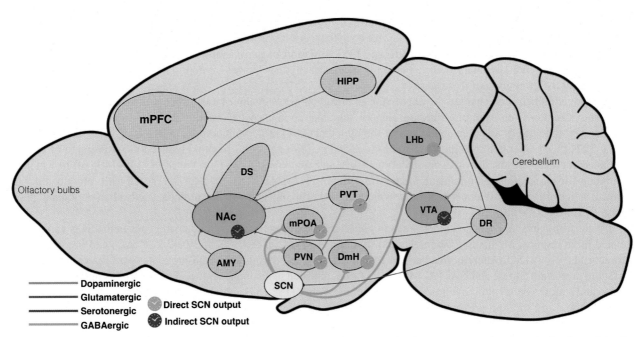

FIG. 3 Connections between the master circadian pacemaker and reward circuitry. The circadian timing system regulates reward-related neural circuitry at the cellular level through the expression and function of molecular clock machinery and at the circuit level through direct or indirect connections from the SCN. The SCN projects to other hypothalamic areas which then send projections throughout the brain. Other direct projections include the medial preoptic area (mPOA) and lateral habenula (LHb). Both indirect and direct outputs of the SCN send projections to the ventral tegmental area (VTA) and nucleus accumbens (NAc), which are major neural substrates of reward and motivation. Direct SCN outputs receive monosynaptic afferents and/or diffusible signals (e.g., prokineticin 2, PK2). *AMY*, amygdala; *DmH*, dorsomedial hypothalamus; *DR*, dorsal raphe nucleus; *DS*, dorsal striatum; *HIPP*, hippocampus; *mPFC*, medial prefrontal cortex; *PVN*, paraventricular nucleus; *PVT*, paraventricular thalamus.

drugs of abuse. For example, pretreating SCN slices with alcohol or methamphetamine prevented glutamatergic and serotoninergic-induced phaseshifts of neural firing [162,167], supporting the notion the SCN develops rapid tolerance to acute drug rather than sensitization. The repeated effects of alcohol or methamphetamine on the SCN neural firing responses to phase-shifting stimuli have yet to be studied. Withdrawal from chronic administration of drugs of abuse produced long-lasting alterations to the SCN firing, indicating repeated drug administration does induce the SCN plasticity [168]. Furthermore, the ability of chronic intermittent alcohol vapor exposure to shorten free-running locomotor rhythms in mice depended on the number of alcohol exposures and the sensitivity of the mice to the intoxicating effects of alcohol [143]. The concept of rapid tolerance of the SCN to acute drug exposure is intriguing since these findings imply the SCN may not respond to other inputs and therefore may be taken "offline" by drugs of abuse. Understanding the mechanisms underlying acute tolerance, especially whether these changes persist or are further exacerbated by chronic exposure, may reveal novel actions of drugs of abuse on the circadian timing system and downstream pathways.

Developmental Exposure to Drugs of Abuse Impact the Circadian System

Drugs of abuse have a deleterious impact on fetal development and enhance the risk for the emergence of diseases later in life [169–174]. Prenatal drug exposure has been linked to substance use during adolescence and adulthood [175,176]. Children with fetal alcohol spectrum disorder (FASD) are at an increased risk for later alcohol abuse and comorbid sleep, circadian, and stress disorders [175,177–179]. Developmental effects of prenatal drug exposure could contribute to why sleep and circadian disturbances often precede the emergence of psychiatric disorders. Some insight has been provided by animal studies employing prenatal and postnatal alcohol exposure approaches.

The circadian system is vulnerable to the effects of early developmental alcohol exposure. Alcohol exposure during postnatal days 4–9 (human "equivalent" of the last trimester) shortened the free-running period and altered the ability of the pacemaker to respond to photic and nonphotic stimuli [180–183]. Cocaine also impaired photic phase shifting [184,185]. Effects of prenatal alcohol on entrainment likely involve direct impact on the retina [186,187], optic tracts [188], or raphe [189,190], since alcohol has minimal effects on the SCN development [182]. Alcohol effects on entrainment may also involve the SCN expression of brain-derived neurotrophic factor [191,192]. Alcohol exposure during the third larval stage shortened free-running rhythms in flies, which depended on the dose of ethanol—higher doses resulted in more pronounced shortening [192a]. A consistent finding regardless of developmental stage of alcohol exposure is shortening of free-running rhythm signifying a robust, common effect of alcohol on rhythms.

Developmental effects of alcohol on period depend on the inherent period of the clock and *Per* expression [193]. *Drosophila* reared in alcohol solutions displayed shortened or lengthened period depending on their baseline period. Flies with longer periods tended to shorten, while shorter periods tended to lengthen. Behavioral rhythms correlated with increased *Per* expression in flies with a longer period and reduced expression in flies with a lengthened period following alcohol rearing [193]. Fly *Per* is highly homologous to mouse *Per2* [193a]. In relation to other circadian genes, *Per2* appears to be the most affected by developmental alcohol exposure [194,195], which has significant consequences on brain cell function [129]. Drugs of abuse send the developmental trajectory awry which have long-term health consequences and may serve as a risk factor for a host of diseases and disorders [196]. Speculatively, the consequences of developmental drug exposure on the circadian system underlie the emergence and progression of psychiatric disorders and comorbid diseases.

CO-INHERITANCE OF CIRCADIAN AND ALCOHOL-RELATED PHENOTYPES

Animals selectively bred over many generations for alcohol-related behaviors, such as high or low voluntary alcohol consumption, display divergent circadian phenotypes. Mice bred for high alcohol preference (HAP-1) had a shortened circadian period relative to low alcohol preferring (LAP-1) mice [197]. High drinking in the dark (HDID-1 or 2) mice had a shortened period [198–200], and less coherent rhythmicity when placed into constant light [201]. Shorter free-running periods are also observed in rats bred for being high-preferring (P) or highalcohol drinking (HAD-2) rats. In comparison to lower drinking animals, high-preferring (P) and high alcohol drinking (HAD-2) rats had a shorter period relative to wild-type mice [202]. Interestingly, alcohol nonpreferring (NP) rats respond faster to changing light-dark cycles compared with P rats [202]. Selectively breeding for alcohol withdrawal phenotypes seems to also influence circadian phenotypes. Withdrawal seizure-prone (WSP-1 and 2) and withdrawal seizure-resistant (WSR-1 and 2) mice were selected for high and low handing-induced convulsions during alcohol withdrawal [203,204]. Relative to WSR-2 lines, WSP-2 mice have longer circadian periods [201]. Thus, circadian phenotypes may co-inherited when selecting for extreme alcohol-related phenotypes over many generations, suggesting common underlying inherited genetic mechanisms underlying these phenotypes. Revealing the genetics has been difficult because of their inherent polygenic and pleiotropic effects [201,205,206].

It has been postulated that alterations in the GABAergic system may be a common mechanism. GABAergic transmission is critical for coordinated neuronal firing within the SCN [15,207,208] and certain GABA receptor subunits are important for many alcohol-related behaviors, including tolerance and withdrawal severity [209–211]. Genetic mapping studies have found bidirectional changes within the GABA-A receptor signaling pathways as being important for the differences between the WSP and WSR selected phenotypes, as well as the HDID mice [209,212,213]. New genetic and molecular tools are beginning to identify important interactions between GABAergic signaling, circadian rhythms, and drug-related behaviors. Genome sequencing, high-precision genetic mapping, and new computational tools are providing the foundation for future mechanistic studies.

ENVIRONMENTAL CIRCADIAN PERTURBATIONS IMPACT REWARD NEURAL CIRCUITRY AND BEHAVIOR

Effects of Repeated Circadian Phaseshifting and Photoperiod on Reward-Related Behaviors

Chronic shiftwork has been deemed a public health hazard and an environmental carcinogen. Shift workers are more likely to use substances and illicit drugs and with greater frequency [214,215]. Altered sleep schedules and circadian misalignment, along with light at night, often experienced by shift workers, likely have additive or even synergistic consequences on overall health. Animal studies have begun to tease apart the impact of sleep loss, misalignment, and light at night on disease models. Depending on species, paradigm, and behavior, experimental manipulation of photoperiod or repeated shifting of the light-dark cycle can promote or attenuate reward-related behaviors. Repeated 6-h advances of the light-dark cycles at 3–4 week intervals decreased alcohol drinking in males and increased drinking in females of the HAD-1 line [216]. In standard laboratory rats (Fisher and Lewis), similar repeated circadian shifts reduced alcohol drinking in males and females [217], but increased voluntary drinking in Sprague-Dawleys [218]. In contrast, photoperiod changes and repeated shifting paradigms had little effect on alcohol intake in mice, but constant conditions (i.e., DD (constant darkness) or LL (constant light)) reduced alcohol drinking [219,220]. Thus, when considering species differences, the direct impact of environmental circadian disruption on drug intake remains unclear. Approaches modeling a combination of sleep loss and circadian disruption could provide translational relevance. Sucrose and cocaine self-administration are both increased following acute sleep deprivation [221,222], although the impact of these deprivation paradigms on circadian rhythms needs to be explored further. Additive effects of sleep and circadian alterations likely influence these behaviors. For example, shorter photoperiods, which also impact sleep length and bouts, suppressed cocaine reinstatement, while long photoperiods had no observable effect on behavior [223]. The effects of short photoperiod on reinstatement behavior persisted even when the rats were returned to standard 12:12 light-dark schedules, suggesting long-lasting plasticity within circadian regulated circuits relevant for reward. The impact of photoperiod on behavior may or may not be SCN dependent [224].

Glucocorticoid pathways in the brain and peripheral tissues mediate the phase resetting and re-entrainment of the circadian system to environmental lighting schedules [127,225,226]. Circadian oscillations of glucocorticoids control time of day dependent variations of neural cortical plasticity [227]. Circadian-regulated plasticity may be directly modulated by "neurotransmitter switching." For example, shorter photoperiods (less "stressful" for nocturnal rodents) increased the number of dopaminergic neurons in the hypothalamus, while longer photoperiods (more stressful) induced a subset of hypothalamic neurons to switch from primarily producing DA to SST [228]. These "switching" neurons send direct projections to corticotropin releasing factor (CRF)-producing neurons [229], and may be responsible for the ability of longer photoperiods to promote CRF release and increase anxiety and depressive behaviors [228]. Shorter photoperiods elicited the opposite behavioral effect, and dopaminergic to SST switching neurons were mostly responsible for these bidirectional modulatory consequences on behavior. These findings reveal direct links between photoperiod and neural plasticity with consequences on mood-related behaviors and provide a basis for seasonal variation of anhedonia and motivation.

Effects of Circadian Misalignment on Reward, Motivation, and Substance Use: A Focus on Human Adolescence and Young Adulthood

Beginning around puberty through the early 1920s, sleep timing and circadian rhythms gradually delay [230,231]. Throughout adolescence, sleep timing shifts later with parallel shifts in circadian preference [232] and slower accumulation of homeostatic sleep drive [233]. Adolescents tend to experience reduced sleep duration, increased daytime sleepiness, and increased sleep disturbances [233]. Decreased sleep duration and increased sleep disturbance may be a consequence of delayed sleep timing and conflicts with early school start times, resulting in circadian misalignment and sleep loss [232,234]. Initiation of substance use often occurs during adolescence with an emerging epidemiological pattern suggesting adolescence as vulnerable period of substance use and risky behavior [235]. Circadian misalignment and sleep loss during adolescence are associated with substance use. Short sleep duration was associated with an increased use of caffeine, nicotine, alcohol, and illicit drugs [236–238]. Weekday-weekend sleep differences have been associated with increased risk-taking behavior, substance use, and depressed mood [239–242].

Positive affect and reward activation varies across the day and may be a pathway by which sleep loss and circadian misalignment affect reward sensitivity [243–245]. Later bedtimes and shorter sleep durations contribute to reduced reward responsiveness, blunted positive affect rhythms, impulsivity, and depression [243,244,246–252]. Chronotype and sleep timing are associated with altered neural responses to reward outcome in adolescents. For example, glucose metabolic response to monetary reward in the striatum and prefrontal cortex displayed diurnal variation—increasing in the afternoon/evening relative to the morning [243,253,254]. Reward circuitry is sensitive to circadian disturbance and sleep loss. Larger weekday-weekend differences in sleep timing (i.e., social jet lag) were associated with lower prefrontal reactivity to anticipation and receipt of monetary reward in young adolescents [255]. Later sleep timing was also associated with reduced prefrontal cortical and enhanced ventral striatal responses to monetary reward [256,257]. Young adults (20-year-old males) who were evening chronotypes exhibited lower prefrontal responses to reward anticipation and higher ventral striatal response to reward outcome relative morning chronotypes, which predicted greater alcohol consumption and dependence [258].

Sleep deprivation also disrupts reward circuitry. Sleep deprivation promoted ventral striatal response and reduced prefrontal response in young adults (18–25 years old) [259] and impacts decision making in adults [260]. Brain reactivity to sleep deprivation is differentially affected in adolescents than adults, possibly related to the greater neural and behavioral sensitivity to reward during adolescence [261]. Striatal reactivity was enhanced in young adults but blunted in adolescents in response to reward outcomes [262,263]. Less effective fronto-striatal function during reward combined with behavioral compensation for blunted motivation may be a mechanism for risky behaviors in adolescents with shorter sleep or sleep loss [264].

Vulnerability to substance use during adolescence may therefore be a consequence of a combination of normative development of reward circuits, sleep, and circadian timing, which are impacted by environmental factors, such as work and school demands. Very little is known about the underlying neurobiology of the impact of sleep and circadian rhythms disruptions on reward function. Sleep deprivation may impact DA receptor expression and function in the ventral striatum [265]. Future studies will need to further clarify the directionality of these effects. For instance, sleep and circadian disturbances impact reward function and behavior, but more sensation-seeking adolescents may be more motivated to stay up later and engage in rewarding, risky activities.

HUMAN CHRONOTYPE AND REWARD

A combination of genetic, biological, and environmental factors contributes to the emergence of chronotype during adolescence and adulthood. Robust markers of circadian phase are critical to reliably measuring individual circadian phase and interpreting associated findings. Circadian phase typically relies on measures extracted from self-report, rather than objective biological measures, although more rigorous approaches are being employed and continually improved upon [266]. Recruitment of participants usually categorically dichotomizes into either "morning" or "evening" preferring [267]. Morning chronotypes prefer daytime activities accompanied by peak performance during morning hours, while evening chronotypes prefer activities later in the day and evening. Mutations within circadian genes are linked to chronotype. For example, the 3111T/C variant of the *CLOCK* was associated with evening preference [268] and variants with *PER* genes were associated with morning preference [268a]. Somewhat unsurprisingly, mutations of core circadian genes are linked to extreme chronotypes. Chronotype is a complex trait that changes across development and more advanced genetic and phenotyping approaches may reveal novel mechanisms [46]. A few of these variants are associated with major depression, bipolar disorder, and substance use [268b,47,75].

Shared genetic substrates, yet to be identified, potentially impact common underlying mechanisms across several psychiatric disorders, especially when considering the empirical and clinical relationships between chronotype, depression, and substance use [249,269]. Chronotype predicts the responses to rewarding stimuli within mood and reward-related neural circuits and pathophysiology of mood disturbances [270,271]. Evening preference was more prevalent in those with bipolar or depression disorders [245,272–274]. Furthermore, evening chronotype and delayed rhythms were found to be associated with poorer emotional adjustment and substance use in clinically depressed patients [275]. Using positron emission tomography (PET) imaging of glucose uptake within the brain, evening chronotypes exhibited a delayed and reduced diurnal response of positive affect correlating with dampened diurnal response to reward stimuli within the prefrontal cortex and striatum [255]. Decreased prefrontal cortex (PFC) response to reward has also been shown in those carrying a specific *PER2* mutation (rs2304672) [256]. PFC activation was negatively correlated with the ventral striatal response during reward outcome in those with the mutation and those with later sleep midpoints [256]. Thus, genotype and chronotype are related to corticostriatal circuit activation to reward outcome [276,277], and continued efforts to investigate the interplay between sleep loss and circadian misalignment and whether chronotype predicts risk for health-related outcomes has the potential for immediate behavioral interventions and possibly other therapeutics.

LOSS OF CIRCADIAN RHYTHMS OF DRUG SELF-ADMINISTRATION: A MODEL FOR "LOSS OF CONTROL"

In humans and animal experimental systems, drug self-administration usually follows a robust diurnal pattern. For example, voluntary intake of alcohol and cocaine self-administration during extended access has shown intake is primarily restricted to the dark phase in nocturnal rodents [278–285]. Cocaine self-administration displayed a free-running rhythm under constant conditions, suggesting endogenous circadian-dependent regulation of drug intake and seeking behaviors [286]. Interestingly, alcohol consumption peaks during the early evening in humans, while "craving" and administration of first drink shifts to the early morning shortly after awakening in alcohol-dependent individuals [287]. Circadian "gating" of rewardseeking or motivation may be important for temporal regulation of goal-directed behaviors. Loss of rhythm, for example, of self-administration of drug may be indicative of the transition from use or abuse to dependence.

Dependence is marked by the transition from sporadic patterns of use to more regular use followed by intense bingeing and intense cravings. The loss of control is the hallmark of addiction. Modeling these phenomena for translational relevance is usually implemented by providing intermittent access or periods of forced deprivation over longer periods of time. Intermittent access to several drugs of abuse usually escalates seeking and taking of alcohol and illicit substances [97,284,288–290]. A few studies have investigated the diurnal patterns of intake using similar approaches with prolonged access. Extended continuous access to nicotine reduced the overall amplitude of self-administration behavior during the night and increased intake during the day, revealing a loss of diurnal self-administration patterns [291]. Under prolonged intermittent access paradigms, the daily rhythm of cocaineself-administration was completely lost [284], similar to the alcohol deprivation effect on voluntary alcohol self-administration [97,292]. These studies are obviously labor and resource intensive, so few venture into experimental investigation of these phenomena, especially since loss of rhythmicity usually emerges following months of continued or intermittent access. Despite, designing therapeutics targeting the circadian system could be clinically effective. Pharmacological therapeutics known to target components of the molecular clock reduced overall drug self-administration by restoring the circadian pattern of drug intake [97,292]. Timing of therapeutic intervention may therefore be important for response or these treatments may be used in combination with other therapies to reduce drug intake.

FUTURE DIRECTIONS

There is growing skepticism as to whether circadian rhythms are the primary mechanisms driving substance use and addiction despite emerging evidence demonstrating the contrary. This may be due in part to the overall lack of research directly and causally relating whether actual rhythms within the proposed molecular, physiological, or biological clocks are truly important for the rhythm of the behavioral phenotypes, and whether or not these circadian-dependent mechanisms even have a major role once someone becomes dependent. The onus on the field lies within the necessity to design pragmatic experimental models which clearly define the impact of circadian disruption on the vulnerability to substance use separated from the role of rhythms on craving, dependence, and withdrawal. For example, circadian gene mutations may contribute to the risk for substance use through circadian-dependent mechanisms or genetic pleiotropy [293]. In addition, light may impact mood and reward circuitry and behavior via mechanisms that are seemingly independent from "classical" circadian pathways [6,224,294].

Site-specific knockdown or overexpression of circadian genes in the brain has been useful for clarifying their roles in reward-related behaviors, yet these do not readily define whether the temporal pattern or action of these genes has a direct impact, only whether these genes have any role. Tools for better spatial and temporal control of circadian gene expression could be an approach for delineating these pleiotropic effects on behavior. Chronic drug administration impacts molecular clock function, which, in turn, disrupts CCG transcription. Some of these genes may be directly involved in neurotransmission, while others are involved in epigenetic modifications. For example, the histone deacetylase activity of SIRT1 is circadiandependent and SIRT1 directly controls CLOCK/BMAL1-mediated transcription [295–297]. Notably, SIRT1 activity in the NAc has been shown to be important for regulating cocaine and morphine reward [298–300]. Epigenetic modifications by drugs of abuse are also important for neural plasticity and long-term adaptations mediating reward, dependence, and craving [301,302]. Connecting molecular clocks to epigenetic regulation of drug-induced neural plasticity is an active area of research and will undoubtedly advance our understanding of the cellular and molecular mechanisms underlying dependence. Large-scale and high-throughput "omics" approaches have yet to be systematically applied to experimental investigation of circadian rhythms and drugs of abuse, but these approaches integrated with advanced behavioral models should prove useful for further demonstrating the connection between rhythms and drug reward and motivation.

Identifying novel mechanisms connecting these phenomena are also feasible with the newly available mouse populations of the collaborative cross [303] and diversity outbred [304], and the heterogeneous stock (HS) rats [305]. These are the most genetically heterogeneous and phenotypically diverse rodent populations available. High-density genetic mapping of complex behaviors is readily possible with these resources [306]. Multiinstitutional and collaborative efforts are underway for identifying novel genetic mechanisms underlying circadian and addiction-related phenotypes. Furthermore, improved spatial and temporal control of neural circuits during drug self-administration is feasible using approaches such as pharmacogenetics (e.g., DREADDs) and optogenetic tools [12]. These tools can also be employed to investigate circuits downstream from circadian pacemakers and to reveal network properties that may vary across a circadian timescale. Phasecoupling, or coherence, of neural circuits may be important for mood and reward behaviors. Coherence may vary across the day or with chronic administration of drugs of abuse, as has been demonstrated for chronic stress, and drug-induced disruptions in coherence may decouple circuits or clocks within circuits, contributing to the pathophysiology of addiction. The advent of newer technologies and approaches will enhance our ability to dissect the spatial and temporal function of cellular and molecular pathways. Combining molecular and translational approaches will be necessary to further elucidate the role of the circadian system in addiction and to determine whether chronotherapeutics are effective treatments for substance use and comorbid disorders.

REFERENCES

[1] Logan RW, Williams III WP, McClung CA. Circadian rhythms and addiction: mechanistic insights and future directions. Behav Neurosci 2014;128(3):387–412. https://doi.org/10.1037/a0036268.

[2] Vosko AM, Colwell CS, Avidan AY. Jet lag syndrome: circadian organization, pathophysiology, and management strategies. Nat Sci Sleep 2010;2:187–98. https://doi.org/10.2147/NSS.S6683.

[3] Wang XS, Armstrong ME, Cairns BJ, Key TJ, Travis RC. Shift work and chronic disease: the epidemiological evidence. Occup Med (Lond) 2011;61(2):78–89. https://doi.org/10.1093/occmed/kqr001.

[4] Moore RY, Eichler VB. Loss of a circadian adrenal corticosterone rhythm following suprachiasmatic lesions in the rat. Brain Res 1972;42(1):201–6.

[5] Stephan FK, Zucker I. Circadian rhythms in drinking behavior and locomotor activity of rats are eliminated by hypothalamic lesions. Proc Natl Acad Sci U S A 1972;69(6):1583–6.

[6] Hattar S, Kumar M, Park A, Tong P, Tung J, Yau KW, Berson DM. Central projections of melanopsin-expressing retinal ganglion cells in the mouse. J Comp Neurol 2006;497(3):326–49. https://doi.org/10.1002/cne.20970.

[7] Hattar S, Liao HW, Takao M, Berson DM, Yau KW. Melanopsin-containing retinal ganglion cells: architecture, projections, and intrinsic photosensitivity. Science 2002;295(5557):1065–70. https://doi.org/10.1126/science.1069609.

[8] Hattar S, Lucas RJ, Mrosovsky N, Thompson S, Douglas RH, Hankins MW, Lem J, Biel M, Hofmann F, Foster RG, Yau KW. Melanopsin and rod-cone photoreceptive systems account for all major accessory visual functions in mice. Nature 2003;424(6944):76–81. https://doi.org/10.1038/nature01761.

[9] Rosenwasser AM, Turek FW. Neurobiology of circadian rhythm regulation. Sleep Med Clin 2015;10(4):403–12. https://doi.org/10.1016/j.jsmc.2015.08.003.

[10] Mieda M, Ono D, Hasegawa E, Okamoto H, Honma K, Honma S, Sakurai T. Cellular clocks in AVP neurons of the SCN are critical for interneuronal coupling regulating circadian behavior rhythm. Neuron 2015;85(5):1103–16. https://doi.org/10.1016/j.neuron.2015.02.005.

[11] Barca-Mayo O, Pons-Espinal M, Follert P, Armirotti A, Berdondini L, De Pietri Tonelli D. Astrocyte deletion of Bmal1 alters daily locomotor activity and cognitive functions via GABA signalling. Nat Commun 2017;8. https://doi.org/10.1038/ncomms14336.

[12] Brancaccio M, Maywood ES, Chesham JE, Loudon AS, Hastings MH. A Gq-Ca^{2+} axis controls circuit-level encoding of circadian time in the suprachiasmatic nucleus. Neuron 2013;78(4):714–28. https://doi.org/10.1016/j.neuron.2013.03.011.

[13] Tso CF, Simon T, Greenlaw AC, Puri T, Mieda M, Herzog ED. Astrocytes regulate daily rhythms in the suprachiasmatic nucleus and behavior. Curr Biol 2017;27(7):1055–61. https://doi.org/10.1016/j.cub.2017.02.037.

[14] Flourakis M, Kula-Eversole E, Hutchison AL, Han TH, Aranda K, Moose DL, White KP, Dinner AR, Lear BC, Ren D, Diekman CO, Raman IM, Allada R. A conserved bicycle model for circadian clock control of membrane excitability. Cell 2015;162(4):836–48. https://doi.org/10.1016/j.cell.2015.07.036.

[15] Freeman Jr. GM, Krock RM, Aton SJ, Thaben P, Herzog ED. GABA networks destabilize genetic oscillations in the circadian pacemaker. Neuron 2013;78(5):799–806. https://doi.org/10.1016/j.neuron.2013.04.003.

[16] Lundkvist GB, Kwak Y, Davis EK, Tei H, Block GD. A calcium flux is required for circadian rhythm generation in mammalian pacemaker neurons. J Neurosci 2005;25(33):7682–6. https://doi.org/10.1523/JNEUROSCI.2211-05.2005.

[17] Dziema H, Oatis B, Butcher GQ, Yates R, Hoyt KR, Obrietan K. The ERK/MAP kinase pathway couples light to immediate-early gene expression in the suprachiasmatic nucleus. Eur J Neurosci 2003;17(8):1617–27.

[18] Herzog ED, Takahashi JS, Block GD. Clock controls circadian period in isolated suprachiasmatic nucleus neurons. Nat Neurosci 1998;1(8):708–13. https://doi.org/10.1038/3708.

[19] Liu C, Weaver DR, Strogatz SH, Reppert SM. Cellular construction of a circadian clock: period determination in the suprachiasmatic nuclei. Cell 1997;91(6):855–60.

[20] Yoo SH, Yamazaki S, Lowrey PL, Shimomura K, Ko CH, Buhr ED, Siepka SM, Hong HK, Oh WJ, Yoo OJ, Menaker M, Takahashi JS. PERIOD2:: LUCIFERASE real-time reporting of circadian dynamics reveals persistent circadian oscillations in mouse peripheral tissues. Proc Natl Acad Sci U S A 2004;101(15):5339–46. https://doi.org/10.1073/pnas.0308709101.

[21] Buijs RM, Scheer FA, Kreier F, Yi C, Bos N, Goncharuk VD, Kalsbeek A. Organization of circadian functions: interaction with the body. Prog Brain Res 2006;153:341–60. https://doi.org/10.1016/S0079-6123(06)53020-1.

[22] Kalsbeek A, Yi CX, Cailotto C, la Fleur SE, Fliers E, Buijs RM. Mammalian clock output mechanisms. Essays Biochem 2011;49(1):137–51. https://doi.org/10.1042/BSE0480137.

[23] Bae K, Jin X, Maywood ES, Hastings MH, Reppert SM, Weaver DR. Differential functions of mPer1, mPer2, and mPer3 in the SCN circadian clock. Neuron 2001;30(2):525–36.

[24] Jin X, Shearman LP, Weaver DR, Zylka MJ, de Vries GJ, Reppert SM. A molecular mechanism regulating rhythmic output from the suprachiasmatic circadian clock. Cell 1999;96(1):57–68.

[25] Takahashi JS. Transcriptional architecture of the mammalian circadian clock. Nat Rev Genet 2017;18(3):164–79. https://doi.org/10.1038/nrg.2016.150.

[26] Zhang R, Lahens NF, Ballance HI, Hughes ME, Hogenesch JB. A circadian gene expression atlas in mammals: implications for biology and medicine. Proc Natl Acad Sci U S A 2014;111(45):16219–24. https://doi.org/10.1073/pnas.1408886111.

[27] Lowrey PL, Shimomura K, Antoch MP, Yamazaki S, Zemenides PD, Ralph MR, Menaker M, Takahashi JS. Positional syntenic cloning and functional characterization of the mammalian circadian mutation tau. Science 2000;288(5465):483–92.

[28] Mohawk JA, Takahashi JS. Cell autonomy and synchrony of suprachiasmatic nucleus circadian oscillators. Trends Neurosci 2011;34(7):349–58. https://doi.org/10.1016/j.tins.2011.05.003.

[29] Ueda HR, Hayashi S, Chen W, Sano M, Machida M, Shigeyoshi Y, Iino M, Hashimoto S. System-level identification of transcriptional circuits underlying mammalian circadian clocks. Nat Genet 2005;37(2):187–92. https://doi.org/10.1038/ng1504.

[30] Bertolucci C, Cavallari N, Colognesi I, Aguzzi J, Chen Z, Caruso P, Foá A, Tosini G, Bernardi F, Pinotti M. Evidence for an overlapping role of CLOCK and NPAS2 transcription factors in liver circadian oscillators. Mol Cell Biol 2008;28(9):3070–5. https://doi.org/10.1128/MCB.01931-07.

[31] DeBruyne JP, Weaver DR, Reppert SM. CLOCK and NPAS2 have overlapping roles in the suprachiasmatic circadian clock. Nat Neurosci 2007; 10(5):543–5. https://doi.org/10.1038/nn1884.

[32] Garcia JA, Zhang D, Estill SJ, Michnoff C, Rutter J, Reick M, Scott K, Diaz-Arrastia R, McKnight SL. Impaired cued and contextual memory in NPAS2-deficient mice. Science 2000;288(5474):2226–30.

[33] Landgraf D, Wang LL, Diemer T, Welsh DK. NPAS2 compensates for loss of CLOCK in peripheral circadian oscillators. PLoS Genet 2016;12(2): e1005882. https://doi.org/10.1371/journal.pgen.1005882.

[34] Reick M, Garcia JA, Dudley C, McKnight SL. NPAS2: an analog of clock operative in the mammalian forebrain. Science 2001;293(5529):506–9. https://doi.org/10.1126/science.1060699.

[35] Duffield GE, Best JD, Meurers BH, Bittner A, Loros JJ, Dunlap JC. Circadian programs of transcriptional activation, signaling, and protein turnover revealed by microarray analysis of mammalian cells. Curr Biol 2002;12(7):551–7.

[36] Panda S, Hogenesch JB, Kay SA. Circadian rhythms from flies to human. Nature 2002;417(6886):329–35. https://doi.org/10.1038/417329a.

[37] Storch KF, Lipan O, Leykin I, Viswanathan N, Davis FC, Wong WH, Weitz CJ. Extensive and divergent circadian gene expression in liver and heart. Nature 2002;417(6884):78–83. https://doi.org/10.1038/nature744.

[38] Comasco E, Nordquist N, Gokturk C, Aslund C, Hallman J, Oreland L, Nilsson KW. The clock gene PER2 and sleep problems: association with alcohol consumption among Swedish adolescents. Ups J Med Sci 2010;115(1):41–8. https://doi.org/10.3109/03009731003597127.

[39] Dong L, Bilbao A, Laucht M, Henriksson R, Yakovleva T, Ridinger M, Desrivieres S, Clarke TK, Lourdusamy A, Smolka MN, Cichon S, Blomeyer D, Treutlein J, Perreau-Lenz S, Witt S, Leonardi-Essmann F, Wodarz N, Zill P, Soyka M, Albrecht U, Rietschel M, Lathrop M, Bakalkin G, Spanagel R, Schumann G. Effects of the circadian rhythm gene period 1 (per1) on psychosocial stress-induced alcohol drinking. Am J Psychiatry 2011;168(10):1090–8. https://doi.org/10.1176/appi.ajp.2011.10111579.

[40] Spanagel R, Pendyala G, Abarca C, Zghoul T, Sanchis-Segura C, Magnone MC, Lascorz J, Depner M, Holzberg D, Soyka M, Schreiber S, Matsuda F, Lathrop M, Schumann G, Albrecht U. The clock gene Per2 influences the glutamatergic system and modulates alcohol consumption. Nat Med 2005;11(1):35–42. https://doi.org/10.1038/nm1163.

[41] Kovanen L, Saarikoski ST, Haukka J, Pirkola S, Aromaa A, Lonnqvist J, Partonen T. Circadian clock gene polymorphisms in alcohol use disorders and alcohol consumption. Alcohol Alcohol 2010;45(4):303–11. https://doi.org/10.1093/alcalc/agq035.

[42] Sjoholm LK, Kovanen L, Saarikoski ST, Schalling M, Lavebratt C, Partonen T. CLOCK is suggested to associate with comorbid alcohol use and depressive disorders. J Circadian Rhythms 2010;8:1. https://doi.org/10.1186/1740-3391-8-1.

[43] Barclay NL, Eley TC, Mill J, Wong CC, Zavos HM, Archer SN, Gregory AM. Sleep quality and diurnal preference in a sample of young adults: associations with 5HTTLPR, PER3, and CLOCK 3111. Am J Med Genet B Neuropsychiatr Genet 2011;156B(6):681–90. https://doi.org/10.1002/ajmg.b.31210.

[44] Benedetti F, Dallaspezia S, Fulgosi MC, Lorenzi C, Serretti A, Barbini B, Colombo C, Smeraldi E. Actimetric evidence that CLOCK 3111 T/C SNP influences sleep and activity patterns in patients affected by bipolar depression. Am J Med Genet B Neuropsychiatr Genet 2007;144B(5):631–5. https://doi.org/10.1002/ajmg.b.30475.

[45] Benedetti F, Serretti A, Colombo C, Barbini B, Lorenzi C, Campori E, Smeraldi E. Influence of CLOCK gene polymorphism on circadian mood fluctuation and illness recurrence in bipolar depression. Am J Med Genet B Neuropsychiatr Genet 2003;123B(1):23–6. https://doi.org/10.1002/ajmg.b.20038.

[46] McCarthy MJ, Nievergelt CM, Kelsoe JR, Welsh DK. A survey of genomic studies supports association of circadian clock genes with bipolar disorder spectrum illnesses and lithium response. PLoS One 2012;7(2):e32091. https://doi.org/10.1371/journal.pone.0032091.

[47] Soria V, Martinez-Amoros E, Escaramis G, Valero J, Perez-Egea R, Garcia C, Gutiérrez-Zotes A, Puigdemont D, Bayés M, Crespo JM, Martorell L, Vilella E, Labad A, Vallejo J, Pérez V, Menchón JM, Estivill X, Gratacòs M, Urretavizcaya M. Differential association of circadian genes with mood disorders: CRY1 and NPAS2 are associated with unipolar major depression and CLOCK and VIP with bipolar disorder. Neuropsychopharmacology 2010;35(6):1279–89. https://doi.org/10.1038/npp.2009.230.

[48] Tsuchimine S, Yasui-Furukori N, Kaneda A, Kaneko S. The CLOCK C3111T polymorphism is associated with reward dependence in healthy Japanese subjects. Neuropsychobiology 2013;67(1):1–5. https://doi.org/10.1159/000342383.

[49] Zhang J, Liao G, Liu C, Sun L, Liu Y, Wang Y, Jiang Z, Wang Z. The association of CLOCK gene T3111C polymorphism and hPER3 gene 54-nucleotide repeat polymorphism with Chinese Han people schizophrenics. Mol Biol Rep 2011;38(1):349–54. https://doi.org/10.1007/s11033-010-0114-2.

[50] Ozburn AR, Purohit K, Parekh PK, Kaplan GN, Falcon E, Mukherjee S, Cates HM, McClung CA. Functional implications of the CLOCK 3111T/C single-nucleotide polymorphism. Front Psych 2016;7:67. https://doi.org/10.3389/fpsyt.2016.00067.

[51] Ozburn AR, Falcon E, Mukherjee S, Gillman A, Arey R, Spencer S, McClung CA. The role of clock in ethanol-related behaviors. Neuropsychopharmacology 2013;38(12):2393–400. https://doi.org/10.1038/npp.2013.138.

[52] McClung CA, Sidiropoulou K, Vitaterna M, Takahashi JS, White FJ, Cooper DC, Nestler EJ. Regulation of dopaminergic transmission and cocaine reward by the Clock gene. Proc Natl Acad Sci U S A 2005;102(26):9377–81. https://doi.org/10.1073/pnas.0503584102.

[53] Ozburn AR, Larson EB, Self DW, McClung CA. Cocaine self-administration behaviors in ClockDelta19 mice. Psychopharmacology (Berl) 2012;223(2):169–77. https://doi.org/10.1007/s00213-012-2704-2.

[54] Arey R, McClung CA. An inhibitor of casein kinase 1 epsilon/delta partially normalizes the manic-like behaviors of the ClockDelta19 mouse. Behav Pharmacol 2012;23(4):392–6. https://doi.org/10.1097/FBP.0b013e32835651fd.

[55] Arey RN, Enwright III JF, Spencer SM, Falcon E, Ozburn AR, Ghose S, Tamminga C, McClung CA. An important role for Cholecystokinin, a CLOCK target gene, in the development and treatment of manic-like behaviors. Mol Psychiatry 2013; https://doi.org/10.1038/mp.2013.47.

[56] Coque L, Mukherjee S, Cao JL, Spencer S, Marvin M, Falcon E, Sidor MM, Birnbaum SG, Graham A, Neve RL, Gordon E, Ozburn AR, Goldberg MS, Han MH, Cooper DC, McClung CA. Specific role of VTA dopamine neuronal firing rates and morphology in the reversal of anxiety-related, but not depression-related behavior in the ClockDelta19 mouse model of mania. Neuropsychopharmacology 2011;36 (7):1478–88. https://doi.org/10.1038/npp.2011.33.

[57] Dzirasa K, Coque L, Sidor MM, Kumar S, Dancy EA, Takahashi JS, CA MC, Nicolelis MA. Lithium ameliorates nucleus accumbens phase-signaling dysfunction in a genetic mouse model of mania. J Neurosci 2010;30(48):16314–23. https://doi.org/10.1523/JNEUROSCI.4289-10.2010.

[58] Mukherjee S, Coque L, Cao JL, Kumar J, Chakravarty S, Asaithamby A, Graham A, Gordon E, Enwright III JF, RJ DL, Birnbaum SG, Cooper DC, McClung CA. Knockdown of clock in the ventral tegmental area through RNA interference results in a mixed state of mania and depression-like behavior. Biol Psychiatry 2010;68(6):503–11. https://doi.org/10.1016/j.biopsych.2010.04.031.

[59] Parekh PK, Becker-Krail D, Sundaravelu P, Ishigaki S, Okado H, Sobue G, Huang Y, McClung CA. Altered GluA1 (Gria1) function and accumbal synaptic plasticity in the clockdelta19 model of bipolar mania. Biol Psychiatry 2017; https://doi.org/10.1016/j.biopsych.2017.06.022.

[60] Roybal K, Theobold D, Graham A, DiNieri JA, Russo SJ, Krishnan V, Chakravarty S, Peevey J, Oehrlein N, Birnbaum S, Vitaterna MH, Orsulak P, Takahashi JS, Nestler EJ, Carlezon Jr. WA, McClung CA. Mania-like behavior induced by disruption of CLOCK. Proc Natl Acad Sci U S A 2007; 104(15):6406–11. https://doi.org/10.1073/pnas.0609625104.

[61] Sidor MM, Spencer SM, Dzirasa K, Parekh PK, Tye KM, Warden MR, Arey RN, Enwright III JF, Jacobsen JP, Kumar S, Remillard EM, Caron MG, Deisseroth K, McClung CA. Daytime spikes in dopaminergic activity drive rapid mood-cycling in mice. Mol Psychiatry 2015;20(11):1406–19. https://doi.org/10.1038/mp.2014.167.

[62] Spencer S, Torres-Altoro MI, Falcon E, Arey R, Marvin M, Goldberg M, Bibb JA, McClung CA. A mutation in CLOCK leads to altered dopamine receptor function. J Neurochem 2012;123(1):124–34. https://doi.org/10.1111/j.1471-4159.2012.07857.x.

[63] King DP, Vitaterna MH, Chang AM, Dove WF, Pinto LH, Turek FW, Takahashi JS. The mouse Clock mutation behaves as an antimorph and maps within the W19H deletion, distal of Kit. Genetics 1997;146(3):1049–60.

[64] King DP, Zhao Y, Sangoram AM, Wilsbacher LD, Tanaka M, Antoch MP, Steeves TD, Vitaterna MH, Kornhauser JM, Lowrey PL, Turek FW, Takahashi JS. Positional cloning of the mouse circadian clock gene. Cell 1997;89(4):641–53.

[65] Vitaterna MH, King DP, Chang AM, Kornhauser JM, Lowrey PL, McDonald JD, Dove WF, Pinto LH, Turek FW, Takahashi JS. Mutagenesis and mapping of a mouse gene, Clock, essential for circadian behavior. Science 1994;264(5159):719–25.

[66] Shumay E, Fowler JS, Wang GJ, Logan J, Alia-Klein N, Goldstein RZ, Maloney T, Wong C, Volkow ND. Repeat variation in the human PER2 gene as a new genetic marker associated with cocaine addiction and brain dopamine D2 receptor availability. Transl Psychiatry 2012;2. https://doi.org/10.1038/tp.2012.11.

[67] Volkow ND, Fowler JS, Wang GJ, Baler R, Telang F. Imaging dopamine's role in drug abuse and addiction. Neuropharmacology 2009; 56(Suppl 1):3–8. https://doi.org/10.1016/j.neuropharm.2008.05.022.

[68] Blomeyer D, Buchmann AF, Lascorz J, Zimmermann US, Esser G, Desrivieres S, Schmidt MH, Banaschewski T, Schumann G, Laucht M. Association of PER2 genotype and stressful life events with alcohol drinking in young adults. PLoS One 2013;8(3):e59136. https://doi.org/10.1371/journal.pone.0059136.

[69] Archer SN, Robilliard DL, Skene DJ, Smits M, Williams A, Arendt J, von Schantz M. A length polymorphism in the circadian clock gene Per3 is linked to delayed sleep phase syndrome and extreme diurnal preference. Sleep 2003;26(4):413–5.

[70] Artioli P, Lorenzi C, Pirovano A, Serretti A, Benedetti F, Catalano M, Smeraldi E. How do genes exert their role? Period 3 gene variants and possible influences on mood disorder phenotypes. Eur Neuropsychopharmacol 2007;17(9):587–94. https://doi.org/10.1016/j.euroneuro.2007.03.004.

[71] Benedetti F, Radaelli D, Bernasconi A, Dallaspezia S, Falini A, Scotti G, Lorenzi C, Colombo C, Smeraldi E. Clock genes beyond the clock: CLOCK genotype biases neural correlates of moral valence decision in depressed patients. Genes Brain Behav 2008;7(1):20–5. https://doi.org/10.1111/j.1601-183X.2007.00312.x.

[72] Kripke DF, Nievergelt CM, Joo E, Shekhtman T, Kelsoe JR. Circadian polymorphisms associated with affective disorders. J Circadian Rhythms 2009;7:2. https://doi.org/10.1186/1740-3391-7-2.

[73] Malison RT, Kranzler HR, Yang BZ, Gelernter J. Human clock, PER1 and PER2 polymorphisms: lack of association with cocaine dependence susceptibility and cocaine-induced paranoia. Psychiatr Genet 2006;16(6):245–9. https://doi.org/10.1097/01.ypg.0000242198.59020.ca.

[74] Viola AU, Archer SN, James LM, Groeger JA, Lo JC, Skene DJ, von Schantz M, Dijk DJ. PER3 polymorphism predicts sleep structure and waking performance. Curr Biol 2007;17(7):613–8. https://doi.org/10.1016/j.cub.2007.01.073.

[75] Wang X, Mozhui K, Li Z, Mulligan MK, Ingels JF, Zhou X, Hori RT, Chen H, Cook MN, Williams RW, Lu L. A promoter polymorphism in the Per3 gene is associated with alcohol and stress response. Transl Psychiatry 2012;2:e73. https://doi.org/10.1038/tp.2011.71.

[76] Perreau-Lenz S, Zghoul T, de Fonseca FR, Spanagel R, Bilbao A. Circadian regulation of central ethanol sensitivity by the mPer2 gene. Addict Biol 2009;14(3):253–9. https://doi.org/10.1111/j.1369-1600.2009.00165.x.

[77] Zghoul T, Abarca C, Sanchis-Segura C, Albrecht U, Schumann G, Spanagel R. Ethanol self-administration and reinstatement of ethanol-seeking behavior in Per1(Brdm1) mutant mice. Psychopharmacology (Berl) 2007;190(1):13–9. https://doi.org/10.1007/s00213-006-0592-z.

[78] Brager A, Prosser RA, Glass JD. Acamprosate-responsive brain sites for suppression of ethanol intake and preference. Am J Physiol Regul Integr Comp Physiol 2011;301(4):R1032–1043. https://doi.org/10.1152/ajpregu.00179.2011.

[79] Brager AJ, Prosser RA, Glass JD. Circadian and acamprosate modulation of elevated ethanol drinking in mPer2 clock gene mutant mice. Chronobiol Int 2011;28(8):664–72. https://doi.org/10.3109/07420528.2011.601968.

[80] Gamsby JJ, Templeton EL, Bonvini LA, Wang W, Loros JJ, Dunlap JC, Green AI, Gulick D. The circadian Per1 and Per2 genes influence alcohol intake, reinforcement, and blood alcohol levels. Behav Brain Res 2013;249:15–21. https://doi.org/10.1016/j.bbr.2013.04.016.

[81] Pohl JB, Ghezzi A, Lew LK, Robles RB, Cormack L, Atkinson NS. Circadian genes differentially affect tolerance to ethanol in Drosophila. Alcohol Clin Exp Res 2013;37(11):1862–71. https://doi.org/10.1111/acer.12173.

[82] Liao J, Seggio JA, Ahmad ST. Mutations in the circadian gene period alter behavioral and biochemical responses to ethanol in Drosophila. Behav Brain Res 2016;302:213–9. https://doi.org/10.1016/j.bbr.2016.01.041.

[83] Abarca C, Albrecht U, Spanagel R. Cocaine sensitization and reward are under the influence of circadian genes and rhythm. Proc Natl Acad Sci U S A 2002;99(13):9026–30. https://doi.org/10.1073/pnas.142039099.

[84] Andretic R, Chaney S, Hirsh J. Requirement of circadian genes for cocaine sensitization in Drosophila. Science 1999;285(5430):1066–8.

[85] Uz T, Arslan AD, Kurtuncu M, Imbesi M, Akhisaroglu M, Dwivedi Y, Pandey GN, Manev H. The regional and cellular expression profile of the melatonin receptor MT1 in the central dopaminergic system. Brain Res Mol Brain Res 2005;136(1–2):45–53. https://doi.org/S0169-328X(05)00011-2 [pii].

[86] Uz T, Akhisaroglu M, Ahmed R, Manev H. The pineal gland is critical for circadian period1 expression in the striatum and for circadian cocaine sensitization in mice. Neuropsychopharmacology 2003;28(12):2117–23. https://doi.org/10.1038/sj.npp.1300254.

[87] Halbout B, Perreau-Lenz S, Dixon CI, Stephens DN, Spanagel R. Per1(Brdm1) mice self-administer cocaine and reinstate cocaine-seeking behaviour following extinction. Behav Pharmacol 2011;22(1):76–80. https://doi.org/10.1097/FBP.0b013e328341e9ca.

[88] Oyefeso A, Sedgwick P, Ghodse H. Subjective sleep-wake parameters in treatment-seeking opiate addicts. Drug Alcohol Depend 1997;48(1):9–16.

[89] Caille S, Espejo EF, Koob GF, Stinus L. Dorsal and median raphe serotonergic system lesion does not alter the opiate withdrawal syndrome. Pharmacol Biochem Behav 2002;72(4):979–86.

[90] Stinus L, Robert C, Karasinski P, Limoge A. Continuous quantitative monitoring of spontaneous opiate withdrawal: locomotor activity and sleep disorders. Pharmacol Biochem Behav 1998;59(1):83–9.

[91] Ammon S, Mayer P, Riechert U, Tischmeyer H, Hollt V. Microarray analysis of genes expressed in the frontal cortex of rats chronically treated with morphine and after naloxone precipitated withdrawal. Brain Res Mol Brain Res 2003;112(1–2):113–25.

[92] Ammon-Treiber S, Hollt V. Morphine-induced changes of gene expression in the brain. Addict Biol 2005;10(1):81–9. https://doi.org/10.1080/13556210412331308994.

[93] Hood S, Cassidy P, Mathewson S, Stewart J, Amir S. Daily morphine injection and withdrawal disrupt 24-h wheel running and PERIOD2 expression patterns in the rat limbic forebrain. Neuroscience 2011;186:65–75. https://doi.org/10.1016/j.neuroscience.2011.04.045.

[94] Perreau-Lenz S, Sanchis-Segura C, Leonardi-Essmann F, Schneider M, Spanagel R. Development of morphine-induced tolerance and withdrawal: involvement of the clock gene mPer2. Eur Neuropsychopharmacol 2010;20(7):509–17. https://doi.org/10.1016/j.euroneuro.2010.03.006.

[95] Ando H, Ushijima K, Kumazaki M, Eto T, Takamura T, Irie S, Kaneko S, Fujimura A. Associations of metabolic parameters and ethanol consumption with messenger RNA expression of clock genes in healthy men. Chronobiol Int 2010;27(1):194–203. https://doi.org/10.3109/07420520903398617.

[96] Huang MC, Ho CW, Chen CH, Liu SC, Chen CC, Leu SJ. Reduced expression of circadian clock genes in male alcoholic patients. Alcohol Clin Exp Res 2010;34(11):1899–904. https://doi.org/10.1111/j.1530-0277.2010.01278.x.

[97] Perreau-Lenz S, Vengeliene V, Noori HR, Merlo-Pich EV, Corsi MA, Corti C, Spanagel R. Inhibition of the casein-kinase-1-epsilon/delta prevents relapse-like alcohol drinking. Neuropsychopharmacology 2012;37(9):2121–31. https://doi.org/10.1038/npp.2012.62.

[98] Bryant CD, Graham ME, Distler MG, Munoz MB, Li D, Vezina P, Sokoloff G, Palmer AA. A role for casein kinase 1 epsilon in the locomotor stimulant response to methamphetamine. Psychopharmacology (Berl) 2009;203(4):703–11. https://doi.org/10.1007/s00213-008-1417-z.

[99] Bryant CD, Parker CC, Zhou L, Olker C, Chandrasekaran RY, Wager TT, Bolivar VJ, Loudon AS, Vitaterna MH, Turek FW, Palmer AA. Csnk1e is a genetic regulator of sensitivity to psychostimulants and opioids. Neuropsychopharmacology 2012;37(4):1026–35. https://doi.org/10.1038/npp.2011.287.

[100] Eide EJ, Woolf MF, Kang H, Woolf P, Hurst W, Camacho F, Vielhaber EL, Giovanni A, Virshup DM. Control of mammalian circadian rhythm by CKIepsilon-regulated proteasome-mediated PER2 degradation. Mol Cell Biol 2005;25(7):2795–807. https://doi.org/10.1128/MCB.25.7.2795-2807.2005.

[101] Meng QJ, Logunova L, Maywood ES, Gallego M, Lebiecki J, Brown TM, Sládek M, Semikhodskii AS, NRJ G, Piggins HD, Chesham JE, Bechtold DA, Yoo SH, Takahashi JS, Virshup DM, Boot-Handford RP, Hastings MH, Loudon ASI. Setting clock speed in mammals: the CK1 epsilon tau mutation in mice accelerates circadian pacemakers by selectively destabilizing PERIOD proteins. Neuron 2008;58(1):78–88. https://doi.org/10.1016/j.neuron.2008.01.019.

[102] Cheng MY, Leslie FM, Zhou QY. Expression of prokineticins and their receptors in the adult mouse brain. J Comp Neurol 2006;498(6):796–809. https://doi.org/10.1002/cne.21087.

[103] Jhou TC, Fields HL, Baxter MG, Saper CB, Holland PC. The rostromedial tegmental nucleus (RMTg), a GABAergic afferent to midbrain dopamine neurons, encodes aversive stimuli and inhibits motor responses. Neuron 2009;61(5):786–800. https://doi.org/10.1016/j.neuron.2009.02.001.

[104] Luo AH, Aston-Jones G. Circuit projection from suprachiasmatic nucleus to ventral tegmental area: a novel circadian output pathway. Eur J Neurosci 2009;29(4):748–60. https://doi.org/10.1111/j.1460-9568.2008.06606.x.

[105] Colavito V, Fabene PF, Grassi-Zucconi G, Pifferi F, Lamberty Y, Bentivoglio M, Bertini G. Experimental sleep deprivation as a tool to test memory deficits in rodents. Front Syst Neurosci 2013;7:106. https://doi.org/10.3389/fnsys.2013.00106.

[106] Matzeu A, Zamora-Martinez ER, Martin-Fardon R. The paraventricular nucleus of the thalamus is recruited by both natural rewards and drugs of abuse: recent evidence of a pivotal role for orexin/hypocretin signaling in this thalamic nucleus in drug-seeking behavior. Front Behav Neurosci 2014;8:117. https://doi.org/10.3389/fnbeh.2014.00117.

[107] Neumann PA, Wang Y, Yan Y, Wang Y, Ishikawa M, Cui R, Huang YH, Sesack SR, Schlüter OM, Dong Y. Cocaine-induced synaptic alterations in thalamus to nucleus accumbens projection. Neuropsychopharmacology 2016;41(9):2399–410. https://doi.org/10.1038/npp.2016.52.

[108] Alamilla J, Aguilar-Roblero R. Glutamate and GABA neurotransmission from the paraventricular thalamus to the suprachiasmatic nuclei in the rat. J Biol Rhythms 2010;25(1):28–36. https://doi.org/10.1177/0748730409357771.

[109] Tataroglu O, Aksoy A, Yilmaz A, Canbeyli R. Effect of lesioning the suprachiasmatic nuclei on behavioral despair in rats. Brain Res 2004;1001 (1-2):118–24. https://doi.org/10.1016/j.brainres.2003.11.063 S0006899303041398 [pii].

[110] Tuma J, Strubbe JH, Mocaer E, Koolhaas JM. Anxiolytic-like action of the antidepressant agomelatine (S 20098) after a social defeat requires the integrity of the SCN, Eur Neuropsychopharmacol 2005;15(5):545–55. https://doi.org/S0924-977X(05)00040-4 [pii].

[111] Luo AH, Georges FE, Aston-Jones GS. Novel neurons in ventral tegmental area fire selectively during the active phase of the diurnal cycle. Eur J Neurosci 2008;27(2):408–22. https://doi.org/10.1111/j.1460-9568.2007.05985.x.

[112] Abe M, Herzog ED, Yamazaki S, Straume M, Tei H, Sakaki Y, Menaker M, Block GD. Circadian rhythms in isolated brain regions. J Neurosci 2002;22(1):350–6.

[113] Logan RW, Edgar N, Gillman AG, Hoffman D, Zhu X, McClung CA. Chronic stress induces brain region-specific alterations of molecular rhythms that correlate with depression-like behavior in mice. Biol Psychiatry 2015;78(4):249–58. https://doi.org/10.1016/j.biopsych.2015.01.011.

[114] Sleipness EP, Sorg BA, Jansen HT. Diurnal differences in dopamine transporter and tyrosine hydroxylase levels in rat brain: dependence on the suprachiasmatic nucleus. Brain Res 2007;1129(1):34–42. https://doi.org/10.1016/j.brainres.2006.10.063.

[115] Sleipness EP, Sorg BA, Jansen HT. Contribution of the suprachiasmatic nucleus to day:night variation in cocaine-seeking behavior. Physiol Behav 2007;91(5):523–30. https://doi.org/10.1016/j.physbeh.2007.02.013.

[116] Cheng MY, Bullock CM, Li C, Lee AG, Bermak JC, Belluzzi J, Weaver DR, Leslie FM, Zhou QY. Prokineticin 2 transmits the behavioural circadian rhythm of the suprachiasmatic nucleus. Nature 2002;417(6887):405–10. https://doi.org/10.1038/417405a.

[117] Masumoto KH, Nagano M, Takashima N, Hayasaka N, Hiyama H, Matsumoto S, Inouye ST, Shigeyoshi Y. Distinct localization of prokineticin 2 and prokineticin receptor 2 mRNAs in the rat suprachiasmatic nucleus. Eur J Neurosci 2006;23(11):2959–70. https://doi.org/10.1111/j.1460-9568. 2006.04834.x.

[118] Li JD, Hu WP, Zhou QY. Disruption of the circadian output molecule prokineticin 2 results in anxiolytic and antidepressant-like effects in mice. Neuropsychopharmacology 2009;34(2):367–73. https://doi.org/10.1038/npp.2008.61.

[119] Webb IC, Baltazar RM, Lehman MN, Coolen LM. Bidirectional interactions between the circadian and reward systems: is restricted food access a unique zeitgeber? Eur J Neurosci 2009;30(9):1739–48. https://doi.org/10.1111/j.1460-9568.2009.06966.x.

[120] Hampp G, Ripperger JA, Houben T, Schmutz I, Blex C, Perreau-Lenz S, Brunk I, Spanagel R, Ahnert-Hilger G, Meijer JH, Albrecht U. Regulation of monoamine oxidase A by circadian-clock components implies clock influence on mood. Curr Biol 2008;18(9):678–83. https://doi.org/10.1016/j. cub.2008.04.012.

[120a] Hokfelt T, Rehfeld JF, Skirboll L, Ivemark B, Goldstein M, Markey K. Evidence for coexistence of dopamine and CCK in meso-limbic neurones. Nature 1980;285:476–8.

[121] Schade R, Vick K, Ott T, Sohr R, Pfister C, Bellach J, Golor G, Lemmer B. Circadian rhythms of dopamine and cholecystokinin in nucleus accumbens and striatum of rats—influence on dopaminergic stimulation. Chronobiol Int 1995;12(2):87–99.

[122] Ferris MJ, Espana RA, Locke JL, Konstantopoulos JK, Rose JH, Chen R, Jones SR. Dopamine transporters govern diurnal variation in extracellular dopamine tone. Proc Natl Acad Sci U S A 2014;111(26):2751–9.

[123] Chung S, Lee EJ, Yun S, Choe HK, Park SB, Son HJ, Hwang O, Son GH, Kim K. Impact of circadian nuclear receptor REV-ERBalpha on midbrain dopamine production and mood regulation. Cell 2014;157(4):858–68. https://doi.org/10.1016/j.cell.2014.03.039.

[124] Piechota M, Korostynski M, Sikora M, Golda S, Dzbek J, Przewlocki R. Common transcriptional effects in the mouse striatum following chronic treatment with heroin and methamphetamine. Genes Brain Behav 2012;11(4):404–14. https://doi.org/10.1111/j.1601-183X.2012.00777.x.

[125] Hampp G, Albrecht U. The circadian clock and mood-related behavior. Commun Integr Biol 2008;1(1):1–3.

[126] Cheon S, Park N, Cho S, Kim K. Glucocorticoid-mediated Period2 induction delays the phase of circadian rhythm. Nucleic Acids Res 2013; 41(12):6161–74. https://doi.org/10.1093/nar/gkt307.

[127] Pezuk P, Mohawk JA, Wang LA, Menaker M. Glucocorticoids as entraining signals for peripheral circadian oscillators. Endocrinology 2012; 153(10):4775–83. https://doi.org/10.1210/en.2012-1486.

[128] Reddy TE, Gertz J, Crawford GE, Garabedian MJ, Myers RM. The hypersensitive glucocorticoid response specifically regulates period 1 and expression of circadian genes. Mol Cell Biol 2012;32(18):3756–67. https://doi.org/10.1128/MCB.00062-12.

[129] Agapito M, Mian N, Boyadjieva NI, Sarkar DK. Period 2 gene deletion abolishes beta-endorphin neuronal response to ethanol. Alcohol Clin Exp Res 2010;34(9):1613–8. https://doi.org/10.1111/j.1530-0277.2010.01246.x.

[130] Sarkar DK. Circadian genes, the stress axis, and alcoholism. Alcohol Res 2012;34(3):362–6.

[131] Spanagel R, Kiefer F. Drugs for relapse prevention of alcoholism: ten years of progress. Trends Pharmacol Sci 2008;29(3):109–15. https://doi.org/ 10.1016/j.tips.2007.12.005.

[132] Spanagel R, Rosenwasser AM, Schumann G, Sarkar DK. Alcohol consumption and the body's biological clock. Alcohol Clin Exp Res 2005;29 (8):1550–7.

[133] Ozburn AR, Falcon E, Twaddle A, Nugent AL, Gillman AG, Spencer SM, Arey RN, Mukherjee S, Lyons-Weiler J, Self DW, McClung, A C. Direct regulation of diurnal Drd3 expression and cocaine reward by NPAS2. Biol Psychiatry 2015;77(5):425–33. https://doi.org/10.1016/j.biopsych. 2014.07.030.

[134] Falcon E, Ozburn A, Mukherjee S, Roybal K, McClung CA. Differential regulation of the period genes in striatal regions following cocaine exposure. PLoS One 2013;8(6):e66438. https://doi.org/10.1371/journal.pone.0066438.

[135] McCarthy MJ, Fernandes M, Kranzler HR, Covault JM, Welsh DK. Circadian clock period inversely correlates with illness severity in cells from patients with alcohol use disorders. Alcohol Clin Exp Res 2013; https://doi.org/10.1111/acer.12106.

[136] Chen CP, Kuhn P, Advis JP, Sarkar DK. Chronic ethanol consumption impairs the circadian rhythm of pro-opiomelanocortin and period genes mRNA expression in the hypothalamus of the male rat. J Neurochem 2004;88(6):1547–54.

[137] Filiano AN, Millender-Swain T, Johnson Jr. R, Young ME, Gamble KL, Bailey SM. Chronic ethanol consumption disrupts the core molecular clock and diurnal rhythms of metabolic genes in the liver without affecting the suprachiasmatic nucleus. PLoS One 2013;8(8):e71684. https://doi.org/10.1371/journal.pone.0071684.

[138] Antle MC, Silver R. Orchestrating time: arrangements of the brain circadian clock. Trends Neurosci 2005;28(3):145–51. https://doi.org/10.1016/j.tins.2005.01.003.

[139] Madeira MD, Andrade JP, Lieberman AR, Sousa N, Almeida OF, Paula-Barbosa MM. Chronic alcohol consumption and withdrawal do not induce cell death in the suprachiasmatic nucleus, but lead to irreversible depression of peptide immunoreactivity and mRNA levels. J Neurosci 1997;17(4):1302–19.

[140] Madeira MD, Paula-Barbosa MM. Effects of alcohol on the synthesis and expression of hypothalamic peptides. Brain Res Bull 1999;48(1):3–22.

[141] Aton SJ, Colwell CS, Harmar AJ, Waschek J, Herzog ED. Vasoactive intestinal polypeptide mediates circadian rhythmicity and synchrony in mammalian clock neurons. Nat Neurosci 2005;8(4):476–83. https://doi.org/10.1038/nn1419.

[142] Yan L, Karatsoreos I, Lesauter J, Welsh DK, Kay S, Foley D, Silver R. Exploring spatiotemporal organization of SCN circuits. Cold Spring Harb Symp Quant Biol 2007;72:527–41. https://doi.org/10.1101/sqb.2007.72.037.

[143] Logan RW, McCulley III WD, Seggio JA, Rosenwasser AM. Effects of withdrawal from chronic intermittent ethanol vapor on the level and circadian periodicity of running-wheel activity in C57BL/6J and C3H/HeJ mice. Alcohol Clin Exp Res 2012;36(3):467–76. https://doi.org/10.1111/j.1530-0277.2011.01634.x.

[144] Rosenwasser AM, Logan RW, Fecteau ME. Chronic ethanol intake alters circadian period-responses to brief light pulses in rats. Chronobiol Int 2005;22(2):227–36.

[145] Seggio JA, Fixaris MC, Reed JD, Logan RW, Rosenwasser AM. Chronic ethanol intake alters circadian phase shifting and free-running period in mice. J Biol Rhythms 2009;24(4):304–12. https://doi.org/10.1177/0748730409338449.

[146] Seggio JA, Logan RW, Rosenwasser AM. Chronic ethanol intake modulates photic and non-photic circadian phase responses in the Syrian hamster. Pharmacol Biochem Behav 2007;87(3):297–305. https://doi.org/10.1016/j.pbb.2007.05.001.

[147] Iijima M, Nikaido T, Akiyama M, Moriya T, Shibata S. Methamphetamine-induced, suprachiasmatic nucleus-independent circadian rhythms of activity and mPer gene expression in the striatum of the mouse. Eur J Neurosci 2002;16(5):921–9. https://doi.org/2140 [pii].

[148] Li SX, Wang ZR, Li J, Peng ZG, Zhou W, Zhou M, Lu L. Inhibition of period1 gene attenuates the morphine-induced ERK-CREB activation in frontal cortex, hippocampus, and striatum in mice. Am J Drug Alcohol Abuse 2008;34(6):673–82. https://doi.org/10.1080/00952990802308197.

[149] Nikaido T, Akiyama M, Moriya T, Shibata S. Sensitized increase of period gene expression in the mouse caudate/putamen caused by repeated injection of methamphetamine. Mol Pharmacol 2001;59(4):894–900.

[150] Sanchis-Segura C, Jancic D, Jimenez-Minchan M, Barco A. Inhibition of cAMP responsive element binding protein in striatal neurons enhances approach and avoidance responses toward morphine—and morphine withdrawal-related cues. Front Behav Neurosci 2009;3. https://doi.org/10.3389/neuro.08.030.2009.

[151] Wongchitrat P, Mukda S, Phansuwan-Pujito P, Govitrapong P. Effect of amphetamine on the clock gene expression in rat striatum. Neurosci Lett 2013;542:126–30. https://doi.org/10.1016/j.neulet.2013.03.009.

[152] Mohawk JA, Baer ML, Menaker M. The methamphetamine-sensitive circadian oscillator does not employ canonical clock genes. Proc Natl Acad Sci U S A 2009;106(9):3519–24. https://doi.org/10.1073/pnas.0813366106.

[153] Salaberry NL, Mateo M, Mendoza J. The clock gene rev-erbalpha regulates methamphetamine actions on circadian timekeeping in the mouse brain. Mol Neurobiol 2017;54(7):5327–34. https://doi.org/10.1007/s12035-016-0076-z.

[154] Lynch WJ, Girgenti MJ, Breslin FJ, Newton SS, Taylor JR. Gene profiling the response to repeated cocaine self-administration in dorsal striatum: a focus on circadian genes. Brain Res 2008;1213:166–77. https://doi.org/10.1016/j.brainres.2008.02.106.

[155] Akhisaroglu M, Kurtuncu M, Manev H, Uz T. Diurnal rhythms in quinpirole-induced locomotor behaviors and striatal D2/D3 receptor levels in mice. Pharmacol Biochem Behav 2005;80(3):371–7. https://doi.org/10.1016/j.pbb.2004.11.016.

[156] Imbesi M, Yildiz S, Dirim Arslan A, Sharma R, Manev H, Uz T. Dopamine receptor-mediated regulation of neuronal "clock" gene expression. Neuroscience 2009;158(2):537–44. https://doi.org/10.1016/j.neuroscience.2008.10.044.

[157] McElroy B, Zakaria A, Glass JD, Prosser RA. Ethanol modulates mammalian circadian clock phase resetting through extrasynaptic GABA receptor activation. Neuroscience 2009;164(2):842–8. https://doi.org/10.1016/j.neuroscience.2009.08.020.

[158] Prosser RA, Glass JD. The mammalian circadian clock exhibits acute tolerance to ethanol. Alcohol Clin Exp Res 2009;33(12):2088–93. https://doi.org/10.1111/j.1530-0277.2009.01048.x.

[159] Brager AJ, Ruby CL, Prosser RA, Glass JD. Chronic ethanol disrupts circadian photic entrainment and daily locomotor activity in the mouse. Alcohol Clin Exp Res 2010;34(7):1266–73. https://doi.org/10.1111/j.1530-0277.2010.01204.x.

[160] Brager AJ, Ruby CL, Prosser RA, Glass JD. Acute ethanol disrupts photic and serotonergic circadian clock phase-resetting in the mouse. Alcohol Clin Exp Res 2011;35(8):1467–74. https://doi.org/10.1111/j.1530-0277.2011.01483.x.

[161] Acuna-Goycolea C, Obrietan K, van den Pol AN. Cannabinoids excite circadian clock neurons. J Neurosci 2010;30(30):10061–6. https://doi.org/10.1523/JNEUROSCI.5838-09.2010.

[162] Biello SM, Dafters RI. MDMA and fenfluramine alter the response of the circadian clock to a serotonin agonist in vitro. Brain Res 2001;920(1-2):202–9.

[163] Yang JJ, Wang YT, Cheng PC, Kuo YJ, Huang RC. Cholinergic modulation of neuronal excitability in the rat suprachiasmatic nucleus. J Neurophysiol 2010;103(3):1397–409. https://doi.org/10.1152/jn.00877.2009.

[164] Liu C, Gillette MU. Cholinergic regulation of the suprachiasmatic nucleus circadian rhythm via a muscarinic mechanism at night. J Neurosci 1996;16(2):744–51.

[165] Ferguson SA, Kennaway DJ, Moyer RW. Nicotine phase shifts the 6-sulphatoxymelatonin rhythm and induces c-Fos in the SCN of rats. Brain Res Bull 1999;48(5):527–38.

[166] Pietila K, Laakso I, Ahtee L. Chronic oral nicotine administration affects the circadian rhythm of dopamine and 5-hydroxytryptamine metabolism in the striata of mice. Naunyn Schmiedebergs Arch Pharmacol 1995;353(1):110–5.

[167] Ruby CL, Prosser RA, DePaul MA, Roberts RJ, Glass JD. Acute ethanol impairs photic and nonphotic circadian phase resetting in the Syrian hamster. Am J Physiol Regul Integr Comp Physiol 2009;296(2):R411–418. https://doi.org/10.1152/ajpregu.90782.2008.

[168] Cutler DJ, Mundey MK, Mason R. Electrophysiological effects of opioid receptor activation on Syrian hamster suprachiasmatic nucleus neurones in vitro. Brain Res Bull 1999;50(2):119–25.

[169] Behnke M, Smith VC, Committee on Substance Abuse, Committee on Fetus and Newborn. Prenatal substance abuse: short- and long-term effects on the exposed fetus. Pediatrics 2013;131(3):e1009–1024. https://doi.org/10.1542/peds.2012-3931.

[170] Molteno CD, Jacobson JL, Carter RC, Dodge NC, Jacobson SW. Infant emotional withdrawal: a precursor of affective and cognitive disturbance in fetal alcohol spectrum disorders. Alcohol Clin Exp Res 2014;38(2):479–88. https://doi.org/10.1111/acer.12240.

[171] O'Connor MJ, Kasari C. Prenatal alcohol exposure and depressive features in children. Alcohol Clin Exp Res 2000;24(7):1084–92.

[172] Pei J, Denys K, Hughes J, Rasmussen C. Mental health issues in fetal alcohol spectrum disorder. J Ment Health 2011;20(5):438–48. https://doi.org/10.3109/09638237.2011.577113.

[173] Streissguth AP, Bookstein FL, Barr HM, Sampson PD, O'Malley K, Young JK. Risk factors for adverse life outcomes in fetal alcohol syndrome and fetal alcohol effects. J Dev Behav Pediatr 2004;25(4):228–38.

[174] Treit S, Lebel C, Baugh L, Rasmussen C, Andrew G, Beaulieu C. Longitudinal MRI reveals altered trajectory of brain development during childhood and adolescence in fetal alcohol spectrum disorders. J Neurosci 2013;33(24):10098–109. https://doi.org/10.1523/JNEUROSCI.5004-12.2013.

[175] Alati R, Al Mamun A, Williams GM, O'Callaghan M, Najman JM, Bor W. In utero alcohol exposure and prediction of alcohol disorders in early adulthood: a birth cohort study. Arch Gen Psychiatry 2006;63(9):1009–16. https://doi.org/10.1001/archpsyc.63.9.1009.

[176] Baer JS, Barr HM, Bookstein FL, Sampson PD, Streissguth AP. Prenatal alcohol exposure and family history of alcoholism in the etiology of adolescent alcohol problems. J Stud Alcohol 1998;59(5):533–43.

[177] Chen ML, Olson HC, Picciano JF, Starr JR, Owens J. Sleep problems in children with fetal alcohol spectrum disorders. J Clin Sleep Med 2012;8(4):421–9. https://doi.org/10.5664/jcsm.2038.

[178] Troese M, Fukumizu M, Sallinen BJ, Gilles AA, Wellman JD, Paul JA, Brown ER, Hayes MJ. Sleep fragmentation and evidence for sleep debt in alcohol-exposed infants. Early Hum Dev 2008;84(9):577–85. https://doi.org/10.1016/j.earlhumdev.2008.02.001.

[179] Verma P, Hellemans KG, Choi FY, Yu W, Weinberg J. Circadian phase and sex effects on depressive/anxiety-like behaviors and HPA axis responses to acute stress. Physiol Behav 2010;99(3):276–85. https://doi.org/10.1016/j.physbeh.2009.11.002.

[180] Allen GC, Farnell YZ, Maeng JU, West JR, Chen WJ, Earnest DJ. Long-term effects of neonatal alcohol exposure on photic reentrainment and phase-shifting responses of the activity rhythm in adult rats. Alcohol 2005;37(2):79–88. https://doi.org/10.1016/j.alcohol.2005.11.003.

[181] Allen GC, West JR, Chen WJ, Earnest DJ. Neonatal alcohol exposure permanently disrupts the circadian properties and photic entrainment of the activity rhythm in adult rats. Alcohol Clin Exp Res 2005;29(10):1845–52.

[182] Farnell YZ, West JR, Chen WJ, Allen GC, Earnest DJ. Developmental alcohol exposure alters light-induced phase shifts of the circadian activity rhythm in rats. Alcohol Clin Exp Res 2004;28(7):1020–7.

[183] Sakata-Haga H, Dominguez HD, Sei H, Fukui Y, Riley EP, Thomas JD. Alterations in circadian rhythm phase shifting ability in rats following ethanol exposure during the third trimester brain growth spurt. Alcohol Clin Exp Res 2006;30(5):899–907. https://doi.org/10.1111/j.1530-0277.2006.00105.x.

[184] Shang EH, Zhdanova IV. The circadian system is a target and modulator of prenatal cocaine effects. PLoS One 2007;2(7):e587. https://doi.org/10.1371/journal.pone.0000587.

[185] Strother WN, Vorhees CV, Lehman MN. Long-term effects of early cocaine exposure on the light responsiveness of the adult circadian timing system. Neurotoxicol Teratol 1998;20(5):555–64.

[186] Chmielewski CE, Hernandez LM, Quesada A, Pozas JA, Picabea L, Prada FA. Effects of ethanol on the inner layers of chick retina during development. Alcohol 1997;14(4):313–7.

[187] Tenkova T, Young C, Dikranian K, Labruyere J, Olney JW. Ethanol-induced apoptosis in the developing visual system during synaptogenesis. Invest Ophthalmol Vis Sci 2003;44(7):2809–17.

[188] Parson SH, Sojitra NM. Loss of myelinated axons is specific to the central nervous system in a mouse model of the fetal alcohol syndrome. J Anat 1995;187(Pt 3):739–48.

[189] Sari Y, Zhou FC. Prenatal alcohol exposure causes long-term serotonin neuron deficit in mice. Alcohol Clin Exp Res 2004;28(6):941–8.

[190] Sliwowska JH, Song HJ, Bodnar T, Weinberg J. Prenatal alcohol exposure results in long-term serotonin neuron deficits in female rats: modulatory role of ovarian steroids. Alcohol Clin Exp Res 2013; https://doi.org/10.1111/acer.12224.

[191] Kim YI, Choi HJ, Colwell CS. Brain-derived neurotrophic factor regulation of N-methyl-D-aspartate receptor-mediated synaptic currents in suprachiasmatic nucleus neurons. J Neurosci Res 2006;84(7):1512–20. https://doi.org/10.1002/jnr.21063.

[192] Liang FQ, Allen G, Earnest D. Role of brain-derived neurotrophic factor in the circadian regulation of the suprachiasmatic pacemaker by light. J Neurosci 2000;20(8):2978–87.

[192a] Seggio JA, Possidente B, Ahmad ST. Larval ethanol exposure alters adult circadian free-running locomotor activity rhythm in *Drosophila melanogaster*, Chronobiol Int 2012;29(1):75–81. https://doi.org/10.3109/07420528.2011.635236.

[193] Ahmad ST, Steinmetz SB, Bussey HM, Possidente B, Seggio JA. Larval ethanol exposure alters free-running circadian rhythm and per Locus transcription in adult *D. melanogaster* period mutants. Behav Brain Res 2013;241:50–5. https://doi.org/10.1016/j.bbr.2012.11.035.

[193a] Shearman LP, Zylka MJ, Weaver DR, Kolakowski Jr. LF, Reppert SM. Two period Homologs: circadian expression and photic regulation in the suprachiasmatic nuclei. Neuron 1997;19(6):1261–9. https://doi.org/10.1016/S0896-6273(00)80417-1.

[194] Agapito MA, Barreira JC, Logan RW, Sarkar DK. Evidence for possible period 2 gene mediation of the effects of alcohol exposure during the postnatal period on genes associated with maintaining metabolic signaling in the mouse hypothalamus. Alcohol Clin Exp Res 2013, 263–269; https://doi.org/10.1111/j.1530-0277.2012.01871.x.

[195] Chen CP, Kuhn P, Advis JP, Sarkar DK. Prenatal ethanol exposure alters the expression of period genes governing the circadian function of beta-endorphin neurons in the hypothalamus. J Neurochem 2006;97(4):1026–33. https://doi.org/10.1111/j.1471-4159.2006.03839.x.

[196] Brooks E, Canal MM. Development of circadian rhythms: role of postnatal light environment. Neurosci Biobehav Rev 2013;37(4):551–60. https://doi.org/10.1016/j.neubiorev.2013.02.012.

[197] Hofstetter JR, Grahame NJ, Mayeda AR. Circadian activity rhythms in high-alcohol-preferring and low-alcohol-preferring mice. Alcohol 2003; 30(1):81–5.

[198] Crabbe JC, Metten P, Rhodes JS, Yu CH, Brown LL, Phillips TJ, Finn DA. A line of mice selected for high blood ethanol concentrations shows drinking in the dark to intoxication. Biol Psychiatry 2009;65(8):662–70. https://doi.org/10.1016/j.biopsych.2008.11.002.

[199] Crabbe JC, Spence SE, Brown LL, Metten P. Alcohol preference drinking in a mouse line selectively bred for high drinking in the dark. Alcohol 2011;45(5):427–40. https://doi.org/10.1016/j.alcohol.2010.12.001.

[200] Rhodes JS, Best K, Belknap JK, Finn DA, Crabbe JC. Evaluation of a simple model of ethanol drinking to intoxication in C57BL/6J mice. Physiol Behav 2005;84(1):53–63. https://doi.org/10.1016/j.physbeh.2004.10.007.

[201] McCulley III WD, Ascheid S, Crabbe JC, Rosenwasser AM. Selective breeding for ethanol-related traits alters circadian phenotype. Alcohol 2013;47(3):187–94. https://doi.org/10.1016/j.alcohol.2013.01.001.

[202] Rosenwasser AM, Fecteau ME, Logan RW, Reed JD, Cotter SJ, Seggio JA. Circadian activity rhythms in selectively bred ethanol-preferring and nonpreferring rats. Alcohol 2005;36(2):69–81. https://doi.org/10.1016/j.alcohol.2005.07.001.

[203] Crabbe JC, Phillips TJ. Selective breeding for alcohol withdrawal severity. Behav Genet 1993;23(2):171–7.

[204] Kosobud A, Crabbe JC. Ethanol withdrawal in mice bred to be genetically prone or resistant to ethanol withdrawal seizures. J Pharmacol Exp Therap 1986;238(1):170–7.

[205] Crabbe JC, Phillips TJ, Buck KJ, Cunningham CL, Belknap JK. Identifying genes for alcohol and drug sensitivity: recent progress and future directions. Trends Neurosci 1999;22(4):173–9.

[206] Shimomura K, Low-Zeddies SS, King DP, Steeves TD, Whiteley A, Kushla J, Zemenides PD, Lin A, Vitaterna MH, Churchill GA, Takahashi JS. Genome-wide epistatic interaction analysis reveals complex genetic determinants of circadian behavior in mice. Genome Res 2001;11(6):959–80. https://doi.org/10.1101/gr.171601.

[207] Belenky MA, Yarom Y, Pickard GE. Heterogeneous expression of gamma-aminobutyric acid and gamma-aminobutyric acid-associated receptors and transporters in the rat suprachiasmatic nucleus. J Comp Neurol 2008;506(4):708–32. https://doi.org/10.1002/cne.21553.

[208] Moore RY, Speh JC. GABA is the principal neurotransmitter of the circadian system. Neurosci Lett 1993;150(1):112–6.

[209] Buck KJ, Finn DA. Genetic factors in addiction: QTL mapping and candidate gene studies implicate GABAergic genes in alcohol and barbiturate withdrawal in mice. Addiction 2001;96(1):139–49. https://doi.org/10.1080/09652140020017012.

[210] Lovinger DM. Communication networks in the brain: neurons, receptors, neurotransmitters, and alcohol. Alcohol Res Health 2008;31(3):196–214.

[211] Sprow GM, Thiele TE. The neurobiology of binge-like ethanol drinking: evidence from rodent models. Physiol Behav 2012;106(3):325–31. https://doi.org/10.1016/j.physbeh.2011.12.026.

[212] Bergeson SE, Kyle Warren R, Crabbe JC, Metten P, Gene Erwin V, Belknap JK. Chromosomal loci influencing chronic alcohol withdrawal severity. Mamm Genome 2003;14(7):454–63. https://doi.org/10.1007/s00335-002-2254-4.

[213] Iancu OD, Oberbeck D, Darakjian P, Metten P, McWeeney S, Crabbe JC, Hitzemann R. Selection for drinking in the dark alters brain gene coexpression networks. Alcohol Clin Exp Res 2013;37(8):1295–303. https://doi.org/10.1111/acer.12100.

[214] Frone MR. Workplace substance use climate: prevalence and distribution in the U.S. Workforce. J Subst Use 2012;71(1):72–83.

[215] Trinkoff AM, Storr CL. Work schedule characteristics and substance use in nurses. Am J Ind Med 1998;34(3):266–71.

[216] Clark JW, Fixaris MC, Belanger GV, Rosenwasser AM. Repeated light-dark phase shifts modulate voluntary ethanol intake in male and female high alcohol-drinking (HAD1) rats. Alcohol Clin Exp Res 2007;31(10):1699–706. https://doi.org/10.1111/j.1530-0277.2007.00476.x.

[217] Rosenwasser AM, Clark JW, Fixaris MC, Belanger GV, Foster JA. Effects of repeated light-dark phase shifts on voluntary ethanol and water intake in male and female Fischer and Lewis rats. Alcohol 2010;44(3):229–37. https://doi.org/10.1016/j.alcohol.2010.03.002.

[218] Gauvin DV, Baird TJ, Vanecek SA, Briscoe RJ, Vallett M, Holloway FA. Effects of time-of-day and photoperiod phase shifts on voluntary ethanol consumption in rats. Alcohol Clin Exp Res 1997;21(5):817–25.

[219] Rosenwasser AM, Fixaris MC. Chronobiology of alcohol: studies in C57BL/6J and DBA/2J inbred mice. Physiol Behav 2013;110–111:140–7. https://doi.org/10.1016/j.physbeh.2013.01.001.

[220] Trujillo JL, Do DT, Grahame NJ, Roberts AJ, Gorman MR. Ethanol consumption in mice: relationships with circadian period and entrainment. Alcohol 2011;45(2):147–59. https://doi.org/10.1016/j.alcohol.2010.08.016.

[221] Chen B, Wang Y, Liu X, Liu Z, Dong Y, Huang YH. Sleep Regulates Incubation of Cocaine Craving. J Neurosci 2015;35(39):13300–10. https://doi.org/10.1523/JNEUROSCI.1065-15.2015.

[222] Liu Z, Wang Y, Cai L, Li Y, Chen B, Dong Y, Huang YH. Prefrontal cortex to accumbens projections in sleep regulation of reward. J Neurosci 2016;36(30):7897–910. https://doi.org/10.1523/JNEUROSCI.0347-16.2016.

[223] Sorg BA, Stark G, Sergeeva A, Jansen HT. Photoperiodic suppression of drug reinstatement. Neuroscience 2011;176:284–95. https://doi.org/10.1016/j.neuroscience.2010.12.022.

[224] LeGates TA, Altimus CM, Wang H, Lee HK, Yang S, Zhao H, Kirkwood A, Todd Weber E, Hattar S. Aberrant light directly impairs mood and learning through melanopsin-expressing neurons. Nature 2012;491(7425):594–8. https://doi.org/10.1038/nature11673.

[225] Gibson EM, Wang C, Tjho S, Khattar N, Kriegsfeld LJ. Experimental 'jet lag' inhibits adult neurogenesis and produces long-term cognitive deficits in female hamsters. PLoS One 2010;5(12):e15267. https://doi.org/10.1371/journal.pone.0015267.

[226] Kiessling S, Eichele G, Oster H. Adrenal glucocorticoids have a key role in circadian resynchronization in a mouse model of jet lag. J Clin Investig 2010;120(7):2600–9. https://doi.org/10.1172/JCI41192.

[227] Liston C, Cichon JM, Jeanneteau F, Jia Z, Chao MV, Gan WB. Circadian glucocorticoid oscillations promote learning-dependent synapse formation and maintenance. Nat Neurosci 2013;16(6):698–705. https://doi.org/10.1038/nn.3387.

[228] Dulcis D, Jamshidi P, Leutgeb S, Spitzer NC. Neurotransmitter switching in the adult brain regulates behavior. Science 2013;340(6131):449–53. https://doi.org/10.1126/science.1234152.

[229] Kumar U. Colocalization of somatostatin receptor subtypes (SSTR1-5) with somatostatin, NADPH-diaphorase (NADPH-d), and tyrosine hydroxylase in the rat hypothalamus. J Comp Neurol 2007;504(2):185–205. https://doi.org/10.1002/cne.21444.

[230] Randler C. Morningness-eveningness comparison in adolescents from different countries around the world. Chronobiol Int 2008;25(6):1017–28. https://doi.org/10.1080/07420520802551519.

[231] Roenneberg T, Kuehnle T, Pramstaller PP, Ricken J, Havel M, Guth A, Merrow M. A marker for the end of adolescence. Curr Biol 2004;14(24):R1038–1039. https://doi.org/10.1016/j.cub.2004.11.039.

[232] Crowley SJ, Van Reen E, LeBourgeois MK, Acebo C, Tarokh L, Seifer R, Barker DH, Carskadon MA. A longitudinal assessment of sleep timing, circadian phase, and phase angle of entrainment across human adolescence. PLoS One 2014;9(11):e112199. https://doi.org/10.1371/journal.pone.0112199.

[233] Carskadon MA. Sleep in adolescents: the perfect storm. Pediatr Clin North Am 2011;58(3):637–47. https://doi.org/10.1016/j.pcl.2011.03.003.

[234] Touitou Y. Adolescent sleep misalignment: a chronic jet lag and a matter of public health. J Physiol Paris 2013;107(4):323–6. https://doi.org/10.1016/j.jphysparis.2013.03.008.

[235] Chen K, Kandel DB. The natural history of drug use from adolescence to the mid-thirties in a general population sample. Am J Public Health 1995;85(1):41–7.

[236] McKnight-Eily LR, Eaton DK, Lowry R, Croft JB, Presley-Cantrell L, Perry GS. Relationships between hours of sleep and health-risk behaviors in US adolescent students. Prev Med 2011;53(4-5):271–3. https://doi.org/10.1016/j.ypmed.2011.06.020.

[237] Paiva T, Gaspar T, Matos MG. Mutual relations between sleep deprivation, sleep stealers and risk behaviours in adolescents. Sleep Sci 2016;9(1):7–13. https://doi.org/10.1016/j.slsci.2016.02.176.

[238] Sivertsen B, Skogen JC, Jakobsen R, Hysing M. Sleep and use of alcohol and drug in adolescence. A large population-based study of Norwegian adolescents aged 16 to 19 years. Drug Alcohol Depend 2015;149:180–6. https://doi.org/10.1016/j.drugalcdep.2015.01.045.

[239] Hasler BP, Kirisci L, Clark DB. Restless sleep and variable sleep timing during late childhood accelerate the onset of alcohol and other drug involvement. J Stud Alcohol Drugs 2016;77(4):649–55.

[240] Hasler BP, Martin CS, Wood DS, Rosario B, Clark DB. A longitudinal study of insomnia and other sleep complaints in adolescents with and without alcohol use disorders. Alcohol Clin Exp Res 2014;38(8):2225–33. https://doi.org/10.1111/acer.12474.

[241] O'Brien EM, Mindell JA. Sleep and risk-taking behavior in adolescents. Behav Sleep Med 2005;3(3):113–33. https://doi.org/10.1207/s15402010bsm0303_1.

[242] Pasch KE, Laska MN, Lytle LA, Moe SG. Adolescent sleep, risk behaviors, and depressive symptoms: are they linked? Am J Health Behav 2010;34(2):237–48.

[243] Hasler BP, Germain A, Nofzinger EA, Kupfer DJ, Krafty RT, Rothenberger SD, James JA, Buysse DJ. Chronotype and diurnal patterns of positive affect and affective neural circuitry in primary insomnia. J Sleep Res 2012;21(5):515–26. https://doi.org/10.1111/j.1365-2869.2012.01002.x.

[244] Miller MA, Rothenberger SD, Hasler BP, Donofry SD, Wong PM, Manuck SB, Kamarck TW, Roecklein KA. Chronotype predicts positive affect rhythms measured by ecological momentary assessment. Chronobiol Int 2015;32(3):376–84. https://doi.org/10.3109/07420528.2014.983602.

[245] Murray G, Nicholas CL, Kleiman J, Dwyer R, Carrington MJ, Allen NB, Trinder J. Nature's clocks and human mood: the circadian system modulates reward motivation. Emotion 2009;9(5):705–16. https://doi.org/10.1037/a0017080.

[246] Adan A, Natale V, Caci H, Prat G. Relationship between circadian typology and functional and dysfunctional impulsivity. Chronobiol Int 2010;27(3):606–19. https://doi.org/10.3109/07420521003663827.

[247] Caci H, Robert P, Boyer P. Novelty seekers and impulsive subjects are low in morningness. Eur Psychiatry 2004;19(2):79–84. https://doi.org/10.1016/j.eurpsy.2003.09.007.

[248] Drennan MD, Klauber MR, Kripke DF, Goyette LM. The effects of depression and age on the Horne-Ostberg morningness-eveningness score. J Affect Disord 1991;23(2):93–8.

[249] Hasler BP, Allen JJ, Sbarra DA, Bootzin RR, Bernert RA. Morningness-eveningness and depression: preliminary evidence for the role of the behavioral activation system and positive affect. Psychiatry Res 2010;176(2-3):166–73. https://doi.org/10.1016/j.psychres.2009.06.006.

[250] Merikanto I, Kronholm E, Peltonen M, Laatikainen T, Vartiainen E, Partonen T. Circadian preference links to depression in general adult population. J Affect Disord 2015;188:143–8. https://doi.org/10.1016/j.jad.2015.08.061.

[251] Owens JA, Dearth-Wesley T, Lewin D, Gioia G, Whitaker RC. Self-Regulation and sleep duration, sleepiness, and chronotype in adolescents. Pediatrics 2016;138(6). https://doi.org/10.1542/peds.2016-1406.

[252] Tonetti L, Adan A, Caci H, De Pascalis V, Fabbri M, Natale V. Morningness-eveningness preference and sensation seeking. Eur Psychiatry 2010;25 (2):111–5. https://doi.org/10.1016/j.eurpsy.2009.09.007.

[253] Buysse DJ, Nofzinger EA, Germain A, Meltzer CC, Wood A, Ombao H, Kupfer DJ, Moore RY. Regional brain glucose metabolism during morning and evening wakefulness in humans: preliminary findings. Sleep 2004;27(7):1245–54.

[254] Germain A, Nofzinger EA, Meltzer CC, Wood A, Kupfer DJ, Moore RY, Buysse DJ. Diurnal variation in regional brain glucose metabolism in depression. Biol Psychiatry 2007;62(5):438–45. https://doi.org/10.1016/j.biopsych.2006.09.043.

[255] Hasler BP, Dahl RE, Holm SM, Jakubcak JL, Ryan ND, Silk JS, Phillips ML, Forbes EE. Weekend-weekday advances in sleep timing are associated with altered reward-related brain function in healthy adolescents. Biol Psychol 2012;91(3):334–41. https://doi.org/10.1016/j.biopsycho.2012.08.008.

[256] Forbes EE, Dahl RE, Almeida JR, Ferrell RE, Nimgaonkar VL, Mansour H, Sciarrillo SR, Holm SM, Rodriguez EE, Phillips ML. PER2 rs2304672 polymorphism moderates circadian-relevant reward circuitry activity in adolescents. Biol Psychiatry 2012;71(5):451–7. https://doi.org/10.1016/j.biopsych.2011.10.012.

[257] Holm SM, Forbes EE, Ryan ND, Phillips ML, Tarr JA, Dahl RE. Reward-related brain function and sleep in pre/early pubertal and mid/late pubertal adolescents. J Adolesc Health 2009;45(4):326–34. https://doi.org/10.1016/j.jadohealth.2009.04.001.

[258] Hasler BP, Casement MD, Sitnick SL, Shaw DS, Forbes EE. Eveningness among late adolescent males predicts neural reactivity to reward and alcohol dependence two years later. Behav Brain Res 2017;327:112–20. https://doi.org/10.1016/j.bbr.2017.02.024.

[259] Mullin BC, Phillips ML, Siegle GJ, Buysse DJ, Forbes EE, Franzen PL. Sleep deprivation amplifies striatal activation to monetary reward. Psychol Med 2013;43(10):2215–25. https://doi.org/10.1017/S0033291712002875.

[260] Venkatraman V, Chuah YM, Huettel SA, Chee MW. Sleep deprivation elevates expectation of gains and attenuates response to losses following risky decisions. Sleep 2007;30(5):603–9.

[261] Forbes EE, Ryan ND, Phillips ML, Manuck SB, Worthman CM, Moyles DL, Jill A, Tarr MSW, Samantha R, Sciarrillo BA, Dahl RE. Healthy adolescents' neural response to reward: associations with puberty, positive affect, and depressive symptoms. J Am Acad Child Adolesc Psychiatry 2010;49(2):162–72 e161-165.

[262] Franzen PL, Buysse DJ, Forbes EE, Jones NP, Ohlsen TJ, Germain A. Sleep restriction lowers striatal responses to the receipt of monetary reward in adolescents. Sleep 2016;39S:A93.

[263] Germain A, McNamee R, Khan H, McMakin DL, Franzen P, Forbes EE. Sleep restriction amplifies neural response to reward cues compared to normal sleep and sleep deprivation. Sleep 2016;39S:A94.

[264] Owens J, Wang G, Lewin D, Skora E, Baylor A. Association between short sleep duration and risk behavior factors in middle school students. Sleep 2016;.

[265] Volkow ND, Tomasi D, Wang GJ, Telang F, Fowler JS, Logan J, Benveniste H, Kim R, Thanos PK, Ferre S. Evidence that sleep deprivation down-regulates dopamine D2R in ventral striatum in the human brain. J Neurosci 2012;32(19):6711–7. https://doi.org/10.1523/JNEUROSCI.0045-12.2012.

[266] Roenneberg T, Wirz-Justice A, Merrow M. Life between clocks: daily temporal patterns of human chronotypes. J Biol Rhythms 2003;18(1):80–90. https://doi.org/10.1177/0748730402239679.

[267] Horne JA, Ostberg O. A self-assessment questionnaire to determine morningness-eveningness in human circadian rhythms. Int J Chronobiol 1976;4(2):97–110.

[268] Katzenberg D, Young T, Finn L, Lin L, King DP, Takahashi JS, Mignot E. A CLOCK polymorphism associated with human diurnal preference. Sleep 1998;21(6):569–76.

[268a] Johansson C, Willeit M, Smedh C, Ekholm J, Paunio T, Kieseppä T, Lichtermann D, Praschak-Rieder N, Neumeister A, Nilsson LG, Kasper S, Peltonen L, Adolfsson R, Schalling M, Partonen T. Circadian clock-related polymorphisms in seasonal affective disorder and their relevance to diurnal preference. Neuropsychopharmacology 2003;28(4):734–9.

[268b] Kennaway DJ. Clock genes at the heart of depression. J Psychopharmacol 2010;24(2 Suppl.):5–14.

[269] Mansour HA, Monk TH, Nimgaonkar VL. Circadian genes and bipolar disorder. Ann Med 2005;37(3):196–205. https://doi.org/10.1080/07853890510007377.

[270] Forbes EE, Hariri AR, Martin SL, Silk JS, Moyles DL, Fisher PM, Brown SM, Ryan ND, Birmaher B, Axelson DA, Dahl RE. Altered striatal activation predicting real-world positive affect in adolescent major depressive disorder. Am J Psychiatry 2009;166(1):64–73. https://doi.org/10.1176/appi.ajp.2008.07081336.

[271] Nestler EJ, Carlezon Jr. WA. The mesolimbic dopamine reward circuit in depression. Biol Psychiatry 2006;59(12):1151–9. https://doi.org/10.1016/j.biopsych.2005.09.018.

[272] Ahn YM, Chang J, Joo YH, Kim SC, Lee KY, Kim YS. Chronotype distribution in bipolar I disorder and schizophrenia in a Korean sample. Bipolar Disord 2008;10(2):271–5. https://doi.org/10.1111/j.1399-5618.2007.00573.x.

[273] Boivin DB, Czeisler CA, Dijk DJ, Duffy JF, Folkard S, Minors DS, Totterdell P, Waterhouse JM. Complex interaction of the sleep-wake cycle and circadian phase modulates mood in healthy subjects. Arch Gen Psychiatry 1997;54(2):145–52.

[274] Hatonen T, Forsblom S, Kieseppa T, Lonnqvist J, Partonen T. Circadian phenotype in patients with the co-morbid alcohol use and bipolar disorders. Alcohol Alcohol 2008;43(5):564–8. https://doi.org/10.1093/alcalc/agn057.

[275] Giannotti F, Cortesi F, Sebastiani T, Ottaviano S. Circadian preference, sleep and daytime behaviour in adolescence. J Sleep Res 2002;11(3):191–9.

[276] Haber SN, Knutson B. The reward circuit: linking primate anatomy and human imaging. Neuropsychopharmacology 2010;35(1):4–26. https://doi.org/10.1038/npp.2009.129.

[277] Hasler BP, Clark DB. Circadian misalignment, reward-related brain function, and adolescent alcohol involvement. Alcohol Clin Exp Res 2013;37(4):558–65. https://doi.org/10.1111/acer.12003.

[278] Aalto J, Kiianmaa K. Role of brain monoaminergic systems in the increased ethanol drinking caused by REM-sleep deprivation. Alcohol Alcohol Suppl 1987;1:313–7.

[279] Agabio R, Cortis G, Fadda F, Gessa GL, Lobina C, Reali R, Colombo G. Circadian drinking pattern of Sardinian alcohol-preferring rats. Alcohol Alcohol 1996;31(4):385–8.

[280] Boyle AE, Smith BR, Amit Z. A descriptive analysis of the structure and temporal pattern of voluntary ethanol intake within an acquisition paradigm. J Stud Alcohol 1997;58(4):382–91.

[281] Dole VP, Ho A, Gentry RT. Toward an analogue of alcoholism in mice: criteria for recognition of pharmacologically motivated drinking. Proc Natl Acad Sci U S A 1985;82(10):3469–71.

[282] Lynch WJ, Taylor JR. Sex differences in the behavioral effects of 24-h/day access to cocaine under a discrete trial procedure. Neuropsychopharmacology 2004;29(5):943–51. https://doi.org/10.1038/sj.npp.1300389.

[283] Pasley JN, Powell EW, Halberg F. Strain differences in circadian drinking behaviors of ethanol and water in rats. Prog Clin Biol Res 1987;227B:467–71.

[284] Roberts DC, Brebner K, Vincler M, Lynch WJ. Patterns of cocaine self-administration in rats produced by various access conditions under a discrete trials procedure. Drug Alcohol Depend 2002;67(3):291–9.

[285] Trujillo JL, Roberts AJ, Gorman MR. Circadian timing of ethanol exposure exerts enduring effects on subsequent ad libitum consumption in C57 mice. Alcohol Clin Exp Res 2009;33(7):1286–93. https://doi.org/10.1111/j.1530-0277.2009.00954.x.

[286] Bass CE, Jansen HT, Roberts DC. Free-running rhythms of cocaine self-administration in rats held under constant lighting conditions. Chronobiol Int 2010;27(3):535–48. https://doi.org/10.3109/07420521003664221.

[287] Danel T, Jeanson R, Touitou Y. Temporal pattern in consumption of the first drink of the day in alcohol-dependent persons. Chronobiol Int 2003; 20(6):1093–102.

[288] Holter SM, Spanagel R. Effects of opiate antagonist treatment on the alcohol deprivation effect in long-term ethanol-experienced rats. Psychopharmacology (Berl) 1999;145(4):360–9.

[289] Sinclair JD, Senter RJ. Development of an alcohol-deprivation effect in rats. Q J Stud Alcohol 1968;29(4):863–7.

[290] Tornatzky W, Miczek KA. Cocaine self-administration "binges": transition from behavioral and autonomic regulation toward homeostatic dysregulation in rats. Psychopharmacology (Berl) 2000;148(3):289–98.

[291] O'Dell LE, Chen SA, Smith RT, Specio SE, Balster RL, Paterson NE, Markou A, Zorrilla EP, Koob GF. Extended access to nicotine self-administration leads to dependence: Circadian measures, withdrawal measures, and extinction behavior in rats. J Pharmacol Exp Therap 2007;*320*(1):180–93. https://doi.org/10.1124/jpet.106.105270.

[292] Vengeliene V, Noori HR, Spanagel R. The use of a novel drinkometer system for assessing pharmacological treatment effects on ethanol consumption in rats. Alcohol Clin Exp Res 2013;37(Suppl 1):E322–328. https://doi.org/10.1111/j.1530-0277.2012.01874.x.

[293] Rosenwasser AM. Circadian clock genes: non-circadian roles in sleep, addiction, and psychiatric disorders? Neurosci Biobehav Rev 2010;34(8):1249–55. https://doi.org/10.1016/j.neubiorev.2010.03.004.

[294] LeGates TA, Fernandez DC, Hattar S. Light as a central modulator of circadian rhythms, sleep and affect. Nat Rev Neurosci 2014;15(7):443–54. https://doi.org/10.1038/nrn3743.

[295] Asher G, Gatfield D, Stratmann M, Reinke H, Dibner C, Kreppel F, Mostoslavsky R, Alt FW, Schibler U. SIRT1 regulates circadian clock gene expression through PER2 deacetylation. Cell 2008;134(2):317–28. https://doi.org/10.1016/j.cell.2008.06.050.

[296] Nakahata Y, Kaluzova M, Grimaldi B, Sahar S, Hirayama J, Chen D, Guarente LP, Sassone-Corsi P. The NAD+-dependent deacetylase SIRT1 modulates CLOCK-mediated chromatin remodeling and circadian control. Cell 2008;134(2):329–40. https://doi.org/10.1016/j.cell.2008.07.002.

[297] Nakahata Y, Sahar S, Astarita G, Kaluzova M, Sassone-Corsi P. Circadian control of the NAD+ salvage pathway by CLOCK-SIRT1. Science 2009;324(5927):654–7. https://doi.org/10.1126/science.1170803.

[298] Ferguson D, Koo JW, Feng J, Heller E, Rabkin J, Heshmati M, Renthal W, Neve R, Liu X, Shao N, Sartorelli V, Shen L, Nestler EJ. Essential role of SIRT1 signaling in the nucleus accumbens in cocaine and morphine action. J Neurosci 2013;33(41):16088–98. https://doi.org/10.1523/JNEUROSCI.1284-13.2013.

[299] Ferguson D, Shao N, Heller E, Feng J, Neve R, Kim HD, Call T, Magazu S, Shen L, Nestler EJ. SIRT1-FOXO3a regulate cocaine actions in the nucleus accumbens. J Neurosci 2015;35(7):3100–11. https://doi.org/10.1523/JNEUROSCI.4012-14.2015.

[300] Renthal W, Kumar A, Xiao G, Wilkinson M, Covington III HE, Maze I, Sikder D, Robison AJ, La Plant Q, Dietz DM, Russo SJ, Vialou V, Chakravarty S, Kodadek TJ, Stack A, Kabbaj M, Nestler EJ. Genome-wide analysis of chromatin regulation by cocaine reveals a role for sirtuins. Neuron 2009;62(3):335–48. https://doi.org/10.1016/j.neuron.2009.03.026.

[301] Nestler EJ. Epigenetic mechanisms of drug addiction. Neuropharmacology 2014;*76*(*Pt B*):259–68. https://doi.org/10.1016/j.neuropharm.2013.04.004.

[302] Walker DM, Cates HM, Heller EA, Nestler EJ. Regulation of chromatin states by drugs of abuse. Curr Opin Neurobiol 2015;30:112–21. https://doi.org/10.1016/j.conb.2014.11.002.

[303] Threadgill DW, Churchill GA. Ten years of the collaborative Cross. Genetics 2012;190(2):291–4. https://doi.org/10.1534/genetics.111.138032.

[304] Svenson KL, Gatti DM, Valdar W, Welsh CE, Cheng R, Chesler EJ, Palmer AA, McMillan L, Churchill GA. High-resolution genetic mapping using the mouse diversity outbred population. Genetics 2012;190(2):437–47. https://doi.org/10.1534/genetics.111.132597.

[305] Solberg Woods LC, Stelloh C, Regner KR, Schwabe T, Eisenhauer J, Garrett MR. Heterogeneous stock rats: a new model to study the genetics of renal phenotypes. Am J Physiol Renal Physiol 2010;298(6):F1484–1491. https://doi.org/10.1152/ajprenal.00002.2010.

[306] Logan RW, Robledo RF, Recla JM, Philip VM, Bubier JA, Jay JJ, Harwood C, Wilcox T, Gatti DM, Bult CJ, Churchill GA, Chesler EJ. High-precision genetic mapping of behavioral traits in the diversity outbred mouse population. Genes Brain Behav 2013;12(4):424–37. https://doi.org/10.1111/gbb.12029.

Chapter 14

Role of Oxytocin in Countering Addiction-Associated Behaviors Exacerbated by Stress

Jacqueline F. McGinty*, Courtney E. King†, Casey E. O'Neill*, and Howard C. Becker†

*Department of Neuroscience, Medical University of South Carolina, Charleston, SC, United States, †Department of Psychiatry and Behavioral Sciences, Medical University of South Carolina, Charleston, SC, United States

Substance use disorder (SUD) and alcohol use disorder (AUD) are substantial, disabling public health concerns, with lifetime prevalence rates of 12.5%–16% reported in the United States. Drug and alcohol dependence is characterized by loss of control of intake, cycles of excessive consumption combined with periods of abstinence and frequent return to heavy use [1]. Despite the significant personal, social, and economic burden SUDs and AUDs present, there are few effective pharmacotherapies available to help curb either addiction.

Recently, the neuropeptide oxytocin (OXT) has shown promise as a novel therapeutic target in the treatment of SUD and AUD. OXT is an endogenous neuropeptide that modulates the adaptive processes associated with reward, tolerance, memory, and stress responses [2]. OXT is synthesized in the paraventricular (PVN), supraoptic (SON), and accessory nuclei (AN) of the hypothalamus. The magnocellular neurons in PVN and SON project to the posterior pituitary where they release OXT into the peripheral circulation. In addition, the magnocellular neurons are the predominant source of OXT projections to forebrain areas (e.g., cortical, limbic, basal ganglia structures) where they mediate an array of motivationally relevant behaviors [3–5]. Parvocellular OXT neurons project to the hindbrain and spinal cord where they mediate autonomic functions and pain perception [6]. However, a more complex characterization of OXT neurons based on chemogenetics and gene expression is emerging [7]. CNS OXT neurons interact with one known OXT receptor that can couple to multiple G proteins [8,9]. The expression of OXT receptors is tissue-, species-, and sex steroid hormone-dependent [10–14]. OXT receptors are abundant in many brain regions that are associated with emotional triggers of drug seeking as well as stress, including the prefrontal cortex, hypothalamus, bed nucleus of the stria terminalis (BNST), ventral tegmental area, and amygdala [15,16].

Aside from its known hormonal role in parturition and maternal behaviors, the OXT nonapeptide has well-documented roles in mammalian (pro) social behavior (e.g., pair-bonding, social reward processing, aggression) that have been largely attributed to overlapping organization and projection sites in the midbrain dopaminergic system and direct modulation of dopamine neurotransmission by OXT [17,18]. In addition, OXT has been shown to exert potent anxiolytic, antifear, and antistress effects in humans [19,20] and in animals [21–23]. Based on these and other findings, intra-nasal OXT has been promoted as a potential early intervention to prevent the development of PTSD after trauma in humans [24,25]. Similarly, OXT has emerged as a potential treatment for SUDs. Activation of OXT in convergent pathways within the mesocorticolimbic dopamine system involved in the processing of motivationally relevant stimuli may extend to influencing the rewarding and the addictive aspects of alcohol and other substances of abuse [17,26–29].

OXT and AUD. A growing body of literature suggests that OXT plays a significant role in alcohol addiction. Preclinical evidence indicates that OXT influences a number of behavioral and physiological effects of alcohol [30]. Both peripheral and central OXT administration blocks the development of tolerance to hypothermic and hypnotic effects of alcohol in rodents and modulates the severity of withdrawal symptoms [28,31,32]. Systemic OXT administration has been shown to reduce alcohol preference and intake in a variety of drinking models in rats [28,33,34] and mice [35,36]. MacFadyen et al. [34] reported that acute administration of OXT to rats in a range of doses (0.1–0.5 mg/kg, ip) 30 min prior to testing, reduced alcohol self-administration in two different paradigms. In alcohol-preferring C57BL/6J mice, systemic OXT when administered 30 min prior to each session, decreased alcohol self-administration and binge-like alcohol consumption in a

Neural Mechanisms of Addiction. https://doi.org/10.1016/B978-0-12-812202-0.00014-2
Copyright © 2019 Elsevier Inc. All rights reserved.

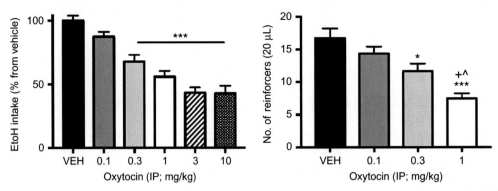

FIG. 1 Effects of OXT treatment on binge-like ethanol (EtOH) drinking and self-administration. Left. Oxytocin reduced binge-like EtOH intake (shown as % change from vehicle) in a dose-related manner. Values are mean ± SEM ($N = 10$/group). ***$P < .001$ vs VEH. Right. Oxytocin reduced the number of ethanol reinforcers earned in a 20-min self-administration session in a dose-related manner. Ethanol reinforcers were earned on a fixed-ratio (FR)-4 schedule of responding and correspond to g/kg intake. Values are mean ± SEM ($N = 13$–15/group). *$P < .05$, ***$P < .001$ vs VEH; +$P < .01$ vs 0.1 mg/kg OXT; ^($P < .05$) vs 0.3 mg/kg OXT.

dose-related manner (0.3, 1, 3, or 10 mg/kg, i.p.), as well as reduced motivation for alcohol reward as measured by a progressive-ratio schedule of responding ([35], Fig. 1). Enduring effects of systemic OXT administration have also been demonstrated. McGregor and Bowen [28] found that a single dose of OXT (1 mg/kg, i.p.) produced a long-lasting reduction in preference for an alcoholic beverage compared with a nonalcoholic sweet solution. Additionally, treatment with OXT for 2 weeks prior to the induction of a two-bottle free choice paradigm resulted in lower alcohol preference in OXT-treated rats compared to controls. Further, when adolescent rats were injected daily for 10 days with 1 mg/kg, i.p. OXT (PND 33-42) and offered ad libitum beer consumption 30 days later, beer, but not water, consumption in OXT-pretreated rats was significantly less than in vehicle-treated rats [33].

OXT and SUD. In addition to the effects of OXT on alcohol-motivated behavior, OXT has also been shown to regulate motivation for other drugs of abuse. For example, systemic administration of OXT inhibits METH-induced hyperactivity and conditioned place preference [37–39] and dose-dependently decreases the motivation to self-administer METH on a progressive ratio schedule [38]. Furthermore, OXT decreases drug seeking in male and female rats induced by conditioned cues, a drug prime, or stress in rats that had a history of cocaine or METH self-administration [38,40–43]. Because OXT decreases locomotor activity for up to 30 min after injection (0.3–1 mg/kg, i.p.) and decreases cocaine-induced locomotor activity in female, not male rats [42], care must be taken when interpreting behavioral data when OXT is injected acutely on the day of experimentation. In contrast, repeated OXT injections prior to METH self-administration in adolescent females [44] or adult males or females [41,45] decreases reinstatement of drug seeking >4 weeks after treatment ended. These data indicate that OXT administration has enduring physiological and behavioral effects that suggest it has therapeutic promise in treating relapse in chronic SUDs. These studies reinforce the idea that mechanisms underlying drug seeking are attenuated in an enduring manner by systemic administration of OXT.

OXT and co-morbid stress and SUDs. OXT has been shown to exert potent anxiolytic, antifear, and antistress effects in humans [19,20] and in animals [21–23]. Based on these and other findings, intra-nasal OXT has been promoted as a potential early intervention to prevent the development of PTSD after trauma in humans [24,25]. A history of stress can alter responses to substances of abuse in preclinical studies as covered in multiple reviews [46–48]. For example, footshock-induced stress increased alcohol consumption [49] and maternal separation in early life increased METH self-administration [50]. Another PTSD model, single prolonged stress, enhanced locomotor sensitization to cocaine but did not alter cocaine SA [51] and had mixed effects on METH- or amphetamine-induced behavioral sensitization [52,53]. Predator odors are ecologically relevant stressors that reliably induce hyperarousal, a long-lasting avoidance of trauma-related cues, and dysregulation of the hypothalamic-pituitary-adrenal (HPA) axis [54–56]. Rats exhibiting persistent high avoidance of predator odor-related stimuli also exhibit long-lasting poststress increases in alcohol drinking [57]. Mice with a history of TMT exposure show an exacerbated stress-induced reinstatement of alcohol seeking compared to mice with a history of saline exposure [58]. Further, OXT treatment attenuated stress-induced reinstatement in the TMT-preexposed group. These results implicate OXT in decreasing stress-induced reinstatement of alcohol seeking.

Repeated exposure to TMT in one environment (activity chambers) exacerbated the response of rats to reinstatement of drug seeking induced by conditioned cues or re-experiencing the predator odor in the operant chambers previously associated with METH self-administration [41]. A single injection of OXT 30 min before reinstatement suppressed METH seeking in both saline- and TMT preexposed rats. Further, OXT injections in male rats for 10 d prior to METH

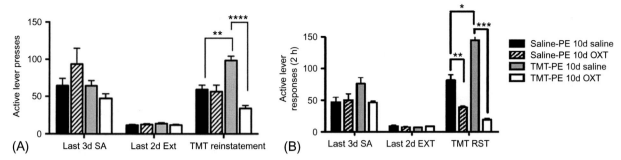

FIG. 2 Effects of 10 daily OXT injections on TMT-induced reinstatement in TMT-preexposed (PE) and Saline-PE male (A) and female (B) rats. In each graph, average number of active lever presses during METH SA (left), last 2 d of extinction (middle), and the TMT-induced reinstatement test (right) are displayed. (A) (Sal-PE 10 d Sal vs **P < .01 Sal-PE 10 d Sal vs TMT-PE 10 d Sal; ****P < .0001 TMT-PE 10 d Sal vs TMT-PE 10 d OXT) Mean ± SEM. (B) (*P < .05 Sal-PE 10 d Sal vs TMT-PE 10 d Sal; **P < .01 Sal-PE 10 d Sal vs Sal-PE 10 d OXT;***P < .0001. TMT-PE 10 d Sal vs TMT-PE 10 d OXT) Mean ± SEM; n = 7–8 per group.

self-administration blocked only the stress-induced exacerbation of drug seeking in TMT preexposed rats ([41], Fig. 2A). However, OXT injections in female rats for 10 d prior to METH self-administration blocked both the stress-induced exacerbation of drug seeking by TMT preexposure and drug seeking in saline preexposed rats ([45], Fig. 2B). These data are consistent with an exacerbated response of females vs males to predator odors [59,59a] and demonstrate sex differences in OXT's effects on stress and substance use vulnerability.

Clinical studies. Studies that have investigated the effects of OXT on SUD and AUD are limited but encouraging. Pedersen and colleagues demonstrated that acute intranasal OXT treatment vs placebo reduced craving, anxiety, and withdrawal symptoms in treatment-seeking alcohol-dependent patients [60]. More recently, a study found that acute intranasal OXT treatment reduced craving for alcohol in more anxious individuals and increased craving in less anxious individuals with AUD [61]. Further, human postmortem studies have shed some light on the effects of alcohol consumption on the OXT system. In male patients with AUD compared to controls, pyknosis, and pericellular edema in the PVN of OXT and vasopressin magnocellular neurons in the hypothalamus was observed along with a decrease in OXT peptide immunoreactivity (IR) in SON and an increase in surviving PVN neurons [62]. These observations are in agreement with the previous evidence of magnocellular neuronal degeneration in severe alcoholic males [63] and in rodents [64], suggesting that chronic alcohol exposure is toxic to these neurons. More recently, a significant increase in *Oxt* mRNA in the prefrontal cortex of human postmortem AUD subjects compared to controls was reported [65]. However, the functional significance of *Oxt* mRNA in cells outside of the hypothalamus and whether it is translated into protein is unclear.

With regard to clinical interactions between stress and addictive substances, few reports have been published and all were small studies that used a single dose of OXT. OXT (40 IU) reduced craving for marijuana and social stress but not anxiety in cannabis-dependent subjects [65a]. In cocaine-addicted individuals with a history of childhood adversity, OXT reduced a social stress-induced cortisol response [65b]. In contrast, OXT increased subjects' desire to use cocaine but had no effect on cue-induced cocaine craving [65c]. Certainly larger, carefully designed studies using repeated OXT should be performed that take into consideration context- and sex-specific responses.

OXT mechanisms of action in AUD and SUD. While there is clear evidence to suggest a role for OXT in AUD and SUD, the mechanisms by which OXT reduces alcohol or drug consumption and seeking are not fully understood. In a recent preclinical study, direct (icv) administration of OXT reduced alcohol consumption and alcohol-induced dopamine efflux in the nucleus accumbens [66]. Additionally, OXT receptors have been identified on neurons in the ventral tegmental area projecting to the nucleus accumbens [15]. Thus, it is possible that OXT reduces ethanol self-administration by altering mesolimbic dopamine activity that ordinarily signals alcohol-related reward although few studies have examined the role of OXT receptors in mediating the neuropeptide's effect on the motivational effects of alcohol. However, recent reports showed that viral-mediated overexpression of OXT receptors in the nucleus accumbens reduced alcohol-induced conditioned place preference and alcohol consumption in mice [67,68]. Additionally, the ability of OXT to reduce binge-like alcohol drinking was blocked by pretreatment with an OXT receptor antagonist, suggesting that OXT reduced alcohol consumption via interaction with its own receptor [35]. However, although OXT receptors appear to be involved in the rewarding properties of alcohol, it is important to note that Bowen and colleagues showed that the attenuation of sedative and ataxic effects of alcohol by central administration OXT was mediated by GABA$_A$ receptors [31].

With regard to other addictive substances, chronic METH, cocaine, and morphine administration alters OXT receptor levels in the hypothalamus and amygdala and increases OXT receptor levels in the hippocampus [29,69,69a]; however,

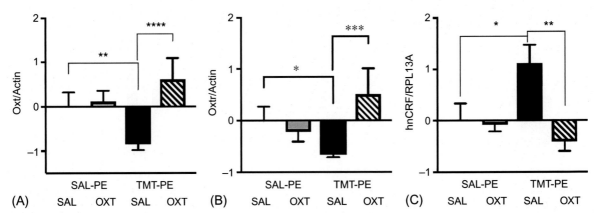

FIG. 3 Changes in mRNA expression in PVN of male rats with a METH self-administration history after TMT-induced reinstatement with or without 10 d of OXT administration. Significant decreases in (A) *Oxt* and (B) *Oxtr* mRNA were seen in the PVN of TMT-PE rats vs Sal-PE rats (*$P < .05$, **$P < .01$) that were prevented by 10 d OXT treatment (***$P < .001$, ****$P < .0001$); $n = 6–8$ per group. (C) An increase in *Crh* hnRNA in PVN of TMT-PE vs Sal-PE rats ($P < .09$) that were prevented by 10 d OXT treatment (**$P < .01$); $n = 3–5$ per group.

METH self-administration decreased OXT fiber density in the nucleus accumbens core [70]. Reinstatement of drug seeking is thought to depend on disturbed glutamate transmission in the prefrontal cortex projection to the nucleus accumbens [70a]. Thus, it is possible that administration of OXT acts via OXTRs in brain regions such as the prefrontal cortex or nucleus accumbens [71] to suppress reinstatement of METH seeking. In fact, OXTRs in the prefrontal cortex of mice are expressed exclusively in GABA-somatostatin interneurons and bath application of OXT in prefrontal cortical slices increases the firing of these neurons [71a] that suppresses pyramidal neuron activity. Further, OXT infusion into the medial prefrontal cortex blocks restraint stress-induced reinstatement of METH-induced conditioned place preference in a glutamate-dependent fashion [38,72].

Addiction has been hypothesized to result from deficits in reward experience coupled with excess stress [73]. Many forms of stress change the expression of corticotropin-releasing hormone (*Crh*) in the CNS [74,75], including stress-induced reinstatement of drug seeking [76]. Thus, it is possible that the effects of exogenous OXT administration decreases drug seeking by suppressing CRH expression in stress-related brain areas. Intriguingly, although *Crh* gene expression in PVN neurons is decreased by the canonical negative feedback from acute stress-induced release of glucocorticoids, *Crh* expression in neurons of the BNST is increased simultaneously [77,78]. Our preliminary data using rtPCR indicate that OXT injections for 10-d augmented *Oxt* mRNA in the PVN and prevented an increase in *Crh* mRNA in BNST of male rats with a history of TMT-induced stress and METH self-administration (Fig. 3). These data suggest that *Crh* neurons in the BNST are chronically engaged by prior exposure to TMT in rats with a history of METH self-administration and that OXT administered immediately after the chronic TMT stress reduces their activation.

Outstanding questions. While recent studies have illuminated a generally inhibitory role for OXT in alcohol and drug taking and seeking, there are some inconsistencies in the literature. The robustness and duration of OXT effects appear to be different depending on the species, age, route of administration, dose, and alcohol or drug consumption paradigm [28,34,35,41,44,79]. Another controversial issue surrounding systemic OXT administration concerns its penetrance into the brain [30,80]. There are several reports that the effects of systemic OXT on behavior are blocked by intracerebral microinfusion of OXT receptor antagonists [71,81,82]. After intranasal or i.p. administration to rodents, exogenous OXT causes an increase in extracellular OXT levels in brain microdialysates [83] and presumably stimulates OXT receptors in the CNS to exert its behavioral actions [21]. Although studies have shown increased cerebrospinal fluid OXT levels following delivery by the intranasal route in humans [84] and nonhuman primates [11], it is unclear whether elevated levels of CSF OXT are the result of exogenous OXT entering the CSF or whether there is a feed-forward mechanism stimulated by the administered peptide that increases endogenous OXT levels. Additional studies are necessary to further investigate these questions. Nevertheless, the growing literature demonstrating the effects of OXT on the neurobehavioral response to addictive drugs and alcohol points to the OXT system as a promising target for therapeutic intervention in AUD and SUDs regardless of its mechanism of action.

FUNDING SOURCES

This research was supported by DoD grant W81XWH-12-2-0048 Subaward 8A-293, T32 AA007474, and F31 AA026483-01.

REFERENCES

[1] Koob GF, Volkow ND. Neurobiology of addiction: a neurocircuitry analysis. Lancet 2016;3:760–73.

[2] Lee MR, Rohn MC, Tanda G, Leggio L. Targeting the oxytocin system to treat addictive disorders: rationale and progress to date. CNS Drugs 2016;30:109–23.

[3] Grinevich V, Knobloch-Bollmann HS, Eliava M, Busnelli M, Chini B. Assembling the puzzle: pathways of oxytocin signaling in the brain. Biol Psychiatry 2016;7:155–64.

[4] Kiss A, Mikkelsen JD. Oxytocin-anatomy and functional assignments: a minireview. Endocr Regul 2005;39:97–105.

[5] Knobloch HS, Charlet A, Hoffmann LC, Khrulev S, Cetin AH, Osten P, Schwarz MK, Seeburg PH, Stoop R, Grinevich V. Evoked axonal oxytocin release in the central amygdala attenuates fear response. Neuron 2012;73:553–66.

[6] Eliava M, Melchior M, Knobloch-Bollmann HS, Stoop R, Charlet A, Grinevich V. A new population of parvocellular oxytocin neurons controlling magnocellular neuron activity and inflammatory pain processing. Neuron 2016;89:1291–304.

[7] Althammer F, Grinevich V. Diversity of oxytocin neurons: beyond magno- and parvocellular cell types? J Neuroendocrinol 2017. https://doi.org/10.1111/jne.12549.

[8] Gimpl G, Farenholz F. The oxytocin receptor system: structure, function, and regulation. Physiol Rev 2001;81:629–83.

[9] Gimpl G, Reitz J, Brauer S, Trosson C. Oxytocin receptors: ligand binding, signalling and cholesterol dependence. Prog Brain Res 2008;170:193–204.

[10] Dumais KM, Veenema AH. Vasopression and oxytocin receptor systems in the brain: sex differences and sex-specific regulation of social behavior. Front Neuroendocrinol 2016;40:1–23.

[11] Freeman S, Young LJ. Comparative perspectives on oxytocin and vasopressin receptor research in rodents and primates: translational implications. J Neuroendocrinol 2016;28. https://doi.org/10.1111/jne.12382.

[12] Nair HP, Gutman AR, Davis M, Young LJ. Central oxytocin, vasopressin, and corticotropin-releasing factor receptor densities in the basal forebrain predict isolation potentiated startle in rats. J Neurosci 2005;25:11479–88.

[13] Smith CJ, Poeklmann ML, Li S, Ratnaseelan AM, Bredewold R, Veenema AH. Age and sex differences in oxytocin and vasopression V1a receptor binding densities in the rat brain: focus on the social decision-making network. Brain Struct Funct 2017;222:981–1006.

[14] Stoop R, Hegoburo C, Van den Burg E. New opportunities in vasopressin and oxytocin research: a perspective from the amygdala. Ann Rev Neurosci 2015;3:369–88.

[15] Peris J, MacFadyen K, Smith JA, De Kloet AD, Wang L, Krause EG. Oxytocin receptors are expressed on dopamine and glutamate neurons in the mouse ventral tegmental area that project to nucleus accumbens and other mesolimbic targets. J Comp Neurol 2017;525:1094–108.

[16] Smeltzer MD, Curtis JT, Aragona BJ, Wang Z. Dopamine, oxytocin, and vasopressin receptor binding in the medial prefrontal cortex of monogamous and promiscuous voles. Neurosci Lett 2006;394:146–51.

[17] Love TM. Oxytocin, motivation and the role of dopamine. Pharmacol Biochem Behav 2014;119:49–60.

[18] Xiao L, Priest MF, Nasenbeny J, Lu T, Kozorovitskiy Y. Biased oxytocinergic modulation of midbrain dopamine systems. Neuron 2017;95:368–84.

[19] Heinrichs M, Domes G. Neuropeptides and social behaviour: effects of oxytocin and vasopressin in humans. Prog Brain Res 2008;170:337–50.

[20] Kirsch P, Esslinger C, Chen Q, Mier D, Lis S, Siddhanti S, Gruppe H, Mattay VS, Gallhofer B, Meyer-Lindenberg A. Oxytocin modulates neural circuitry for social cognition and fear in humans. J Neurosci 2005;25:11489–93.

[21] Neumann ID, Landgraf R. Balance of brain oxytocin and vasopressin: implications for anxiety, depression, and social behaviors. Trends Neurosci 2012;35:649–59.

[22] Ring RH, Malberg JE, Potestio L, Ping J, Boikess S, Luo B, Schechter LE, Rizzo S, Rahman Z, Rosenzweig-Lipson S. Anxiolytic-like effects of oxytocin in male mice: behavioral and autonomic evidence, therapeutic implications. Psychopharmacol 2006;185:218–25.

[23] Viviani D, Charlet A, van den Burg E, Robinet C, Hurni N, Abatis M, Magara F, Stoop R. Oxytocin selectively gates fear responses through distinct outputs from the central amygdala. Science 2011;333:104–7.

[24] Frijling JL, van Zuiden M, Koch SB, Nawijn L, Goslings JC, Luitse JS, Biesheuvel TH, Honig A, Bakker FC, Denys D, Veltman DJ, Olff M. Efficacy of oxytocin administration early after psychotrauma in preventing the development of PTSD: Study protocol of a randomized controlled trial. BMC Psychiatry 2014;14:92.

[25] Frijling JL, van Zuiden M, Koch SB, Nawijn L, Veltman DJ, Olff M. Intranasal oxytocin affects amygdala functional connectivity after trauma script-driven imagery in distressed recently trauma-exposed individuals. Neuropsychopharmacology 2015;41:1286–96.

[26] Baskerville TA, Douglas AJ. Dopamine and oxytocin interactions underlying behaviors: Potential contributions to behavioral disorders. CNS Neurosci Ther 2010;1:e92–e123.

[27] Burkett JP, Young LJ. The behavioral, anatomical, and pharmacological parallels between social attachment, love and addiction. Psychopharmacology 2012;224:1–26.

[28] McGregor IS, Bowen MT. Breaking the loop: oxytocin as a potential treatment for drug addiction. Horm Behav 2012;61:331–9.

[29] Sarnyai Z, Kovacs GL. Oxytocin in learning and addiction. Pharmacol Biochem Behav 2014;119:3–9.

[30] Lee MR, Weerts EM. Oxytocin for the treatment of drug and alcohol use disorders. Behav Pharmacol 2016;27:640–8.

[31] Bowen MT, Peters ST, Absalom N, Chebib M, Neumann ID, McGregor IS. Oxytocin prevents ethanol actions at delta subunit GABA$_A$ receptors and attenuates ethanol-induced motor impairments in rats. Proc Natl Acad Sci U S A 2015;112:3104–9.

[32] Szabo G, Kovacs GL, Szekeli S, Telegdy G. The effects of neurohypophyseal hormones on tolerance to the hypothermic effect of ethanol. Alcohol 1985;2:567–74.

[33] Bowen MT, Carson DS, Spiro A, Arnold JC, McGregor IS. Adolescent oxytocin exposure causes persistent reductions in anxiety and alcohol consumption and enhances sociability in rats. PloS One 2011;6:e27237.

[34] MacFadyen K, Loveless R, DeLucca B, Wardley K, Deogan S, Thomas C, Peris J. Peripheral oxytocin administration reduces ethanol consumption in rats. Pharmacol Biochem Behav 2016;140:27–32.

[35] King CE, Griffin WC, Luderman LN, Kates MM, McGinty JF, Becker HC. Oxytocin reduces ethanol self-administration in mice. Alcohol Clin Exp Res 2017;41:955–64.

[36] Peters S, Slattery DA, Flor PJ, Neumann ID, Reber SO. Differential effects of baclofen and oxytocin on the increased ethanol consumption following chronic psychosocial stress in mice. Addict Biol 2013;18:66–77.

[37] Baracz SJ, Rourke PI, Pardey MC, Hunt GE, McGregor IS, Cornish JL. Oxytocin directly administered into the nucleus accumbens core or subthalamic nucleus attenuates methamphetamine-induced conditioned place preference. Behav Brain Res 2012;228:185–93.

[38] Carson DS, Cornish JL, Guastella AJ, Hunt GE, McGregor IS. Oxytocin decreases methamphetamine self-administration, methamphetamine hyperactivity, and relapse to methamphetamine-seeking behaviour in rats. Neuropharmacology 2010;58:38–43.

[39] Qi J, Yang JY, Wang F, Zhao YN, Song M, Wu CF. Effects of oxytocin on methamphetamine-induced conditioned place preference and the possible role of glutamatergic neurotransmission in the medial prefrontal cortex of mice in reinstatement. Neuropharmacology 2009;56:856–65.

[40] Cox BM, Young AB, See RE, Reichel CM. Sex differences in methamphetamine seeking in rats: Impact of oxytocin. Psychoneuroendocrinology 2013;38:2343–53.

[41] Ferland CL, Reichel CM, McGinty JF. Effects of oxytocin on methamphetamine-seeking exacerbated by predator odor pre-exposure in rats. Psychopharmacology (Berl) 2016;233:1015–24.

[42] Leong KC, Zhou L, See RE, Ghee S, Reichel CM. Oxytocin dereases cocaine taking, cocaine seeking, and locomotor activity in female rats. Exp Clin Psychopharmacol 2015;24:55–64.

[43] Zhou L, Ghee SM, See RE, Reichel CM. Oxytocin differentially affects sucrose taking and seeking in male and female rats. Behav Brain Res 2015;283:184–90.

[44] Hicks C, Cornish JL, Baracz SJ, Suraev A, McGregor IS. Adolescent pre-treatment with oxytocin protects against adult methamphetamine-seeking behavior in female rats. Addict Biol 2014;21:304–15.

[45] O'Neill CE, Newsom RJ, McGinty JF. Effects of oxytocin following traumatic stress on methamphetamine seeking in female rats. Soc Neurosci Abst 2016;457:07.

[46] Gilpin NW, Weiner JL. Neurobiology of co-morbid post-traumatic stress disorder & alcohol use disorder. Genes Brain Behav 2017;16:15–43.

[47] Zhou Y, Proudnikov D, Yuferov V, Kreek MJ. Drug-induced and genetic alterations in stress-responsive systems: implications for specific addictive diseases. Brain Res 2010;1314:235–52.

[48] Zorilla EP, Logrip ML, Koob GF. Corticotropin releasing factor: a key role in the neurobiology of addiction. Front Neuroendocrinol 2014;35:234–2244.

[49] Meyer EM, Long V, Fanselow MS, Spigelman I. Stress increases voluntary alcohol intake, but does not alter established drinking habits in a rat model of posttraumatic stress disorder. Alcohol Clin Exp Res 2013;37:566–74.

[50] Lewis CR, Staudinger K, Scheck L, Olive MF. The effects of maternal separation on adult methamphetamine self-administration, extinction, reinstatement, and MeCP2 immunoreactivity in the nucleus accumbens. Front Psychiatry 2015;4:55.

[51] Eagle AL, Singh R, Kohler RJ, Friedman AL, Liebowitz CP, Galloway MP, Enman NM, Jutkiewicz EM, Perrine SA. Single prolonged stress effects on sensitization to cocaine and cocaine self-administration in rats. Behav Brain Res 2015;284:218–24.

[52] Eagle AL, Perrine SA. Methamphetamine-induced behavioral sensitization in a rodent model of posttraumatic stress disorder. Drug Alcohol Depend 2013;131:36–43.

[53] Toledano D, Tassin JP, Gisquet-Verrier P. Traumatic stress in rats induces noradrenergic-dependent long-term behavioral sensitization: role of individual differences and similarities with dependence on drugs of abuse. Psychopharmacology (Berl) 2013;230:465–76.

[54] Fendt M, Endres T, Lowry CA, Apfelbach R, McGregor IS. TMT-induced autonomic and behavioral changes and the neural basis of its processing. Neurosci Biobehav Rev 2005;29:1145–56.

[55] Ojo JO, Greenberg MB, Leary P, Mouzon B, Bachmeier C, Mullan M, Diamond DM, Crawford F. Neurobehavioral, neuropathological and biochemical profiles in a novel mouse model of co-morbid post-traumatic stress disorder and mild traumatic brain injury. Front Behav Neurosci 2014;8:213.

[56] Thomas RM, Urban JH, Peterson DA. Acute exposure to predator odor elicits a robust increase in corticosterone and a decrease in activity without altering proliferation in the adult rat hippocampus. Exp Neurol 2006;201:308–15.

[57] Edwards S, Baynes BB, Carmichael CY, Zamora-Martinez ER, Barrus M, Koob GF, Gilpin NW. Traumatic stress reactivity promotes excessive alcohol drink and alters the balance of prefrontal cortex-amygdala activity. Transl Psychiatry 2013;3:e296.

[58] King CE, Griffin WC, McGinty JF, Becker HC. Oxytocin attenuates stress-induced reinstatement of alcohol seeking in mice with a history of trauma. Soc Neurosci Abst 2016;453:13.

[59] Campbell T, Lin S, DeVries C, Lambert K. Coping strategies in male and female rats exposed to multiple stressors. Physiol Behav 2003;78:495–504.

[59a] Vandruscolo LF, Vandruscolo JCM, Terenina-Rigaldie E, Raba F, Ramos A, Takahashi RN, Mormede P. Genetic influences on behavioral and neuroendocrine responses to predator odor stress in rats. Neurosci Lett 2017;405:89–94.

[60] Pedersen CA, Smedley KL, Leserman J, Jarskog LF, Rau SW, Kampov-Polevoi A, Casey RL, Fender T, Garbutt JC. Intranasal oxytocin blocks alcohol withdrawal in human subjects. Alcohol Clin Exp Res 2013;37:484–9.

[61] Mitchell JM, Arcuni PA, Weinstein D, Wooley JD. Intranasal oxytocin selectively modulates social perception, craving, and approach behavior in subjects with alcohol use disorder. J Addict Med 2016;10:182–9.

[62] Sivukhina EV, Dolzhikov AA, Morozov lE, Jinkowski GF, Grinevich V. Effects of chronic alcoholic disease on magnocellular and parvocellular hypothalamic neurons in men. Horm Metab Res 2006;38:382–90.

[63] Harding AJ, Halliday GM, Ng JL, Harper CG, Krill JJ. Loss of vasopressin-immunoreactive neruons in alcoholics is dose-related and time-dependent. Neuroscience 1996;72:699–708.

[64] Madeira M, Lieberman A, aula-Barbosa M. Effects of chronic alcohol consumption and dehydration on the supraoptic nucleus of adult male and female rats. Neuroscience 1993;56:657–72.

[65] Lee MR, Schwandt ML, Sankar V, Suchankova P, Sun H, Leggio L. Effect of alcohol use disorder on oxytocin peptide and receptor mRNA expression in human brain: a post-mortem case-control study. Psychoneuroendocrinology 2017;85:14–9.

[65a] McRae-Clark AL, Baker NL, Moran-Santa Maria M, Brady KT. Effect of oxytocin on craving and stress response in marijuana-dependent individuals: a pilot study. Psychopharmacology 2013;228:623–31.

[65b] Flanagan JC, Baker NL, McRae-Clark AL, Brady KT, Moran-Santa Maria MM. Effects of adverse childhood experiences on the association between intranasal oxytocin and social stress reactivity among individuals with cocaine dependence. Psychiatry Res 2015;229:94–100.

[65c] Lee MR, Rohn MC, Tanda G, Leggio L. Targeting the oxytocin system to treat addictive disorders: rationale and progress to date. CNS Drugs 2016;30:109–23.

[66] Peters ST, Bowen MT, Bohrer K, McGregor IS, Neumann ID. Oxytocin inhibits ethanol consumption and ethanol-induced dopamine release in the nucleus accumbens. Addict Biol 2016;22:702–11.

[67] Bahi A. The oxytocin receptor impairs ethanol reward in mice. Physiol Behav 2015;139:321–7.

[68] Bahi A, Al Mansouri S, Al Maamari E. Nucleus accumbens lentiviral-mediated gain of function of the oxytocin receptor regulates anxiety- and ethanol-related behaviors in adult mice. Physiol Behav 2016;164(Pt A):249–58.

[69] Zanos P, Wright SR, Georgiou P, Yoo JH, Ledent C, Hourani SM, Kitchen I, Winsky-Sommerer R, Bailey A. Chronic methamphetamine treatment induces oxytocin receptor up-regulation in the amygdala and hypothalamus via an adenosine A2A receptor-independent mechanism. Pharmacol Biochem Behav 2014;119:72–9.

[69a] Georgiou P, Zanos P, Hourani S, Kitchen I, Bailey A. Cocaine abstinence induces emotional impairment and brain region-specific upregulation of the oxytocin receptor binding. Eur J Neurosci 2015;44:2446–54.

[70] Baracz SJ, Parker LM, Suraev AS, Everett NA, Goodchild AK, McGregor IS, Cornish JL. Chronic methamphetamine self-administration dysregulates oxytocin plasma levels and oxytocin receptor fibre density in the nucleus accumbens core and subthalamic nucleus of the rat. J Neuroendocrinol 2016;28. https://doi.org/10.1111/jne.12337.

[70a] Parsegian A, See RE. Dysregulation of dopamine and glutamate release in the prefrontal cortex and nucleus accumbens following methamphetamine self-administration and during reinstatement in rats. Neuropsychopharmacol 2014;39:811–22.

[71] Baracz SJ, Everett NA, McGregor IS, Cornish JL. Oxytocin in the nucleus accumbens core reduces reinstatement of methamphetamine-seeking behaviour in rats. Addict Biol 2014;21:316–25.

[71a] Nakajima M, Gorlich A, Heintz N. Oxytocin modulates female sociosexual behavior through a specific class of prefrontal cortical interneurons. Cell 2014;159:295–305.

[72] Han WY, Du P, Fu SY, Wang F, Song M, Wu CF, Yang JY. Oxytocin via its receptor affects restraint stress-induced methamphetamine CPP reinstatement in mice: Involvement of the medial prefrontal cortex and dorsal hippocampus glutamatergic system. Pharmacol Biochem Behav 2014;11:80–7.

[73] Koob GF, Buck CL, Cohen A, Edwards S, Park PE, Schlosburg JE, Schmeichel B, Vendruscolo LF, Wade CL, Whitfield Jr. TW, George O. Addiction as a stress surfeit disorder. Neuropharmacology 2014;76:370–82.

[74] Herman JP, Ostrander MM, Mueller NK, Figueiredo HF. Limbic system mechanisms of stress regulation: hypothalamo-pituitary-adrenocortical ais. Prog Neuro-Psychopharmacol Biol Psychiatry 2005;29:1201–13.

[75] Makino S, Smith MA, Gold PW. Increased expression of corticotropin releasing hormone and vasopressin messenger ribonucleic acid (mRNA) in the hypothalamic paraventricular nucleus during repeated stress: association with reduction in glucocorticoid receptor mRNA levels. Endocrinology 1995;136:3299–309.

[76] McReynolds JR, Vranjkovic O, Thao M, Baker DA, Makky K, Lim Y, Mantsch JR. Beta-2-adrenergic receptors mediate stress-evoked reinstatement of cocaine-induced conditioned place preference and increases in CRF mRNA in the bed nucleus of the stria terminalis. Psychopharmacology (Berl) 2014;231:3953–63.

[77] Makino S, Gold PW, Schulkin J. Effects of corticosterone on CRH mRNA and content in the bed nucleus of the stria terminalis; comparison with the effects in the central nucleus of the amygdala and the paraventricular nucleus of the hypothalamus. Brain Res 1994;657:141–9.

[78] Watts AG, Sanchez-Watts G. Region-specific regulation of neuropeptide mRNAs in rat limbic forebrain neurons by aldosterone and corticosterone. J Physiol 1995;484:721–36.

[79] Stevenson JR, Wenner SM, Freestone DM, Romaine CC, Parian MC, Christian SM, Bohidar AE, Ndem JR, Vogel IR, O'Kane CM. Oxytocin reduces alcohol consumption in praire voles. Physiol Behav 2017;179:411–21.

[80] Leng G, Ludwig M. Intranasal oxytocin: myths and delusions. Biol Psychiatry 2016;79:243–50.

[81] Cox BM, Bentzley BS, Regen-Tuero H, See RE, Reichel CM, Aston-Jones G. Oxytocin acts in nucleus accumbens to attenuate methamphetamine seeking and demand. Biol Psychiatry 2017;81:949–58.

[82] Sabihi S, Durosko NE, Dong SM, Leuner B. Oxytocin in the prelimbic medial prefrontal cortex reducs anxiety-like behavior in female and male rats. Psychoneuroendocrinology 2014;45:31–42.

[83] Neumann ID, Maloumby R, Beiderbeck DI, Lukas M, Landgraf R. Increased brain and plasma oxytocin after nasal and peripheral administration in rats and mice. Psychoneuroendocrinology 2013;38:1985–93.

[84] Striepens N, Kendrick KM, Hanking V, Landgraf R, Wullner U, Maier W, et al. Elevated cerebrospinal fluid and blood concentrations of oxytocin following its intranasal administration in humans. Sci Rep 2013;3:3440.

Chapter 15

The Role of Norepinephrine in Drug Addiction: Past, Present, and Future

Stephanie L. Foster and David Weinshenker

Department of Human Genetics, Emory University School of Medicine, Atlanta, GA, United States

INTRODUCTION

Dopamine (DA) has classically been the neurotransmitter at the forefront of addiction research, largely owing to the well-established role of the mesolimbic system in reward and reinforcement [1]. However, other neurotransmitters, such as norepinephrine (NE), influence the activity of brain regions implicated in addiction as well as addiction-related behaviors including drug-associated reward and drug seeking. Given that the decades-long pursuit of DA-based therapies for addiction has not yielded promising results, it is important to consider alternative targets like the noradrenergic system for the development of new addiction therapies.

NE is a catecholamine that is derived from DA by the enzyme dopamine β-hydroxylase (DBH). The major noradrenergic afferents in the brain can be divided into the dorsal and ventral noradrenergic bundles. The dorsal noradrenergic bundle originates in the locus coeruleus (LC) and projects extensively throughout the brain to the cerebellum, hippocampus, forebrain, and amygdala. The ventral noradrenergic bundle, which originates in A1 and A2, also projects to the amygdala, as well as the hypothalamus, and midbrain [2,3]. NE is critical for mediating arousal states and stress responses, both of which play a role in addiction by shaping reward and the salience of reward-associated cues, as well as susceptibility to stressors that can trigger relapse.

NE signals through a class of G-protein-coupled receptors known as adrenergic receptors (ARs). This receptor family consists of three receptor types: α1, α2, and β [4]. Each receptor type has three subtypes, resulting in a total of nine AR subtypes encoded by distinct genes (α1a, α1b, α1d, α2a, α2b, α2c, β1, β2, β3). Downstream signaling through α1, α2, and β ARs is typically mediated by the $G\alpha_q$, and $G\alpha_i$, and $G\alpha_s$ pathways, but noncanonical pathways also exist. While most of the noradrenergic drugs mentioned in this review exhibit binding specificity for a particular receptor type, it is important to note that many are nonspecific regarding receptor subtype, which could explain some of the conflicting findings between studies that will be discussed later in this chapter.

The location of ARs at the level of the synapse is also critical for understanding mechanisms of action of noradrenergic drugs. ARs can be presynaptic, postsynaptic, or extra-synaptic, with receptor types and distribution varying greatly by brain region [5]. Postsynaptic and presynaptic ARs on NE-target neurons (i.e., heteroreceptors) modulate neural activity via downstream signaling cascades, while presynaptic ARs on noradrenergic neurons function as inhibitory autoreceptors and thereby reduce NE transmission. Because the overarching goal for many addiction pharmacotherapies is to block the actions of NE, compounds of interest generally either antagonize heteroreceptors (α1 and β) or stimulate autoreceptors (α2).

There is a substantial body of literature implicating the noradrenergic system in the pathophysiology of addiction, as well as promising results in both animal and human studies regarding the use of noradrenergic therapies for drug dependence. This chapter will provide an overview of the current state of research by (1) briefly reviewing the literature implicating the noradrenergic system in addiction, (2) highlighting the findings of preclinical and clinical studies using noradrenergic therapies, and (3) proposing ways in which current neuroscientific tools can be used to further probe the role of NE in the neural circuitry of addiction or even eventually directly control noradrenergic activity in a clinical setting.

THE ROLE OF THE NORADRENERGIC SYSTEM IN ADDICTION

Recognition of NE's role in addiction has fluctuated substantially over time. This section of the chapter focuses on how NE influences the midbrain and prefrontal cortex (PFC), two brain regions involved in addiction, as well as how psychostimulants, opioids, and alcohol dysregulate noradrenergic transmission. For a more comprehensive history of the role of NE in addiction, see the previous review on this topic [6].

Noradrenergic Regulation of Mesolimbic DA

The clearest evidence for the role of NE in addiction is through its influence on the mesolimbic DA system, which is comprised of the ventral tegmental area (VTA) and its projections to the nucleus accumbens (NAc). It is well established that the release of DA in the NAc is critical for mediating the rewarding effects of stimuli in both animals and humans [7,8]. As such, noradrenergic modulation of midbrain DA neuron activity is significant because this could shape behaviors related to drug reward. For example, burst firing activity of DA neurons, which drives higher DA release than tonic firing, is thought to assist with learning by encoding the salience of a stimulus [9]. This process appears to underlie incentive salience, a conceptualization of addiction in which abused substances alter reward circuitry to promote drug seeking over a desire for natural rewards [1,10]. Therefore, the way in which NE alters the excitability of midbrain DA neurons could impact the development of addiction.

Multiple studies indicate that NE is indeed capable of driving midbrain DA neuron activity, both directly and indirectly (via glutamate and GABA). In vivo pressure injection of the α1AR agonist phenylephrine into the VTA increased pacemaker (i.e., spontaneous) activity and burst firing of DA neurons, and NE iontophoretically applied into the VTA modulated glutamate-evoked excitation of these cells [11,12]. Furthermore, direct in vivo electrical stimulation of noradrenergic nuclei, such as the LC, evoked excitatory responses from midbrain DA neurons [13]. Conversely, blocking NE signaling in rats via the α1AR antagonist prazosin blunted the stimulatory effects of the LC on midbrain DA neurons. Collectively, this work shows that NE influences the activity of VTA DA neurons, and that NE transmission generally provides excitatory drive via the α1AR, although NE can also inhibit these cells via the α2AR or local GABA circuits [14,15].

Current research indicates that NE does not merely alter dopaminergic neuron activity, but also impacts DA release in VTA projection areas. Lesioning LC fibers with the selective noradrenergic neurotoxin N-(2-chloroethyl)-N-ethyl-2-bromobenzylamine (DSP-4) impaired DA release potential in the NAc of rats, and DBH knockout mice that cannot synthesize NE similarly have reduced striatal DA overflow [16,17]. Furthermore, a human PET study showed that blocking α1AR signaling with prazosin increased binding of the D2/D3-PET ligand [11C]-(+)-4-propyl-3,4,4a,5,6,10b-hexahydro-2H-naphtho[1,2-b][1,4]oxazin-9-ol ([^{11}C]-(+)-PHNO) in the dorsal caudate, suggesting reduced receptor occupancy by endogenous extracellular DA [18].

Noradrenergic Effects Beyond the Mesolimbic System

The role of NE in addiction extends beyond its direct influence on mesolimbic DA; NE also mediates addiction-like behavior by altering neuronal activity in the PFC, another brain region central to addiction and relapse. Chronic drug use induces adaptations of dendritic spines and neuronal activity in the PFC, NAc, and other brain regions [19]. In the PFC, increased spine growth is believed to heighten the salience of drug-associated cues and thereby promote drug seeking and relapse [1]. Because this region receives significant noradrenergic projections from the LC, recent work has addressed the influence of noradrenergic signaling on drug-induced plasticity in the medial PFC (mPFC). One study found that blocking βAR signaling with propranolol attenuated cocaine-induced plasticity in the mPFC of rats, as well cocaine-associated memory retrieval [20]. More specifically, βAR blockade during memory reactivation reversed cocaine-induced changes in spontaneous synaptic transmission, paired-pulse facilitation, and AMPA receptor plasticity, and ultimately impaired the retrieval of cocaine-associated memory. Other work has shown that NE depletion in the mPFC impaired natural- and drug-induced motivational salience and reward, and that α1AR signaling in this region was required for cocaine-primed reinstatement of cocaine seeking, a model of relapse-like behavior [21,22]. Collectively, this work indicates that noradrenergic signaling alters behaviorally relevant drug-induced changes in the mPFC.

Other work indicates that drugs of abuse may also have DA-independent consequences in the brain by causing an uncoupling of noradrenergic and serotonergic signaling [23]. One study found that repeated administration of addictive substances, including cocaine, morphine, and ethanol in mice resulted in sensitization of the noradrenergic and serotonergic systems as indicated by increased extracellular NE and serotonin (5-HT) when dosed with d-amphetamine or para-chloroamphetamine, respectively [24]. Importantly, while the canonical D1 receptor antagonist SCH23390 was able to block this sensitization, its effect was mediated by the concurrent 5-HT$_{2C}$ agonist property of the compound and not through its effects on DA. This study illustrates the importance of NE in addiction-related neural adaptations that can occur independent of the influence of DA.

Substance-Induced Dysregulation of NE

The role of NE in addiction is apparent when one considers how abused substances inherently dysregulate the noradrenergic system. Psychostimulants increase synaptic concentrations of DA, 5-HT, and NE [25]. In the case of NE, psychostimulants block the reuptake of extracellular NE by the NE transporter (NET) and induce vesicular release of NE through vesicular monoamine transporter 2 (VMAT2) [26]. A host of studies suggest that psychostimulants dysregulate noradrenergic tone, which contributes to addiction-like behaviors.

Self-administration studies show that while NE does not appear to mediate regular, established drug taking, it is critically important for relapse-like behavior. For example, maintenance of cocaine self-administration in rats is not affected by blocking NE synthesis with DBH inhibitors, or by lesioning noradrenergic afferents with 6-hydroxydopamine (6-OHDA) [27,28]. However, NE is required for the full expression of drug-primed, cue-induced, and stress-induced reinstatement of drug seeking, indicating that reducing NE signaling may be beneficial for preventing relapse. This potential therapeutic approach is supported by other work showing that blockade of α1ARs prevents the escalation of cocaine self-administration following preexposure or extended access to the drug [29,30]. In locomotor sensitization studies, rats that were sensitized to amphetamine by repeated exposure exhibited a higher spontaneous LC neuron firing rate than control animals, while treatment with the α2AR agonist clonidine, which reduces LC tone by acting on inhibitory autoreceptors, attenuated the previously observed locomotor sensitization [31]. These results suggest that psychostimulant exposure induces changes in noradrenergic transmission, and that correcting NE dysregulation could provide an important pathway for addiction intervention. For more in-depth discussion of the role of NE in psychostimulant-induced behaviors, see the following reviews [4,32].

The role of NE in opioid addiction is particularly complex because NE mediates both the rewarding effects of opioid intake and the aversive effects of opioid withdrawal. Previous work shows that NE transmission is necessary for mice to form a morphine-induced conditioned place preference; DBH knockout mice originally showed no preference for a morphine-paired environment but developed a preference when NE synthesis was restored [33]. Moreover, viral restoration of DBH in the nucleus tractus solitarius (NTS; A2), but not the LC, restored morphine-conditioned place preference, indicating that noradrenergic activity in the NTS is necessary for opioid reward.

While NE is required for opioid reward, it also contributes to withdrawal. LC neurons abundantly express μ-receptors, and as such, they are susceptible to opioid-induced changes in noradrenergic tone [34]. While acute exposure attenuates intrinsic LC activity, chronic exposure is associated with normal LC activity levels due to compensatory upregulation of NE biosynthetic capacity and the cAMP pathway [35–37]. This functional change in the LC becomes evident during detoxification, where the lack of opioid intake reveals LC hyperactivity and sympathetic symptoms associated with opioid withdrawal, including elevated pulse rate, sweating, tremors, and irritability/anxiety [38,39]. Likewise, neuroanatomical tracer studies show that the other noradrenergic nuclei, A1 and A2, also become highly active during withdrawal. Combined, these noradrenergic projections likely release excess NE in downstream target regions. One such target of interest has been the bed nucleus of the stria terminalis (BNST), a component of the extended amygdala. Tracer studies show that BNST-projecting cells from A1 and A2 are active during opioid withdrawal, while the LC is not [40]. Moreover, microinjections of βAR antagonists into the BNST markedly reduced conditioned place aversion and somatic signs of withdrawal, indicating that noradrenergic signaling in this region is critical for mediating withdrawal symptoms. Overall, these findings show that NE is dysregulated by opioids and provide additional support for treating addiction with therapies that modulate noradrenergic signaling.

Noradrenergic dysregulation also plays a role in the recovery process, as it confers stress sensitivity that increases the likelihood of relapse. Upon completing detoxification, opioid-dependent individuals often transition to maintenance therapies, such as the agonist methadone, the partial agonist buprenorphine, or the antagonist naltrexone to minimize withdrawal symptoms and prevent opioid abuse [41]. However, a large proportion of individuals struggle with treatment adherence, and often relapse [42,43]. Preclinical studies in rats showed that chronic morphine administration sensitizes LC NE neurons to become hyperactive in response to even minor stressors, which might explain how stressful life events trigger relapse in recovering addicts [44]. As opioid abuse clearly dysregulates NE, it is intrinsically involved in the addiction process and presents an important target for addiction therapies.

Alcohol abuse likewise induces lasting changes in the noradrenergic system. In vitro work in neuroblastoma cultures showed that alcohol upregulated expression of the NE-synthesizing enzyme DBH, as well as releasable NE [45].

Additionally, rats that had received chronic alcohol had elevated plasma NE levels 5 weeks after the last alcohol exposure, indicating that alcohol-associated increases in noradrenergic activity persist even following abstinence [46]. In humans, active alcoholics exhibit higher plasma concentrations of NE and altered DBH activity compared to healthy controls [47,48]. Together, these findings suggest that chronic alcohol abuse causes a dysregulation of NE levels, and that this dysregulation persists for some time following abstinence.

Overall, current evidence suggests a complex role for NE in the pathophysiology of addiction. NE shapes various aspects of addiction from reward to relapse behavior, and chronic drug exposure likewise alters noradrenergic transmission. In general, it appears that substance abuse causes an increase in LC noradrenergic activity as well as NE release. Given that addiction dysregulates NE in this way, the most promising therapies for addiction may involve compounds that decrease NE levels or AR signaling.

TRANSLATIONAL RESEARCH ON NORADRENERGIC THERAPIES

Based on the results described in the previous section, there is reason to believe that interfering with NE production and/or signaling could be an effective treatment for addiction. As such, a number of studies have investigated the potential therapeutic effect of adrenergic drugs. The following section reviews preclinical and clinical studies testing the effects of noradrenergic manipulation on addiction-related behaviors. Study findings are discussed among three major classes of commonly abused substances: psychostimulants, opioids, and alcohol, and further subdivided by mechanism of action (i.e., inhibition of NE synthesis, α2AR agonism, βAR antagonism, etc.). A summary of various pharmacotherapeutic approaches to modulate NE transmission is shown in Fig. 1.

Psychostimulants

Perhaps the most direct way to attenuate all noradrenergic signaling is to interfere with the production of NE. Disulfiram is one such drug with this pharmacologic capability. Originally used for the purpose of discouraging alcohol abuse, disulfiram blocks alcohol metabolism by inhibiting the enzyme aldehyde dehydrogenase [49]. As a result, the alcohol metabolite acetaldehyde accumulates and induces noxious symptoms, such as flushing, headache, and vomiting [50]. Interestingly, studies found that disulfiram reduced cocaine use in individuals with and without comorbid alcohol addiction, suggesting another possible use for this drug [51,52]. Later work revealed that disulfiram also inhibits DBH, the enzyme responsible for the conversion of DA to NE, thus decreasing central NE and increasing DA [53,54]. Disulfiram's mechanism of action is similar to other DBH inhibitors in that its primary metabolite chelates copper and thereby deprives DBH of the co-factor necessary for its activity [55–57]. When tested in rats, disulfiram attenuated cocaine-primed reinstatement,

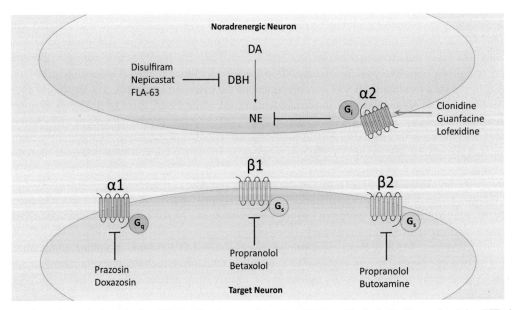

FIG. 1 Summary of noradrenergic therapies for addiction. The drugs used to treat addiction collectively block norepinephrine (NE) signaling by (1) inhibiting NE production, (2) agonizing α2 autoreceptors to attenuate NE transmission, or (3) antagonizing α1 and β heteroreceptors on target cells.

indicating that blocking NE production may reduce relapse-like behavior [28]. A similar study in nonhuman primates showed that disulfiram impaired cocaine-induced DA overflow in the NAc but did not replicate the inhibition of cocaine-primed reinstatement seen in rats [58].

The effects of disulfiram in human studies are likewise mixed in their findings. A 12-week long, randomized, placebo-controlled trial found that disulfiram reduced cocaine use among cocaine-dependent individuals, and that disulfiram mediated its effects directly on cocaine use rather than through alcohol intake [51]. Disulfiram treatment was also effective in reducing the frequency and quantity of cocaine use among cocaine-dependent individuals who were on methadone maintenance for co-morbid dependence, as well as decreasing frequency of cocaine use at a 1 year follow-up [59,60]. One clinical study, while small in size, evaluated the effect of disulfiram on intravenous cocaine and found that treatment attenuated subjective reports of a "high" or "rush," indicating that perhaps disulfiram decreases the rewarding aspect of cocaine [61]. However, other studies have found only modest reductions in cocaine use with disulfiram treatment when compared to baseline use [62]. It is important to note that clinical results may vary based on the duration of treatment with DBH inhibitors. Because chronic DBH inhibition increases the aversive effects of cocaine, increased treatment duration could more effectively reduce cocaine use [63].

Future disulfiram studies will also need to account for other factors that may introduce variability in study outcomes. Meta-analyses have noted gender differences in efficacy, with disulfiram being less efficacious in women [64,65]. Moreover, other work has shown that certain DBH gene variants are associated with lower levels of DBH, which may influence disulfiram efficacy [66]. As a whole, these results indicate that disulfiram may be of some benefit for a subset of patients, but studies need to be carried out in larger, more diverse cohorts, and across several types of psychostimulants to better characterize disulfiram's therapeutic profile.

Given that disulfiram's therapeutic effects on cocaine addiction are attributed to inhibition of DBH, researchers have also pursued the selective DBH inhibitor, nepicastat, which blocks DBH activity by interfering with the active site rather than by chelating copper, as a potential therapeutic agent [67]. Previous work has shown that nepicastat attenuates the reinforcing properties of cocaine and suppresses all modalities of relapse-like behavior in rats, including stress, cue, and drug-primed reinstatement, although the exact mechanism is a subject of some debate [68]. One study showed that DBH inhibition attenuated cocaine-primed reinstatement by inducing excessive DA release from noradrenergic terminals in the PFC, while another provided evidence that reduced signaling via PFC α1ARs is critical [21,69]. In either case, despite promising preclinical findings with nepicastat, detailed human data are lacking. There is one published study testing nepicastat among inpatients with cocaine use disorder, which found that DBH inhibition attenuated the positive subjective effects of cocaine similar to disulfiram clinical studies [70]. While this work is promising, more studies employing larger cohorts and an outpatient design are needed to more comprehensively evaluate the effectiveness of nepicastat. A large, multisite, double blind, placebo-controlled Phase II trial of nepicastat for cocaine dependence was recently completed. Initial outcome measures indicated no significant benefit, but full results have yet to be published (https://www.clinicaltrials.gov/ct2/show/NCT01704196).

Additional recent work has focused on the α2AR agonist clonidine, which stimulates presynaptic α2 inhibitory autoreceptors and reduces NE release [71]. Clonidine attenuated cue-induced reinstatement of cocaine seeking in rats, as well as drug-primed reinstatement of cocaine seeking in squirrel monkeys [72,73]. One human study found that treatment with oral clonidine 3 h prior to viewing stress or drug-cue auditory-imagery scripts significantly reduced subjective drug craving for stress scripts at 0.1 and 0.2 mg doses of clonidine [74]. However, subjective drug craving elicited by drug-cue scripts was only decreased at the higher dose. Together, these studies indicate that noradrenergic signaling specific to the α2AR may protect against relapse by decreasing psychostimulant craving.

The effect of α2AR agonism on factors leading to relapse among cocaine users has been further underscored in separate studies using the α2AR agonist guanfacine. Guanfacine dose-dependently attenuated impulsive and premature responding during the five choice serial reaction time task in rats, suggesting an ability to reduce cocaine-related cognitive inflexibility [75]. Additionally, repeated administration of guanfacine during withdrawal from cocaine decreased anxiety-like behavior in the elevated plus-maze and shock-probe burying task and attenuated stress-induced reinstatement [76]. Combined, this work suggests that blocking noradrenergic signaling could reduce impulsive and stress-related behaviors that increase risk for relapse. Guanfacine has also shown promise in human studies. Notably, cocaine-dependent individuals who had been abstinent for 21 days and were treated with guanfacine reported decreased cocaine craving when presented with cocaine-related imagery [77].

Previous and ongoing studies in both animals and humans provide strong support for the therapeutic effect of α1AR blockade in psychostimulant addiction. Mice lacking the α1bAR did not form a place preference in response to treatment with either cocaine or d-amphetamine, indicating that signaling from this receptor subtype is necessary for the rewarding effects of psychostimulants [78]. These findings are further strengthened by corroborating pharmacologic work showing that prazosin pretreatment attenuated both acute and sensitized locomotor effects in rats given either cocaine or amphetamine [79]. So far, data from human studies also appear to support the use of α1AR antagonists. A pilot study found that doxazosin treatment was effective in reducing the number of cocaine-positive urine samples from study participants [80]. The reason for this effect might be explained by a separate Phase I study, which found that doxazosin decreased participants' ratings of subjective indices, such as "stimulation" and "likeliness to use cocaine," while on cocaine [81]. Doxazosin did not, however, significantly affect ratings of euphoria, consistent with a study that tested the effect of this drug on the subjective effects of the stimulant 3,4-methylenedioxymethamphetamine [82]. Thus, it appears that doxazosin decreases desire or "craving" for cocaine without affecting the subjective experience of cocaine intoxication. These findings indicate that α1AR antagonists may be effective in reducing relapse. Since α1AR blockade has been shown to attenuate cocaine-primed reinstatement in rats, there is already preclinical support for this use of doxazosin [83]. A Phase I clinical trial was recently completed to test the effectiveness of doxazosin in preventing relapse but results are not yet published (https://clinicaltrials.gov/ct2/show/NCT00880997). Doxazosin has also shown differential efficacy among certain gene variants of the α1AR gene ADRA1A; a Phase II clinical trial is currently underway to determine whether the R492C polymorphism is associated with greater efficacy in cocaine-dependent patients (https://clinicaltrials.gov/ct2/show/NCT01953432).

The role of βARs has also been investigated in the context of psychostimulant addiction. In preclinical studies, βAR blockade reduced anxiety-like behavior in mice following acute cocaine exposure, as well as in rats undergoing withdrawal from chronic cocaine [84,85]. These observations were corroborated in a randomized, placebo-controlled clinical study in which the nonselective βAR antagonist propranolol reduced withdrawal symptoms compared to placebo [86]. While another study evaluating the effect of combined treatment with propranolol and the indirect DA agonist amantadine seemed to contradict these findings, a secondary analysis accounting for medication adherence among study subjects showed that propranolol was superior to placebo at managing withdrawal symptoms [87].

Other studies have noted protective effects of βAR antagonism on measures of craving and relapse. Preclinical work showed that while forced swim stress reinstated cocaine-seeking in wild-type mice, it was ineffective in βAR-deficient animals [88]. Likewise, administration of either β1AR or β2AR antagonists prevented forced swim stress-induced reinstatement of cocaine seeking, indicating that βAR signaling mediates stress-induced relapse-like behavior. However, results from human studies have been less compelling. When cocaine-dependent patients were given propranolol or placebo after exposure to a cocaine cue, propranolol attenuated craving in the 24 h after the cue exposure, but was no different from placebo in controlling craving over the following week [89]. Moreover, another study actually found greater drug-cue reactivity in propranolol-treated polydrug users (methadone-maintained opioid-dependent individuals who used cocaine) compared to those receiving placebo [90]. So far, βAR antagonism does not appear to be as promising in the clinic as it was in animal studies.

Opioids

Agonists of the α2AR have shown effectiveness for managing both withdrawal symptoms and relapse. A Phase III clinical trial among patients undergoing medically supervised detoxification found that replacing morphine with the α2AR agonist lofexidine during initial withdrawal significantly decreased withdrawal symptoms and improved retention in treatment compared to placebo [91]. A more recent Phase III trial confirmed that lofexidine-only treatment during withdrawal effectively reduced symptoms and was associated with longer retention in treatment and a higher proportion of treatment completion [92]. This work is particularly intriguing because it provides support for a nonopioid based treatment for opioid withdrawal; currently no such treatment is available. In addition, a study using a combination therapy of clonidine and low-dose naltrexone noted greater attenuation of withdrawal symptoms compared to either medication alone, and that subjects on this regimen were more likely to complete detoxification [93]. Therefore, α2AR agonists present a promising option for improving the success of detoxification treatment by managing withdrawal symptoms.

Other clinical trials have investigated whether supplementing maintenance regimens with α2AR agonists improves long-term outcomes with respect to relapse. Preliminary findings in a study among opioid-dependent patients indicated that concurrent treatment with naltrexone and lofexidine increased abstinence rates, improved relapse outcomes, and attenuated cue-induced craving compared to naltrexone-placebo patients [94]. Another trial, which allowed subjects to report their mood and activities in real time, found that while buprenorphine maintenance with adjunct clonidine increased the length of time before lapse (one-time use) compared to placebo, it did not have any effect on relapse [95]. Similarly, clonidine treatment decoupled daily life stress from drug craving, but did not affect cue-induced craving. Other work using adjunct guanfacine found that α2AR agonism was not sufficient to improve adherence to naltrexone maintenance therapy or opioid abstinence [96]. At this point in time, α2AR agonists appear to be most effective when used during opioid detoxification, and less so during later stages of recovery from opioid addiction. While all of these studies employed noradrenergic therapies that act on α2ARs, disparate efficacy between craving and relapse outcomes suggests that the mechanism behind the noradrenergic effect on opioid abstinence is complex and requires more preclinical investigation. Differences in drug potency, receptor affinity, pharmacokinetics, and site of action (presynaptic versus postsynaptic) likely contribute to these observed variability in efficacy and need to be compared in order to identify the most suitable therapeutic candidate and delivery regimen.

Antagonism of βARs also appears to have beneficial effects on various aspects of opioid addiction. Studies in mice posited a role for β2ARs by showing that pharmacological blockade with butoxamine reversed morphine tolerance and reduced physical dependence as measured by calcitonin gene-related peptide and substance P expression following naloxone-precipitated withdrawal [97]. Additionally, mice lacking β2ARs did not develop morphine tolerance, suggesting that β2AR signaling is necessary for mediating some aspects of morphine dependence. Interestingly, other work has found that rats with increased reactivity to novelty, a predictor of vulnerability to drug abuse, exhibited increased β1AR expression in the BNST [98]. During morphine withdrawal, these animals exhibited more severe somatic signs associated with noradrenergic activity, and microinjection of the β1AR antagonist betaxolol attenuated these symptoms. Despite the favorable findings in preclinical studies, the only report of βAR antagonists being tested in humans in the context of addiction was a small, outpatient study in 1976, which reported no effect of propranolol on withdrawal symptoms among opioid-dependent patients [99]. Given the compelling evidence for βAR antagonism in rodent studies and lack of recent investigation in humans, these compounds should be evaluated in more nonhuman primate and human studies.

Alcohol

Unlike psychostimulant and opioid addiction studies, there has been little research on the inhibition of NE synthesis for the treatment of alcohol dependence. Disulfiram has been used to treat alcohol abuse for many years with mixed results, and its efficacy has been attributed solely to the aversive effects associated with inhibition of aldehyde dehydrogenase [50,100]. However, preclinical studies indicate that DBH inhibition reduces voluntary alcohol intake in rodents, suggesting that the copper chelating effect of disulfiram on NE production could also contribute to its therapeutic effects [101]. For example, pharmacologic or genetic inhibition of DBH decreased alcohol intake in the two bottle choice test and/or operant self-administration, indicating that inhibition of NE production alone can indeed influence alcohol intake [102,103]. This work is further supported by pharmacokinetic studies showing that the DBH inhibitor FLA-63 decreased alcohol intake in rats independent of aldehyde dehydrogenase, as it neither increased blood acetaldehyde levels nor altered aldehyde dehydrogenase activity [104]. While preclinical findings suggest that DBH inhibition may be an effective means of decreasing alcohol intake in humans, these compounds have not been tested in clinical trials.

α2AR agonists have not been extensively pursued in alcohol addiction research. Previous studies showed that clonidine and lofexidine both decreased alcohol consumption in rats, and lofexidine also attenuated stress-induced reinstatement of alcohol self-administration [105,106]. But despite their effectiveness in animal studies, α2AR agonists were only briefly tested in small human studies in the late 1980s for their effect on alcohol withdrawal symptoms, in which these agents proved only marginally better than placebo [107,108].

The majority of preclinical and clinical work regarding treatments for alcohol abuse has focused on α1AR antagonists, such as prazosin. Blockade of α1AR signaling with prazosin decreased withdrawal-associated operant self-administration of alcohol, as well as alcohol intake during the two bottle choice testing in rats [109,110]. Apart from modulating alcohol-taking behavior, prazosin also blocked relapse-like alcohol seeking among rats in two modalities of stress-induced reinstatement: yohimbine and intermittent footshock [111]. The effects of prazosin in human studies are not as consistent. A 6-week long, double-blinded pilot study found no differences between prazosin and placebo groups in mean number of drinking days or drinks per week [112]. However, when only the last 3 weeks of the study were evaluated, prazosin-treated subjects decreased their average total number of drinking days compared to placebo-treated subjects. These results were encouraging, but the study was limited by its small size, short timeframe, and lack of female participants. More recently, clinical trials have been conducted to evaluate the effect of prazosin on veterans with comorbid PTSD and alcohol dependence. While neither study found an effect of prazosin on PTSD-related outcomes, one found that prazosin decreased drinking frequency and severity [113,114].

Even though prazosin only provides a marginal benefit for maintaining initial alcohol abstinence in humans, it may be favorable for preventing longer-term relapse. A study among alcohol-dependent individuals who had been abstinent for 3 weeks found that prazosin reduced both cue- and stress-induced alcohol craving [115]. The trial also found that prazosin treatment decreased anxiety, which corroborated preclinical work showing a decrease in anxiety-like behavior among prazosin-treated rats deprived of alcohol [116]. These findings suggest that while prazosin might not have a strong effect on reducing alcohol intake, its ability to temper anxiety may be an important means of preventing relapse among individuals who are already abstinent. It is important to note that this study was performed in a locked inpatient treatment facility, so further studies will need to be performed to evaluate prazosin's effect under typical outpatient conditions.

METHODS FOR INVESTIGATING THE NEURAL CIRCUITRY OF ADDICTION

While many studies demonstrate a relationship between noradrenergic signaling and addiction-related behaviors, there is still much to learn about the circuit-level changes that occur in the addicted brain. Despite the numerous gene knockout and pharmacological manipulations summarized here that implicate the noradrenergic system in drug addiction, many of these studies employed conventional "whole body" genetic modifications or systemic agonist/antagonist administration that lack brain region and/or receptor subtype specificity. Recent improvements in neuroscientific tools have expanded our ability to understand neural circuits by allowing for spatial and temporal control over activity as never before. This section provides an overview of cutting-edge tools for neuronal stimulation and visualization that could be employed to identify the noradrenergic circuitry responsible for modulating addiction-related behaviors.

Optogenetics

The introduction of optogenetics to the neuroscience field has made a significant impact on the study of neural circuits. This technique, which involves expressing light-sensitive ion channels (opsins) in neurons, usually via viral vectors, renders the activity of specific neuronal populations sensitive to light [117,118]. Different opsins can be delivered based on the desired neuronal effect, with some opsins (e.g., channelrhodopsin) causing depolarization when triggered by light, and others (e.g., halorhodopsin) causing hyperpolarization. By implanting a fiber optic ferrule over the brain region of interest and using a laser or light emitting diode to deliver light at biologically relevant frequencies, genetically identified and targeted neuronal populations, or even specific projections, can be excited or inhibited with temporal and regional specificity. Furthermore, pairing optogenetics with electrophysiological recordings or behavioral assays allows for systematic evaluation of how one brain region or cell type controls the activity of downstream target regions and resulting behavioral changes. This approach could be applied to examine the effects of noradrenergic activity on addiction-like behaviors in real time. For example, rats could undergo viral transfection in order to drive expression of an excitatory or inhibitory opsin in noradrenergic neurons. Then, noradrenergic terminals in the VTA could be selectively stimulated while recording from the VTA during a behavioral assay, such as a progressive ratio task to examine (1) how NE influences VTA activity and (2) how this pattern of activity correlates with motivation for drug. The ability to control neural activity and observe a corresponding behavioral outcome in this way provides many opportunities to investigate the contributions of specific neural populations within larger circuits implicated in addiction. Furthermore, investigation of NE's role in addictive behavior could be extended to LC afferents. The nucleus paragigantocellularis is among the structures that projects to the LC, and these projections consist of both glutamatergic and GABAergic neurons [119]. By selectively expressing an excitatory opsin in one of these cell types and stimulating its terminals in the LC, one could better elucidate how shifts in excitatory or inhibitory input to this brain region shape addiction-related behavior.

Designer Receptors Exclusively Activated by Designer Drugs

For situations in which sustained inhibition or excitation of particular brain regions is desired, designer receptors exclusively activated by designer drugs (DREADDs) provide an alternative to optogenetics. DREADDs are a chemogenetic tool that allows for the control of neural activity through the use of altered G-protein-coupled receptors [120]. DREADDs are mutated human muscarinic receptors that have been rendered insensitive to their natural ligand acetylcholine, and instead made sensitive to an exogenous ligand that can be administered by the experimenter. As a result, DREADDs can be introduced into the brains of animals using transgenic/knock-in or viral vector-based approaches and will exhibit no functional activity until the ligand is administered, allowing for selective, and even cell-type specific, control over neural activity.

The purported benefit of using DREADDs was that these receptors would only be active in response to the exogenous compound clozapine-N-oxide (CNO). However, the recent discovery that DREADDs are in fact activated by clozapine generated from reverse metabolism of CNO—rather than CNO itself—has generated controversy around the technique [121]. Clozapine is an atypical antipsychotic that has a wide range of activity and binds to dopaminergic, serotonergic, and muscarinic receptors [122]. As such, there is concern that this off-target activity could elicit behavioral effects, thereby confounding study results. Despite this surprising finding, it is important to note that DREADDs can still serve as a viable method for studying neural circuitry as long as appropriate controls are implemented. If behavioral effects are observed in experimental (DREADD-expressing+CNO) animals, but not control (CNO only) animals, this would suggest that the off-target effects of clozapine are not sufficient to elicit behavioral changes, and that observed effects in experimental animals can be attributed to the DREADD-induced change in neural activity.

Because CNO-induced DREADD activation can last for several hours in rodents and does not require a wired connection, DREADDs may be preferable to optogenetics in cases where when a sustained change in neural activity is needed, or when a complex behavioral task that would be confounded by the presence of a head-attached cable is desired [123]. Much like the use of different opsins in optogenetics to oppositely control neural activity, the use of Gq/Gs versus Gi DREADDs allows for specific activation or silencing of neuronal firing, respectively [124].

In the context of addiction research, DREADDs could be used to explore the well-documented paradoxical effect of acute versus chronic NE suppression. As noted in the studies above, treatments in rodents that acutely block NE production or signaling can suppress drug-seeking behavior and drug-related reward with respect to psychostimulants [68,125]. However, DBH knockout mice with chronic NE deficiency, or repeated treatment with DBH inhibitors, paradoxically induce hypersensitivity to psychostimulants [17,54,126–128]. Considering that DBH inhibitors like disulfiram have shown promise in treating cocaine addiction, a better understanding of the effects conferred by chronic NE suppression is needed [28]. One potential approach might be to introduce an inhibitory DREADD to noradrenergic neurons that normally provide excitatory drive onto dopaminergic VTA neurons. By regularly administering CNO, one could mimic the effect of chronically decreased noradrenergic input and determine the length of time needed to induce the previously observed compensatory changes in DA receptor function in addiction-associated brain regions such as the NAc [17,126]. Such an approach would provide more insight about the time course under which these neural adaptations occur. Moreover, pairing these experiments with electrophysiological recordings could also reveal how these perturbations affect neuronal activity in other behaviorally relevant regions.

Genetically Encoded Indicators of Neural Activity

One of the greatest challenges in systems neuroscience has been obtaining large-scale recordings of neuronal populations. Traditionally, such data were collected by performing iterative intracellular patch clamp recordings of individual neurons or extracellular recordings of local field potentials that could only provide information about one or a few cells at a time. While multielectrode arrays allow for recordings from many more neurons at once, the scale of data collection is still small compared to that needed to evaluate regional brain activity. Moreover, although neuroimaging methods such as fMRI provide insight about activity on a larger scale, changes in blood oxygen levels lack the temporal resolution needed to provide information about individual action potentials [129].

The aforementioned technical limitations have been at least partially alleviated by the advent of genetically encoded indicators of neural activity (GINAs). GINAs are fluorescent biosensors comprised of two components: an analyte-sensing or -binding domain and a fluorescent protein-based reporter element [130]. In the presence of the analyte of interest, a GINA will experience a change in fluorescence intensity, which can then be detected through microscopy methods. Because calcium events are the best proxy for action potentials in neurons, the most commonly used GINAs are calcium indicators. However, other GINAs have been developed to report alternative aspects of neural activity, such as synaptic transmission and voltage or pH changes.

There are several major advantages of using GINAs to characterize the activity of a neural population. First and foremost, GINAs enable recording of neuronal activity to a degree that surpasses other available techniques [131]. The data gathered from such large populations of neurons have proven particularly important for informing optogenetic approaches, as this information helps researchers define physiologically relevant optical stimulation parameters [132]. Second, GINAs can be incorporated into viral vectors containing drivers for a specific cell type to ensure that GINA expression is limited to a discrete neuronal population. Finally, GINAs allow for repeated measurements of the same cell population [133]. As such, this technique has become a powerful tool for investigating neural circuitry.

Future studies could use GINAs to investigate the role of NE in addiction. There is very little information on the activity of noradrenergic neurons during addiction-like behaviors; most electrophysiological recordings have been performed in brain slices taken from drug-treated animals or in an anesthetized preparation. Calcium indicators could be expressed in noradrenergic LC neurons and used to visualize the activity of these neurons during conditioned place preference or self-administration sessions over a period of time. While this work would be ambitious, recording these neurons over time would provide great insight about how chronic use of abused substances alter noradrenergic tone. Because these experiments would help characterize the LC activity patterns in an "addicted" brain, future pharmacological therapies could be evaluated by their capacity to restore normal LC tone. Overall, the use of GINAs in addiction research could yield great insight regarding addiction-induced changes in neural activity.

Inducible Tango

Yet another groundbreaking technique for investigating neural circuits is the development of the inducible Tango (*i*Tango) system, a method of labeling cells exposed to a particular ligand through corresponding expression of a reporter gene [134]. The technique is so named because two proteins must "partner," or associate, to stimulate a response. Tango involves coupling a transcription factor for a reporter gene to a membrane-bound receptor via a linker that contains a specific cleavage site. Activation of the receptor of interest triggers protease activity via a fused signaling peptide, which cuts the linker, releases the transcription factor, and allows transcription of the reporter gene. As such, one can visualize neurons whose receptors were activated by a specific ligand with high temporal specificity. Moreover, this technique utilizes a novel signaling pathway to activate the reporter, preventing it from being confounded by other concurrent cell-signaling processes [134].

The newly described *i*Tango method, which combines Tango with optogenetics, allows one to investigate the effects of neuromodulators on the neural activity of target regions by making them responsive to a specific wavelength of light [135]. The technique relies on a double-gated system, such that expression of a marker gene correlating with neuronal activity will only occur when a neuron receives coincident stimulation by light and a ligand of interest. By using *i*Tango to specifically stimulate VTA terminals in the NAc, the authors of a recent study were able to characterize the neuromodulatory effect of DA on the NAc in vivo while mice navigated a ball maze in order to earn a water reward [135].

*i*Tango is currently possible for several receptor types including GPCRs, steroid hormone receptors, and receptor tyrosine kinases. Because NE mediates numerous functions related to neuronal activity such as stress and arousal through its GPCR-linked neuromodulatory actions, this technique could foreseeably by applied to the noradrenergic system [136]. For example, researchers could better explore the interaction between noradrenergic projections from the LC, A1, and A2 cell groups and their effect on the VTA. By optically stimulating LC terminals derived from neuroanatomically defined noradrenergic nuclei and altering NE release in the VTA, one could explore how NE circuitry modulates VTA activity while also acquiring a visual record of activity via the expression of marker genes.

Drugs Acutely Restricted by Tethering

Apart from neuroscientific advances in controlling neuronal activity, there has also been substantial improvement in cell-specific drug delivery. Currently, intracranial and intraventricular infusions allow for drug delivery to the brain, but diffusion away from the injection site and nondiscriminate action across other cells that expresses the drug target hinder the ability to study cell-type specific pharmacological effects. The newly developed technique, drugs acutely restricted by tethering (DART), circumvents these issues by sequestering the drug to a particular area where it will act preferentially on a predetermined cell type [137]. DART involves fusing an engineered bacterial enzyme, HaloTag, to a transmembrane domain whose expression is driven by a cell-type specific promoter, ensuring that the fusion protein will only be present on the cell type of interest. An additional fusion molecule is generated consisting of HaloTag's substrate, HaloTag ligand, and the experimental drug. When the HaloTag ligand-drug component is infused into the brain, HaloTag covalently binds HaloTag ligand, effectively sequestering it. Accumulation of the drug induced by HaloTag and HaloTag ligand binding increases the drug concentration in close proximity to target receptors on the cell type of interest, allowing for preferential action of the drug on the specified neurons. The binding between HaloTag and its ligand is highly efficient, such that very low doses of the drug can be administered, further minimizing off-target effects. This technique offers a major advantage in that temporal specificity is maintained without having to modify the pharmacological target.

DART could conceivably be applied to the noradrenergic system to evaluate cell-type specific effects of noradrenergic transmission during reinstatement. The circuitry involved in stress-induced reinstatement has already been characterized, but less is known about the circuit underlying drug-primed reinstatement [138]. Recent work indicates that activation of α1ARs in the mPFC, but not the VTA or NAc, is necessary for cocaine-primed reinstatement in rats [21]. α1ARs are found on multiple types of neurons in the mPFC (e.g., glutamatergic, GABAergic, etc.), and it is not clear which population mediates α1AR-sensitive effects [139]. Using DART, an α1AR antagonist could be linked to HaloTag ligand and infused into the mPFC of rats that express HaloTag only on glutamatergic or GABAergic cells, ensuring that the antagonist would preferentially act on a specific cell population during reinstatement testing.

CONCLUSION

There is strong preclinical evidence that multiple classes of abused substances including psychostimulants, opioids, and alcohol dysregulate noradrenergic tone. Disruptions to the noradrenergic system are significant, as NE plays a key role in the process of addiction through both direct and indirect modulation of the mesolimbic dopaminergic system, which shapes the rewarding aspect of drug intake. Additionally, NE modulates other neuroanatomical and neurobiological substrates that impact relapse-like behaviors. Because many classes of abused substances appear to induce noradrenergic hyperactivity, therapeutic approaches have focused on dampening excessive NE signaling.

Presently, clinical studies regarding noradrenergic therapies for addiction have not been as consistent as animal studies. Despite this variability, the more promising candidates appear to be α1AR antagonists for psychostimulant abuse and α2AR agonists for opioid dependence. While disulfiram has been a mainstay for alcohol abuse treatment for some time, conflicting results regarding its efficacy suggest that other therapeutic options are needed. α1AR antagonists such as prazosin have been shown to decrease drinking frequency, so it is possible that this drug class will be more heavily pursued in the future for alcoholism. Ultimately, larger studies with more diverse cohorts should be conducted in order to test the most promising noradrenergic therapies. The current collection of clinical trials for addiction has been limited to small, inpatient, largely male cohorts, which impairs assessment of the "real-world" efficacy of these therapies.

Advancements in neuroscientific techniques present new options for increasingly complex dissection of addiction neurocircuitry. Increased specificity in both temporal and spatial control of neuronal and pharmacological activity will allow neuroscientists to develop a better understanding of these circuits and thereby direct the development of new treatments. In fact, some of the techniques discussed above may have applications that extend beyond the bench to the bedside. For example, α2AR agonists have been effective in some clinical studies for addiction, but a caveat of these drugs is that patients frequently suffer from hypotension as a side effect [92]. If techniques such as DART are eventually introduced into human studies, drug activity could be limited to addiction-related circuits, reducing off-target side effects and enhancing the tolerability of this treatment. Additionally, optogenetics may someday be applied to humans to help correct the activity of brain regions that become dysregulated by substance abuse. After all, it is not a huge leap from the current use of deep brain stimulation as a treatment for degenerative brain diseases, such as Parkinson's disease. Optogenetics is already being considered in the field of cardiology as a potential means to regulate heart rate among patients with serious arrhythmias, indicating the potential for this technique beyond addiction and across other specialties of medicine [140].

In summary, the noradrenergic system plays a critical role in the pathophysiology of addiction and is an important therapeutic target. Continued examination of the noradrenergic system and its ability to shape addiction-related behaviors using cutting-edge techniques will be critical for the development of new therapies.

REFERENCES

[1] Koob GF, Volkow ND. Neurocircuitry of addiction. Neuropsychopharmacology 2010;35(1):217–38. https://doi.org/10.1038/npp.2009.110.

[2] Kostowski W. Two noradrenergic systems in the brain and their interactions with other monoaminergic neurons. Pol J Pharmacol Pharm 1979;31 (4):425–36.

[3] Ungerstedt U. Stereotaxic mapping of the monoamine pathways in the rat brain. Acta Physiol Scand Suppl 1971;367:1–48.

[4] Schmidt KT, Weinshenker D. Adrenaline rush: the role of adrenergic receptors in stimulant-induced behaviors. Mol Pharmacol 2014;85(4):640–50. https://doi.org/10.1124/mol.113.090118.

[5] Molinoff PB. Alpha- and beta-adrenergic receptor subtypes properties, distribution and regulation. Drugs 1984;28(Suppl 2):1–15.

[6] Weinshenker D, Schroeder JP. There and back again: a tale of norepinephrine and drug addiction. Neuropsychopharmacology 2007;32(7):1433–51. https://doi.org/10.1038/sj.npp.1301263.

[7] Imperato A, Di Chiara G. Preferential stimulation of dopamine release in the nucleus accumbens of freely moving rats by ethanol. J Pharmacol Exp Ther 1986;239(1):219–28.

[8] Volkow ND, Fowler JS, Wang GJ, Baler R, Telang F. Imaging dopamine's role in drug abuse and addiction. Neuropharmacology 2009;56(Suppl 1):3–8. https://doi.org/10.1016/j.neuropharm.2008.05.022.

[9] Robinson TE, Berridge KC. Incentive-sensitization and addiction. Addiction 2001;96(1):103–14. https://doi.org/10.1080/09652140020016996.

[10] Valyear MD, Villaruel FR, Chaudhri N. Alcohol-seeking and relapse: a focus on incentive salience and contextual conditioning. Behav Processes 2017. https://doi.org/10.1016/j.beproc.2017.04.019.

[11] Almodovar-Fabregas LJ, Segarra O, Colon N, Dones JG, Mercado M, Mejias-Aponte CA, Vazquez R, Abreu R, Vazquez E, Williams JT, Jimenez-Rivera CA. Effects of cocaine administration on VTA cell activity in response to prefrontal cortex stimulation. Ann N Y Acad Sci 2002;965:157–71.

[12] Goertz RB, Wanat MJ, Gomez JA, Brown ZJ, Phillips PE, Paladini CA. Cocaine increases dopaminergic neuron and motor activity via midbrain alpha1 adrenergic signaling. Neuropsychopharmacology 2015;40(5):1151–62. https://doi.org/10.1038/npp.2014.296.

[13] Grenhoff J, Nisell M, Ferre S, Aston-Jones G, Svensson TH. Noradrenergic modulation of midbrain dopamine cell firing elicited by stimulation of the locus coeruleus in the rat. J Neural Transm Gen Sect 1993;93(1):11–25.

[14] Jimenez-Rivera CA, Figueroa J, Vazquez-Torres R, Velez-Hernandez ME, Schwarz D, Velasquez-Martinez MC, Arencibia-Albite F. Presynaptic inhibition of glutamate transmission by alpha2 receptors in the VTA. Eur J Neurosci 2012;35(9):1406–15. https://doi.org/10.1111/j.1460-9568.2012.08029.x.

[15] Velasquez-Martinez MC, Vazquez-Torres R, Rojas LV, Sanabria P, Jimenez-Rivera CA. Alpha-1 adrenoreceptors modulate GABA release onto ventral tegmental area dopamine neurons. Neuropharmacology 2015;88:110–21. https://doi.org/10.1016/j.neuropharm.2014.09.002.

[16] Haidkind R, Kivastik T, Eller M, Kolts I, Oreland L, Harro J. Denervation of the locus coeruleus projections by treatment with the selective neurotoxin DSP-4 [N (2-chloroethyl)-N-ethyl-2-bromobenzylamine] reduces dopamine release potential in the nucleus accumbens shell in conscious rats. Neurosci Lett 2002;332(2):79–82.

[17] Schank JR, Ventura R, Puglisi-Allegra S, Alcaro A, Cole CD, Liles LC, Seeman P, Weinshenker D. Dopamine beta-hydroxylase knockout mice have alterations in dopamine signaling and are hypersensitive to cocaine. Neuropsychopharmacology 2006;31(10):2221–30. https://doi.org/10.1038/sj.npp.1301000.

[18] Le Foll B, Thiruchselvam T, Lu SX, Mohammed S, Mansouri E, Lagzdins D, Nakajima S, Wilson AA, Graff-Guerrero A, Di Ciano P, Boileau I. Investigating the effects of norepinephrine alpha1 receptor blockade on dopamine levels: A pilot PET study with [11 C]-(+)-PHNO in controls. Synapse 2017;https://doi.org/10.1002/syn.21968.

[19] Robinson TE, Kolb B. Structural plasticity associated with exposure to drugs of abuse. Neuropharmacology 2004;47(Suppl 1):33–46. https://doi.org/10.1016/j.neuropharm.2004.06.025.

[20] Otis JM, Mueller D. Reversal of cocaine-associated synaptic plasticity in prefrontal cortex. Neuropsychopharmacology 2017;https://doi.org/10.1038/npp.2017.90.

[21] Schmidt KT, Schroeder JP, Foster SL, Squires K, Smith BM, Pitts EG, Epstein MP, Weinshenker D. Norepinephrine regulates cocaine-primed reinstatement via alpha1-adrenergic receptors in the medial prefrontal cortex. Neuropharmacology 2017;119:134–40. https://doi.org/10.1016/j.neuropharm.2017.04.005.

[22] Ventura R, Morrone C, Puglisi-Allegra S. Prefrontal/accumbal catecholamine system determines motivational salience attribution to both reward- and aversion-related stimuli. Proc Natl Acad Sci U S A 2007;104(12):5181–6. https://doi.org/10.1073/pnas.0610178104.

[23] Salomon L, Lanteri C, Glowinski J, Tassin JP. Behavioral sensitization to amphetamine results from an uncoupling between noradrenergic and serotonergic neurons. Proc Natl Acad Sci U S A 2006;103(19):7476–81. https://doi.org/10.1073/pnas.0600839103.

[24] Lanteri C, Salomon L, Torrens Y, Glowinski J, Tassin JP. Drugs of abuse specifically sensitize noradrenergic and serotonergic neurons via a non-dopaminergic mechanism. Neuropsychopharmacology 2008;33(7):1724–34. https://doi.org/10.1038/sj.npp.1301548.

[25] White FJ, Kalivas PW. Neuroadaptations involved in amphetamine and cocaine addiction. Drug Alcohol Depend 1998;51(1–2):141–53.

[26] Sofuoglu M, Sewell RA. Norepinephrine and stimulant addiction. Addict Biol 2009;14(2):119–29. https://doi.org/10.1111/j.1369-1600.2008.00138.x.

[27] Roberts DC, Corcoran ME, Fibiger HC. On the role of ascending catecholaminergic systems in intravenous self-administration of cocaine. Pharmacol Biochem Behav 1977;6(6):615–20.

[28] Schroeder JP, Cooper DA, Schank JR, Lyle MA, Gaval-Cruz M, Ogbonmwan YE, Pozdeyev N, Freeman KG, Iuvone PM, Edwards GL, Holmes PV, Weinshenker D. Disulfiram attenuates drug-primed reinstatement of cocaine seeking via inhibition of dopamine beta-hydroxylase. Neuropsychopharmacology 2010;35(12):2440–9. https://doi.org/10.1038/npp.2010.127.

[29] Wee S, Mandyam CD, Lekic DM, Koob GF. Alpha 1-noradrenergic system role in increased motivation for cocaine intake in rats with prolonged access. Eur Neuropsychopharmacol 2008;18(4):303–11. https://doi.org/10.1016/j.euroneuro.2007.08.003.

[30] Zhang XY, Kosten TA. Previous exposure to cocaine enhances cocaine self-administration in an alpha 1-adrenergic receptor dependent manner. Neuropsychopharmacology 2007;32(3):638–45. https://doi.org/10.1038/sj.npp.1301120.

[31] Doucet EL, Bobadilla AC, Houades V, Lanteri C, Godeheu G, Lanfumey L, Sara SJ, Tassin JP. Sustained impairment of alpha2A-adrenergic auto-receptor signaling mediates neurochemical and behavioral sensitization to amphetamine. Biol Psychiatry 2013;74(2):90–8. https://doi.org/10.1016/j.biopsych.2012.11.029.

[32] Zaniewska M, Filip M, Przegalinski E. The involvement of norepinephrine in behaviors related to psychostimulant addiction. Curr Neuropharmacol 2015;13(3):407–18.

[33] Olson VG, Heusner CL, Bland RJ, During MJ, Weinshenker D, Palmiter RD. Role of noradrenergic signaling by the nucleus tractus solitarius in mediating opiate reward. Science 2006;311(5763):1017–20. https://doi.org/10.1126/science.1119311.

[34] Van Bockstaele EJ, Menko AS, Drolet G. Neuroadaptive responses in brainstem noradrenergic nuclei following chronic morphine exposure. Mol Neurobiol 2001;23(2–3):155–71.

[35] Guitart X, Hayward M, Nisenbaum LK, Beitner-Johnson DB, Haycock JW, Nestler EJ. Identification of MARPP-58, a morphine- and cyclic AMP-regulated phosphoprotein of 58 kDa, as tyrosine hydroxylase: evidence for regulation of its expression by chronic morphine in the rat locus coeruleus. J Neurosci 1990;10(8):2649–59.

[36] Mazei-Robison MS, Nestler EJ. Opiate-induced molecular and cellular plasticity of ventral tegmental area and locus coeruleus catecholamine neurons. Cold Spring Harb Perspect Med 2012;2(7). https://doi.org/10.1101/cshperspect.a012070.

[37] Rasmussen K, Beitner-Johnson DB, Krystal JH, Aghajanian GK, Nestler EJ. Opiate withdrawal and the rat locus coeruleus: behavioral, electrophysiological, and biochemical correlates. J Neurosci 1990;10(7):2308–17.

[38] Aghajanian GK. Tolerance of locus coeruleus neurones to morphine and suppression of withdrawal response by clonidine. Nature 1978;276(5684):186–8.

[39] Wesson DR, Ling W. The clinical opiate withdrawal scale (COWS). J Psychoactive Drugs 2003;35(2):253–9. https://doi.org/10.1080/02791072.2003.10400007.

[40] Delfs JM, Zhu Y, Druhan JP, Aston-Jones G. Noradrenaline in the ventral forebrain is critical for opiate withdrawal-induced aversion. Nature 2000;403(6768):430–4. https://doi.org/10.1038/35000212.

[41] Ward J, Hall W, Mattick RP. Role of maintenance treatment in opioid dependence. Lancet 1999;353(9148):221–6. https://doi.org/10.1016/s0140-6736(98)05356-2.

[42] Broers B, Giner F, Dumont P, Mino A. Inpatient opiate detoxification in Geneva: follow-up at 1 and 6 months. Drug Alcohol Depend 2000;58 (1–2):85–92.

[43] Gossop M, Green L, Phillips G, Bradley B. Lapse, relapse and survival among opiate addicts after treatment. A prospective follow-up study. Br J Psychiatry 1989;154:348–53.

[44] Xu GP, Van Bockstaele E, Reyes B, Bethea T, Valentino RJ. Chronic morphine sensitizes the brain norepinephrine system to corticotropin-releasing factor and stress. J Neurosci 2004;24(38):8193–7. https://doi.org/10.1523/jneurosci.1657-04.2004.

[45] Hassan S, Duong B, Kim KS, Miles MF. Pharmacogenomic analysis of mechanisms mediating ethanol regulation of dopamine beta-hydroxylase. J Biol Chem 2003;278(40):38860–9. https://doi.org/10.1074/jbc.M305040200.

[46] Rasmussen DD, Wilkinson CW, Raskind MA. Chronic daily ethanol and withdrawal: 6. Effects on rat sympathoadrenal activity during "abstinence" Alcohol 2006;38(3):173–7. https://doi.org/10.1016/j.alcohol.2006.06.007.

[47] Kohnke MD, Zabetian CP, Anderson GM, Kolb W, Gaertner I, Buchkremer G, Vonthein R, Schick S, Lutz U, Kohnke AM, Cubells JF. A genotype-controlled analysis of plasma dopamine beta-hydroxylase in healthy and alcoholic subjects: Evidence for alcohol-related differences in noradrenergic function. Biol Psychiatry 2002;52(12):1151–8.

[48] Patkar AA, Marsden CA, Naik PC, Kendall DA, Gopalakrishnan R, Vergare MJ, Weinstein SP. Differences in peripheral noradrenergic function among actively drinking and abstinent alcohol-dependent individuals. Am J Addict 2004;13(3):225–35. https://doi.org/10.1080/10550490490459898.

[49] Deitrich RA, Erwin VG. Mechanism of the inhibition of aldehyde dehydrogenase in vivo by disulfiram and diethyldithiocarbamate. Mol Pharmacol 1971;7(3):301–7.

[50] Hald J, Jacobsen E. A drug sensitizing the organism to ethyl alcohol. Lancet 1948;2(6539):1001–4.

[51] Carroll KM, Fenton LR, Ball SA, Nich C, Frankforter TL, Shi J, Rounsaville BJ. Efficacy of disulfiram and cognitive behavior therapy in cocaine-dependent outpatients: a randomized placebo-controlled trial. Arch Gen Psychiatry 2004;61(3):264–72. https://doi.org/10.1001/archpsyc.61.3.264.

[52] George TP, Chawarski MC, Pakes J, Carroll KM, Kosten TR, Schottenfeld RS. Disulfiram versus placebo for cocaine dependence in buprenorphine-maintained subjects: a preliminary trial. Biol Psychiatry 2000;47(12):1080–6.

[53] Bourdelat-Parks BN, Anderson GM, Donaldson ZR, Weiss JM, Bonsall RW, Emery MS, Liles LC, Weinshenker D. Effects of dopamine beta-hydroxylase genotype and disulfiram inhibition on catecholamine homeostasis in mice. Psychopharmacology (Berl) 2005;183(1):72–80. https://doi.org/10.1007/s00213-005-0139-8.

[54] Gaval-Cruz M, Liles LC, Iuvone PM, Weinshenker D. Chronic inhibition of dopamine beta-hydroxylase facilitates behavioral responses to cocaine in mice. PLoS One 2012;7(11). https://doi.org/10.1371/journal.pone.0050583.

[55] Brodde OE, Nagel M, Schumann HJ. Mechanism of inhibition of dopamine beta-hydroxylase evoked by FLA-63. An in vitro study. Arch Int Pharmacodyn Ther 1977;228(2):184–90.

[56] Goldstein M. Inhibition of norepinephrine biosynthesis at the dopamine-beta-hydroxylation stage. Pharmacol Rev 1966;18(1):77–82.

[57] Torronen R, Marselos M. Changes in the hepatic copper conent after treatment with foreign compounds. Arch Toxicol Suppl 1978;(1):247–9.

[58] Cooper DA, Kimmel HL, Manvich DF, Schmidt KT, Weinshenker D, Howell LL. Effects of pharmacologic dopamine beta-hydroxylase inhibition on cocaine-induced reinstatement and dopamine neurochemistry in squirrel monkeys. J Pharmacol Exp Ther 2014;350(1):144–52. https://doi.org/10.1124/jpet.113.212357.

[59] Carroll KM, Nich C, Ball SA, McCance E, Frankforter TL, Rounsaville BJ. One-year follow-up of disulfiram and psychotherapy for cocaine-alcohol users: Sustained effects of treatment. Addiction 2000;95(9):1335–49.

[60] Petrakis IL, Carroll KM, Nich C, Gordon LT, McCance-Katz EF, Frankforter T, Rounsaville BJ. Disulfiram treatment for cocaine dependence in methadone-maintained opioid addicts. Addiction 2000;95(2):219–28.

[61] Baker JR, Jatlow P, McCance-Katz EF. Disulfiram effects on responses to intravenous cocaine administration. Drug Alcohol Depend 2007;87 (2–3):202–9. https://doi.org/10.1016/j.drugalcdep.2006.08.016.

[62] Shorter D, Nielsen DA, Huang W, Harding MJ, Hamon SC, Kosten TR. Pharmacogenetic randomized trial for cocaine abuse: disulfiram and alpha1A-adrenoceptor gene variation. Eur Neuropsychopharmacol 2013;23(11):1401–7. https://doi.org/10.1016/j.euroneuro.2013.05.014.

[63] Gaval-Cruz M, Weinshenker D. Mechanisms of disulfiram-induced cocaine abstinence: antabuse and cocaine relapse. Mol Interv 2009;9 (4):175–87. https://doi.org/10.1124/mi.9.4.6.

[64] DeVito EE, Babuscio TA, Nich C, Ball SA, Carroll KM. Gender differences in clinical outcomes for cocaine dependence: randomized clinical trials of behavioral therapy and disulfiram. Drug Alcohol Depend 2014;145:156–67. https://doi.org/10.1016/j.drugalcdep.2014.10.007.

[65] Nich C, McCance-Katz EF, Petrakis IL, Cubells JF, Rounsaville BJ, Carroll KM. Sex differences in cocaine-dependent individuals' response to disulfiram treatment. Addict Behav 2004;29(6):1123–8. https://doi.org/10.1016/j.addbeh.2004.03.004.

[66] Kosten TR, Wu G, Huang W, Harding MJ, Hamon SC, Lappalainen J, Nielsen DA. Pharmacogenetic randomized trial for cocaine abuse: disulfiram and dopamine beta-hydroxylase. Biol Psychiatry 2013;73(3):219–24. https://doi.org/10.1016/j.biopsych.2012.07.011.

[67] Kapoor A, Shandilya M, Kundu S. Structural insight of dopamine beta-hydroxylase, a drug target for complex traits, and functional significance of exonic single nucleotide polymorphisms. PLoS One 2011;6(10):e26509. https://doi.org/10.1371/journal.pone.0026509.

[68] Schroeder JP, Epps SA, Grice TW, Weinshenker D. The selective dopamine beta-hydroxylase inhibitor nepicastat attenuates multiple aspects of cocaine-seeking behavior. Neuropsychopharmacology 2013;38(6):1032–8. https://doi.org/10.1038/npp.2012.267.

[69] Devoto P, Flore G, Saba P, Frau R, Gessa GL. Selective inhibition of dopamine-beta-hydroxylase enhances dopamine release from noradrenergic terminals in the medial prefrontal cortex. Brain Behav 2015;5(10). https://doi.org/10.1002/brb3.393.

[70] De La Garza 2nd R, Bubar MJ, Carbone CL, Moeller FG, Newton TF, Anastasio NC, Harper TA, Ware DL, Fuller MA, Holstein GJ, Jayroe JB, Bandak SI, Reiman KZ, Neale AC, Pickford LB, Cunningham KA. Evaluation of the dopamine beta-hydroxylase (DbetaH) inhibitor nepicastat in participants who meet criteria for cocaine use disorder. Prog Neuropsychopharmacol Biol Psychiatry 2015;59:40–8. https://doi.org/10.1016/j.pnpbp.2015.01.009.

[71] Starke K, Montel H, Gayk W, Merker R. Comparison of the effects of clonidine on pre- and postsynaptic adrenoceptors in the rabbit pulmonary artery. Alpha-sympathomimetic inhibition of Neurogenic vasoconstriction. Naunyn Schmiedebergs Arch Pharmacol 1974;285(2):133–50.

[72] Platt DM, Rowlett JK, Spealman RD. Noradrenergic mechanisms in cocaine-induced reinstatement of drug seeking in squirrel monkeys. J Pharmacol Exp Ther 2007;322(2):894–902. https://doi.org/10.1124/jpet.107.121806.

[73] Smith RJ, Aston-Jones G. Alpha(2) adrenergic and imidazoline receptor agonists prevent cue-induced cocaine seeking. Biol Psychiatry 2011;70 (8):712–9. https://doi.org/10.1016/j.biopsych.2011.06.010.

[74] Jobes ML, Ghitza UE, Epstein DH, Phillips KA, Heishman SJ, Preston KL. Clonidine blocks stress-induced craving in cocaine users. Psychopharmacology (Berl) 2011;218(1):83–8. https://doi.org/10.1007/s00213-011-2230-7.

[75] Terry Jr AV, Callahan PM, Schade R, Kille NJ, Plagenhoef M. Alpha 2A adrenergic receptor agonist, guanfacine, attenuates cocaine-related impairments of inhibitory response control and working memory in animal models. Pharmacol Biochem Behav 2014;126:63–72. https://doi.org/10.1016/j.pbb.2014.09.010.

[76] Buffalari DM, Baldwin CK, See RE. Treatment of cocaine withdrawal anxiety with guanfacine: relationships to cocaine intake and reinstatement of cocaine seeking in rats. Psychopharmacology (Berl) 2012;223(2):179–90. https://doi.org/10.1007/s00213-012-2705-1.

[77] Fox H, Sinha R. The role of guanfacine as a therapeutic agent to address stress-related pathophysiology in cocaine-dependent individuals. Adv Pharmacol 2014;69:217–65. https://doi.org/10.1016/b978-0-12-420118-7.00006-8.

[78] Drouin C, Darracq L, Trovero F, Blanc G, Glowinski J, Cotecchia S, Tassin JP. Alpha1b-adrenergic receptors control locomotor and rewarding effects of psychostimulants and opiates, J Neurosci 2002;22(7):2873–84. https://doi.org/20026237.

[79] Drouin C, Blanc G, Villegier AS, Glowinski J, Tassin JP. Critical role of alpha1-adrenergic receptors in acute and sensitized locomotor effects of D-amphetamine, cocaine, and GBR 12783: influence of preexposure conditions and pharmacological characteristics. Synapse 2002;43(1):51–61. https://doi.org/10.1002/syn.10023.

[80] Shorter D, Lindsay JA, Kosten TR. The alpha-1 adrenergic antagonist doxazosin for treatment of cocaine dependence: a pilot study. Drug Alcohol Depend 2013;131(1–2):66–70. https://doi.org/10.1016/j.drugalcdep.2012.11.021.

[81] Newton, T. F., De La Garza, R., 2nd, Brown, G., Kosten, T. R., Mahoney, J. J., 3rd, & Haile, C. N. (2012). Noradrenergic alpha(1) receptor antagonist treatment attenuates positive subjective effects of cocaine in humans: a randomized trial. PLoS One, 7(2), e30854. doi:https://doi.org/10.1371/journal.pone.0030854

[82] Hysek CM, Fink AE, Simmler LD, Donzelli M, Grouzmann E, Liechti ME. Alpha(1)-adrenergic receptors contribute to the acute effects of 3,4-methylenedioxymethamphetamine in humans. J Clin Psychopharmacol 2013;33(5):658–66. https://doi.org/10.1097/JCP.0b013e3182979d32.

[83] Zhang XY, Kosten TA. Prazosin, an alpha-1 adrenergic antagonist, reduces cocaine-induced reinstatement of drug-seeking. Biol Psychiatry 2005;57 (10):1202–4. https://doi.org/10.1016/j.biopsych.2005.02.003.

[84] Rudoy CA, Van Bockstaele EJ. Betaxolol, a selective beta(1)-adrenergic receptor antagonist, diminishes anxiety-like behavior during early withdrawal from chronic cocaine administration in rats. Prog Neuropsychopharmacol Biol Psychiatry 2007;31(5):1119–29. https://doi.org/10.1016/j.pnpbp.2007.04.005.

[85] Schank JR, Liles LC, Weinshenker D. Norepinephrine signaling through beta-adrenergic receptors is critical for expression of cocaine-induced anxiety. Biol Psychiatry 2008;63(11):1007–12. https://doi.org/10.1016/j.biopsych.2007.10.018.

[86] Kampman KM, Volpicelli JR, Mulvaney F, Alterman AI, Cornish J, Gariti P, Cnaan A, Poole S, Muller E, Acosta T, Luce D, O'Brien C. Effectiveness of propranolol for cocaine dependence treatment may depend on cocaine withdrawal symptom severity. Drug Alcohol Depend 2001;63 (1):69–78.

[87] Kampman KM, Dackis C, Lynch KG, Pettinati H, Tirado C, Gariti P, Sparkman T, Atzram M, O'Brien CP. A double-blind, placebo-controlled trial of amantadine, propranolol, and their combination for the treatment of cocaine dependence in patients with severe cocaine withdrawal symptoms. Drug Alcohol Depend 2006;85(2):129–37. https://doi.org/10.1016/j.drugalcdep.2006.04.002.

[88] Vranjkovic O, Hang S, Baker DA, Mantsch JR. β-Adrenergic receptor mediation of stress-induced reinstatement of extinguished cocaine-induced conditioned place preference in mice: Roles for β1 and β2 adrenergic receptors. J Pharmacol Exp Ther 2012;342(2):541–51. https://doi.org/10.1124/jpet.112.193615.

[89] Saladin ME, Gray KM, McRae-Clark AL, Larowe SD, Yeatts SD, Baker NL, Hartwell KJ, Brady KT. A double blind, placebo-controlled study of the effects of post-retrieval propranolol on reconsolidation of memory for craving and cue reactivity in cocaine dependent humans. Psychopharmacology (Berl) 2013;226(4):721–37. https://doi.org/10.1007/s00213-013-3039-3.

[90] Jobes ML, Aharonovich E, Epstein DH, Phillips KA, Reamer D, Anderson M, Preston KL. Effects of Prereactivation propranolol on cocaine craving elicited by imagery script/cue sets in opioid-dependent Polydrug users: a randomized study. J Addict Med 2015;9(6):491–8. https://doi.org/10.1097/adm.0000000000000169.

[91] Yu E, Miotto K, Akerele E, Montgomery A, Elkashef A, Walsh R, Montoya I, Fischman MW, Collins J, McSherry F, Boardman K, Davies DK, O'Brien CP, Ling W, Kleber H, Herman BH. A phase 3 placebo-controlled, double-blind, multi-site trial of the alpha-2-adrenergic agonist, lofexidine, for opioid withdrawal. Drug Alcohol Depend 2008;97(1–2):158–68. https://doi.org/10.1016/j.drugalcdep.2008.04.002.

[92] Gorodetzky CW, Walsh SL, Martin PR, Saxon AJ, Gullo KL, Biswas K. A phase III, randomized, multi-center, double blind, placebo controlled study of safety and efficacy of lofexidine for relief of symptoms in individuals undergoing inpatient opioid withdrawal. Drug Alcohol Depend 2017;176:79–88. https://doi.org/10.1016/j.drugalcdep.2017.02.020.

[93] Mannelli P, Peindl K, Wu LT, Patkar AA, Gorelick DA. The combination very low-dose naltrexone-clonidine in the management of opioid withdrawal. Am J Drug Alcohol Abuse 2012;38(3):200–5. https://doi.org/10.3109/00952990.2011.644003.

[94] Sinha R, Kimmerling A, Doebrick C, Kosten TR. Effects of lofexidine on stress-induced and cue-induced opioid craving and opioid abstinence rates: preliminary findings. Psychopharmacology (Berl) 2007;190(4):569–74. https://doi.org/10.1007/s00213-006-0640-8.

[95] Kowalczyk WJ, Phillips KA, Jobes ML, Kennedy AP, Ghitza UE, Agage DA, Schmittner JP, Epstein DH, Preston KL. Clonidine maintenance prolongs opioid abstinence and decouples stress from craving in daily life: a randomized controlled trial with ecological momentary assessment. Am J Psychiatry 2015;172(8):760–7. https://doi.org/10.1176/appi.ajp.2014.14081014.

[96] Krupitsky E, Zvartau E, Blokhina E, Verbitskaya E, Tsoy M, Wahlgren V, Burakov A, Masalov D, Romanova TN, Palatkin V, Tyurina A, Yaroslavtseva T, Sinha R, Kosten TR. Naltrexone with or without guanfacine for preventing relapse to opiate addiction in St.-Petersburg, Russia. Drug Alcohol Depend 2013;132(3):674–80. https://doi.org/10.1016/j.drugalcdep.2013.04.021.

[97] Liang DY, Shi X, Li X, Li J, Clark JD. The beta2 adrenergic receptor regulates morphine tolerance and physical dependence. Behav Brain Res 2007;181(1):118–26. https://doi.org/10.1016/j.bbr.2007.03.037.

[98] Cecchi M, Capriles N, Watson SJ, Akil H. Beta1 adrenergic receptors in the bed nucleus of stria terminalis mediate differential responses to opiate withdrawal. Neuropsychopharmacology 2007;32(3):589–99. https://doi.org/10.1038/sj.npp.1301140.

[99] Resnick RB, Kestenbaum RS, Schwartz LK, Smith A. Evaluation of propranolol in opiate dependence. Arch Gen Psychiatry 1976;33(8):993–7.

[100] Zindel LR, Kranzler HR. Pharmacotherapy of alcohol use disorders: seventy-five years of progress. J Stud Alcohol Drugs Suppl 2014;75(Suppl 17):79–88.

[101] Winslow BT, Onysko M, Hebert M. Medications for alcohol use disorder. Am Fam Physician 2016;93(6):457–65.

[102] Colombo G, Maccioni P, Vargiolu D, Loi B, Lobina C, Zaru A, Carai MA, Gessa GL. The dopamine beta-hydroxylase inhibitor, nepicastat, reduces different alcohol-related behaviors in rats. Alcohol Clin Exp Res 2014;38(9):2345–53. https://doi.org/10.1111/acer.12520.

[103] Weinshenker D, Rust NC, Miller NS, Palmiter RD. Ethanol-associated behaviors of mice lacking norepinephrine. J Neurosci 2000;20(9):3157–64.

[104] Tottmar O, Hellstrom E, Holmberg K, Lindros KO. Effects of dopamine-beta-hydroxylase inhibitors FLA-57 and FLA-63 on ethanol metabolism and aldehyde dehydrogenase activity in rats. Acta Pharmacol Toxicol (Copenh) 1982;51(3):198–202.

[105] Le AD, Harding S, Juzytsch W, Funk D, Shaham Y. Role of alpha-2 adrenoceptors in stress-induced reinstatement of alcohol seeking and alcohol self-administration in rats. Psychopharmacology (Berl) 2005;179(2):366–73. https://doi.org/10.1007/s00213-004-2036-y.

[106] Rasmussen DD, Alexander L, Malone J, Federoff D, Froehlich JC. The alpha2-adrenergic receptor agonist, clonidine, reduces alcohol drinking in alcohol-preferring (P) rats. Alcohol 2014;48(6):543–9. https://doi.org/10.1016/j.alcohol.2014.07.002.

[107] Cushman Jr P, Forbes R, Lerner W, Stewart M. Alcohol withdrawal syndromes: clinical management with lofexidine. Alcohol Clin Exp Res 1985; 9(2):103–8.

[108] Cushman Jr P, Sowers JR. Alcohol withdrawal syndrome: clinical and hormonal responses to alpha 2-adrenergic agonist treatment. Alcohol Clin Exp Res 1989;13(3):361–4.

[109] Rasmussen DD, Alexander LL, Raskind MA, Froehlich JC. The alpha1-adrenergic receptor antagonist, prazosin, reduces alcohol drinking in alcohol-preferring (P) rats. Alcohol Clin Exp Res 2009;33(2):264–72. https://doi.org/10.1111/j.1530-0277.2008.00829.x.

[110] Walker BM, Rasmussen DD, Raskind MA, Koob GF. Alpha1-noradrenergic receptor antagonism blocks dependence-induced increases in responding for ethanol. Alcohol 2008;42(2):91–7. https://doi.org/10.1016/j.alcohol.2007.12.002.

[111] Le AD, Funk D, Juzytsch W, Coen K, Navarre BM, Cifani C, Shaham Y. Effect of prazosin and guanfacine on stress-induced reinstatement of alcohol and food seeking in rats. Psychopharmacology (Berl) 2011;218(1):89–99. https://doi.org/10.1007/s00213-011-2178-7.

[112] Simpson TL, Saxon AJ, Meredith CW, Malte CA, McBride B, Ferguson LC, Gross CA, Hart KL, Raskind M. A pilot trial of the alpha-1 adrenergic antagonist, prazosin, for alcohol dependence. Alcohol Clin Exp Res 2009;33(2):255–63. https://doi.org/10.1111/j.1530-0277.2008.00807.x.

[113] Petrakis IL, Desai N, Gueorguieva R, Arias A, O'Brien E, Jane JS, Sevarino K, Southwick S, Ralevski E. Prazosin for veterans with post-traumatic stress disorder and comorbid alcohol dependence: A clinical trial. Alcohol Clin Exp Res 2016;40(1):178–86. https://doi.org/10.1111/acer.12926.

[114] Simpson TL, Malte CA, Dietel B, Tell D, Pocock I, Lyons R, Varon D, Raskind M, Saxon AJ. A pilot trial of prazosin, an alpha-1 adrenergic antagonist, for comorbid alcohol dependence and posttraumatic stress disorder. Alcohol Clin Exp Res 2015;39(5):808–17. https://doi.org/10.1111/acer.12703.

[115] Fox HC, Anderson GM, Tuit K, Hansen J, Kimmerling A, Siedlarz KM, Morgan PT, Sinha R. Prazosin effects on stress- and cue-induced craving and stress response in alcohol-dependent individuals: Preliminary findings. Alcohol Clin Exp Res 2012;36(2):351–60. https://doi.org/10.1111/j.1530-0277.2011.01628.x.

[116] Rasmussen DD, Kincaid CL, Froehlich JC. Prazosin prevents increased anxiety behavior that occurs in response to stress during alcohol deprivations. Alcohol Alcohol 2017;52(1):5–11. https://doi.org/10.1093/alcalc/agw082.

[117] Deisseroth K. Optogenetics: 10 years of microbial opsins in neuroscience. Nat Neurosci 2015;18(9):1213–25. https://doi.org/10.1038/nn.4091.

[118] Grosenick L, Marshel JH, Deisseroth K. Closed-loop and activity-guided optogenetic control. Neuron 2015;86(1):106–39. https://doi.org/10.1016/j.neuron.2015.03.034.

[119] Weinshenker D, Holmes PV. Regulation of neurological and neuropsychiatric phenotypes by locus coeruleus-derived galanin. Brain Res 2016;1641 (Pt B):320–37. https://doi.org/10.1016/j.brainres.2015.11.025.

[120] Armbruster BN, Li X, Pausch MH, Herlitze S, Roth BL. Evolving the lock to fit the key to create a family of G protein-coupled receptors potently activated by an inert ligand. Proc Natl Acad Sci U S A 2007;104(12):5163–8. https://doi.org/10.1073/pnas.0700293104.

[121] Gomez JL, Bonaventura J, Lesniak W, Mathews WB, Sysa-Shah P, Rodriguez LA, Ellis RJ, Richie CT, Harvey BK, Dannals RF, Pomper MG, Bonci A, Michaelides M. Chemogenetics revealed: DREADD occupancy and activation via converted clozapine. Science 2017;357(6350):503–7. https://doi.org/10.1126/science.aan2475.

[122] Meltzer HY. An overview of the mechanism of action of clozapine. J Clin Psychiatry 1994;55(Suppl B):47–52.

[123] Bender D, Holschbach M, Stocklin G. Synthesis of n.c.a. carbon-11 labelled clozapine and its major metabolite clozapine-N-oxide and comparison of their biodistribution in mice. Nucl Med Biol 1994;21(7):921–5.

[124] Roth BL. DREADDs for neuroscientists. Neuron 2016;89(4):683–94. https://doi.org/10.1016/j.neuron.2016.01.040.

[125] Mantsch JR, Weyer A, Vranjkovic O, Beyer CE, Baker DA, Caretta H. Involvement of noradrenergic neurotransmission in the stress- but not cocaine-induced reinstatement of extinguished cocaine-induced conditioned place preference in mice: role for beta-2 adrenergic receptors. Neuropsychopharmacology 2010;35(11):2165–78. https://doi.org/10.1038/npp.2010.86.

[126] Gaval-Cruz M, Goertz RB, Puttick DJ, Bowles DE, Meyer RC, Hall RA, Ko D, Paladini CA, Weinshenker D. Chronic loss of noradrenergic tone produces beta-arrestin2-mediated cocaine hypersensitivity and alters cellular D2 responses in the nucleus accumbens. Addict Biol 2016;21 (1):35–48. https://doi.org/10.1111/adb.12174.

[127] Haile CN, During MJ, Jatlow PI, Kosten TR, Kosten TA. Disulfiram facilitates the development and expression of locomotor sensitization to cocaine in rats. Biol Psychiatry 2003;54(9):915–21.

[128] Weinshenker D, Miller NS, Blizinsky K, Laughlin ML, Palmiter RD. Mice with chronic norepinephrine deficiency resemble amphetamine-sensitized animals. Proc Natl Acad Sci U S A 2002;99(21):13873–7. https://doi.org/10.1073/pnas.212519999.

[129] Logothetis NK. What we can do and what we cannot do with fMRI. Nature 2008;453(7197):869–78. https://doi.org/10.1038/nature06976.

[130] Broussard GJ, Liang R, Tian L. Monitoring activity in neural circuits with genetically encoded indicators. Front Mol Neurosci 2014;7. https://doi.org/10.3389/fnmol.2014.00097.

[131] Knopfel T. Genetically encoded optical indicators for the analysis of neuronal circuits. Nat Rev Neurosci 2012;13(10):687–700. https://doi.org/10.1038/nrn3293.

[132] Zalocusky KA, Ramakrishnan C, Lerner TN, Davidson TJ, Knutson B, Deisseroth K. Nucleus accumbens D2R cells signal prior outcomes and control risky decision-making. Nature 2016;531(7596):642–6. https://doi.org/10.1038/nature17400.

[133] Tian L, Hires SA, Mao T, Huber D, Chiappe ME, Chalasani SH, Petreanu L, Akerboom J, McKinney SA, Schreiter ER, Bargmann CI, Jayaraman V, Svoboda K, Looger LL. Imaging neural activity in worms, flies and mice with improved GCaMP calcium indicators. Nat Methods 2009;6 (12):875–81. https://doi.org/10.1038/nmeth.1398.

[134] Barnea G, Strapps W, Herrada G, Berman Y, Ong J, Kloss B, Axel R, Lee KJ. The genetic design of signaling cascades to record receptor activation. Proc Natl Acad Sci U S A 2008;105(1):64–9. https://doi.org/10.1073/pnas.0710487105.

[135] Lee D, Creed M, Jung K, Stefanelli T, Wendler DJ, Oh WC, Mignocchi NL, Luscher C, Kwon HB. Temporally precise labeling and control of neuromodulatory circuits in the mammalian brain. Nat Methods 2017;14(5):495–503. https://doi.org/10.1038/nmeth.4234.

[136] van den Brink RL, Pfeffer T, Warren CM, Murphy PR, Tona KD, van der Wee NJ, Giltay E, van Noorden MS, Rombouts SA, Donner TH, Nieuwenhuis S. Catecholaminergic neuromodulation shapes intrinsic MRI functional connectivity in the human brain. J Neurosci 2016;36 (30):7865–76. https://doi.org/10.1523/jneurosci.0744-16.2016.

[137] Shields BC, Kahuno E, Kim C, Apostolides PF, Brown J, Lindo S, Mensh BD, Dudman JT, Lavis LD, Tadross MR. Deconstructing behavioral neuropharmacology with cellular specificity. Science 2017;356(6333). https://doi.org/10.1126/science.aaj2161.

[138] Leri F, Flores J, Rodaros D, Stewart J. Blockade of stress-induced but not cocaine-induced reinstatement by infusion of noradrenergic antagonists into the bed nucleus of the stria terminalis or the central nucleus of the amygdala. J Neurosci 2002;22(13):5713–8. https://doi.org/20026536.

[139] Mitrano DA, Schroeder JP, Smith Y, Cortright JJ, Bubula N, Vezina P, Weinshenker D. Alpha-1 adrenergic receptors are localized on presynaptic elements in the nucleus accumbens and regulate mesolimbic dopamine transmission. Neuropsychopharmacology 2012;37(9):2161–72. https://doi.org/10.1038/npp.2012.68.

[140] Crocini C, Ferrantini C, Pavone FS, Sacconi L. Optogenetics gets to the heart: a guiding light beyond defibrillation. Prog Biophys Mol Biol 2017. https://doi.org/10.1016/j.pbiomolbio.2017.05.002.

Chapter 16

Glial Dysregulation in Addiction

Evan Hess, Aric Madayag, Matthew Hearing, and David A. Baker
Department of Biomedical Sciences, Marquette University, Milwaukee, WI, United States

In high-income countries, addiction is a leading cause of disease burden [1], a fact, which helps to establish understanding brain function as one of the greatest priorities in modern science. Although preclinical and clinical research has long implicated dysregulation of key cortico-striatal circuits in addiction [2–6], our incomplete understanding of the human brain continues to be a fundamental barrier in the development of disease-modifying therapeutics. Revealing the molecular and cellular basis of intercellular communication is central to this endeavor, especially since neurotransmitter receptors are one of the most accessible targets for central nervous system (CNS) therapeutics. However, this is challenging because of the remarkable complexity of mammalian brains.

The evolution of nervous systems provides useful insights into the organizational and functional principles of the human brain [7]. In species ranging from nematodes to humans, behavioral capabilities are limited by the processing or computational capacity of intercellular communication. Phylogenetic studies reveal that several classical vertebrate neurotransmitters including glutamate, gamma-aminobutyric acid (GABA), and monoamines are present in at least some of the basal metazoan clades [8–12], which help to establish synaptic transmission as a key form of intercellular signaling that has been present throughout the evolution of nervous systems. However, in response to increasingly complex internal states and external environments, evolution enriched this ancient form of signaling by accumulating new genes that increased the computational capacity of intercellular signaling to the enabling of sophisticated behaviors [13–15]. This included the appearance of receptors, transporters, and release mechanisms for these ancient neurotransmitter systems, which enable diverse forms of intercellular communication [16].

A key evolutionary event in the expansion of intercellular signaling was the recruitment of glia to the processing, integration, and storage of information in complex nervous systems [16–19]. While multiple types of glia are being revealed as critical contributors to these processes [20], this chapter will focus specifically on astrocytes since these cells are capable of (a) monitoring the activity of thousands to millions of synapses, (b) integrating this information throughout the spatial domain of an individual astrocyte or a network of astrocytes that are physically coupled by gap junctions, and (c) communicating to neurons through the release of neuroactive substances [18,21,22].

CHAPTER OVERVIEW

While our limited understanding of the molecular and cellular regulation of neural circuits implicated in addiction poses a fundamental barrier in establishing the etiology of these diseases, the role of astrocytes in addiction is a new concept that has not been extensively studied. However, interest in these cells is increasing because astrocytes have been found to contribute to virtually every aspect of brain function, including neural development, synapse formation, synapse pruning, synaptic plasticity, glutamate signaling, GABA signaling, and gating of neural circuits (for excellent reviews, see [18,23–27]). Hence, the primary approach taken here is to establish the enormous potential for these cells to be a key in understanding addiction. To do this, we focus on providing a broad overview of the diverse ways in which astrocyte-neuron signaling are known to impact mammalian brain function, and highlight, where possible, when drugs of abuse appear to alter astrocyte-neuron signaling. In doing this, we seek to review current findings while also establishing a framework for future efforts to better understand this important area of addiction biology.

Neural Mechanisms of Addiction. https://doi.org/10.1016/B978-0-12-812202-0.00016-6
Copyright © 2019 Elsevier Inc. All rights reserved.

THE STRUCTURAL BASIS FOR NEURON-ASTROCYTE SIGNALING

The expanded computational processing capacity of the vertebrate brain included the appearance of a network of glutamate receptors, transporters, and release mechanisms that are expressed by neurons and astrocytes, which collectively enable diverse forms of intercellular communication [16]. This includes signaling between neurons and astrocytes that occurs outside of the synaptic cleft [28–30]. The potential for extrasynaptic signaling to increase the computational capacity of the mammalian brain is evident from estimates that the synaptic cleft represents only 2% of the total interstitial space available to support signaling [31]. The existence of synaptic and extrasynaptic signaling indicates the need to create functionally distinct, specialized signaling zones in the interstitial space of the brain. Hence, we will begin by discussing how the structure and composition of the interstitial compartment is dependent on astrocytes and neurons, and impacts intercellular signaling.

Many mammalian synapses are composed of a presynaptic neuron, postsynaptic neuron, and perisynaptic astrocytic process (PAP), which collectively are termed as the tripartite synapse [32]. However, the fraction of synapses containing a PAP varies across brain regions; 90% of synapses in the somatosensory cortex contain a PAP, whereas this is reduced to 57% of synapses in the hippocampus [25,33]. Structural heterogeneity is further evident by the observation that the extent of ensheathment of the synapse by a PAP can vary from full to partial, and that partially ensheathed synapses can display asymmetry in which the PAP is positioned near the postsynaptic neuron with minimal coverage of the presynaptic neuron [33,34]. The structural diversity between neurons and astrocytes poses the possibility of functional heterogeneity between these cells. For example, the capacity for astrocytes to regulate presynaptic and postsynaptic cell function may be maximal at synapses fully ensheathed by PAPs. Further, structural plasticity may dictate the molecular basis of neuron-astrocyte signaling in the event that synapses lacking PAPs rely primarily on molecules capable of volume transmission, such as peptides released by neurons and/or astrocytes, whereas fully ensheathed synapses may use a range of neurotransmitters since astrocytes express many of the same receptors as neurons [21].

Extrasynaptic signaling is also dependent on the structural relationships between neurons and astrocytes. As the two most abundant cells in the human brain, the cellular membranes of astrocytes and neurons are critical components of the physical barriers of the interstitial space, which consists of a series of asymmetrical channels 20–60 nm in width that form the external environment for every brain cell [35,36]. Asymmetric channels formed by the membranes of astrocytes and neurons are presumed to be interconnected throughout the brain thereby creating a global extracellular network supporting volume transmission. However, recent work indicates that the network of asymmetrical channels may be highly organized in a manner that restricts volume diffusion within and between brain regions. In support, the flow of a tracer injected into the thalamus can be restricted to the anatomical division of the thalamus, whereas administration of a tracer into the nearby caudate displays diffusion directed toward the ipsilateral frontotemporal cortex [37]. If this represents a ubiquitous feature of the interstitial space, then the structural relationship of neurons and astrocytes may form a previously unrecognized form of specialized communication within and across brain regions.

Key Features of the Interstitial Space and Intercellular Signaling

Features of the interstitial space that influence intercellular signaling are volume fraction and tortuosity. Volume fraction refers to the proportion of brain tissue occupied by the interstitial compartment. Tortuosity indicates the degree to which elements of the interstitial space hinder the free diffusion of extracellular molecules. This is typically reflected as a function of the relative diffusion characteristics of a molecule in pure water relative to the interstitial compartment. While the volume fraction and tortuosity of the brain are broadly estimated at 0.2 (or 20% of the total volume of the brain) and 1.6, respectively, each parameter is subject to regulation by neurons and astrocytes, and hence can vary greatly across brain regions. For example, the volume fraction in the rat cerebral cortex ranges between 19% and 23% while tortuosity in rat cortex is estimated to range between 1.51 and 1.65. In the CA1 region of the hippocampus, the volume fraction and tortuosity are estimated at 0.24 and 1.45, respectively [38,39]. The degree to which these parameters are dynamically regulated is evident from the finding that the volume fraction of the cortex increases by more than 60% during sleep [40].

The efficiency of synaptic transmission is impacted by the volume fraction and tortuosity of the synaptic cleft [41]. In a synaptic event that releases 5000 molecules of glutamate, the concentration of glutamate increases from the nM to the mM range in a typical synaptic cleft with a height of 20 nm and a diameter of 150 nm. However, this peak concentration is only reached in the part of the cleft closest to the release sites [41]. Likely as a result, the distribution of postsynaptic glutamate receptors is organized with low-affinity α-amino-3-hydroxy-5-methyl-4-isoazole-proprionic acid (AMPA) receptors being expressed on the postsynaptic membrane that is nearest to the release sites whereas high-affinity N-methyl-D-aspartic acid

(NMDA) and metabotropic glutamate receptors are often more lateral. As a result, a change to the volume fraction or tortuosity of the cleft would alter synaptic transmission, especially as encoded by AMPA receptors.

Signaling between neurons and astrocytes can impact the volume fraction and tortuosity of the synaptic cleft. For example, the volume fraction and tortuosity are influenced by PAPs coverage of the synapse, which can be regulated by neurons [33,42]. Notably, neuron-stimulated PAP motility contributes to synapse stability and synaptic plasticity [33], which is significant since addiction is linked to changes in synaptic density [43,44].

The morphology of astrocytes and neurons will also impact the volume fraction and tortuosity of the nonsynaptic interstitial space, and in doing so, will likely also impact the capacity to support specialized signaling zones that may be essential for extrasynaptic signaling [45]. In support, extrasynaptic NMDA receptors expressed by neurons are often concentrated at points of contact with adjacent axons or astrocytes [46]. These receptors may exist in specialized signaling microdomains since inhibition of glutamate transporters increases the activation of these receptors, which reveals the existence of active mechanisms that limit access to these receptors by extrasynaptic glutamate [28,47]. Given that the extrasynaptic compartment can be as narrow as the synaptic cleft, changes to the morphology of astrocytes or neurons may alter extrasynaptic signaling.

Extracellular Matrix as a Key Component of the Interstitial Space

The extracellular matrix is another key determinant of the volume fraction and tortuosity of the entire interstitial space [45,48]. The extracellular matrix is a formed by a lattice-like network of collagen, glycosaminoglycans, proteoglycans, and glycoproteins, which collectively form the three major components of the extracellular matrix: the basement membrane, perineuronal nets (PNNs), and neuronal interstitial matrix [49]. The spacing and physical connectivity of neurons and astrocytes can be influenced by the extracellular matrix since each of these cells express extracellular matrix receptors and proteins (e.g., integrins and ephrins) that interface with the extracellular matrix or cellular adhesion proteins present on adjacent cells [50]. These proteins can also serve as anchors for synaptogenesis which organize presynaptic and postsynaptic machinery to create a functional synapse. For example, thrombospondins and hevin are synthesized and released by astrocytes during CNS development and into adulthood. Their role is to enhance initial glutamatergic synapse formation through physically coupling the presynaptic terminal to a postsynaptic target. The protein hevin bridges presynaptic transmembrane neurexins to postsynaptic neuroligins to promote the formation of synaptic contacts. Inside the cell, neurexins trigger the mobilization of presynaptic glutamate release machinery to the membrane while neuroligins stabilize AMPA receptors to the postsynaptic membrane, putting them in an advantaged location to receive signals, thus forming a mature synapse [24,51,52].

Beyond contributing to the physical structure of the interstitial compartment, the extracellular matrix regulates the diffusion of neurotransmitters, gliotransmitters, and their receptors largely although the release of chondroitin sulfate proteoglycans (CSPGs) from astrocytes. CSPGs are the major subunits of the PNN which can physically support and insulate neurons and maintain cellular spacing. Owing to their negatively charged residues, they can control the diffusion of charged molecules in the extracellular milieu such as glutamate [53]. Indeed, overexpression of extracellular matrix proteins can physically hinder the flow of signaling molecules in the interstitial space (an increase in tortuosity) and cause pathological signaling [45,54], whereas a loss of the CSPG tenascin results in a lower tortuosity and volume fraction due to its role in maintaining spacing between adjacent cells [55]. PNNs can also cause the clustering of glutamate receptors such as AMPA receptors through physically impairing their ability to laterally diffuse across the membrane, which can cause rapid desensitization of the receptor [56,57]. NMDA receptors are under a similar form of regulation, but through their interaction with integrins [58]. Importantly, while the extracellular matrix may limit diffusion of transmitters and receptors to inhibit synaptic plasticity, enzymes that cleave extracellular matrix interactions can be secreted into the interstitial space to allow for synaptic remodeling. As an example, NR1 diffusion is enhanced when matrix metalloproteinase-9 (MMP-9), an astrocyte secreted enzyme that can cleave extracellular matrix/adhesion associations, interacts with integrin $\beta 1$ [58].

Pathophysiology of Addiction and the Interstitial Space

Disruptions to corticostriatal circuits contribute to drug seeking in many preclinical models of addiction [59,60]. Imaging studies have extended these findings by discovering that the severity of drug craving in humans correlates to the activity in relevant cortical structures [2,6,61]. A potential molecular and cellular basis of aberrant activity may include factors regulating the volume fraction or tortuosity of the interstitial space. Notably, drugs of abuse are known to alter several of these factors including the number of synapses containing a PAP, density of glutamate transporters and receptors, glutamate release by astrocytes into the extrasynaptic compartment (which would alter the synaptic to extrasynaptic gradient), and the molecular composition of the extracellular matrix [62–65]. Hence, there is a need to evaluate the possibility that aberrant neural circuit activity in addiction stems from drug-induced changes altering the volume fraction and tortuosity of the interstitial space. Consistent with this possibility, increases in MMP activity restores AMPA receptor function during reinstatement testing [65].

NEURON TO ASTROCYTE SIGNALING

The consequence of presynaptic neurotransmitter release is typically defined by changes occurring in the postsynaptic cell, however, astrocytes also express many of the same receptors used by neurons [21]. Early evidence that neurons and astrocytes were functionally linked included findings that hippocampal neuronal activation resulted in calcium waves in astrocyte networks [66]. It is now clear that the stimulation of neurotransmitter receptors on the astrocyte cell membrane can activate intracellular signaling mechanisms, such as intracellular calcium events, that may be locally confined in microdomains of the astrocyte, spread throughout the full extent of the astrocyte, or propagate throughout entire astrocyte networks [25,26,67]. The sensitivity of neuron to astrocyte signaling was established by the observation that astrocytes even detect spontaneous synaptic events and that blocking the activation of astrocytes prevented changes in synaptic transmission [67].

Receptor-Mediated Neuron to Astrocyte Signaling

The molecular basis of neuronal signaling to astrocytes will depend on the receptors expressed by these cells. Astrocytes have been found to express protein or mRNA for numerous neurotransmitters receptors including those for glutamate, monoamines, endocannabinoids, and adenosine [21]. However, there are several issues that need to be resolved. First, many studies used cultured astrocytes, and there is a need to understand in vivo receptor expression. Second, receptor expression in astrocytes may vary across development, which indicates the need to determine expression patterns in adult astrocytes [68]. Third, astrocytes within a region or across regions likely display heterogeneous patterns of receptor expression. Hence, a more complete characterization of the receptor expressing astrocytes will provide key insights into the regional capacity of astrocytes to regulate neural circuits.

Since synaptic glutamate is the primary excitatory neurotransmitter in the brain, astrocytes need to be equipped with glutamate receptors to effectively monitor neurotransmission. Astrocytes express several metabotropic glutamate receptors [69]. Although these receptors vary in their downstream signaling mechanisms, they can often produce identical changes in astrocytes. For example, metabotropic glutamate receptors 3 and 5 signal through $G\alpha_{i/o}$ and $G\alpha_{q/11}$, respectively, but both can upregulate glutamate transporter function [70–72]. This highlights the phenomena that astrocytic receptors coupled to diverse signaling pathways may converge on increases in intracellular Ca^{2+}.

Ionotropic glutamate receptors have also been detected in astrocytes, yet the functional role of these receptors is unclear, as astrocytes are not electrically excitable cells. AMPA receptors have been detected in astrocytes in the thalamus and the nucleus of the solitary tract [73]. The function of these receptors is equivocal, in part because astrocytes are not thought to be exposed to the mM levels of glutamate typically needed to activate these receptors. Revealing the source of glutamate stimulating to these receptors may provide insight into their functional role. For example, expression on PAPs would indicate that these receptors are positioned to detect high levels of synaptic glutamate. In comparison, there is more support for the expression of NMDA receptors by astrocytes, which appear to exhibit reduced Mg^{2+} blockade due to the expression of GluN2C/D [74]. This would be crucial since astrocytes are not depolarized to the extent needed to remove the Mg^{2+} block of NMDA receptors expressing GluN2A/B.

The potential existence of ionotropic receptors permeable to Ca^{2+} and Na^+ is intriguing given the importance of these ions to intracellular signaling in astrocytes [75,76]. Cultured adult and fetal human astrocytes contain NMDA receptors that were found to directly contribute to inward Ca^{2+} signaling [77]. Determining whether ionotropic receptors contribute to inward Ca^{2+} or Na^+ signaling in adult astrocytes in vivo will be key in establishing the molecular basis of neuronal regulation of astrocytes as encoded by glutamate.

Nonreceptor-Mediated Neuron to Astrocyte Signaling

Neuronal signaling to astrocytes may also involve nonreceptor-mediated mechanisms. Electron microscopy studies have detected glutamate transporters on perisynaptic astrocyte processes [78], and the transport of glutamate results in the cotransport of Na^+ molecules, which may serve as a key intracellular signal regulating the role of astrocytes in the trafficking of energetic substances [76]. Similarly, astrocytes express dopamine transporters and monoamine oxidase

[79,80], and the metabolism of intracellular dopamine by monoamine oxidase drives lipid peroxidation and IP3 production resulting in an increase in astrocyte Ca^{2+} levels [81]. Hence, astrocytes can detect glutamate, dopamine, and potentially other neurotransmitters even in the absence of receptors for these neurotransmitters.

Revealing the molecular and functional impact of neuronal signaling to astrocytes will be essential to understanding how these cells are impacted by drugs of abuse. For example, repeated cocaine produces disruptions in synaptic signaling encoded by dopamine and glutamate in the nucleus accumbens. In addition, cocaine produces persistent changes in astrocytes that range from the density of PAPs to the function of glutamate transporters and release mechanisms [62,64,82]. These changes likely contribute to the dysregulation of neural circuits encoding maladaptive behaviors in addiction since manipulations that target astrocytes can normalize drug-related behaviors [62,83]. However, it is unclear if dysregulation of astrocytes stem from aberrant signaling of dopamine, glutamate, or some unidentified neurotransmitter.

INTRACELLULAR SIGNALING IN ASTROCYTES

Intracellular Ca^{2+} is an apparent ubiquitous second messenger system in astrocytes since activation of these cells through diverse signals often converges on this pathway [25,84]. Intracellular Ca^{2+} signaling in astrocytes is thought to be initiated by activation of G-protein-coupled receptors, and signaling pathways, such as phospholipase C, that leads to the activation of inositol triphosphate receptor-mediated release of Ca^{2+} from the endoplasmic reticulum [25]. Local intracellular Ca^{2+} signals can then be amplified through additional mechanisms to the point of spreading throughout the cell [84]. From there, Ca^{2+} signals often propagate to neighboring astrocytes, thereby forming Ca^{2+} waves. The cellular and molecular steps involved in the formation of Ca^{2+} waves between astrocytes involve two primary pathways. The first involves direct communication between the cytosols of two cells adjoined by gap junction channels, while the second involves the release of gliotransmitters [84]. This provides the molecular basis for the existence of astrocyte networks, which in turn has the potential to represent a novel form of long-distance communication integrating the function of multiple neural networks, which could contribute to aberrant neural circuits encoding aspects of addiction.

A fundamental question is whether astrocytes transform the complex extracellular molecular environment created by numerous neurotransmitters into a single event in the form of whole-cell increases in Ca^{2+}. Moreover, current data imply that intracellular calcium encodes both the incoming and outgoing information within astrocytes [85]. Alternatively, the apparent similarities observed when measuring whole-cell Ca^{2+} signaling does not accurately portray what may be a high level of complexity of intracellular Ca^{2+} signaling. The possibility that Ca^{2+} may be more complex than initially revealed was indicated by the recent observations that deleting inositol triphosphate receptors in astrocytes or stimulating transgenic G-protein coupled receptors expressed on astrocytes failed to alter synaptic transmission [86], and that ultrathin PAPs are too narrow to accommodate the endoplasmic reticulum which means that internal calcium sources are roughly 500 nm away from synapses [85].

Newly developed tools are providing key insights into the spatial, temporal, and molecular basis of Ca^{2+} signaling, which reveal that astrocytic signaling is indeed highly complex. Specifically, the use of genetically encoded calcium indicators and two-photon microscopy imaging have enabled for the first time the detection of Ca^{2+} transients in PAPs. Recent findings indicate that the majority of in vivo astrocytic Ca^{2+} signals occur in PAPs, and hence, do not require the mobilization of internal calcium stores [85,87]. These exciting advances should lead to an improved understanding of how astrocytes receive and integrate information, which will be essential in evaluating drug-induced changes in astrocytes that may underlie addiction.

ASTROCYTE TO NEURON SIGNALING

While astrocytes appear to consolidate information about the extracellular milieu into a singular universal Ca^{2+} signal, the outputs that astrocytes produce are numerous and capable of influencing neurotransmission. This can occur through ion buffering, regulation of cerebral blood flow, water transport, glucose uptake, energy substrate production for adjacent neurons, and antioxidant production. In addition, astrocytes can regulate the extracellular levels of neuroactive substances ranging from small molecules (e.g., glutamate, GABA) to trophic factors (e.g., brain derived neurotrophic factor, transforming growth factor beta 1). Notably, the actions of molecules released by astrocytes can be functionally distinct from the identical molecules released by neurons due to spatial and temporal receptor activation parameters. For example, astrocytes may target gliotransmitters to a population of neuronal receptors that are distal from the synapse. Further, these receptors are often embedded in large protein complexes that tailor the physiological changes occurring in the neuron following receptor activation. For example, extrasynaptic NMDA receptors have been detected in highly specialized protein complexes that are often anchored near adjacent astrocyte processes [46,88], and extrasynaptic NMDA receptors can have effects on the neuron that are opposite to synaptic NMDA receptor activation [29]. Beyond the release of gliotransmitters, astrocytes release a diverse array of proteins that compose the extracellular matrix, which can impact synapse formation, neuronal receptor localization, and other aspects of neurotransmission [24,51,52,89].

Astrocyte to Neuron Signaling: Glutamate Homeostasis

Glutamate is the most studied gliotransmitter in drug addiction. Glutamate uptake into astrocytes primarily involves the Na^+-dependent glutamate transporter 1 (GLT-1) which is a critical determinant of synaptic glutamate signaling [90]. As mentioned previously, the cotransport of Na^+ can influence astrocyte swelling, which can alter the volume of the extracellular space. Glutamate cleared by astrocytes can have numerous fates including re-release and metabolism via oxidation to produce adenosine triphosphate, or conversion into glutamine to be utilized by neurons [91]. The expression and function of GLT-1 can be regulated by metabotropic glutamate receptors which suggests that astrocytes sense local synaptic glutamate events and respond by appropriately enhancing uptake [71].

Astrocytes under physiological conditions can release glutamate by channel/transporter- and vesicular-mediated mechanisms [92]. The cystine-glutamate antiporter system x_c^- (Sxc) releases glutamate and in doing so, can activate presynaptic and postsynaptic neuronal receptors in the striatum and hippocampus [93–95]. Similarly, ultrastructural electron microscopy studies provide an evidence for vesicles containing glutamate in hippocampal astrocytes proximal to presynaptic neuronal NMDA receptors which can enhance synaptic glutamate signaling [96,97]. Unresolved questions include the conditions leading to release by each mechanism, and the neuronal function of each release mechanism.

The finding that astrocytes regulate glutamate homeostasis has been one of the key findings implicating these cells in addiction [98]. For example, Sxc activity has been shown to be reduced following self-administration of cocaine in rats resulting in reduced extracellular glutamate levels in the nucleus accumbens. This reduction is thought to produce drug seeking through a metabotropic glutamate receptor 2 dependent mechanism whereby a reduction in its activity produces a disinhibition of glutamate release from cortical afferents; enhancing excitatory drive on accumbal efferents and consequently promoting drug seeking [62]. Similarly, GLT-1 function and expression in the nucleus accumbens is also reduced by cocaine self-administration [99], and this is also required for cocaine reinstatement [83,100]. While more work is needed to reveal the functional impact of impaired Sxc and GLT-1 activity in the nucleus accumbens, one consequence may involve impaired extrasynaptic signaling. As previously noted, extrasynaptic glutamate signaling requires intact glutamate clearance. Hence, the abnormal regulation of the nucleus accumbens circuitry may include impaired fidelity between distinct types of intercellular signaling (e.g., synaptic and extrasynaptic glutamate).

CONCLUSION

Evolution of mammalian brains has rendered intercellular signaling between neurons and astrocytes as a critical component of virtually every aspect of brain function. Hence, our understanding of the human brain in the healthy and diseased states will be limited by our understanding of bidirectional neuron signaling. As a result, the study of bidirectional neuron and astrocyte signaling, including attempts to understand the regulation of glutamate signaling, has the potential to revolutionize our understanding of the human brain, a development that may expedite our efforts to discover treatments for addiction. Consistent with this, preclinical and clinical studies establish the therapeutic potential of targeting astrocyte to neuron signaling. Restoring extracellular glutamate levels using drugs such as N-acetylcysteine or ceftriaxone reduces reinstatement produced by cocaine, heroin, and other drugs of abuse in rodents and attenuates drug craving or use in humans [62,83,101,102]. These studies illustrate the therapeutic potential of targeting astrocytes to normalize neural circuits implicated in drug addiction.

REFERENCES

[1] World Health Organization. The global burden of disease: 2004 update, Geneva: World Health Organization; 2008. http://www.who.int/iris/handle/10665/43942.

[2] Breiter HC, Gollub RL, Weisskoff RM, Kennedy DN, Makris N, Berke JD, Hyman SE. Acute effects of cocaine on human brain activity and emotion. Neuron 1997;19(3):591–611.

[3] Creese I, Iversen SD. The pharmacological and anatomical substrates of the amphetamine response in the rat. Brain Res 1975;83:419–36.

[4] Goeders NE, Smith JE. Cortical dopaminergic involvement in cocaine reinforcement. Science 1983;221(4612):773–5.

[5] Kosten TR, Cheeves C, Palumbo J, Seibyl JP, Price LH, Woods SW. Regional cerebral blood flow during acute and chronic abstinence from combined cocaine-alcohol abuse. Drug Alcohol Depend 1998;50(3):187–95.

[6] Volkow ND, Wang GJ, Fowler JS, Hitzemann R, Angrist B, Gatley SJ, Pappas N. Association of methylphenidate-induced craving with changes in right striato-orbitofrontal metabolism in cocaine abusers: implications in addiction. Am J Psychiatry 1999;156(1):19–26.

[7] Bosch TC, Klimovich A, Domazet-Loso T, Grunder S, Holstein TW, Jekely G, Yuste R. Back to the basics: Cnidarians start to fire. Trends Neurosci 2017;40(2):92–105. https://doi.org/10.1016/j.tins.2016.11.005.

[8] Moroz LL. On the independent origins of complex brains and neurons. Brain Behav Evol 2009;74(3):177–90. https://doi.org/10.1159/000258665.

[9] Moroz LL. Convergent evolution of neural systems in ctenophores. J Exp Biol 2015;218(Pt 4):598–611. https://doi.org/10.1242/jeb.110692.

[10] Moroz LL, Kohn AB. Independent origins of neurons and synapses: insights from ctenophores. Philos Trans R Soc Lond B Biol Sci 2016;371(1685) https://doi.org/10.1098/rstb.2015.0041.

[11] Ryan JF, Chiodin M. Where is my mind? How sponges and placozoans may have lost neural cell types. Philos Trans R Soc Lond B Biol Sci 2015;370 (1684). https://doi.org/10.1098/rstb.2015.0059.

[12] Ryan JF, Pang K, Schnitzler CE, Nguyen AD, Moreland RT, Simmons DK, Baxevanis AD. The genome of the ctenophore Mnemiopsis leidyi and its implications for cell type evolution. Science 2013;342(6164). https://doi.org/10.1126/science.1242592.

[13] Bayes A, Collins MO, Reig-Viader R, Gou G, Goulding D, Izquierdo A, Grant SG. Evolution of complexity in the zebrafish synapse proteome. Nat Commun 2017;8. https://doi.org/10.1038/ncomms14613.

[14] Bayes A, van de Lagemaat LN, Collins MO, Croning MD, Whittle IR, Choudhary JS, Grant SG. Characterization of the proteome, diseases and evolution of the human postsynaptic density. Nat Neurosci 2011;14(1):19–21. https://doi.org/10.1038/nn.2719.

[15] Oberheim Bush NA, Nedergaard M. Do evolutionary changes in astrocytes contribute to the computational power of the hominid brain? Neurochem Res 2017;42(9):2577–87. https://doi.org/10.1007/s11064-017-2363-0.

[16] Min R, Santello M, Nevian T. The computational power of astrocyte mediated synaptic plasticity. Front Comput Neurosci 2012;6. https://doi.org/10.3389/fncom.2012.00093.

[17] Alvarellos-Gonzalez A, Pazos A, Porto-Pazos AB. Computational models of neuron-astrocyte interactions lead to improved efficacy in the performance of neural networks. Comput Math Methods Med 2012;2012:476324. https://doi.org/10.1155/2012/476324.

[18] Oberheim NA, Wang X, Goldman S, Nedergaard M. Astrocytic complexity distinguishes the human brain. Trends Neurosci 2006;29(10):547–53.

[19] Porto-Pazos AB, Veiguela N, Mesejo P, Navarrete M, Alvarellos A, Ibanez O, Araque A. Artificial astrocytes improve neural network performance. PLoS One 2011;6(4). https://doi.org/10.1371/journal.pone.0019109.

[20] Wu Y, Dissing-Olesen L, MacVicar BA, Stevens B. Microglia: dynamic mediators of synapse development and plasticity. Trends Immunol 2015;36 (10):605–13. https://doi.org/10.1016/j.it.2015.08.008.

[21] Cahoy JD, Emery B, Kaushal A, Foo LC, Zamanian JL, Christopherson KS, Barres BA. A transcriptome database for astrocytes, neurons, and oligodendrocytes: a new resource for understanding brain development and function. J Neurosci 2008;28(1):264–78.

[22] Ogata K, Kosaka T. Structural and quantitative analysis of astrocytes in the mouse hippocampus. Neuroscience 2002;113(1):221–33.

[23] Allen NJ, Barres BA. Signaling between glia and neurons: focus on synaptic plasticity. Curr Opin Neurobiol 2005;15(5):542–8. https://doi.org/10.1016/j.conb.2005.08.006.

[24] Christopherson KS, Ullian EM, Stokes CC, Mullowney CE, Hell JW, Agah A, Barres BA. Thrombospondins are astrocyte-secreted proteins that promote CNS synaptogenesis. Cell 2005;120(3):421–33. https://doi.org/10.1016/j.cell.2004.12.020.

[25] Papouin T, Dunphy J, Tolman M, Foley JC, Haydon PG. Astrocytic control of synaptic function. Philos Trans R Soc Lond B Biol Sci 2017;372 (1715)https://doi.org/10.1098/rstb.2016.0154.

[26] Perea G, Sur M, Araque A. Neuron-glia networks: integral gear of brain function. Front Cell Neurosci 2014;8. https://doi.org/10.3389/fncel.2014.00378.

[27] Poskanzer KE, Yuste R. Astrocytic regulation of cortical UP states. Proc Natl Acad Sci U S A 2011;108(45):18453–8.

[28] Le Meur K, Galante M, Angulo MC, Audinat E. Tonic activation of NMDA receptors by ambient glutamate of non-synaptic origin in the rat hippocampus. J Physiol 2007;580(Pt. 2):373–83. https://doi.org/10.1113/jphysiol.2006.123570.

[29] Papouin T, Oliet SH. Organization, control and function of extrasynaptic NMDA receptors. Philos Trans R Soc Lond B Biol Sci 2014;369(1654) https://doi.org/10.1098/rstb.2013.0601.

[30] Wu YW, Grebenyuk S, McHugh TJ, Rusakov DA, Semyanov A. Backpropagating action potentials enable detection of extrasynaptic glutamate by NMDA receptors. Cell Rep 2012;1(5):495–505. https://doi.org/10.1016/j.celrep.2012.03.007.

[31] Rusakov DA, Harrison E, Stewart MG. Synapses in hippocampus occupy only 1–2% of cell membranes and are spaced less than half-micron apart: a quantitative ultrastructural analysis with discussion of physiological implications. Neuropharmacology 1998;37(4–5):513–21.

[32] Araque A, Parpura V, Sanzgiri RP, Haydon PG. Tripartite synapses: glia, the unacknowledged partner. Trends Neurosci 1999;22(5):208–15.

[33] Bernardinelli Y, Randall J, Janett E, Nikonenko I, Konig S, Jones EV, Muller D. Activity-dependent structural plasticity of perisynaptic astrocytic domains promotes excitatory synapse stability. Curr Biol 2014;24(15):1679–88. https://doi.org/10.1016/j.cub.2014.06.025.

[34] Lehre KP, Rusakov DA. Asymmetry of glia near central synapses favors presynaptically directed glutamate escape. Biophys J 2002;83(1):125–34.

[35] Herculano-Houzel S. The remarkable, yet not extraordinary, human brain as a scaled-up primate brain and its associated cost. Proc Natl Acad Sci U S A 2012;109(Suppl. 1):10661–8. https://doi.org/10.1073/pnas.1201895109.

[36] Nicholson C, Kamali-Zare P, Tao L. Brain extracellular space as a diffusion barrier. Comput Vis Sci 2011;14(7):309–25. https://doi.org/10.1007/s00791-012-0185-9.

[37] Shi C, Lei Y, Han H, Zuo L, Yan J, He Q, Xu W. Transportation in the interstitial space of the brain can be regulated by neuronal excitation. Sci Rep 2015;5. https://doi.org/10.1038/srep17673.

[38] Hrabetova S. Extracellular diffusion is fast and isotropic in the stratum radiatum of hippocampal CA1 region in rat brain slices. Hippocampus 2005;15(4):441–50. https://doi.org/10.1002/hipo.20068.

[39] Lehmenkuhler A, Sykova E, Svoboda J, Zilles K, Nicholson C. Extracellular space parameters in the rat neocortex and subcortical white matter during postnatal development determined by diffusion analysis. Neuroscience 1993;55(2):339–51.

[40] Xie L, Kang H, Xu Q, Chen MJ, Liao Y, Thiyagarajan M, Nedergaard M. Sleep drives metabolite clearance from the adult brain. Science 2013;342 (6156):373–7. https://doi.org/10.1126/science.1241224.

[41] Rusakov DA, Savtchenko LP, Zheng K, Henley JM. Shaping the synaptic signal: molecular mobility inside and outside the cleft. Trends Neurosci 2011;34(7):359–69. https://doi.org/10.1016/j.tins.2011.03.002.

[42] Haber M, Zhou L, Murai KK. Cooperative astrocyte and dendritic spine dynamics at hippocampal excitatory synapses. J Neurosci 2006;26 (35):8881–91.

[43] Lai KO, Ip NY. Structural plasticity of dendritic spines: the underlying mechanisms and its dysregulation in brain disorders. Biochim Biophys Acta 2013;1832(12):2257–63. https://doi.org/10.1016/j.bbadis.2013.08.012.

[44] Scofield MD, Heinsbroek JA, Gipson CD, Kupchik YM, Spencer S, Smith AC, Kalivas PW. The nucleus Accumbens: Mechanisms of addiction across drug classes reflect the importance of glutamate homeostasis. Pharmacol Rev 2016;68(3):816–71. https://doi.org/10.1124/pr.116.012484.

[45] Vargova L, Sykova E. Astrocytes and extracellular matrix in extrasynaptic volume transmission. Philos Trans R Soc Lond B Biol Sci 2014;369 (1654). https://doi.org/10.1098/rstb.2013.0608.

[46] Petralia RS, Wang YX, Hua F, Yi Z, Zhou A, Ge L, Wenthold RJ. Organization of NMDA receptors at extrasynaptic locations. Neuroscience 2010;167(1):68–87. https://doi.org/10.1016/j.neuroscience.2010.01.022.

[47] Mulholland PJ, Chandler LJ. Inhibition of glutamate transporters couples to Kv4.2 dephosphorylation through activation of extrasynaptic NMDA receptors. Neuroscience 2010;165(1):130–7.

[48] Murai KK, Nguyen LN, Irie F, Yamaguchi Y, Pasquale EB. Control of hippocampal dendritic spine morphology through ephrin-A3/EphA4 signaling. Nat Neurosci 2003;6(2):153–60. https://doi.org/10.1038/nn994.

[49] Lau LW, Cua R, Keough MB, Haylock-Jacobs S, Yong VW. Pathophysiology of the brain extracellular matrix: a new target for remyelination. Nat Rev Neurosci 2013;14(10):722–9. https://doi.org/10.1038/nrn3550.

[50] Tomaselli KJ, Neugebauer KM, Bixby JL, Lilien J, Reichardt LF. N-cadherin and integrins: two receptor systems that mediate neuronal process outgrowth on astrocyte surfaces. Neuron 1988;1(1):33–43.

[51] Kania A, Klein R. Mechanisms of ephrin-Eph signalling in development, physiology and disease. Nat Rev Mol Cell Biol 2016;17(4):240–56. https://doi.org/10.1038/nrm.2015.16.

[52] Kucukdereli H, Allen NJ, Lee AT, Feng A, Ozlu MI, Conatser LM, Eroglu C. Control of excitatory CNS synaptogenesis by astrocyte-secreted proteins Hevin and SPARC. Proc Natl Acad Sci U S A 2011;108(32):E440–449. https://doi.org/10.1073/pnas.1104977108.

[53] Sykova E. Extrasynaptic volume transmission and diffusion parameters of the extracellular space. Neuroscience 2004;129(4):861–76. https://doi.org/10.1016/j.neuroscience.2004.06.077.

[54] Zamecnik J, Homola A, Cicanic M, Kuncova K, Marusic P, Krsek P, Vargova L. The extracellular matrix and diffusion barriers in focal cortical dysplasias. Eur J Neurosci 2012;36(1):2017–24. https://doi.org/10.1111/j.1460-9568.2012.08107.x.

[55] Sykova E. Glia and volume transmission during physiological and pathological states. J Neural Transm (Vienna) 2005;112(1):137–47. https://doi.org/10.1007/s00702-004-0120-4.

[56] Frischknecht R, Heine M, Perrais D, Seidenbecher CI, Choquet D, Gundelfinger ED. Brain extracellular matrix affects AMPA receptor lateral mobility and short-term synaptic plasticity. Nat Neurosci 2009;12(7):897–904. https://doi.org/10.1038/nn.2338.

[57] Heine M, Groc L, Frischknecht R, Beique JC, Lounis B, Rumbaugh G, Choquet D. Surface mobility of postsynaptic AMPARs tunes synaptic transmission. Science 2008;320(5873):201–5. https://doi.org/10.1126/science.1152089.

[58] Michaluk P, Mikasova L, Groc L, Frischknecht R, Choquet D, Kaczmarek L. Matrix metalloproteinase-9 controls NMDA receptor surface diffusion through integrin beta1 signaling. J Neurosci 2009;29(18):6007–12. https://doi.org/10.1523/jneurosci.5346-08.2009.

[59] Mantsch JR, Baker DA, Funk D, Le AD, Shaham Y. Stress-induced reinstatement of drug seeking: 20 years of progress. Neuropsychopharmacology 2016;41(1):335–56. https://doi.org/10.1038/npp.2015.142.

[60] Schmidt HD, Anderson SM, Famous KR, Kumaresan V, Pierce RC. Anatomy and pharmacology of cocaine priming-induced reinstatement of drug seeking. Eur J Pharmacol 2005;526(1–3):65–76.

[61] Volkow ND, Wang GJ, Ma Y, Fowler JS, Wong C, Ding YS, Kalivas P. Activation of orbital and medial prefrontal cortex by methylphenidate in cocaine-addicted subjects but not in controls: relevance to addiction. J Neurosci 2005;25(15):3932–9.

[62] Baker DA, McFarland K, Lake RW, Shen H, Tang XC, Toda S, Kalivas PW. Neuroadaptations in cystine-glutamate exchange underlie cocaine relapse. Nat Neurosci 2003;6(7):743–9. https://doi.org/10.1038/nn1069.

[63] Mulholland PJ, Chandler LJ, Kalivas PW. Signals from the fourth dimension regulate drug relapse. Trends Neurosci 2016;39(7):472–85. https://doi.org/10.1016/j.tins.2016.04.007.

[64] Scofield MD, Li H, Siemsen BM, Healey KL, Tran PK, Woronoff N, Reissner KJ. Cocaine self-administration and extinction leads to reduced glial Fibrillary acidic protein expression and morphometric features of astrocytes in the nucleus Accumbens Core. Biol Psychiatry 2016;80(3):207–15. https://doi.org/10.1016/j.biopsych.2015.12.022.

[65] Smith AC, Scofield MD, Kalivas PW. The tetrapartite synapse: extracellular matrix remodeling contributes to corticoaccumbens plasticity underlying drug addiction. Brain Res 2015;1628(Pt A):29–39. https://doi.org/10.1016/j.brainres.2015.03.027.

[66] Dani JW, Chernjavsky A, Smith SJ. Neuronal activity triggers calcium waves in hippocampal astrocyte networks. Neuron 1992;8(3):429–40.

[67] Shigetomi E, Bushong EA, Haustein MD, Tong X, Jackson-Weaver O, Kracun S, Khakh BS. Imaging calcium microdomains within entire astrocyte territories and endfeet with GCaMPs expressed using adeno-associated viruses. J Gen Physiol 2013;141(5):633–47. https://doi.org/10.1085/jgp.201210949.

[68] Sun W, McConnell E, Pare JF, Xu Q, Chen M, Peng W, Nedergaard M. Glutamate-dependent neuroglial calcium signaling differs between young and adult brain. Science 2013;339(6116):197–200. https://doi.org/10.1126/science.1226740.

[69] Bradley SJ, Challiss RA. G protein-coupled receptor signalling in astrocytes in health and disease: a focus on metabotropic glutamate receptors. Biochem Pharmacol 2012;84(3):249–59. https://doi.org/10.1016/j.bcp.2012.04.009.

[70] Aronica E, Gorter JA, Ijlst-Keizers H, Rozemuller AJ, Yankaya B, Leenstra S, Troost D. Expression and functional role of mGluR3 and mGluR5 in human astrocytes and glioma cells: opposite regulation of glutamate transporter proteins. Eur J Neurosci 2003;17(10):2106–18.

[71] Devaraju P, Sun MY, Myers TL, Lauderdale K, Fiacco TA. Astrocytic group I mGluR-dependent potentiation of astrocytic glutamate and potassium uptake. J Neurophysiol 2013;109(9):2404–14. https://doi.org/10.1152/jn.00517.2012.

[72] Vermeiren C, Najimi M, Vanhoutte N, Tilleux S, de Hemptinne I, Maloteaux JM, Hermans E. Acute up-regulation of glutamate uptake mediated by mGluR5a in reactive astrocytes. J Neurochem 2005;94(2):405–16. https://doi.org/10.1111/j.1471-4159.2005.03216.x.

[73] McDougal DH, Hermann GE, Rogers RC. Vagal afferent stimulation activates astrocytes in the nucleus of the solitary tract via AMPA receptors: evidence of an atypical neural-glial interaction in the brainstem. J Neurosci 2011;31(39):14037–45. https://doi.org/10.1523/JNEUROSCI.2855-11.2011.

[74] Palygin O, Lalo U, Pankratov Y. Distinct pharmacological and functional properties of NMDA receptors in mouse cortical astrocytes. Br J Pharmacol 2011;163(8):1755–66. https://doi.org/10.1111/j.1476-5381.2011.01374.x.

[75] Araque A, Carmignoto G, Haydon PG. Dynamic signaling between astrocytes and neurons. Ann Rev Physiol 2001;63:795–813. https://doi.org/10.1146/annurev.physiol.63.1.795.

[76] Bernardinelli Y, Magistretti PJ, Chatton JY. Astrocytes generate Na+-mediated metabolic waves. Proc Natl Acad Sci U S A 2004;101(41):14937–42. https://doi.org/10.1073/pnas.0405315101.

[77] Lee MC, Ting KK, Adams S, Brew BJ, Chung R, Guillemin GJ. Characterisation of the expression of NMDA receptors in human astrocytes. PLoS One 2010;5(11). https://doi.org/10.1371/journal.pone.0014123.

[78] Chaudhry FA, Lehre KP, van Lookeren Campagne M, Ottersen OP, Danbolt NC, Storm-Mathisen J. Glutamate transporters in glial plasma membranes: highly differentiated localizations revealed by quantitative ultrastructural immunocytochemistry. Neuron 1995;15(3):711–20.

[79] Gasser PJ, Hurley MM, Chan J, Pickel VM. Organic cation transporter 3 (OCT3) is localized to intracellular and surface membranes in select glial and neuronal cells within the basolateral amygdaloid complex of both rats and mice. Brain Struct Funct 2017;222(4):1913–28. https://doi.org/10.1007/s00429-016-1315-9.

[80] Wu X, Kekuda R, Huang W, Fei YJ, Leibach FH, Chen J, Ganapathy V. Identity of the organic cation transporter OCT3 as the extraneuronal monoamine transporter (uptake2) and evidence for the expression of the transporter in the brain. J Biol Chem 1998;273(49):32776–86.

[81] Vaarmann A, Gandhi S, Abramov AY. Dopamine induces Ca2+ signaling in astrocytes through reactive oxygen species generated by monoamine oxidase. J Biol Chem 2010;285(32):25018–23. https://doi.org/10.1074/jbc.M110.111450.

[82] Roberts-Wolfe DJ, Kalivas PW. Glutamate transporter GLT-1 as a therapeutic target for substance use disorders. CNS Neurol Disord Drug Targets 2015;14(6):745–56.

[83] Reissner KJ, Gipson CD, Tran PK, Knackstedt LA, Scofield MD, Kalivas PW. Glutamate transporter GLT-1 mediates N-acetylcysteine inhibition of cocaine reinstatement. Addict Biol 2015;20(2):316–23. https://doi.org/10.1111/adb.12127.

[84] Scemes E, Giaume C. Astrocyte calcium waves: what they are and what they do. Glia 2006;54(7):716–25. https://doi.org/10.1002/glia.20374.

[85] Rusakov DA. Disentangling calcium-driven astrocyte physiology. Nat Rev Neurosci 2015;16(4):226–33. https://doi.org/10.1038/nrn3878.

[86] Agulhon C, Fiacco TA, McCarthy KD. Hippocampal short- and long-term plasticity are not modulated by astrocyte Ca2+ signaling. Science 2010;327(5970):1250–4. https://doi.org/10.1126/science.1184821.

[87] Zheng K, Bard L, Reynolds JP, King C, Jensen TP, Gourine AV, Rusakov DA. Time-resolved imaging reveals heterogeneous landscapes of Nanomolar Ca(2+) in neurons and Astroglia. Neuron 2015;88(2):277–88. https://doi.org/10.1016/j.neuron.2015.09.043.

[88] Sah P, Hestrin S, Nicoll RA. Tonic activation of NMDA receptors by ambient glutamate enhances excitability of neurons. Science 1989;246(4931):815–8.

[89] Eroglu C, Allen NJ, Susman MW, O'Rourke NA, Park CY, Ozkan E, Barres BA. Gabapentin receptor alpha2delta-1 is a neuronal thrombospondin receptor responsible for excitatory CNS synaptogenesis. Cell 2009;139(2):380–92. https://doi.org/10.1016/j.cell.2009.09.025.

[90] Danbolt NC. Glutamate uptake. Prog Neurobiol 2001;65(1):1–105.

[91] McKenna MC. Glutamate pays its own way in astrocytes. Front Endocrinol (Lausanne) 2013;4. https://doi.org/10.3389/fendo.2013.00191.

[92] Malarkey EB, Parpura V. Mechanisms of glutamate release from astrocytes. Neurochem Int 2008;52(1–2):142–54. https://doi.org/10.1016/j.neuint.2007.06.005.

[93] Baker DA, Xi ZX, Shen H, Swanson CJ, Kalivas PW. The origin and neuronal function of in vivo nonsynaptic glutamate. J Neurosci 2002;22(20):9134–41.

[94] Lewerenz J, Hewett SJ, Huang Y, Lambros M, Gout PW, Kalivas PW, Maher P. The cystine/glutamate antiporter system x(c)(−) in health and disease: from molecular mechanisms to novel therapeutic opportunities. Antioxid Redox Signal 2013;18(5):522–55. https://doi.org/10.1089/ars.2011.4391.

[95] Williams LE, Featherstone DE. Regulation of hippocampal synaptic strength by glial xCT. J Neurosci 2014;34(48):16093–102. https://doi.org/10.1523/JNEUROSCI.1267-14.2014.

[96] Bezzi P, Gundersen V, Galbete JL, Seifert G, Steinhauser C, Pilati E, Volterra A. Astrocytes contain a vesicular compartment that is competent for regulated exocytosis of glutamate. Nat Neurosci 2004;7(6):613–20. https://doi.org/10.1038/nn1246.

[97] Jourdain P, Bergersen LH, Bhaukaurally K, Bezzi P, Santello M, Domercq M, Volterra A. Glutamate exocytosis from astrocytes controls synaptic strength. Nat Neurosci 2007;10(3):331–9. https://doi.org/10.1038/nn1849.

[98] Kalivas PW. The glutamate homeostasis hypothesis of addiction. Nat Rev Neurosci 2009;10(8):561–72. https://doi.org/10.1038/nrn2515.

[99] Knackstedt LA, Melendez RI, Kalivas PW. Ceftriaxone restores glutamate homeostasis and prevents relapse to cocaine seeking. Biol Psychiatry 2010;67(1):81–4.

[100] Trantham-Davidson H, LaLumiere RT, Reissner KJ, Kalivas PW, Knackstedt LA. Ceftriaxone normalizes nucleus accumbens synaptic transmission, glutamate transport, and export following cocaine self-administration and extinction training. J Neurosci 2012;32(36):12406–10. https://doi.org/10.1523/JNEUROSCI.1976-12.2012.

[101] Schmaal L, Berk L, Hulstijn KP, Cousijn J, Wiers RW, van den Brink W. Efficacy of N-acetylcysteine in the treatment of nicotine dependence: a double-blind placebo-controlled pilot study. Eur Addict Res 2011;17(4):211–6. https://doi.org/10.1159/000327682.

[102] Zhou W, Kalivas PW. N-acetylcysteine reduces extinction responding and induces enduring reductions in cue- and heroin-induced drug-seeking. Biol Psychiatry 2008;63(3):338–40. https://doi.org/10.1016/j.biopsych.2007.06.008.

Chapter 17

Role of the Extracellular Matrix in Addiction

Lauren N. Beloate and Peter W. Kalivas

Department of Neuroscience, Medical University of South Carolina, Charleston, SC, United States

INTRODUCTION

There are many contributing factors in drug addiction development, maintenance, and relapse, including genetics [1], physiological effects of drug and drug withdrawal [2], contextual learning [3], and social environment [4]. There is now a large body of literature showing that neuroplasticity in the corticostriatal pathway regulates drug experience-induced behavioral changes [5,6]. More recently, evidence of a role for the "tetrapartite synapse" within this pathway, which includes the presynapse, postsynapse, glia, and extracellular matrix (ECM), in drug addiction and drug-induced neuroplasticity has emerged [7,8].

The ECM is composed of secreted glycoproteins and proteoglycans and provides scaffolding and a network of receptors, cell adhesion molecules (CAMs), and proteases that surround and interact with neurons and glia to influence their signaling and plasticity [9]. The ECM can be categorized into a "loose" ECM present throughout the brain and spinal cord [10], perineuronal nets (PNNs) found more proximal to classical synapses [11], and membrane-bound molecules [12]. Research on the ECM has focused on neural plasticity during development [13], and only recently has the addiction field brought focus to tetrapartite synapses.

In this chapter, we provide an overview of the elements of the ECM (Fig. 1) and their role in the initial contingent or noncontingent drug experience, withdrawal with or without extinction training, and relapse for psychostimulants, opiates, and ethanol. Throughout, we discuss mechanisms regulating how the ECM components interact with drug-induced neural signaling and plasticity, and identify possible future experimental targets.

OVERVIEW OF STRUCTURE

Proteases

Proteases in the ECM catalytically create ligands that can signal within the ECM and into the intracellular neuropil through membrane receptors. Here, we focus on the matrix metalloproteinases (MMPs) and tissue plasminogen activator (tPA), both of which have been classified as "pro-addictive factors." Their expression is increased following drugs of abuse, and has been linked to potentiating drug reward [14].

MMPs are a family of the ECM proteases that promote cellular reorganization [15] and mediate synaptic plasticity [16]. Specifically, they degrade and influence the other ECM components and cell-surface molecules through three mechanisms: regulation of gene transcription, regulation of proenzyme activation, and presence of tissue inhibitors of metalloproteinases (TIMPs) (reviewed in [9]). MMPs are also known to regulate learning and memory, such as in fear conditioning and spatial learning tasks [17,18]. The discovery of regulation of neuroplasticity and learning and memory provided the first indication that MMPs play a role in drug addiction. Furthermore, postmortem tissue from cocaine abusers shows decreased MMP-9 activity in the hippocampus [19], heroin users have significantly higher circulating MMP-2 and -9 [20], and serum MMP-9 levels are increased in human alcohol abusers [21,22]. There are more than 20 genes encoding MMPs; however, the most studied forms in the brain are MMP-2, -3, and -9 [23]. Therefore, these will be highlighted in the current chapter.

tPA is a prominent ECM serine protease that converts plasminogen into plasmin [24], allowing for the cleavage of the other ECM components, such as proteoglycans and laminin [25]. tPA is abundantly expressed through the brain in synaptic vesicles and is released into the extracellular space by depolarization [26]. tPA is also an activation factor for

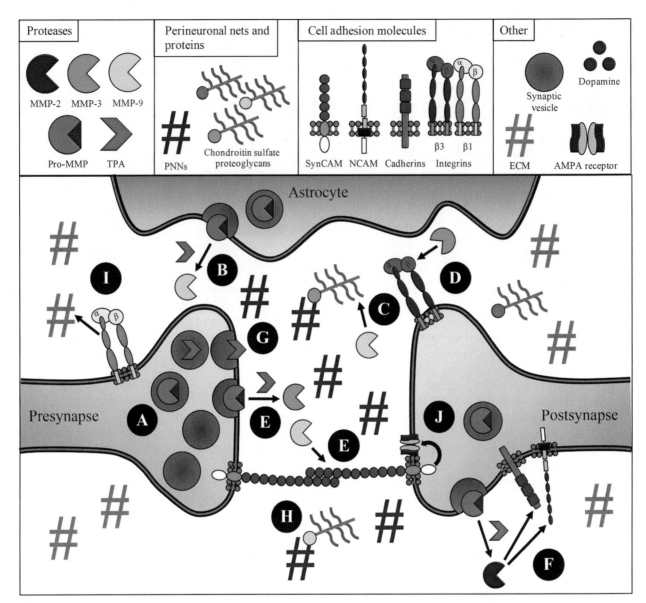

FIG. 1 Extracellular matrix structure. The extracellular matrix (ECM) is made up of the "loose" ECM and proteases, the PNN proteins, and their proteins and membrane-bound molecules, such as CAMs. Pro-MMPs are stored in synaptic vesicles and released from both neuronal and nonneuronal cells (A) and converted to active MMPs by various activation factors, such as tPA (B). In the tetrapartite synapse, MMPs degrade and influence the other ECM factors, such as CSPGs (C), integrins (D), cadherins (E), SynCAMs and NCAMs (F) to play a role in neuroplasticity and learning and memory processes. tPA is also expressed in synaptic vesicles and released into the extracellular space by depolarization (G) and is important for synaptic plasticity and remodeling at glutamatergic and dopaminergic synapses. PNNs are proximal to neurons and glia and are composed of molecules such as CSPGs (H). The integrin family of CAMs support interface between cell membranes and the surrounding ECM (I). The cadherin CAM family, which are important for cell-to-cell inter-actions, become localized at the synaptic membrane following increased neural activity to stabilize AMPA glutamate receptors (AMPARs) at the synaptic membrane (J) to play a role in both LTP and LTD. *Abbreviations*: MMPs = matrix metalloproteinases; tPA = tissue plasminogen activator; PNNs = perineuronal nets; CAMs = cell adhesion molecules; ECM = extracellular matrix.

certain MMPs [27], and plays a role in synaptic plasticity and remodeling [28] in glutamatergic and dopaminergic pathways [29]. Together, these data support a role for the tPA in regulation of addictive behaviors.

PNNs and Proteins

PNNs are the part of the ECM that is most proximal to neurons and glia [11], and they are composed of molecules such as the chondroitin sulfate proteoglycans (CSPGs), hyaluronans, tenascins, and link proteins (reviewed in [30]). In some brain areas, they specifically surround gamma-Aminobutyric acid (GABA)-ergic interneurons [31]. Past research on PNNs has primarily focused on plasticity during critical developmental stages, and the role of PNNs in drug addiction and drug-induced neuroplasticity is a relatively new field of study. However, the expression of PNNs throughout development and in brain areas associated with addiction learning suggest importance in regulating drug-induced behavior [30].

Cell Adhesion Molecules

CAMs are cell-surface molecules that facilitate cell-to-ECM and cell-to-cell interactions through processes such as adhesion, neurite outgrowth, and intracellular signaling [32]. The intracellular domains of CAMs interact with cytoskeletal proteins, while an extracellular domain serves as a target for proteinase activity [9]. Importantly, human CAM expression has been linked to addiction phentotypes [33]. Here, we focus on the calcium-dependent CAMs (integrins and cadherins), as these have been studied in greater detail in the context of drug addiction [34]; however, there are also calcium-independent CAMs [35,36].

The integrin family of CAMs is made up of transmembrane cell adhesion receptors that support interface between cell membranes and the surrounding ECM [37,38]. MMPs cleave protein gelatin sequences to reveal a binding motif that can signal into neurons via integrins [39]. Binding of a ligand to integrin receptors creates a link between the ECM and actin cytoskeleton of cells through intermediate intracellular proteins. The activation of these pathways is one way in which the ECM CAMs regulate changes in cell shape, motility, growth, apoptosis, and gene regulation [35,36].

While integrins are important for cell-to-ECM interactions, the cadherin family of CAMs seem to be more important for cell-to-cell interactions [40], primarily through hemophilic interactions across the synaptic membrane [41]. Furthermore, they associate with α-amino-3-hydroxy-5-methyl-4-isoxazolepropionic acid receptor glutamate receptors (AMPARs) through direct and indirect interactions with receptor subunits [42,43]. Following increased neural activity, cadherins become localized at the synaptic membrane [44], which leads to increased synapse stability and stabilization of AMPARs at the synaptic membrane [45,46]. Due to influences on glutamate receptor expression, it is not surprising that cadherins play an important role in both long-term potentiation and depression (LTP and LTD, respectively) in brain areas such as the hippocampus and ventral tegmental area (VTA) [47,48]. Furthermore, disruption of cadherin interactions in the hippocampus blocks the acquisition of context-dependent memory formation [48]. Lastly, cadherins are widely expressed in dopaminergic neurons in the VTA, which is an area known to regulate drug-related behavior [47], and N-cadherin and E-cadherin are both MMP-9 substrates at synapses [9].

Synaptic cell adhesion molecules (SynCAMs) are immunoglobulin proteins that maintain and instruct formation of excitatory synapses in the brain [49]. SynCAM 1 is expressed throughout the adult brain, including in the striatum [50]. Furthermore, SynCAM adhesion regulates synapse ultrastructure and synapse density, and SynCAM in the hippocampus plays a role in spatial learning [51]. Animal studies discussed below suggest that SynCAM could play a role in some aspects of psychostimulant-induced behavior and synaptic plasticity.

MMP-9 regulates axonal growth and guidance through interaction with neural cell adhesion molecules (NCAMs) [52]. NCAMs also interact with the other ECM components to regulate brain and behavior functions that are related to drugs of abuse. For instance, NCAMs and integrins are thought to act in concert to contribute to separate phases of LTP formation [53,54], and blockade of NCAM or integrin function interferes with both LTP and memory formation [55]. Furthermore, hippocampal tissue from humans with lethal heroin blood levels shows increased amounts of polysialic acid NCAM in both neurons and glia and NCAM levels positively correlate with blood heroin levels, providing evidence from the human literature that NCAM is regulated in opiate addiction [56].

ROLE OF ECM IN THE DEVELOPMENT OF DRUG ADDICTION

Recreational drug use transitions to abuse overtime and requires repeated exposure to a drug [6,57]. We model the initial drug experience and development of drug addiction in laboratory settings using either noncontingent (passive) drug exposure or contingent (active) drug self-administration. Below, we review the role of the ECM in this stage of drug addiction, with a focus on proteases, PNNs, and CAMs.

Proteases

Work thus far on the ECM proteases in the context of drug addiction has been focused primarily on MMPs, and both psychostimulants and opiates increase proteolytic activity following drug experience. For example, noncontingent methamphetamine exposure upregulates MMP-2 and -9 expression in the nucleus accumbens core [58], and 5 days of nicotine conditioned place preference (CPP) conditioning increases MMP-2 and -9 in the hippocampus, but not in prefrontal cortex of adolescent female rats [59]. Furthermore, acute exposure to noncontingent morphine increases both expression and activity (as measured with zymography) of MMP-9 in the midbrain, and pharmacological MMP-9 blockade prevents the development of morphine tolerance following chronic exposure [60]. Thus, acute or chronic experience with either psychostimulants or opiates increases MMP expression, particularly MMP-9, and proteolytic activity in the ECM. Results indicate that effects of ethanol depend on MMP subtype and brain region. For instance, in one study, acute ethanol intoxication decreased MMP-9 activity in the hippocampus, with no effect on MMP-2 [61]. In an ethanol dependence model where animals received chronic vapor preexposure, repeated intracerebroventricular (ICV) infusions of broad spectrum MMP inhibitor, FN-439, blocked escalation of ethanol self-administration, suggesting that MMP activity is increased during initial ethanol experience and regulates subsequent drug behavior [62].

Similar to MMPs, tPA is increased in some brain areas following acute and chronic noncontingent exposure to psychostimulants, opiates, and ethanol. Specifically, a single cocaine injection increases tPA expression in the medial prefrontal cortex, accumbens and insula, and chronic exposure increases tPA expression in the hippocampus [63,64]. Increased tPA expression in the hippocampus also occurs following chronic but not acute methamphetamine [65]. Nicotine exposure during 3 days of CPP conditioning additionally increases tPA expression in the accumbens [66]. Like psychostimulants, both single and repeated exposure to morphine increases tPA in the frontal cortex, accumbens and hippocampus [67], and acute ethanol exposure increases tPA in the accumbens [68]. Furthermore, lentiviral overexpression of tPA increases cocaine-, amphetamine-, and morphine-induced locomotor behavior during initial experience [63,69], while tPA knockout mice display enhanced morphine self-administration. However, pressing under a progressive-ratio schedule for morphine is decreased in tPA knockout mice [70]. Together, these results suggest that tPA expression regulates the reinforcing properties of opiates.

PNNs and Proteins

Although most studies have focused on ethanol, expression of PNNs is increased in response to all classes of drugs of abuse, and their upregulation seems to play a role in the initial drug acquisition and conditioning, as well as future drug-related behaviors. Binge ethanol drinking increases PNNs in brain areas, such as the insula [71], and ethanol exposure increases PNN CSPG expression. Specifically, short-term ethanol exposure increases neurocan in CA1-cultured astrocytes [72], and binge drinking increases aggrecan, brevican, and phosphacan in the insula [73]. These effects seem to be long lasting, as adult mice exposed to ethanol during adolescence display increased expression of the PNN proteins, CSPGs, brevican, and neurocan, in the orbitofrontal cortex [74]. Using the chrondroitanase ABC (Ch-ABC), removal of PNNs in the dorsomedial prefrontal cortex (PFC) and anterior dorsal lateral hypothalamic area (LHAad) during the initial drug experience, can block cocaine CPP and self-administration acquisition [30,71]. Interestingly, unlike most proteins discussed in this chapter, the PNN proteoglycan, syndecan-3, has been proposed as a "resiliency factor" for cocaine addiction, as syndecan-3 null mice display enhanced self-administration and adeno-associated virus (AAV) re-expression normalizes responding for cocaine [75].

Cell Adhesion Molecules

To the authors' knowledge, the role of the ECM integrin signaling in initial drug exposure has not been directly studied. However, measurements made during extinction and withdrawal suggest that integrin subunits are increased in the accumbens during drug experience and regulate future drug behavior (described below). There is an increase in hippocampal cadherin 2 RNA expression in adults but not adolescent mice following oxycodone self-administration [76]. Although more work is needed, this study suggests that cadherins may play a role in the development of opiate addiction.

SynCAM knockout mice display increased locomotor activity during noncontingent cocaine administration; however, they do not differ in development of cocaine CPP compared with wild-type mice [77]. Therefore, SynCAM seems to be protective against responses to psychostimulants although this effect seems to be specific to locomotor behavior. Lastly, rats show increased expression of NCAM in the prefrontal cortex following cocaine exposure [78], and NCAM knockouts find ethanol, morphine, cocaine, and amphetamine less rewarding ([79,80]).

Summary and Conclusions

Noncontingent exposure to psychostimulants increases MMP-2 and -9 expression in areas such as the accumbens core and hippocampus (Fig. 2A), and noncontingent exposure to morphine increases MMP-9 expression in the midbrain (Fig. 2B). While acute ethanol exposure decreases MMP-9 expression in the hippocampus, chronic ethanol exposure is more similar to other drug classes and increases MMP activity (Fig. 2C). Together, these results suggest that MMP-9 upregulation assists

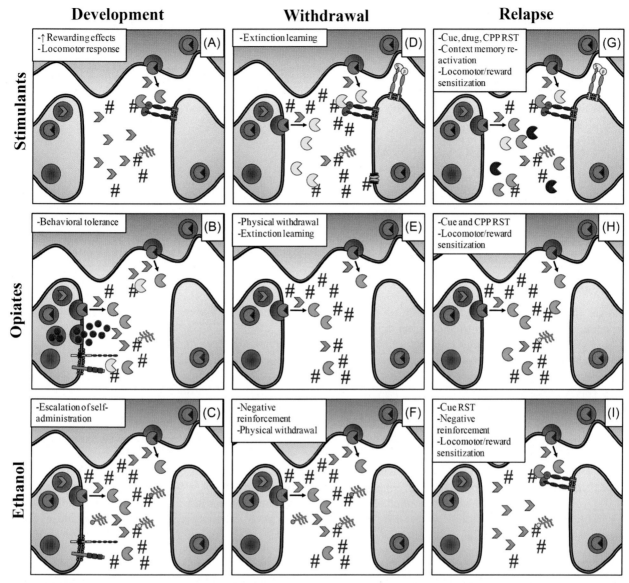

FIG. 2   Role of ECM in drug addiction. Changes in the ECM during the development of (A) stimulant, (B) opiate, and (C) ethanol addiction; withdrawal from (D) stimulants, (E) opiates, or (F) ethanol; and relapse to (G) stimulants, (H) opiates, and (I) ethanol. *Brain areas*: (A) core, HC, mPFC, insula (proteases); dmPFC, LHAad (PNNs); and PFC (CAMs); (B) midbrain, frontal cortex, accumbens, and HC (proteases); (C) HC, NAc (proteases); insula (PNNs); and OFC, HC, insula (CSPGs); (D) core (proteases); amygdala (PNNs); and core (CAMs); (E) SC (proteases); amygdala (PNNs); and mPFC, NAc (CSPGs); (F) core, HC, amygdala (proteases) and OFC (CSPGs); (G) core, mPFC, HC (proteases); amygdala, dmPFC (PNNs); and core, VTA, ventral midbrain (CAMs); (H) amygdala (PNNs); and (I) core (CAMs). See Fig. 1 for definition of symbols.

in the rewarding effects of initial drug experience. Other MMPs and tPA can activate the proteolytic action of MMPs. Therefore, it is no surprise that the patterns of tPA expression closely follow those of MMPs during initial exposure to psychostimulants, opiates, and ethanol (Fig. 2A–C). Furthermore, studies with knockout mice suggest that tPA regulates psychostimulant- and opiate-induced dopamine release in the accumbens [67,70]. Experience leads to a tolerance to opiates, which is defined as an increase in responding for a drug due to a decrease in the rewarding effects [81]. Results from behavioral studies suggest that the increase in tPA and MMPs during opiate administration plays an important role in the development of that tolerance [60,70].

Although PNN expression has been shown to increase in some brain areas and influence behavior following psychostimulant exposure, most work thus far has focused on ethanol experience. Both acute and chronic ethanol experience leads to increased expression of both PNNs and their CSPGs in cortical areas (Fig. 2C) and the increase is long lasting when ethanol exposure occurs in adolescence. Interestingly, one study showed that PNNs surrounding GABA interneurons increase immediately following nicotine exposure in both the VTA and orbitofrontal cortex (OFC), suggesting that psychostimulants remodel PNNs surrounding GABA interneuron to allow for drug-induced neuroplasticity [82]. Future studies will hopefully directly test this hypothesis and extend it to other drug classes.

CAMs, such as SynCAM and NCAM, are also increased in brain areas such as the PFC and hippocampus in response to all three drug classes discussed here, and this upregulation seems to mediate the rewarding effects of those drugs (Fig. 2A–C). MMPs proteolytically process the other ECM components, including CAMs, so it would also be of interest to look more at the interaction between MMPs and CAMs during drug experience.

ROLE OF ECM IN DRUG WITHDRAWAL

The negative effects that often occur during the attempt to abstain from drugs can play a consequential role in vulnerability for drug relapse [2]. The withdrawal stage of drug addiction is modeled using two primary approaches in animal experiments: the abstinence model (remain in home cage for a period of time following drug experience) or extinction training (placed in operant chamber but receive no drug and/or drug-associated discrete cues). Below, we review the ways in which different components of the ECM are changed or play a role during withdrawal.

Proteases

Unlike what is seen following initial drug exposure, MMPs and their activity seem to differentially regulate what is happening during withdrawal from different drug classes.

In rodents extinguished from cocaine and nicotine self-administration, MMP activity is increased in the accumbens core [83]. In the same study, there were enduring increases in MMP-2 expression compared with yoked saline controls. Acute intraaccumbens core inhibition of MMP-2 (but not MMP-9) reversed cocaine extinction-induced increases in MMP activity [83]. This suggests that MMP-2 specifically regulates the role of the ECM in psychostimulant extinction learning. Unlike rodents extinguished from psychostimulant self-administration, MMP activity in the accumbens core is unchanged in rodents extinguished from opiate self-administration [83]. However, one study suggests that MMP expression in the spinal cord regulates morphine withdrawal signs, as spinal administration of exogenous MMP-9 following chronic morphine administration increases diarrhea, paw tremor, tremor, ptosis, and wet dog shakes. Furthermore, targeted mutation of MMP-9 attenuates the behavioral signs of withdrawal [84]. Acute administration of the nonspecific MMP inhibitor, FN-439, during a short withdrawal between ethanol vapor preexposure and ethanol self-administration prevents escalation of responding [62]. This indicates the role of MMPs in negative reinforcement learning during acute withdrawal from depressants. Together, these results suggest that the accumbens core MMP-2 activity regulates psychostimulant extinction learning, spinal MMP-9 activity regulates opiate-induced withdrawal symptoms, and MMP activity in general regulates negative reinforcement learning during acute ethanol withdrawal.

Although tPA is increased following experience with all drug classes discussed in this chapter, it may be particularly important for regulating ethanol withdrawal. For example, 2 weeks into ethanol withdrawal, tPA is increased in the hippocampus and amygdala [85]. Furthermore, tPA knockout mice display a reduction in ethanol withdrawal-induced seizures [85]. Because the expression and role of tPA and MMPs are similar during initial psychostimulant and opiate experiences, we predict that tPA would most likely mimic the pattern of MMPs during withdrawal discussed above.

PNNs and Proteins

As mentioned above, PNNs are upregulated in the brain during drug experience and this increase regulates initial drug intake. Studies suggest that level of PNN expression continues to regulate behavior when the drug is removed. When the Ch-ABC PNN removal strategy is employed in the basolateral amygdala and central amygdala following drug experience, extinction training is enhanced following cocaine and morphine CPP conditioning and morphine and heroin self-administration [86]. While PNN proteoglycans are increased in the OFC in a long-term manner following short- and long-term ethanol exposure, Tenascin R, brevican and hyaloronan mRNA levels are decreased in the mPFC and accumbens core following heroin self-administration. Furthermore, this decrease seems to be a pharmacological effect of heroin exposure and withdrawal, as extinction training and forced abstinence lead to similar levels [87]. Therefore, PNN levels in the amygdala negatively regulate extinction learning following exposure to psychostimulants and opiates. Finally, opiate abstinence and extinction training decrease PNN CSPGs in the PFC and accumbens, and withdrawal from ethanol increases CSPGs in orbitofrontal cortex.

Cell Adhesion Molecules

Most work studying integrin signaling in drug addiction has focused on the accumbens. The β3 integrin subunit is increased in the accumbens core of rodents extinguished from cocaine self-administration or acute, noncontingent exposure, with no change in the β1 subunit [88,89]. Alternatively, the β1 integrin subunit (but not β3) is increased in the accumbens during withdrawal from chronic, noncontingent cocaine exposure [89]. Therefore, cocaine seems to bidirectionally regulate beta integrin subunits, depending on length of initial exposure. Lastly, polysialylated NCAM (PSA-NCAM) was decreased in the ventromedial PFC of rodents with an increased risk for alcohol-related behaviors, and depleting ventromedial PFC PSA-NCAM blocks extinction of alcohol seeking [90]. The role of other CAMs, like cadherins and SynCAM, in drug withdrawal or extinction training remains relatively unknown.

Summary and Conclusions

MMP-2, but not MMP-9, activity is upregulated in the accumbens during psychostimulant extinction training (Fig. 2D), and MMP activity is increased during ethanol withdrawal (Fig. 2F). Although there are no increases in the accumbens MMP expression during opiate extinction, both MMP activity in the spinal cord and tPA expression in the hippocampus and amygdala seem to be important in regulating opiate withdrawal symptoms (Fig. 2E). Because CAM expression is increased during initial exposure to psychostimulants, opiates, and ethanol, one would think they might regulate extinction training following self-administration. However, the only existing work has focused on changes to integrin subunit expression following psychostimulant experience and NCAM expression during ethanol extinction. These studies suggest that, in the accumbens core, the β3 subunit is transiently upregulated during acute exposure while β1 is increased only following chronic exposure (Fig. 2D), and decreased NCAM exposure in the PFC prevents extinction of ethanol-seeking behavior. The current literature suggests that PNNs and some CSPGs remain high in areas such as the PFC and amygdala to maintain drug-seeking behavior or block extinction training for psychostimulants, opiates, and ethanol (Fig. 2D–F). One study shows that during cocaine withdrawal, but not during initial experience, PNNs surrounding glutamatergic projection neurons are increased in some parts of the cerebellum, although it is unknown if this occurs in other brain areas [91]. In addition, the combination of extinction training and the Ch-ABC method of removing amygdala PNNs has been proposed as a treatment for psychostimulant and opiate relapse prevention [86].

ROLE OF ECM IN DRUG RELAPSE OR RE-EXPOSURE

Drug-associated cues and context, stress, and re-exposure to the drug itself can lead to drug relapse in humans and laboratory animals [92]. In both the self-administration and CPP paradigms, relapse is modeled in the reinstatement portion of the assay. The ECM components are changed during reinstatement, and evidence supports a direct role in reinstated behavior.

Proteases

Fifteen minutes of cocaine-, nicotine-, or heroin-associated cue reinstatement transiently increases MMP-9 activity in the accumbens core. Cocaine-primed reinstatement also increases MMP activity, presumably through MMP-9 but not MMP-2; however, this only occurs after 45 min [83]. Cocaine CPP reinstatement also increases MMP-9 expression in the medial PFC [93], but there was no change in MMP-9 or -2 following nicotine CPP [59]. Furthermore, broad spectrum pharmacological inhibition of MMP activity blocks cocaine CPP reinstatement [94]. Again, this is most likely due to MMP-9 activity blockade, as inhibition of the accumbens core MMP-9 (but not MMP-2) prevents both cocaine- and cocaine cue-induced reinstatement [83]. MMP-3 expression is increased in the hippocampus and prefrontal cortex of rodents following a nicotine CPP test. However, this same increase is seen in saline-conditioned control rats. This suggests that MMP-3 regulates the reactivation of contextual memories in general, and not specific to drug-related contextual memories [59]. Furthermore, methamphetamine locomotor and reward sensitization is prevented in MMP-9 or MMP-2 knockout mice [58]. Also, broad spectrum inhibition of MMP activity blocks heroin cue reinstatement [62]. Therefore, MMP activity regulates relapse to opiates and most psychostimulants, except nicotine. MMP-9 and -2 activities may differ between cocaine and methamphetamine. A role for MMPs in relapse to alcohol is unknown at this time.

Knocking out tPA knockout or shRNA-mediated tPA knockdown decreases locomotor sensitization and CPP to methamphetamine, morphine, and ethanol [65,67,95] and decreases CPP for nicotine [66]. Conversely, tPA overexpression increases locomotor sensitization and CPP for cocaine, amphetamine, morphine, and ethanol [63,69,95]. Therefore, tPA expression seems to play a role in relapse to most drug classes.

PNNs and Proteins

The role of PNNs in drug relapse is relatively understudied. However, the amygdala seems to be a promising brain area of study for PNN involvement in the effects of both psychostimulants and opiates. Removal of PNNs in the basolateral amygdala (BLA) and central amygdala following cocaine or morphine conditioning decreases CPP reinstatement [86]. Furthermore, PNN removal in the dorsomedial PFC after cocaine CPP conditioning impaired reconsolidation of CPP reinstatement. This seems to be selective to the dorsomedial PFC, as blockade in ventromedial PFC had no effect [30].

Cell Adhesion Molecules

Genetic deletion of the integrin β1 subunit or the β1 signaling kinase Arg augments cocaine experience-induced locomotor sensitization [96,97]. This is line with the role of the β1 subunit in locomotor sensitization (but not cocaine reward) as discussed above. Administration of the Arg-Gly-Asp (RGD) peptide antagonist blocks the ability of integrins to bind the ECM, and RGD infusion into the accumbens core inhibits cocaine cue-induced reinstatement [88]. Unpublished work in our lab shows that MMP-9 signaling in the accumbens core has its action specifically through the integrin β3 subunit and this pathway is required for the transient synaptic plasticity and cue-induced reinstatement following either heroin or cocaine self-administration (Dr. Constanza Garcia-Keller, unpublished data). Therefore, the β1 integrin subunit seems to specifically regulate locomotor sensitization to cocaine, but the β3 subunit in the accumbens regulates both psychostimulant and opiate cue reinstatement.

Cocaine CPP or locomotor sensitization decreases cadherin13 in the ventral midbrain [98], and stabilizing cadherins at dopaminergic synapses in the VTA reduces cocaine CPP [47]. Furthermore, cadherin13 knockout mice display a shift to the left of the cocaine dose-response curve and sensitize to a lower dose of cocaine [98]. These effects are consistent with cadherin13 genetic alterations in human addicts [98] and suggest that increased cadherin expression regulates psychostimulant relapse. The role of other CAMs, like SynCAM and NCAM, in drug relapse or re-exposure remains relatively unknown although it seems likely given their role in earlier drug addiction stages.

Summary and Conclusions

MMP activity regulates psychostimulant cue-induced and drug-primed reinstatement (accumbens) and CPP reinstatement (PFC) (Fig. 2G), as well as opiate cue-induced reinstatement (Fig. 2H). MMP-9 seems to be the primary regulator of this activity, while MMP-2 activity preferentially regulates psychostimulant locomotor sensitization. The role of tPA and PNN in relapse is akin to MMP activity, as blocking tPA blocks locomotor and reward sensitization for all drug classes discussed in this chapter (Fig. 2G–I). Furthermore, blocking PNNs in PFC and amygdala blocks psychostimulant and opiate reinstatement. Studies suggest that psychostimulant locomotor and reward sensitization requires cadherin expression at VTA dopamine synapses and that psychostimulant and opiate reinstatement require the signaling of MMP activity through β3 integrin signaling in the accumbens core (Fig. 2G). The role of MMP activity, PNN expression, and CAMs in ethanol relapse remains unknown.

OVERALL SUMMARY

In conclusion, there is a growing body of evidence that the components of the ECM, including proteases, PNNs, CSPGs, and CAMs, play a crucial role in the development, withdrawal/abstinence, and relapse of psychostimulants, opiates, and ethanol. As the field of addiction research progresses, it will important to focus on the tetrapartite synapse and how it adapts in response to drugs of abuse. Furthermore, uncovering how the extracellular component interacts with and regulates the components of the classical synapse could provide targets in developing treatments for human drug addiction.

REFERENCES

[1] Nestler EJ. Epigenetic mechanisms of drug addiction. Neuropharmacology 2014;76(*Pt B*):259–68. https://doi.org/10.1016/j.neuropharm.2013.04.004.

[2] Wise RA, Koob GF. The development and maintenance of drug addiction. Neuropsychopharmacology 2014;39(2):254–62. https://doi.org/10.1038/npp.2013.261.

[3] Marchant NJ, Kaganovsky K, Shaham Y, Bossert JM. Role of corticostriatal circuits in context-induced reinstatement of drug seeking. Brain Res 2015;1628(Pt A):219–32. https://doi.org/10.1016/j.brainres.2014.09.004.

[4] Beloate LN, Coolen LM. Influences of social reward experience on behavioral responses to drugs of abuse: review of shared and divergent neural plasticity mechanisms for sexual reward and drugs of abuse. Neurosci Biobehav Rev 2017;83:356–72. https://doi.org/10.1016/j.neubiorev.2017.10.024.

[5] Scofield MD, Heinsbroek JA, Gipson CD, Kupchik YM, Spencer S, Smith AC, Roberts-Wolfe D, Kalivas PW. The nucleus Accumbens: mechanisms of addiction across drug classes reflect the importance of glutamate homeostasis. Pharmacol Rev 2016;68(3):816–71. https://doi.org/10.1124/pr.116.012484.

[6] Volkow ND, Morales M. The brain on drugs: from reward to addiction. Cell 2015;162(4):712–25. https://doi.org/10.1016/j.cell.2015.07.046.

[7] Mulholland PJ, Chandler LJ, Kalivas PW. Signals from the fourth dimension regulate drug relapse. Trends Neurosci 2016;39(7):472–85. https://doi.org/10.1016/j.tins.2016.04.007.

[8] Smith AC, Scofield MD, Kalivas PW. The tetrapartite synapse: extracellular matrix remodeling contributes to corticoaccumbens plasticity underlying drug addiction. Brain Res 2015;1628(Pt A):29–39. https://doi.org/10.1016/j.brainres.2015.03.027.

[9] Wright JW, Harding JW. Contributions of matrix metalloproteinases to neural plasticity, habituation, associative learning and drug addiction. Neural Plast 2009;2009:579382. https://doi.org/10.1155/2009/579382.

[10] Soleman S, Filippov MA, Dityatev A, Fawcett JW. Targeting the neural extracellular matrix in neurological disorders. Neuroscience 2013;253:194–213. https://doi.org/10.1016/j.neuroscience.2013.08.050.

[11] Sorg BA, Berretta S, Blacktop JM, Fawcett JW, Kitagawa H, Kwok JC, Miquel M. Casting a wide net: role of Perineuronal nets in neural plasticity. J Neurosci 2016;36(45):11459–68. https://doi.org/10.1523/JNEUROSCI.2351-16.2016.

[12] Song I, Dityatev A. Crosstalk between glia, extracellular matrix and neurons. Brain Res Bull 2017. https://doi.org/10.1016/j.brainresbull.2017.03.003.

[13] Frischknecht R, Gundelfinger ED. The brain's extracellular matrix and its role in synaptic plasticity. Adv Exp Med Biol 2012;970:153–71. https://doi.org/10.1007/978-3-7091-0932-8_7.

[14] Yamada K, Nabeshima T. Pro- and anti-addictive neurotrophic factors and cytokines in psychostimulant addiction: mini review. Ann N Y Acad Sci 2004;1025:198–204. https://doi.org/10.1196/annals.1316.025.

[15] Himelstein BP, Canete-Soler R, Bernhard EJ, Dilks DW, Muschel RJ. Metalloproteinases in tumor progression: the contribution of MMP-9. Invasion Mestastasis 1994-1995;14(1–6):246–58.

[16] Ethell IM, Ethell DW. Matrix metalloproteinases in brain development and remodeling: synaptic functions and targets. J Neurosci Res 2007;85(13):2813–23. https://doi.org/10.1002/jnr.21273.

[17] Meighan SE, Meighan PC, Choudhury P, Davis CJ, Olson ML, Zornes PA, Wright JW, Harding JW. Effects of extracellular matrix-degrading proteases matrix metalloproteinases 3 and 9 on spatial learning and synaptic plasticity. J Neurochem 2006;96(5):1227–41. https://doi.org/10.1111/j.1471-4159.2005.03565.x.

[18] Nagy V, Bozdagi O, Huntley GW. The extracellular protease matrix metalloproteinase-9 is activated by inhibitory avoidance learning and required for long-term memory. Learn Mem 2007;14(10):655–64. https://doi.org/10.1101/lm.678307.

[19] Mash DC, ffrench-Mullen J, Adi N, Qin Y, Buck A, Pablo J. Gene expression in human hippocampus from cocaine abusers identifies genes which regulate extracellular matrix remodeling. PLoS One 2007;2(11):e1187. https://doi.org/10.1371/journal.pone.0001187.

[20] Kovatsi L, Batzios S, Nikolaou K, Fragou D, Njau S, Tsatsakis A, Karakiulakis G, Papakonstantinou E. Alterations in serum MMP and TIMP concentrations following chronic heroin abuse. Toxicol Mech Methods 2013;23(5):377–81. https://doi.org/10.3109/15376516.2012.758681.

[21] Samochowiec A, Grzywacz A, Kaczmarek L, Bienkowski P, Samochowiec J, Mierzejewski P, Preuss UW, Grochans E, Ciechanowicz A. Functional polymorphism of matrix metalloproteinase-9 (MMP-9) gene in alcohol dependence: family and case control study. Brain Res 2010;1327:103–6. https://doi.org/10.1016/j.brainres.2010.02.072.

[22] Sillanaukee P, Kalela A, Seppa K, Hoyhtya M, Nikkari ST. Matrix metalloproteinase-9 is elevated in serum of alcohol abusers. Eur J Clin Invest 2002;32:225–9.

[23] Verslegers M, Lemmens K, Van Hove I, Moons L. Matrix metalloproteinase-2 and -9 as promising benefactors in development, plasticity and repair of the nervous system. Prog Neurobiol 2013;105:60–78. https://doi.org/10.1016/j.pneurobio.2013.03.004.

[24] Melchor JP, Strickland S. Tissue plasminogen activator in central nervous system physiology and pathology. Thromb Haemost 2005;93(4):655–60.

[25] Werb Z. ECM and cell surface proteolysis: regulating cellular ecology. Cell 1997;97:439–42.

[26] Gualandris A, Jones TE, Strickland S, Tsirka SE. Membrane depolarization induced calcium-dependent secretion of tissue plasminogen activator. J Neurosci 1996;16(7):2220–5.

[27] Sternlicht MD, Werb Z. How matrix metalloproteinases regulate cell behavior. Annu Rev Cell Dev Biol 2001;17:463–516. https://doi.org/10.1146/annurev.cellbio.17.1.463.

[28] Calabresi P, Napolitano M, Centoanze D, Marfia GA, Gubellini P, Teule MA, Bernardi G, Frati L, Tolu M, Gulino A. Tissue plasminogen activator controls multiple forms of synaptic plasticity and memory. Eur J Neurosci 2000;12:1002–12.

[29] Samson AL, Medcalf RL. Tissue-type plasminogen activator: a multifaceted modulator of neurotransmission and synaptic plasticity. Neuron 2006;50 (5):673–8. https://doi.org/10.1016/j.neuron.2006.04.013.

[30] Slaker M, Blacktop JM, Sorg BA. Caught in the net: Perineuronal nets and addiction. Neural Plast 2016;2016:7538208. https://doi.org/10.1155/2016/7538208.

[31] Dityatev A, Bruckner G, Dityateva G, Grosche J, Kleene R, Schachner M. Activity-dependent formation and functions of chondroitin sulfate-rich extracellular matrix of perineuronal nets. Dev Neurobiol 2007;67(5):570–88. https://doi.org/10.1002/dneu.20361.

[32] Fields RD, Itoh K. Neural cell adhesion molecules in activity-dependent development and synaptic plasticity. Trends Neurosci 1996;19:473–80.

[33] Zhong X, Drgonova J, Li CY, Uhl GR. Human cell adhesion molecules: annotated functional subtypes and overrepresentation of addiction-associated genes. Ann N Y Acad Sci 2015;1349:83–95. https://doi.org/10.1111/nyas.12776.

[34] Gourley SL, Taylor JR, Koleske AJ. Cell adhesion signaling pathways: first responders to cocaine exposure? Commun Integr Biol 2011;4(1):30–3. https://doi.org/10.4161/cib.4.1.14083.

[35] Danen EH, Sonnenberg A. Integrins in regulation of tissue development and function. J Pathol 2003;200(4):471–80. https://doi.org/10.1002/path.1416.

[36] Frisch SM, Ruoslahti E. Integrins and anoikis. Curr Opin Cell Biol 1997;9:701–6.

[37] Hynes RO. Integrins: bidirectional, allosteric signaling machines. Cell 2002;110:673–87.

[38] Luo BH, Carman CV, Springer TA. Structural basis of integrin regulation and signaling. Annu Rev Immunol 2007;25:619–47.

[39] Barczyk M, Carracedo S, Gullberg D. Integrins. Cell Tissue Res 2010;339:269–80. https://doi.org/10.1007/s00441-009-0834-6.

[40] Goldbrunner RH, Bernstein JJ, Tonn JC. ECM-mediated glioma cell invasion. Microsc Res Tech 1998;43(3):250–7.

[41] Noe V, Fingleton B, Jacobs K, Crawford HC, Vermeulen S, Steelant W, Bruyneel E, Matrisian LM, Mareel M. Release of an invasion promoter E-cadherin fragment by matrilysin and stromelysin-1. J Cell Sci 2000;114(1):111–8.

[42] Saglietti L, Dequidt C, Kamieniarz K, Rousset MC, Valnegri P, Thoumine O, Beretta F, Fagni L, Choquet D, Sala C, Sheng M, Passafaro M. Extra-cellular interactions between GluR2 and N-cadherin in spine regulation. Neuron 2007;54(3):461–77. https://doi.org/10.1016/j.neuron.2007.04.012.

[43] Silverman JB, Restituito S, Lu W, Lee-Edwards L, Khatri L, Ziff EB. Synaptic anchorage of AMPA receptors by cadherins through neural plakophilin-related arm protein AMPA receptor-binding protein complexes. J Neurosci 2007;27(32):8505–16. https://doi.org/10.1523/JNEUROSCI.1395-07.2007.

[44] Tanaka H, Shan W, Phillips GR, Arndt K, Bozdagi O, Shapiro L, Huntley GW, Benson DL, Colman DR. Molecular modification of N-cadherin in response to synaptic activity. Neuron 2000;25:93–107.

[45] Brigidi GS, Sun Y, Beccano-Kelly D, Pitman K, Mobasser M, Borgland SL, Milnerwood AJ, Bamji SX. Palmitoylation of delta-catenin by DHHC5 mediates activity-induced synapse plasticity. Nat Neurosci 2014;17(4):522–32. https://doi.org/10.1038/nn.3657.

[46] Mendez P, De Roo M, Poglia L, Klauser P, Muller D. N-cadherin mediates plasticity-induced long-term spine stabilization. J Cell Biol 2010;189 (3):589–600. https://doi.org/10.1083/jcb.201003007.

[47] Mills F, Globa AK, Liu S, Cowan CM, Mobasser M, Phillips AG, Borgland SL, Bamji SX. Cadherins mediate cocaine-induced synaptic plasticity and behavioral conditioning. Nat Neurosci 2017;20(4):540–9. https://doi.org/10.1038/nn.4503.

[48] Schrick C, Fischer A, Srivastava DP, Tronson NC, Penzes P, Radulovic J. N-cadherin regulates cytoskeletally-associated IQGAP1/ERK signaling and memory formation. Neuron 2007;55(5):786–98.

[49] Biederer T, Sara Y, Mozhayeva M, Atasoy D, Liu X, Kavalali ET, Sudhof TC. SynCAM, a synaptic adhesion molecule that drives synapse assembly. Science 2002;297(5586):1525–31.

[50] Thomas LA, Akins MR, Biederer T. Expression and adhesion profiles of SynCAM molecules indicate distinct neuronal functions. J Comp Neurol 2008;510(1):47–67. https://doi.org/10.1002/cne.21773.

[51] Robbins EM, Krupp AJ, Perez de Arce K, Ghosh AK, Fogel AI, Boucard A, Südhof TC, Biederer T. SynCAM 1 adhesion dynamically regulates synapse number and impacts plasticity and learning. Neuron 2010;68(5):894–906. https://doi.org/10.1016/j.neuron.2010.11.003.

[52] Niethammer P, Delling M, Sytnyk V, Dityatev A, Fukami K, Schachner M. Cosignaling of NCAM via lipid rafts and the FGF receptor is required for neuritogenesis. J Cell Biol 2002;157(3):521–32. https://doi.org/10.1083/jcb.200109059.

[53] Luthi A, Laurent JP, Figurovt A, Mullert D, Schachner M. Hippocampal long-term potentiation and neural cell adhesion molecules L1 and NCAM. Nature 1994;372:777–9.

[54] Xiao P, Bahr BA, Staubli U, Vanderklish PW, Lynch G. Evidence that matrix recognition contributes to stabilization but not induction of LTP. Neuroreport 1991;2(8):461–4.

[55] Staubli U, Chun D, Lynch G. Time-dependent reversal of long-term potentiation by an integrin antagonist. J Neurosci 1998;18(9):3460–9.

[56] Weber M, Modemann S, Schipper P, Trauer H, Franke H, Illes P, Kleemann …, J W. Increased polysialic acid neural cell adhesion molecule expression in human hippocampus of heroin addicts. Neuroscience 2006;138(4):1215–23. https://doi.org/10.1016/j.neuroscience.2005.11.059.

[57] Everitt BJ, Robbins TW. Drug addiction: updating actions to habits to compulsions ten years on. Annu Rev Psychol 2016;67:23–50. https://doi.org/10.1146/annurev-psych-122414-033457.

[58] Mizoguchi H, Yamada K, Mouri A, Niwa M, Mizuno T, Noda Y, Nabeshima T. Role of matrix metalloproteinase and tissue inhibitor of MMP in methamphetamine-induced behavioral sensitization and reward: implications for dopamine receptor down-regulation and dopamine release. J Neurochem 2007;102(5):1548–60. https://doi.org/10.1111/j.1471-4159.2007.04623.x.

[59] Natarajan R, Harding JW, Wright JW. A role for matrix metalloproteinases in nicotine-induced conditioned place preference and relapse in ado-lescent female rats. J Exp Neurosci 2013;7:1–14. https://doi.org/10.4137/JEN.S11381This.

[60] Nakamoto K, Kawasaki S, Kobori T, Fujita-Hamabe W, Mizoguchi H, Yamada K, Nabeshima T, Tokuyama S. Involvement of matrix metalloproteinase-9 in the development of morphine tolerance. Eur J Pharmacol 2012;683(1–3):86–92. https://doi.org/10.1016/j.ejphar.2012.03.006.

[61] Wright JW, Masino AJ, Reichert JR, Turner GD, Meighan SE, Meighan PC, Harding JW. Ethanol-induced impairment of spatial memory and brain matrix metalloproteinases. Brain Res 2003;963:252–61.

[62] Smith AW, Nealey KA, Wright JW, Walker BM. Plasticity associated with escalated operant ethanol self-administration during acute withdrawal in ethanol-dependent rats requires intact matrix metalloproteinase systems. Neurobiol Learn Mem 2011;96(2):199–206. https://doi.org/10.1016/j.nlm.2011.04.011.

[63] Bahi A, Dreyer JL. Overexpression of plasminogen activators in the nucleus accumbens enhances cocaine-, amphetamine- and morphine-induced reward and behavioral sensitization. Genes Brain Behav 2008;7(2):244–56. https://doi.org/10.1111/j.1601-183X.2007.00346.x.

[64] Hashimoto K, Yasushi K, Nishikawa T. Psychomimetic-induction of tissue plasminogen activator mRNA in corticostriatal neurons in rat brain. Eur J Neurosci 1998;10:3387–99.

[65] Nagai T, Noda Y, Ishikawa K, Miyamoto Y, Yoshimura M, Ito M, Takayanagi M, Takuma K, Yamada K, Nabeshima T. The role of tissue plasminogen activator in methamphetamine-related reward and sensitization. J Neurochem 2005;92(3):660–7. https://doi.org/10.1111/j.1471-4159.2004.02903.x.

[66] Nagai T, Ito M, Nakamichi N, Mizoguchi H, Kamei H, Fukakusa A, Nabeshima T, Takuma K, Yamada K. The rewards of nicotine: regulation by tissue plasminogen activator-plasmin system through protease activated receptor-1. J Neurosci 2006;26(47):12374–83. https://doi.org/10.1523/JNEUROSCI.3139-06.2006.

[67] Nagai T, Yamada K, Yoshimura M, Ishiwaka K, Miyamoto Y, Hashimoto K, Noda Y, Nitta A, Nabeshima T. The tissue plasminogen activator-plasmin system participates in the rewarding effect of morphine by regulating dopamine release. Proc Natl Ann Sci 2004;101(10):3650–5.

[68] Bahi A, Dreyer JL. Involvement of tissue plasminogen activator "tPA" in ethanol-induced locomotor sensitization and conditioned-place preference. Behav Brain Res 2012;226(1):250–8. https://doi.org/10.1016/j.bbr.2011.09.024.

[69] Bahi A, Kusnecov AW, Dreyer JL. Effects of urokinase-type plasminogen activator in the acquisition, expression and reinstatement of cocaine-induced conditioned-place preference. Behav Brain Res 2008;191(1):17–25. https://doi.org/10.1016/j.bbr.2008.03.004.

[70] Yan Y, Yamada K, Mizoguchi H, Noda Y, Nagai T, Nitta A, Nabeshima T. Reinforcing effects of morphine are reduced in tissue plasminogen activator-knockout mice. Neuroscience 2007;146(1):50–9. https://doi.org/10.1016/j.neuroscience.2007.01.011.

[71] Blacktop JM, Todd RP, Sorg BA. Role of perineuronal nets in the anterior dorsal lateral hypothalamic area in the acquisition of cocaine-induced conditioned place preference and self-administration. Neuropharmacology 2017;118:124–36. https://doi.org/10.1016/j.neuropharm.2017.03.018.

[72] Zhang X, Bhattacharyya S, Kusumo H, Goodlett CR, Tobacman JK, Guizzetti M. Arylsulfatase B modulates neurite outgrowth via astrocyte chondroitin-4-sulfate: dysregulation by ethanol. Glia 2014;62(2):259–71. https://doi.org/10.1002/glia.22604.

[73] Chen H, He D, Lasek AW. Repeated binge drinking increases Perineuronal nets in the insular cortex. Alcohol Clin Exp Res 2015;39(10):1930–8. https://doi.org/10.1111/acer.12847.

[74] Coleman Jr. LG, Liu W, Oguz I, Styner M, Crews FT. Adolescent binge ethanol treatment alters adult brain regional volumes, cortical extracellular matrix protein and behavioral flexibility. Pharmacol Biochem Behav 2014;116:142–51. https://doi.org/10.1016/j.pbb.2013.11.021.

[75] Chen J, Repunte-Canonigo V, Kawamura T, Lefebvre C, Shin W, Howell LL, Hemby SE, Harvey BK, Califano A, Morales M, Koob GF, Sanna PP. Hypothalamic proteoglycan syndecan-3 is a novel cocaine addiction resilience factor. Nat Commun 2013;4:1955. https://doi.org/10.1038/ncomms2955.

[76] Zhang Y, Brownstein AJ, Buonora M, Niikura K, Ho A, Correa da Rosa J, Kreek MJ, Ott J. Self administration of oxycodone alters synaptic plasticity gene expression in the hippocampus differentially in male adolescent and adult mice. Neuroscience 2015;285:34–46. https://doi.org/10.1016/j.neuroscience.2014.11.013.

[77] Giza JI, Jung Y, Jeffrey RA, Neugebauer NM, Picciotto MR, Biederer T. The synaptic adhesion molecule SynCAM 1 contributes to cocaine effects on synapse structure and psychostimulant behavior. Neuropsychopharmacology 2013;38(4):628–38. https://doi.org/10.1038/npp.2012.226.

[78] Mackowiak M, Mordalska P, Dudys D, Korostynski M, Bator E, Wedzony K. Cocaine enhances ST8Siall mRNA expression and neural cell adhesion molecule polysialylation in the rat medial prefrontal cortex. Neuroscience 2011;186:21–31.

[79] Ishiguro H, Hall FS, Horiuchi Y, Sakurai T, Hishimoto A, Grumet M, Uhl GR, Onaivi ES, Arinami T. NrCAM-regulating neural systems and addiction-related behaviors. Addict Biol 2014;19(3):343–53. https://doi.org/10.1111/j.1369-1600.2012.00469.x.

[80] Ishiguro H, Liu QR, Gong JP, Hall FS, Ujike H, Morales M, Uhl RG. NrCAM in addiction vulnerability: positional cloning, drug-regulation, haplotype-specific expression, and altered drug reward in knockout mice. Neuropsychopharmacology 2006;31(3):572–84. https://doi.org/10.1038/sj.npp.1300855.

[81] Fields HL, Margolis EB. Understanding opioid reward. Trends Neurosci 2015;38(4):217–25. https://doi.org/10.1016/j.tins.2015.01.002.

[82] Vazquez-Sanroman DB, Monje RD, Bardo MT. Nicotine self-administration remodels perineuronal nets in ventral tegmental area and orbitofrontal cortex in adult male rats. Addict Biol 2017;22(6):1743–55. https://doi.org/10.1111/adb.12437.

[83] Smith AC, Kupchik YM, Scofield MD, Gipson CD, Wiggins A, Thomas CA, Kalivas PW. Synaptic plasticity mediating cocaine relapse requires matrix metalloproteinases. Nat Neurosci 2014;17(12):1655–7. https://doi.org/10.1038/nn.3846.

[84] Liu WT, Han Y, Liu YP, Song AA, Barnes B, Song XJ. Spinal matrix metalloproteinase-9 contributes to physical dependence on morphine in mice. J Neurosci 2010;30(22):7613–23. https://doi.org/10.1523/JNEUROSCI.1358-10.2010.

[85] Pawlak R, Melchor JP, Matys T, Skrzypiec AE, Strickland S. Ethanol-withdrawal seizures are controlled by tissue plasminogen activator via modulation of NR2B-containing NMDA receptors. Proc Natl Ann Sci 2005;102(2):443–8.

[86] Xue YX, Xue LF, Liu JF, He J, Deng JH, Sun SC, Han HB, Luo YX, Xu LZ, Wu P, Lu L. Depletion of perineuronal nets in the amygdala to enhance the erasure of drug memories. J Neurosci 2014;34(19):6647–58. https://doi.org/10.1523/JNEUROSCI.5390-13.2014.

[87] Van den Oever MC, Lubbers BR, Goriounova NA, Li KW, Van der Schors RC, Loos M, Riga D, Wiskerke J, Binnekade R, Stegeman M, Schoffelmeer ANM, Mansvelder HD, Smit AB, De Vries TJ, Spijker S. Extracellular matrix plasticity and GABAergic inhibition of prefrontal cortex pyramidal cells facilitates relapse to heroin seeking. Neuropsychopharmacology 2010;35(10):2120–33. https://doi.org/10.1038/npp.2010.90.

[88] Wiggins A, Smith RJ, Shen HW, Kalivas PW. Integrins modulate relapse to cocaine-seeking. J Neurosci 2011;31(45):16177–84. https://doi.org/10.1523/JNEUROSCI.3816-11.2011.

[89] Wiggins AT, Pacchioni AM, Kalivas PW. Integrin expression is altered after acute and chronic cocaine. Neurosci Lett 2009;450(3):321–3. https://doi.org/10.1016/j.neulet.2008.12.006.

[90] Barker JM, Torregrossa MM, Taylor JR. Low prefrontal PSA-NCAM confers risk for alcoholism-related behavior. Nat Neurosci 2012;15 (10):1356–8. https://doi.org/10.1038/nn.3194.

[91] Vazquez-Sanroman D, Leto K, Cerezo-Garcia M, Carbo-Gas M, Sanchis-Segura C, Carulli D, Rossi F, Miquel M. The cerebellum on cocaine: plasticity and metaplasticity. Addict Biol 2015;20(5):941–55. https://doi.org/10.1111/adb.12223.

[92] Venniro M, Caprioli D, Shaham Y. Animal models of drug relapse and craving: From drug priming-induced reinstatement to incubation of craving after voluntary abstinence. Prog Brain Res 2016;224:25–52. https://doi.org/10.1016/bs.pbr.2015.08.004.

[93] Brown TE, Forquer MR, Harding JW, Wright JW, Sorg BA. Increase in matrix metalloproteinase-9 levels in the rat medial prefrontal cortex after cocaine reinstatement of conditioned place preference. Synapse 2008;62(12):886–9.

[94] Brown TE, Forquer MR, Cocking DL, Jansen HT, Harding JW, Sorg BA. Role of matrix metalloproteinases in the acquisition and reconsolidation of cocaine-induced conditioned place preference. Learn Mem 2007;14(3):214–23. https://doi.org/10.1101/lm.476207.

[95] Bahi A, Dreyer JL. Involvement of nucleus accumbens dopamine D1 receptors in ethanol drinking, ethanol-induced conditioned place preference, and ethanol-induced psychomotor sensitization in mice. Psychopharmacology (Berl) 2012;222(1):141–53. https://doi.org/10.1007/s00213-011-2630-8.

[96] Gourley SL, Olevska A, Warren MS, Taylor JR, Koleske AJ. Arg kinase regulates prefrontal dendritic spine refinement and cocaine-induced plasticity. J Neurosci 2012;32(7):2314–23. https://doi.org/10.1523/JNEUROSCI.2730-11.2012.

[97] Warren MS, Bradley WD, Gourley SL, Lin YC, Simpson MA, Reichardt LF, Greer CA, Taylor JR, Koleske AJ. Integrin beta1 signals through Arg to regulate postnatal dendritic arborization, synapse density, and behavior. J Neurosci 2012;32(8):2824–34. https://doi.org/10.1523/JNEUROSCI.3942-11.2012.

[98] Uhl GR, Drgonova J, Hall FS. Curious cases: altered dose-response relationships in addiction genetics. Pharmacol Ther 2014;141(3):335–46. https://doi.org/10.1016/j.pharmthera.2013.10.013.

Chapter 18

Striatal Cell-Type Specific Plasticity in Addiction

Michel Engeln*,a, T. Chase Francis†,‡,a, and Mary Kay Lobo*

*Department of Anatomy and Neurobiology, University of Maryland School of Medicine, Baltimore, MD, United States, †Intramural Research Program, National Institute on Drug Abuse, US National Institutes of Health, Baltimore, MD, United States, ‡Biomedical Research Center, US National Institutes of Health, Baltimore, MD, United States

INTRODUCTION

The striatum, a key brain region in the basal ganglia network, receives dense dopaminergic projections from the substantia nigra *pars compacta* (SNc) and the ventral tegmental area (VTA). These inputs make up the nigrostriatal and mesolimbic circuit, which are involved in motor versus limbic functions, and are critical to different aspects of substance abuse. The ventral striatum, or nucleus accumbens (NAc), is divided in two compartments known as shell and core. The shell mediates the reinforcing effects of drugs and has a role in the development of addiction [1], while the core supports acquired drug-related behaviors and their long-term execution [2]. The dorsal striatum (DStr) is involved in motor aspects of reward but its activity is modulated by reward expectation [3]. The DStr is thus considered to be involved in habitual responding for drug [4]. Therefore, drug addiction relies on a transitional recruitment of the ventral to the DStr [5], possibly through spiraling pathways [6], where the drug progressively loses its rewarding properties leading to habitual drug-seeking behavior [7].

The major projection neurons (>95%) of the striatum are medium spiny neurons (MSNs), which are differentiated by their dopamine (DA) receptor expression, either dopamine 1 (D1) receptor or dopamine 2 (D2) receptor. Anatomically, overlap between these two neuron subtypes and their projections is minimal [8]. MSNs also differ in their receptor (R) and peptide expression, where D1-MSNs are enriched in muscarinic 4 receptor expression, dynorphin, and substance P, while D2-MSNs have enrichment of G-protein-coupled receptor (GPCR) 6 and enkephalin. In the DStr, MSN projections are segregated into a direct pathway, projecting to the globus pallidus internal (GPi) and substantia nigra *pars reticulata* (SNr), and an indirect pathway, projecting to the globus pallidus external (GPe), with very little overlap in projections. In the NAc, MSN projections are not as segregated. D1-MSNs project to the GPi and midbrain structures, SNr and VTA, as well as the ventral pallidum (VP) and lateral hypothalamus. D2-MSNs project to the VP. These projections underlie the differential roles for MSN subtypes functionally through distinct circuits, which underlies differential behavioral output (for review, see [9]).

Consistent with their differential projections in the basal ganglia, D1- and D2-MSNs seem to have different roles in drug-related behaviors. Studies using cell-type specific approaches demonstrate that optogenetic stimulation of D1-MSNs enhances the rewarding properties of cocaine and morphine, while D2-MSN stimulation has the opposite effect [10–12]. Conversely, optogenetic or chemogenetic inhibition as well as transmission-blocking techniques in D1-MSNs reduces while chemogenetic inhibition of D2-MSNs increases psychostimulant-induced behaviors [13–17].

PSYCHOSTIMULANTS ABUSE

Physiology and Synaptic Plasticity

Nearly all drugs of abuse directly or indirectly alter DA release throughout the brain [18]. Cocaine directly enhances DA levels within the brain by acting on the DA transporter to prevent reuptake of DA. DA receptors are coupled with G-proteins and can oppositely regulate MSNs activity. D1-R activation stimulates adenylyl cyclase through $G_{s/olf}$ protein, increasing

---

a. Contributed equally.

Copyright © 2019 Elsevier Inc. All rights reserved.

cAMP function, causing D1-MSNs to activate their output structures. Conversely, D2-R activation inhibits adenylyl cyclase and cAMP formation through G_i protein, causing D2-MSNs to inhibit their output structures. DA receptor signaling alters excitatory and inhibitory plasticity on MSNs, and the psychomotor effects of cocaine strongly depend on altering MSN activity. Cocaine promotes dissociable effects on MSN subtypes. In general, enhancing the activity of D1-MSNs drives cocaine seeking and conditioned reinforcement, while stimulation of D2-MSNs attenuates both of these behaviors [11–16] (Fig. 1). Coordinated DA receptor activity and excitatory input drive cocaine-mediated alterations in MSN-subtype activity through long-term plasticity changes in MSNs [19]. Effects of cocaine on excitatory plasticity can vary by administration procedures and inputs. Most studies demonstrate repeated cocaine potentiates global excitatory transmission on D1-MSNs and does not affect transmission on D2-MSNs [13,20–22]. In fact, a single injection of cocaine is sufficient to occlude long-term potentiation (LTP) on D1-MSNs, but not D2-MSNs, for up to 1 week following treatment [21]. Potentiation within the striatum requires NMDA-R activation and D1-R activation in D1-MSNs [19], suggesting enhanced DA within the striatum plays a role in driving excitatory plasticity.

Optogenetic-mediated de-potentiation of prefrontocortical excitatory inputs to D1-MSNs is sufficient to restore locomotor behavior from the sensitized state in mice treated with cocaine noncontingently [21]. In mice that self-administer cocaine, a 1 Hz long-term depression (LTD) of ventral hippocampus (vHipp) or 13 Hz of medial prefrontal cortex (mPFC) inputs was sufficient to block seeking for cocaine by reversing α-amino-3-hydroxy-5-methyl-4-isoxazolepropionic acid (AMPA)-R/N-methyl-D-aspartate (NMDA)-R enhancements or inward rectification by GluR2-lacking AMPA-R insertion, respectively [23]. The increase in GluR2-lacking AMPA-Rs correlates with increased cocaine craving [24], occurs at silent synapses [20,25], and can occur on mPFC-D1-MSN synapses at normal cocaine concentrations and amygdala-D2-MSNs at higher cocaine concentrations [22]. High-frequency stimulation of putative D1-MSNs (i.e., the NAc-VTA projecting neurons) attenuates locomotor sensitization and occludes conditioned place preference (CPP), suggesting downstream D1-MSN signaling promotes cocaine-mediated reward [26]. Furthermore, blocking cocaine-mediated potentiation of D1-MSN inhibitory transmission, within the VP, by 1 Hz LTD restored locomotor sensitization [27]. These results strongly suggest that the NAc D1-MSNs promote rewarding effects of cocaine via excitatory transmission that drives downstream signaling within the NAc target regions.

Coordinated activity and individual differences between MSN subtypes play a crucial role in the expression of cocaine seeking. Lateral inhibition of D1-MSNs by D2-MSNs is suppressed following cocaine administration via a D2-R mechanism and is required for locomotor sensitization [28], suggesting that disinhibition of D1-MSNs plays a role in mediating behavioral outcomes to cocaine. D2-MSN activity on the VP projections suggests a different role for these neurons with cocaine exposure. Combined measurements of perseverance and motivation for cocaine seeking after extended access cocaine negatively correlate with AMPA-R/NMDA-R ratios in D2-MSNs, but not D1-MSNs. Consistent with this, chemogenetic inhibition of D2-MSNs enhanced motivation to obtain cocaine [13]. A similar effect is observed downstream

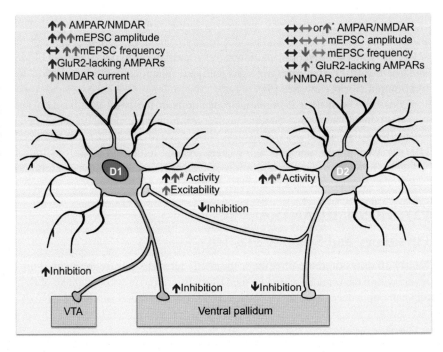

FIG. 1 Alterations in electrophysiological properties of medium spiny neuron (MSN) subtypes during withdrawal from drugs of abuse: when examining of global excitation, inhibition, or activity, striatal D1-MSNs display increased excitation and decreased inhibition which facilitates output, while D2-MSNs exhibit decreased activity or excitation, or no changes. *: Thalamic inputs and #: c-fos labeling for putative activity changes. *Red arrow*: cocaine; *blue arrow*: opiates; and *green arrow*: alcohol.

of the NAc, since restoring D2-MSN inhibitory transmission in the VP by 10 Hz LTP caused recovery from motivational deficits [27]. Furthermore, optogenetic inhibition of the NAc to VP afferents was sufficient to block reinstatement to cocaine seeking in rats [29]. Cocaine specifically blocks normal inhibitory LTD in the D2-MSN to VP synapses [16,27]. Together, these results imply the D2-MSN to VP activity drives motivational changes to cocaine administration through D1-MSN disinhibition of local collaterals and downstream VP.

MSN-subtype dendritic morphology changes, produced by cocaine, often correlate to changes in excitatory plasticity. Cocaine significantly enhances spine density of both the NAc MSN subtypes 2 days following repeated cocaine administration [30]. Furthermore, 1 day following repeated administration of cocaine, AMPA-R/NMDA-R ratios remain unchanged in the NAc MSN subtypes, but miniature excitatory postsynaptic currents (mEPSC) frequency is significantly enhanced in the NAc D1-MSNs and decreased in D2-MSNs. At this same time point, only density of D1-MSN spines was enhanced [31]. Furthermore, silent synapses are formed on D1-MSNs, but not on D2-MSNs, early after cocaine. However, after 21 or 30 days of withdrawal, only D1-MSNs display enhanced spine density [30,32], and synapses become "unsilenced" after extended withdrawal [20], likely through GluR2-lacking AMPA-R insertion.

Differential activity of MSN subtypes drives or suppresses cocaine reward. Optogenetic activation of the NAc D1-MSNs enhances CPP to cocaine and D2-MSN activation significantly attenuates CPP [11]. New in vivo techniques have shed light on activity of MSNs during cocaine administration and cocaine-motivated tasks. In the DStr, MSN subtypes show similar in vivo calcium activity dynamics and neuronal activity clustering in response to cocaine injections and locomotor activity [33]. In the NAc, and in accordance with electrophysiological and optogenetic findings, activity of MSN subtypes dichotomizes. Confirming early pharmacological work on DA action on D1 versus D2 receptors [34,35], in vivo calcium activity increased in D1-MSNs and decreased in D2-MSNs following a single cocaine injection. Following cocaine CPP, D1-MSNs display enhanced calcium transients to the cocaine-paired chamber, while D2-MSNs display diminished activity [36].

Molecular Adaptations

Activity of MSN subtypes occurs through differential molecular changes, which drives opposing functional and behavioral outcomes (Fig. 2). It is likely the effects on overall MSN-subtype activity [36] are mediated by enhanced signaling on sparsely overlapping DA receptors found on the NAc MSN subtypes, among other cell-type-specific modulatory

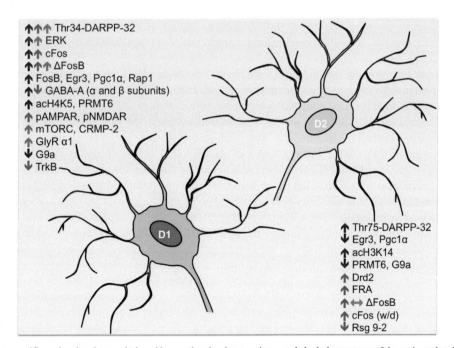

FIG. 2 MSN subtype-specific molecular changes induced by psychostimulants, opiates, and alcohol: summary of the main molecular changes occurring D1- and D2-MSNs after drug exposure. All three classes of drugs alter the expression of immediate early genes, epigenetic marks, and signaling molecules in D1-MSNs. Some of these genes are oppositely regulated in D2-MSNs. *Red arrow:* cocaine; *blue arrow:* opiates; *green arrow:* alcohol; and *w/d:* changes observed during withdrawal.

mechanisms. DA D1 and D2 receptors oppositely stimulate adenylyl cyclase and cAMP production mediating protein kinase A (PKA) activity. Cocaine increases PKA phosphorylation of the DA- and cyclic AMP-regulated phosphoprotein of 32 kDa (DARPP-32) at Thr34 and reduces phosphorylation at Thr75 in D1-MSNs while having the opposite effect in D2-MSNs. Increased phosphorylation of DARPP-32 at Thr34 inhibits protein phosphatase 1 (PP-1), further impacting the expression of various genes to support the response to cocaine [37]. The deletion of DARPP-32 from D1-MSNs decreases cocaine-induced locomotion, while its deletion from D2-MSNs has the opposite effect [38]. Cocaine also modulates extracellular signal-regulated kinase (ERK) phosphorylation by DARPP-32 in D1-MSNs [39]. Thus, acute cocaine administration rapidly leads to ERK expression in D1-MSNs, further regulating various neuronal functions underlying cocaine reward [40]. These observations are supported by work showing that the D1-R knock-out (KO) has decreased expression of ERK1/2 in the NAc after D-amphetamine injection [41]. Repeated cocaine administration induces pERK, and leads to a downregulation of its direct nuclear target mitogen- and stress-activated kinase 1 (pMSK1) in both the DStr and NAc D1-MSNs, likely accounting for the rewarding effect of cocaine [42,43]. However, activation of DA-R signaling can also be independent of cAMP. For instance, D2-Rs activate protein kinase C (PKC) in the NAc shell leading to reduced AMPA-R signaling and decreased excitatory synaptic efficacy to promote cocaine seeking [44]. Moreover TrkB, the brain-derived neurotrophic factor (BDNF) receptor, activates ERK signaling and TrkB deletion from D1-MSNs increases cocaine CPP, while having the opposite effect in D2-MSNs [11]. The activation of ERK further modulates the expression of multiple transcription factors regulating long-lasting neuronal changes. The expression of immediate early genes such as cFos or zif268 (Egr1) is increased in the NAc D1-MSNs following repeated cocaine injection [42], which is supported by in vivo calcium imaging [36]. Conversely, fluorescence-activated cell sorting (FACS) of cFos-expressing striatal neurons show enriched prodynorphin expression (preferentially found in D1-MSNs) in this population after sensitization to cocaine [45]. Acute cocaine injection leads to FosB expression in striatal D1-MSNs [46], while repeated cocaine injections induce the accumulation of its truncated splice variant, ΔFosB, in D1-MSNs only [47]. Overexpressing ΔFosB in D1-MSNs enhances behavioral response to cocaine associated with increased stubby spine density and altered excitatory synaptic function. These effects are not found when ΔFosB is overexpressed in D2-MSNs [48]. Work from our group recently used the RiboTag methodology to investigate the cell-type-specific mRNA levels of the transcription factor Egr3 after cocaine. We investigated Egr3 since this transcription factor is a key target of cocaine-mediated signaling molecules such as DA and BDNF. In the NAc, Egr3 expression was increased in D1- and decreased in D2-MSNs following repeated cocaine exposure. These opposite changes were confirmed by showing that Egr3 knock down in D1-MSNs and overexpression in D2-MSNs both lead to increased cocaine locomotor sensitization and CPP [49]. In addition, Egr3 regulates the expression of the transcriptional coactivator peroxisome proliferator-activated receptor gamma coactivator (PGC)-1α. We found that PGC-1α was increased in the NAc D1-MSNs and reduced in D2-MSNs after cocaine treatment. Overexpression of PGC-1α in D1-MSNs increased locomotor sensitization and CPP for cocaine while its overexpression in D2-MSNs had the opposite effect [50].

Cocaine alters dendritic and spine morphology (see Chapter 10). Molecules involved in neuronal outgrowth also show cell-type-specific alterations following cocaine exposure. A protein involved in cellular morphology, Tiam1, is downregulated following cocaine injections or repeated optogenetic stimulation of D1-MSNs [14]. Optogenetic inhibition of D1-MSNs blunted cocaine-induced locomotor sensitization and reversed the downregulation of Tiam 1. In addition silencing p11, a protein regulating receptor and neurotransmitter transport to the plasma membrane, in the NAc D1-MSNs increases cocaine CPP [51]. However, knocking out WAVE1, a protein involved in de novo actin polymerization, from D1-MSNs (but not from D2-MSNs) reduces cocaine-induced CPP and locomotor sensitization. WAVE1 KO in D1-MSNs is associated with spine alterations and decreased glutamatergic function, suggesting that WAVE1 is involved in cocaine-induced synaptic plasticity in D1-MSNs [52]. In addition, thin, immature spines have been associated with bidirectional alteration of small GTPase Rap1b signaling after cocaine, supporting synaptic reconfiguration after repeated drug exposure [53]. Knockout of Rap1 in the NAc D1-MSNs decreases neuronal firing rate and behavioral response to cocaine, while activation of Rap1 in these cells produces the opposite effect [54]. Moreover, mutation of NMDA-R obligatory subunit GluN1 in D1-MSNs blocks cocaine-induced locomotor sensitization without altering normal ambulation [55]. Similarly, while genetic deletion of GluN1 from either D1- or D2-MSNs has no effect on cocaine CPP acquisition and extinction, its deletion from D1-MSN exclusively alters cocaine-primed reinstatement of CPP, suggesting that NMDA-R signaling in D1-MSNs is involved in relapse [56]. Altogether these data suggest that cocaine induces profound transcriptional changes in D1-MSNs altering key neuronal functions such as neurotransmitter release, receptor signaling, dendritic spine formation that ultimately lead to altered synaptic plasticity, and behavioral response to the drug.

Along with glutamate-related alterations, cocaine also modifies GABAergic function. Using a translating ribosome affinity purification (TRAP) methodology to assess gene expression in each MSN subtype, a study found that chronic cocaine injections increased the expression of the GABA-A α3, α4, and β3 subunits in D1-MSNs associated with increased miniature inhibitory postsynaptic current (mIPSC) frequency [57]. Another study showed, however, that deletion of the GABA-Aα4 subunit from D1-MSNs increases cocaine CPP, while its ablation from D2-MSNs has no effect on cocaine-mediated behaviors [58].

Finally, a growing number of studies are focusing on epigenetic changes elicited by psychostimulants. Among them, only a few have assessed these changes in a cell-type-specific manner. For instance, both acute and chronic cocaine increase phosphorylation of histone 3 on Ser-10 in striatal D1-MSNs, likely through ERK activation [42]. Both acute and chronic cocaine increase acetylation of H4K$_5$ in D1-MSNs with only transient effects observed in D2-MSNs. In contrast, acetylation on H3K$_{14}$ is transient in D1-MSNs and is only elicited by chronic treatment in D2-MSNs [59]. Recent work showed that protein arginine methyltransferase-6 (PRMT6) expression was decreased in the NAc of rodent and humans after cocaine exposure. Cell-type-specific assessment revealed that PRMT6 was increased in D1-MSNs and decreased in D2-MSNs. Surprisingly, overexpressing PRMT6 in D1-MSNs decreases cocaine CPP while this manipulation in D2-MSNs increases it. Cocaine also decreased PRMT6-associated histone mark H3R2me2a, leading to reduced transcriptional repression on target genes such as Scrin1 and overexpression of Scrin1 in D2-MSNs decreased cocaine CPP. Together, epigenetic mechanisms in D2-MSNs could constitute a homeostatic response necessary to oppose cocaine's behavioral effects [60]. In addition, cell-type-specific mRNA studies using the TRAP methodology found that the repressive histone methyltransferase G9a was decreased in striatal D1- and D2-MSNs following repeated cocaine. Interestingly, knocking out G9a from D2-MSNs increased cocaine-induced locomotor sensitization and CPP while G9a knockout from D1-MSNs had the opposite effect. Increased sensitivity to cocaine effects after G9a KO from D2-MSNs was associated with molecular alterations including partial phenotypic switch, making D2-MSNs more similar to D1-MSNs [61]. Focusing on the NAc, our group showed recently that chronic cocaine decreases G9a expression in D1-MSNs only, likely through decreased binding of Egr3 on the G9a promoter [49]. Moreover, recent work proposed that cocaine alters the three-dimensional (3D) genome organization. Repeated cocaine exposure induces chromatin alterations, including DNA methylation, at two gene loci Auts2 and Caln1 in D2-MSNs selectively. Increased DNA methylation prevents the binding of the chromosomal scaffolding protein, CCCTC-binding factor (CTCF), and lead to opening of the Auts2-Caln1 loop as well as histone methylation to promote the expression of these two genes. The resulting induction of the Auts2 and Caln1 in D2-MSNs augments response to reward [62]. Finally, in addition to transcriptional studies, posttranscriptional modifications are being assessed in cocaine addiction. Ablation of Argonaute 2 (Ago2), a protein involved in RNA interference through miRNA-mediated gene silencing, from D2-MSNs decreases cocaine self-administration. Moreover, in D2-MSNs, Ago2 regulates the expression of many other miRNAs following acute cocaine, suggesting that posttranscriptional modifications play an important role in altered gene expression associated with cocaine addiction [63].

OPIATE ABUSE

Physiology and Synaptic Plasticity

Excitatory transmission on MSN subtypes is altered through mu-opioid receptor (MOR) modulation, particularly on presynaptic sites, driving long-term changes observed presynaptically and postsynaptically (Fig. 1). As with other drugs discussed, differential excitatory plasticity is observed on MSN subtypes, mostly explored within the ventromedial NAc shell. Similar effects occur despite the route of administration, but rely on the length of administration and the time in which effects are observed after withdrawal.

Spontaneous withdrawal from repeated morphine exposure significantly enhances AMPA-R/NMDA-R ratios and mEPSC amplitude and frequency in D1-MSNs, while in D2-MSNs, frequency is diminished [64]. Repeated self-administration of remifentanil attenuated [D-Ala2, N-MePhe4, Gly-ol]-enkephalin (DAMGO)-mediated inhibition of mEPSC frequency on D1-MSNs, but not D2-MSNs [65]. Optogenetic activation of the NAc D1-MSNs, but not D2-MSNs, accelerates morphine tolerance [66]. Moreover, activation of D1-MSNs enhances morphine reward, while activation of D2-MSNs has the opposite effect. Interestingly, knockout of TrkB receptors from D1-MSNs significantly attenuates spontaneous IPSC (sIPSC) amplitude on D1-MSNs, mimicking a reduction of sIPSCs by morphine exposure and promoting morphine CPP [10]. These results suggest repeated opioid exposure enhances global excitation of D1-MSNs and potentially D1-MSN output via various mechanisms. However, input specificity to MSN subtypes often reveals a different story.

Optogenetic stimulation has provided some clue as to what inputs drive opioid-mediated alterations on these subtypes. In particular, the thalamus provides a strong aversive stimulus to the NAc. De-potentiation or inhibition of paraventricular thalamic input on D2-MSNs is sufficient to suppress morphine withdrawal symptoms and place aversion induced by withdrawal [67]. Thalamostriatal synapses are robustly depressed by DAMGO. Treatment with oxycodone, a potent opioid drug and MOR-agonist, may occlude endocannabinoid LTD by DAMGO, suggesting opioid drugs may act on thalamostriatal synapses to inhibit input to the striatum [68]. Naloxone precipitated withdrawal in morphine-dependent animals promotes cFos expression in all the NAc MSNs, with a greater effect in D2-MSNs [69]. These results suggest D2-MSN output, driven by thalamic input, promotes withdrawal symptoms. Withdrawal from morphine significantly enhances the number of silent synapses on D2-MSNs following 1 day of withdrawal, which are likely eliminated over extended withdrawal via synaptic pruning [20]. Acute morphine promotes the opposite change, enhancing cFos in putative D1-MSNs [69]. Therefore, the incubation time for abstinence and length of administration is important in understanding the effects on subpopulations of striatal cells. Future studies are needed to understand large-scale excitatory changes on D1-MSNs following morphine withdrawal, at varying time points of withdrawal, to determine input specificity driving observed changes in excitatory input.

Molecular Adaptations

The opioidergic and dopaminergic systems closely interact and both D1- and D2-MSNs express MORs [70]. Thus, various treatment regimens or manipulations lead to alterations in D1-R and D2-R expression and function (Fig. 2). For instance in the NAc, Drd2 protein increases after 7 days of morphine treatment, while Drd1 protein levels are unchanged [71]. Restoring MOR in striatal D1-MSNs of MOR KO mice restores opiate reward with no effect on analgesia or withdrawal [70]. In addition, other receptors might be involved in opiate reward or withdrawal. Knocking out the adenosine 2a receptor (A2A-R), selectively expressed in D2-MSNs, has deleterious effect on rewarding properties of morphine. In A2A KO mice, while naloxone fails to induce place aversion in morphine-dependent animals [72], morphine CPP is abolished and morphine self-administration as well as break point are decreased [73]. At the molecular level, striatal A2A activation increases cAMP during morphine withdrawal. This increase allows PKA activation, leading to increased activator of G-protein 3 (AGS3) expression. In turn, AGS3 binds to $G\alpha_i$-guanosine diphosphate (GDP) to inactivate $G\alpha_i$ signaling to further increase cAMP. High cAMP/PKA levels activate PKC/phospholipase C (PLC) signaling and stimulate adenylyl cyclase 5 and 7, producing cAMP superactivation [74]. Along with D1-R and D2-R, A2A-R interacts with DARPP-32. Interestingly, acute morphine increases the state of phosphorylation of DARPP-32 at Thr34, which can be blocked by D1-R antagonism. These effects are restricted to the primary effects of morphine as DARPP-32 KO and T34A DARPP-32 mutant mice show low acute hyperlocomotor effect of morphine but are able to develop locomotor sensitization and CPP [75]. In later phases of morphine action (2 h post morphine), PKA activation by D1-R stimulation increases AMPA-R and NMDA-R phosphorylation leading to synaptic potentiation and Ca^{2+}-dependent events that dephosphorylate Thr34-DARPP-32. Increased glutamatergic function through mGluR5 receptors activates Cdk5 and phosphorylates DARPP-32 at Thr75. This, in turn, reduces PKA activity and D1-R-dependent phosphorylation events that together might support long-term adaptation to morphine [76]. Finally, after chronic morphine, the AMPA-R subunit GluR1 is decreased at the membrane of the NAc shell D1-MSNs and on putative D2-MSNs of the NAc core [77].

In addition to modifications in DA-R function and related signaling pathways, opiates also activate transcription factors. Fos-related antigens (FRA) expression is increased in the NAc D2-MSNs following naltrexone-precipitated withdrawal [78]. Similar results are seen with cFos expression after a single morphine injection in D1-MSNs of both the NAc core and shell. After naloxone-precipitated morphine withdrawal, while cFos is activated in both the NAc D1- and D2-MSNs, its expression is predominant in D2-MSNs [69]. In addition, morphine subcutaneous implant and heroin self-administration increases ΔFosB expression in both the NAc and Dstr D1- and D2-MSNs [47]. Interestingly, rewarding properties and physical dependence of morphine rely on ΔFosB-mediated inhibition of dynorphin, a neuropeptide mostly released by D1-MSNs [79]. Moreover, other molecular alterations targeting GPCRs are associated with opiate abuse. The expression of the Rsg9-2 gene is drastically decreased in the NAc D2-MSNs following chronic morphine. When overexpressed in the NAc, Rsg9-2 blocks morphine CPP and reduces physical signs of withdrawal, likely through an effect on D2-MSNs [66]. In addition, in cultured striatal D1-MSNs, MORs regulate Ras signaling pathway through neurofibromin 1 (NF1), a direct effector of GPCR subunit Gβγ. Activation of Ras pathway could then underlie the long-lasting neuronal changes induced by morphine and support sensitization of psychomotor effect of the drug [80]. Finally, chronic morphine reduces the expression of the TrkB receptor in the NAc D1-MSNs and is accompanied by a downregulation of various GABA-A subunits, suggesting that morphine reward relies on altered GABA-A response and TrkB signaling. Knocking out TrkB receptor from D1-MSNs recapitulates these findings with enhanced morphine reward, reduced sIPSCs amplitude, and decreased GABA-A subunit expression [10].

Although some data suggest that D1-MSNs support the positive rewarding properties of opiates while D2-MSNs are involved in the negative properties of opiate withdrawal, recent findings suggest that this dichotomy might be less strict than previously thought. Information on cell-type-specific alterations is needed to delineate D1- and D2-MSNs respective role in opiate intake and withdrawal.

ALCOHOL ABUSE

Physiology and Synaptic Plasticity

Ethanol administration also alters plasticity on MSN subtypes (Fig. 1). Acute administration of ethanol on striatal slices impairs NMDA-R-mediated potentiation of population spikes and promotes LTD at higher ethanol concentrations, which is dependent on D2-R activation and endocannabinoid signaling [81]. Chronic intermittent ethanol (CIE) vapor exposure resulted in potentiation in excitatory transmission in the NAc D1-MSNs in a protocol that normally results in LTD [82]. Similarly, in the dorsomedial striatum D1-MSNs, alcohol consumption significantly enhances mEPSC amplitude and frequency as well as AMPA-mediated current by enhancing dendritic branching and spine density [83]. Interestingly, excitatory transmission on D2-MSNs displayed depression with the same protocol, which normally produces no effect [82]. This effect appears to disappear after 2 weeks following CIE [84]. CIE also promotes increased sEPSC frequency in D1-MSNs, increased inward rectifying AMPA-Rs, and an increased NMDA-R/AMPA-R ratio [82], suggesting baseline effects of repeated ethanol administration occur on D1-MSNs and NMDA-R-mediated plasticity is altered on both MSN subtypes.

Molecular Adaptations

Similar to psychostimulants and opiates, alcohol increases DA release in the NAc and activates the reward circuit. Alcohol increases adenosine levels and, interestingly, most of the brain's A2A-Rs are present in the striatum on D2-MSNs. Studies in cell culture have shown that activation of A2A-R increases cAMP levels leading to activation of PKA which, in turn, increases gene transcription. Moreover, A2A-R and D2-R act synergistically through $G_{i/o}$ βγ dimers when exposed to ethanol [85]. Thus, both A2A-R and D2-R antagonists attenuate ethanol self-administration in rats, suggesting that D2-MSNs could play an important role in alcohol abuse [86]. Work evaluating striatal DA-R function in humans showed that alcohol-dependent subjects have a reduction in D2-R receptor availability, implying a role for D2-MSNs in alcohol abuse [87]. In rats, overexpression of D2-R cDNA in the NAc of animals previously trained to self-administer cocaine decreased alcohol preference and intake. This effect progressively returned to basal, high levels of intake when D2-R overexpression fades out, suggesting that high expression levels of D2-R protects against alcohol abuse [88]. However, recent work overexpressing the NAc D2-R preferentially in D2-MSNs had no effect on alcohol consumption, suggesting that D2-R alterations in alcohol dependence may not be confined to the indirect pathway [89]. Knocking out D1-Rs decreases ethanol consumption without altering water intake in D1-null mice, illustrating an important role of D1-Rs in motivation for ethanol [90]. Indeed, a single exposure to ethanol can produce a long-lasting increase of excitatory transmission onto D1-MSNs in the NAc shell through mTORC1 activation [91]. Interestingly, it has recently been shown that mTORC1 activation increases the phosphorylation of the microtubule-associated protein Collapsin Response Mediator Protein-2 (CRMP-2) in the NAc [92]. These mechanisms could likely support the increased dendritic complexity observed in striatal D1- but not in D2-MSNs following alcohol intake [83]. In addition, ethanol consumption increases glycine receptor (Gly-R) subunit α1 expression in the NAc D1-MSNs and glycine receptor-mediated tonic currents, suggesting that the modulation of the excitability of D1-MSNs by alcohol also occurs through Gly-R [93].

Because alcohol acts on the same pathways as psychostimulants and opiates, it is likely that similar molecular alterations occur in alcoholism (Fig. 2). Thus, ethanol administration increases phospho-ERK expression in the NAc. This increase can be blocked by the administration of a D1-R antagonist, suggesting that alterations of the ERK pathway in D1-MSNs are involved in the deleterious effect of ethanol on the reward system [94]. Mice exhibiting locomotor sensitization to ethanol show higher D1-R agonist-induced phosphor-Thr34-DARPP-32 levels in the NAc compared with animals showing no sensitization or control mice. These results highlight that repeated alcohol exposure could sensitize intracellular pathways in D1-MSNs underlying behavioral adaptation to the drug [95]. Finally, ethanol exposure preferentially increases ΔFosB expression in both the dorsal and ventral striatum D1-MSNs [47].

Altogether, alcohol seems to alter many of the main molecular functions in D1- and D2-MSNs. However, more cell-type-specific studies are needed to reveal dysfunctions specific to each cell-type.

CONCLUSIONS

Drugs of abuse exert plasticity on MSN subtypes differentially through multiple presynaptic and postsynaptic signaling mechanisms to ultimately drive reward and future drug use. While D1- and D2-MSNs can have opposite function in mediating drug-related behaviors, they can also play a role in different aspects of these behaviors. It is thus crucial to use cell-type-specific methodologies to uncover adaptations specific to each MSN subtype that could be, otherwise, over-shadowed when assessed in total tissue. Moreover, these plastic changes occur at different stages of addiction. Thus, future studies should aim to continue utilizing MSN-subtype examination, including in vivo techniques such as calcium imaging, to better define temporal aspects of drug seeking and addiction.

REFERENCES

[1] Ikemoto S. Brain reward circuitry beyond the mesolimbic dopamine system: a neurobiological theory. Neurosci Biobehav Rev 2010;35(2):129–50. https://doi.org/10.1016/j.neubiorev.2010.02.001.

[2] Di Chiara G. Nucleus accumbens shell and core dopamine: differential role in behavior and addiction. Behav Brain Res 2002;137(1–2):75–114.

[3] Isomura Y, Takekawa T, Harukuni R, Handa T, Aizawa H, Takada M, Fukai T. Reward-modulated motor information in identified striatum neurons. J Neurosci 2013;33(25):10209–20. https://doi.org/10.1523/JNEUROSCI.0381-13.2013.

[4] Balleine BW, Liljeholm M, Ostlund SB. The integrative function of the basal ganglia in instrumental conditioning. Behav Brain Res 2009;199 (1):43–52. https://doi.org/10.1016/j.bbr.2008.10.034.

[5] Murray JE, Dilleen R, Pelloux Y, Economidou D, Dalley JW, Belin D, Everitt BJ. Increased impulsivity retards the transition to dorsolateral striatal dopamine control of cocaine seeking. Biol Psychiatry 2014;76(1):15–22. https://doi.org/10.1016/j.biopsych.2013.09.011.

[6] Haber SN, Fudge JL, McFarland NR. Striatonigrostriatal pathways in primates form an ascending spiral from the shell to the dorsolateral striatum. J Neurosci 2000;20(6):2369–82.

[7] Everitt BJ, Robbins TW. Neural systems of reinforcement for drug addiction: from actions to habits to compulsion. Nat Neurosci 2005;8(11):1481–9. https://doi.org/10.1038/nn1579.

[8] Kupchik YM, Brown RM, Heinsbroek JA, Lobo MK, Schwartz DJ, Kalivas PW. Coding the direct/indirect pathways by D1 and D2 receptors is not valid for accumbens projections. Nat Neurosci 2015;18(9):1230–2. https://doi.org/10.1038/nn.4068.

[9] Engeln M, Lobo MK. Cocaine and striatal projection neuron subtype mechanisms. In: Preedy VR, editor. The Neuroscience of Cocaine. 1st ed. London, UK: Elsevier; 2017. p. 297–305.

[10] Koo JW, Lobo MK, Chaudhury D, Labonte B, Friedman A, Heller E, Peña CJ, Han MH, Nestler EJ. Loss of BDNF signaling in D1R-expressing NAc neurons enhances morphine reward by reducing GABA inhibition. Neuropsychopharmacology 2014;39(11):2646–53. https://doi.org/10.1038/npp.2014.118.

[11] Lobo MK, Covington III HE, Chaudhury D, Friedman AK, Sun H, Damez-Werno D, Dietz DM, Zaman S, Koo JW, Kennedy PJ, Mouzon E, Mogri M, Neve RL, Deisseroth K, Han MH, Nestler EJ. Cell type-specific loss of BDNF signaling mimics optogenetic control of cocaine reward. Science 2010;330(6002):385–90. https://doi.org/10.1126/science.1188472.

[12] Song SS, Kang BJ, Wen L, Lee HJ, Sim HR, Kim TH, Yoon S, Yoon BJ, Augustine GJ, Baik JH. Optogenetics reveals a role for accumbal medium spiny neurons expressing dopamine D2 receptors in cocaine-induced behavioral sensitization. Front Behav Neurosci 2014;8. https://doi.org/10.3389/fnbeh.2014.00336.

[13] Bock R, Shin JH, Kaplan AR, Dobi A, Markey E, Kramer PF, Christine H Christensen, Martin F Adrover, Alvarez, V. A.. Strengthening the accumbal indirect pathway promotes resilience to compulsive cocaine use. Nat Neurosci 2013;16(5):632–8. https://doi.org/10.1038/nn.3369.

[14] Chandra R, Lenz JD, Gancarz AM, Chaudhury D, Schroeder GL, Han MH, Cheer JF, Dietz DM, Lobo MK. Optogenetic inhibition of D1R containing nucleus accumbens neurons alters cocaine-mediated regulation of Tiam1. Front Mol Neurosci 2013;6. https://doi.org/10.3389/fnmol.2013.00013.

[15] Ferguson SM, Eskenazi D, Ishikawa M, Wanat MJ, Phillips PE, Dong Y, Roth BL, Neumaier JF. Transient neuronal inhibition reveals opposing roles of indirect and direct pathways in sensitization. Nat Neurosci 2011;14(1):22–4. https://doi.org/10.1038/nn.2703.

[16] Heinsbroek JA, Neuhofer DN, Griffin III WC, Siegel GS, Bobadilla AC, Kupchik YM, Kalivas PW. Loss of plasticity in the D2-accumbens pallidal pathway promotes cocaine seeking. J Neurosci 2017;37(4):757–67. https://doi.org/10.1523/JNEUROSCI.2659-16.2016.

[17] Hikida T, Yawata S, Yamaguchi T, Danjo T, Sasaoka T, Wang Y, Nakanishi S. Pathway-specific modulation of nucleus accumbens in reward and aversive behavior via selective transmitter receptors. Proc Natl Acad Sci U S A 2013;110(1):342–7. https://doi.org/10.1073/pnas.1220358110.

[18] Di Chiara G, Imperato A. Drugs abused by humans preferentially increase synaptic dopamine concentrations in the mesolimbic system of freely moving rats. Proc Natl Acad Sci U S A 1988;85(14):5274–8.

[19] Shen W, Flajolet M, Greengard P, Surmeier DJ. Dichotomous dopaminergic control of striatal synaptic plasticity. Science 2008;321(5890):848–51. https://doi.org/10.1126/science.1160575.

[20] Graziane NM, Sun S, Wright WJ, Jang D, Liu Z, Huang YH, Nestler EJ, Wang YT, Schlüter OM, Dong Y. Opposing mechanisms mediate morphine- and cocaine-induced generation of silent synapses. Nat Neurosci 2016;19(7):915–25. https://doi.org/10.1038/nn.4313.

[21] Pascoli V, Turiault M, Luscher C. Reversal of cocaine-evoked synaptic potentiation resets drug-induced adaptive behaviour. Nature 2011;481 (7379):71–5. https://doi.org/10.1038/nature10709.

[22] Terrier J, Luscher C, Pascoli V. Cell-type specific insertion of GluA2-lacking AMPARs with cocaine exposure leading to sensitization, cue-induced seeking, and incubation of craving. Neuropsychopharmacology 2016;41(7):1779–89. https://doi.org/10.1038/npp.2015.345.

[23] Pascoli V, Terrier J, Espallergues J, Valjent E, O'Connor EC, Luscher C. Contrasting forms of cocaine-evoked plasticity control components of relapse. Nature 2014;509(7501):459–64. https://doi.org/10.1038/nature13257.

[24] Conrad KL, Tseng KY, Uejima JL, Reimers JM, Heng LJ, Shaham Y, Marinelli M, Wolf ME. Formation of accumbens GluR2-lacking AMPA receptors mediates incubation of cocaine craving. Nature 2008;454(7200):118–21. https://doi.org/10.1038/nature06995.

[25] Lee BR, Ma YY, Huang YH, Wang X, Otaka M, Ishikawa M, Neumann PA, Graziane NM, Brown TE, Suska A, Guo C, Lobo MK, Sesack SR, Wolf ME, Nestler EJ, Shaham Y, Schlüter OM, Dong Y. Maturation of silent synapses in amygdala-accumbens projection contributes to incubation of cocaine craving. Nat Neurosci 2013;16(11):1644–51. https://doi.org/10.1038/nn.3533.

[26] Bocklisch C, Pascoli V, Wong JC, House DR, Yvon C, de Roo M, Tan KR, Luscher C. Cocaine disinhibits dopamine neurons by potentiation of GABA transmission in the ventral tegmental area. Science 2013;341(6153):1521–5. https://doi.org/10.1126/science.1237059.

[27] Creed M, Ntamati NR, Chandra R, Lobo MK, Luscher C. Convergence of reinforcing and Anhedonic cocaine effects in the ventral pallidum. Neuron 2016;92(1):214–26. https://doi.org/10.1016/j.neuron.2016.09.001.

[28] Dobbs LK, Kaplan AR, Lemos JC, Matsui A, Rubinstein M, Alvarez VA. Dopamine regulation of lateral inhibition between striatal neurons gates the stimulant actions of cocaine. Neuron 2016;90(5):1100–13. https://doi.org/10.1016/j.neuron.2016.04.031.

[29] Stefanik MT, Kupchik YM, Brown RM, Kalivas PW. Optogenetic evidence that pallidal projections, not nigral projections, from the nucleus accumbens core are necessary for reinstating cocaine seeking. J Neurosci 2013;33(34):13654–62. https://doi.org/10.1523/JNEUROSCI.1570-13.2013.

[30] Lee KW, Kim Y, Kim AM, Helmin K, Nairn AC, Greengard P. Cocaine-induced dendritic spine formation in D1 and D2 dopamine receptor-containing medium spiny neurons in nucleus accumbens. Proc Natl Acad Sci U S A 2006;103(9):3399–404. https://doi.org/10.1073/pnas.0511244103.

[31] Kim J, Park BH, Lee JH, Park SK, Kim JH. Cell type-specific alterations in the nucleus accumbens by repeated exposures to cocaine. Biol Psychiatry 2011;69(11):1026–34. https://doi.org/10.1016/j.biopsych.2011.01.013.

[32] Dobi A, Seabold GK, Christensen CH, Bock R, Alvarez VA. Cocaine-induced plasticity in the nucleus accumbens is cell specific and develops without prolonged withdrawal. J Neurosci 2011;31(5):1895–904. https://doi.org/10.1523/JNEUROSCI.5375-10.2011.

[33] Barbera G, Liang B, Zhang L, Gerfen CR, Culurciello E, Chen R, Li Y, Lin DT. Spatially compact neural clusters in the dorsal striatum encode locomotion relevant information. Neuron 2016;92(1):202–13. https://doi.org/10.1016/j.neuron.2016.08.037.

[34] Baker DA, Fuchs RA, Specio SE, Khroyan TV, Neisewander JL. Effects of intraaccumbens administration of SCH-23390 on cocaine-induced loco-motion and conditioned place preference. Synapse 1998;30(2):181–93. https://doi.org/10.1002/(SICI)1098-2396(199810)30:2<181::AID-SYN8>3.0.CO;2-8.

[35] Nazarian A, Russo SJ, Festa ED, Kraish M, Quinones-Jenab V. The role of D1 and D2 receptors in the cocaine conditioned place preference of male and female rats. Brain Res Bull 2004;63(4):295–9. https://doi.org/10.1016/j.brainresbull.2004.03.004.

[36] Calipari ES, Bagot RC, Purushothaman I, Davidson TJ, Yorgason JT, Pena CJ, Walker DM, Pirpinias ST, Guise KG, Ramakrishnan C, Deisseroth K, Nestler EJ. In vivo imaging identifies temporal signature of D1 and D2 medium spiny neurons in cocaine reward. Proc Natl Acad Sci U S A 2016;113 (10):2726–31. https://doi.org/10.1073/pnas.1521238113.

[37] Bateup HS, Svenningsson P, Kuroiwa M, Gong S, Nishi A, Heintz N, Greengard P. Cell type-specific regulation of DARPP-32 phosphorylation by psychostimulant and antipsychotic drugs. Nat Neurosci 2008;11(8):932–9. https://doi.org/10.1038/nn.2153.

[38] Bateup HS, Santini E, Shen W, Birnbaum S, Valjent E, Surmeier DJ, Fisone G, Nestler EJ, Greengard P. Distinct subclasses of medium spiny neurons differentially regulate striatal motor behaviors. Proc Natl Acad Sci U S A 2010;107(33):14845–50. https://doi.org/10.1073/pnas.1009874107.

[39] Lu L, Koya E, Zhai H, Hope BT, Shaham Y. Role of ERK in cocaine addiction. Trends Neurosci 2006;29(12):695–703. https://doi.org/10.1016/j.tins.2006.10.005.

[40] Valjent E, Corvol JC, Pages C, Besson MJ, Maldonado R, Caboche J. Involvement of the extracellular signal-regulated kinase cascade for cocaine-rewarding properties. J Neurosci 2000;20(23):8701–9.

[41] Gerfen CR, Paletzki R, Worley P. Differences between dorsal and ventral striatum in Drd1a dopamine receptor coupling of dopamine- and cAMP-regulated phosphoprotein-32 to activation of extracellular signal-regulated kinase. J Neurosci 2008;28(28):7113–20. https://doi.org/10.1523/JNEUROSCI.3952-07.2008.

[42] Bertran-Gonzalez J, Bosch C, Maroteaux M, Matamales M, Herve D, Valjent E, Girault JA. Opposing patterns of signaling activation in dopamine D1 and D2 receptor-expressing striatal neurons in response to cocaine and haloperidol. J Neurosci 2008;28(22):5671–85. https://doi.org/10.1523/JNEUROSCI.1039-08.2008.

[43] Brami-Cherrier K, Valjent E, Herve D, Darragh J, Corvol JC, Pages C, Arthur SJ, Girault JA, Caboche J. Parsing molecular and behavioral effects of cocaine in mitogen- and stress-activated protein kinase-1-deficient mice. J Neurosci 2005;25(49):11444–54. https://doi.org/10.1523/JNEUROSCI.1711-05.2005.

[44] Ortinski PI, Briand LA, Pierce RC, Schmidt HD. Cocaine-seeking is associated with PKC-dependent reduction of excitatory signaling in accumbens shell D2 dopamine receptor-expressing neurons. Neuropharmacology 2015;92:80–9. https://doi.org/10.1016/j.neuropharm.2015.01.002.

[45] Guez-Barber D, Fanous S, Golden SA, Schrama R, Koya E, Stern AL, Bossert JM, Harvey BK, Picciotto MR, Hope BT. FACS identifies unique cocaine-induced gene regulation in selectively activated adult striatal neurons. J Neurosci 2011;31(11):4251–9. https://doi.org/10.1523/JNEUROSCI.6195-10.2011.

[46] Berretta S, Robertson HA, Graybiel AM. Dopamine and glutamate agonists stimulate neuron-specific expression of Fos-like protein in the striatum. J Neurophysiol 1992;68(3):767–77.

[47] Lobo MK, Zaman S, Damez-Werno DM, Koo JW, Bagot RC, DiNieri JA, Nugent A, Finkel E, Chaudhury D, Chandra R, Riberio E, Rabkin J, Mouzon E, Cachope R, Cheer JF, Han MH, Dietz DM, Self DW, Hurd YL, Vialou V, Nestler EJ. DeltaFosB induction in striatal medium spiny neuron subtypes in response to chronic pharmacological, emotional, and optogenetic stimuli. J Neurosci 2013;33(47):18381–95. https://doi.org/10.1523/JNEUROSCI.1875-13.2013.

[48] Grueter BA, Robison AJ, Neve RL, Nestler EJ, Malenka RC. FosB differentially modulates nucleus accumbens direct and indirect pathway function. Proc Natl Acad Sci U S A 2013;110(5):1923–8. https://doi.org/10.1073/pnas.1221742110.

[49] Chandra R, Francis TC, Konkalmatt P, Amgalan A, Gancarz AM, Dietz DM, Lobo MK. Opposing role for Egr3 in nucleus accumbens cell subtypes in cocaine action. J Neurosci 2015;35(20):7927–37. https://doi.org/10.1523/JNEUROSCI.0548-15.2015.

[50] Chandra R, Engeln M, Francis TC, Konkalmatt P, Patel D, Lobo MK. A role for peroxisome proliferator-activated receptor gamma coactivator-1alpha in nucleus accumbens neuron subtypes in cocaine action. Biol Psychiatry 2016. https://doi.org/10.1016/j.biopsych.2016.10.024.

[51] Arango-Lievano M, Schwarz JT, Vernov M, Wilkinson MB, Bradbury K, Feliz A, Marongiu R, Gelfand Y, Warner-Schmidt J, Nestler EJ, Greengard P, Russo SJ, Kaplitt MG. Cell-type specific expression of p11 controls cocaine reward. Biol Psychiatry 2014;76(10):794–801. https://doi.org/10.1016/j.biopsych.2014.02.012.

[52] Ceglia I, Lee KW, Cahill ME, Graves SM, Dietz D, Surmeier DJ, Nestler EJ, Nairn AC, Greengard P, Kim Y. WAVE1 in neurons expressing the D1 dopamine receptor regulates cellular and behavioral actions of cocaine. Proc Natl Acad Sci U S A 2017;114(6):1395–400. https://doi.org/10.1073/pnas.1621185114.

[53] Cahill ME, Bagot RC, Gancarz AM, Walker DM, Sun H, Wang ZJ, Heller EA, Feng J, Kennedy PJ, Koo JW, Cates HM, Neve RL, Shen L, Dietz DM, Nestler EJ. Bidirectional synaptic structural plasticity after chronic cocaine administration occurs through Rap1 small GTPase signaling. Neuron 2016;89(3):566–82. https://doi.org/10.1016/j.neuron.2016.01.031.

[54] Nagai T, Nakamuta S, Kuroda K, Nakauchi S, Nishioka T, Takano T, Zhang X, Tsuboi D, Funahashi Y, Nakano T, Yoshimoto J, Kobayashi K, Uchigashima M, Watanabe M, Miura M, Nishi A, Kobayashi K, Yamada K, Amano M, Kaibuchi K. Phosphoproteomics of the dopamine pathway enables discovery of Rap1 activation as a reward signal in vivo. Neuron 2016;89(3):550–65. https://doi.org/10.1016/j.neuron.2015.12.019.

[55] Heusner CL, Palmiter RD. Expression of mutant NMDA receptors in dopamine D1 receptor-containing cells prevents cocaine sensitization and decreases cocaine preference. J Neurosci 2005;25(28):6651–7. https://doi.org/10.1523/JNEUROSCI.1474-05.2005.

[56] Joffe ME, Vitter SR, Grueter BA. GluN1 deletions in D1- and A2A-expressing cell types reveal distinct modes of behavioral regulation. Neuropharmacology 2017;112(Pt A):172–80. https://doi.org/10.1016/j.neuropharm.2016.03.026.

[57] Heiman M, Schaefer A, Gong S, Peterson JD, Day M, Ramsey KE, Suárez-Fariñas M, Schwarz C, Stephan DA, Surmeier DJ, Greengard P, Heintz N. A translational profiling approach for the molecular characterization of CNS cell types. Cell 2008;135(4):738–48. https://doi.org/10.1016/j.cell.2008.10.028.

[58] Maguire EP, Macpherson T, Swinny JD, Dixon CI, Herd MB, Belelli D, Stephens DN, King SL, Lambert JJ. Tonic inhibition of accumbal spiny neurons by extrasynaptic alpha4betadelta GABAA receptors modulates the actions of psychostimulants. J Neurosci 2014;34(3):823–38. https://doi.org/10.1523/JNEUROSCI.3232-13.2014.

[59] Jordi E, Heiman M, Marion-Poll L, Guermonprez P, Cheng SK, Nairn AC, Greengard P, Girault JA. Differential effects of cocaine on histone post-translational modifications in identified populations of striatal neurons. Proc Natl Acad Sci U S A 2013;110(23):9511–6. https://doi.org/10.1073/pnas.1307116110.

[60] Damez-Werno DM, Sun H, Scobie KN, Shao N, Rabkin J, Dias C, Calipari ES, Maze I, Pena CJ, Walker DM, Cahill ME, Chandra R, Gancarz A, Mouzon E, Landry JA, Cates H, Lobo MK, Dietz D, Allis CD, Guccione E, Turecki G, Defilippi P, Neve RL, Hurd YL, Shen L, Nestler EJ. Histone arginine methylation in cocaine action in the nucleus accumbens. Proc Natl Acad Sci U S A 2016;113(34):9623–8. https://doi.org/10.1073/pnas.1605045113.

[61] Maze I, Chaudhury D, Dietz DM, Von Schimmelmann M, Kennedy PJ, Lobo MK, Sillivan SE, Miller ML, Bagot RC, Sun H, Turecki G, Neve RL, Hurd YL, Shen L, Han MH, Schaefer A, Nestler EJ. G9a influences neuronal subtype specification in striatum. Nat Neurosci 2014;17(4):533–9. https://doi.org/10.1038/nn.3670.

[62] Engmann O, Labonte B, Mitchell A, Bashtrykov P, Calipari ES, Rosenbluh C, Loh YE, Walker DM, Burek D, Hamilton PJ, Issler O, Neve RL, Turecki G, Hurd Y, Chess A, Shen L, Mansuy I, Jeltsch A, Akbarian S, Nestler EJ. Cocaine-induced chromatin modifications associate with increased expression and three-dimensional looping of Auts2. Biol Psychiatry 2017. https://doi.org/10.1016/j.biopsych.2017.04.013.

[63] Schaefer A, Im HI, Veno MT, Fowler CD, Min A, Intrator A, Kjems J, Kenny PJ, O'Carroll D, Greengard P. Argonaute 2 in dopamine 2 receptor-expressing neurons regulates cocaine addiction. J Exp Med 2010;207(9):1843–51. https://doi.org/10.1084/jem.20100451.

[64] Hearing MC, Jedynak J, Ebner SR, Ingebretson A, Asp AJ, Fischer RA, Schmidt C, Larson EB, Thomas MJ. Reversal of morphine-induced cell-type-specific synaptic plasticity in the nucleus accumbens shell blocks reinstatement. Proc Natl Acad Sci U S A 2016;113(3):757–62. https://doi.org/10.1073/pnas.1519248113.

[65] James AS, Chen JY, Cepeda C, Mittal N, Jentsch JD, Levine MS, Evans CJ, Walwyn W. Opioid self-administration results in cell-type specific adaptations of striatal medium spiny neurons. Behav Brain Res 2013;256:279–83. https://doi.org/10.1016/j.bbr.2013.08.009.

[66] Gaspari S, Papachatzaki MM, Koo JW, Carr FB, Tsimpanouli ME, Stergiou E, Bagot RC, Ferguson D, Mouzon E, Chakravarty S, Deisseroth K, Lobo MK, Zachariou V. Nucleus accumbens-specific interventions in RGS9-2 activity modulate responses to morphine. Neuropsychopharmacology 2014;39(8):1968–77. https://doi.org/10.1038/npp.2014.45.

[67] Zhu Y, Wienecke CF, Nachtrab G, Chen X. A thalamic input to the nucleus accumbens mediates opiate dependence. Nature 2016;530(7589):219–22. https://doi.org/10.1038/nature16954.

[68] Atwood BK, Kupferschmidt DA, Lovinger DM. Opioids induce dissociable forms of long-term depression of excitatory inputs to the dorsal striatum. Nat Neurosci 2014;17(4):540–8. https://doi.org/10.1038/nn.3652.

[69] Enoksson T, Bertran-Gonzalez J, Christie MJ. Nucleus accumbens D2- and D1-receptor expressing medium spiny neurons are selectively activated by morphine withdrawal and acute morphine, respectively. Neuropharmacology 2012;62(8):2463–71. https://doi.org/10.1016/j.neuropharm.2012.02.020.

[70] Cui Y, Ostlund SB, James AS, Park CS, Ge W, Roberts KW, Mittal N, Murphy NP, Cepeda C, Kieffer BL, Levine MS, Jentsch JD, Walwyn WM, Sun YE, Evans CJ, Maidment NT, Yang XW. Targeted expression of mu-opioid receptors in a subset of striatal direct-pathway neurons restores opiate reward. Nat Neurosci 2014;17(2):254–61. https://doi.org/10.1038/nn.3622.

[71] Garcia-Perez D, Nunez C, Laorden ML, Milanes MV. Regulation of dopaminergic markers expression in response to acute and chronic morphine and to morphine withdrawal. Addict Biol 2016;21(2):374–86. https://doi.org/10.1111/adb.12209.

[72] Castane A, Wells L, Soria G, Hourani S, Ledent C, Kitchen I, Opacka-Juffry J, Maldonado R, Valverde O. Behavioural and biochemical responses to morphine associated with its motivational properties are altered in adenosine A(2A) receptor knockout mice. Br J Pharmacol 2008;155(5):757–66. https://doi.org/10.1038/bjp.2008.299.

[73] Brown RM, Short JL, Cowen MS, Ledent C, Lawrence AJ. A differential role for the adenosine A2A receptor in opiate reinforcement vs opiate-seeking behavior. Neuropsychopharmacology 2009;34(4):844–56. https://doi.org/10.1038/npp.2008.72.

[74] Fan P, Jiang Z, Diamond I, Yao L. Up-regulation of AGS3 during morphine withdrawal promotes cAMP superactivation via adenylyl cyclase 5 and 7 in rat nucleus accumbens/striatal neurons. Mol Pharmacol 2009;76(3):526–33. https://doi.org/10.1124/mol.109.057802.

[75] Borgkvist A, Usiello A, Greengard P, Fisone G. Activation of the cAMP/PKA/DARPP-32 signaling pathway is required for morphine psychomotor stimulation but not for morphine reward. Neuropsychopharmacology 2007;32(9):1995–2003. https://doi.org/10.1038/sj.npp.1301321.

[76] Scheggi S, Crociani A, De Montis MG, Tagliamonte A, Gambarana C. Dopamine D1 receptor-dependent modifications in the dopamine and cAMP-regulated phosphoprotein of Mr 32 kDa phosphorylation pattern in striatal areas of morphine-sensitized rats. Neuroscience 2009;163(2):627–39. https://doi.org/10.1016/j.neuroscience.2009.06.053.

[77] Glass MJ, Lane DA, Colago EE, Chan J, Schlussman SD, Zhou Y, Kreek MJ, Pickel VM. Chronic administration of morphine is associated with a decrease in surface AMPA GluR1 receptor subunit in dopamine D1 receptor expressing neurons in the shell and non-D1 receptor expressing neurons in the core of the rat nucleus accumbens. Exp Neurol 2008;210(2):750–61. https://doi.org/10.1016/j.expneurol.2008.01.012.

[78] Walters CL, Aston-Jones G, Druhan JP. Expression of fos-related antigens in the nucleus accumbens during opiate withdrawal and their attenuation by a D2 dopamine receptor agonist. Neuropsychopharmacology 2000;23(3):307–15. https://doi.org/10.1016/S0893-133X(00)00113-5.

[79] Zachariou V, Bolanos CA, Selley DE, Theobald D, Cassidy MP, Kelz MB, Shaw-Lutchman T, Berton O, Sim-Selley LJ, Dileone RJ, Kumar A, Nestler EJ. An essential role for DeltaFosB in the nucleus accumbens in morphine action. Nat Neurosci 2006;9(2):205–11. https://doi.org/10.1038/nn1636.

[80] Xie K, Colgan LA, Dao MT, Muntean BS, Sutton LP, Orlandi C, Boye SL, Boye SE, Shih CC, Li Y, Xu B, Smith RG, Yasuda R, Martemyanov KA. NF1 is a direct G protein effector essential for opioid signaling to Ras in the striatum. Curr Biol 2016;26(22):2992–3003. https://doi.org/10.1016/j.cub.2016.09.010.

[81] Yin HH, Park BS, Adermark L, Lovinger DM. Ethanol reverses the direction of long-term synaptic plasticity in the dorsomedial striatum. Eur J Neurosci 2007;25(11):3226–32. https://doi.org/10.1111/j.1460-9568.2007.05606.x.

[82] Renteria R, Maier EY, Buske TR, Morrisett RA. Selective alterations of NMDAR function and plasticity in D1 and D2 medium spiny neurons in the nucleus accumbens shell following chronic intermittent ethanol exposure. Neuropharmacology 2017;112(Pt A):164–71. https://doi.org/10.1016/j.neuropharm.2016.03.004.

[83] Wang J, Cheng Y, Wang X, Roltsch Hellard E, Ma T, Gil H, Hamida SB, Ron D. Alcohol elicits functional and structural plasticity selectively in dopamine D1 receptor-expressing neurons of the dorsomedial striatum. J Neurosci 2015;35(33):11634–43. https://doi.org/10.1523/JNEUROSCI.0003-15.2015.

[84] Jeanes ZM, Buske TR, Morrisett RA. Cell type-specific synaptic encoding of ethanol exposure in the nucleus accumbens shell. Neuroscience 2014;277:184–95. https://doi.org/10.1016/j.neuroscience.2014.06.063.

[85] Yao L, Arolfo MP, Dohrman DP, Jiang Z, Fan P, Fuchs S, Janak PH, Gordon AS, Diamond I. Betagamma dimers mediate synergy of dopamine D2 and adenosine A2 receptor-stimulated PKA signaling and regulate ethanol consumption. Cell 2002;109(6):733–43.

[86] Arolfo MP, Yao L, Gordon AS, Diamond I, Janak PH. Ethanol operant self-administration in rats is regulated by adenosine A2 receptors. Alcohol Clin Exp Res 2004;28(9):1308–16.

[87] Volkow ND, Wang GJ, Fowler JS, Logan J, Hitzemann R, Ding YS, Pappas N, Shea C, Piscani K. Decreases in dopamine receptors but not in dopamine transporters in alcoholics. Alcohol Clin Exp Res 1996;20(9):1594–8.

[88] Thanos PK, Volkow ND, Freimuth P, Umegaki H, Ikari H, Roth G, Ingram DK, Hitzemann R. Overexpression of dopamine D2 receptors reduces alcohol self-administration. J Neurochem 2001;78(5):1094–103.

[89] Gallo EF, Salling MC, Feng B, Moron JA, Harrison NL, Javitch JA, Kellendonk C. Upregulation of dopamine D2 receptors in the nucleus accumbens indirect pathway increases locomotion but does not reduce alcohol consumption. Neuropsychopharmacology 2015;40(7):1609–18. https://doi.org/10.1038/npp.2015.11.

[90] El-Ghundi M, George SR, Drago J, Fletcher PJ, Fan T, Nguyen T, Liu C, Sibley DR, Westphal H, O'Dowd BF. Disruption of dopamine D1 receptor gene expression attenuates alcohol-seeking behavior. Eur J Pharmacol 1998;353(2–3):149–58.

[91] Beckley JT, Laguesse S, Phamluong K, Morisot N, Wegner SA, Ron D. The first alcohol drink triggers mTORC1-dependent synaptic plasticity in nucleus accumbens dopamine D1 receptor neurons. J Neurosci 2016;36(3):701–13. https://doi.org/10.1523/JNEUROSCI.2254-15.2016.

[92] Liu F, Laguesse S, Legastelois R, Morisot N, Ben Hamida S, Ron D. mTORC1-dependent translation of collapsin response mediator protein-2 drives neuroadaptations underlying excessive alcohol-drinking behaviors. Mol Psychiatry 2017;22(1):89–101. https://doi.org/10.1038/mp.2016.12.

[93] Forstera B, Munoz B, Lobo MK, Chandra R, Lovinger DM, Aguayo LG. Presence of ethanol sensitive glycine receptors in medium spiny neurons in the mouse nucleus accumbens. J Physiol 2017. https://doi.org/10.1113/JP273767.

[94] Ibba F, Vinci S, Spiga S, Peana AT, Assaretti AR, Spina L, Longoni R, Acquas E. Ethanol-induced extracellular signal regulated kinase: role of dopamine D1 receptors. Alcohol Clin Exp Res 2009;33(5):858–67. https://doi.org/10.1111/j.1530-0277.2009.00907.x.

[95] Abrahao KP, Goeldner FO, Souza-Formigoni ML. Individual differences in ethanol locomotor sensitization are associated with dopamine D1 receptor intra-cellular signaling of DARPP-32 in the nucleus accumbens. PLoS One 2014;9(2):e98296. https://doi.org/10.1371/journal.pone.0098296.

Chapter 19

Harnessing Circuits for the Treatment of Addictive Disorders

Yann Pelloux and Christelle Baunez

Institut de Neurosciences de la Timone, UMR7289, CNRS and Aix Marseille Université, Marseille, France

INTRODUCTION

Addicted individuals compulsively use drugs despite their negative consequences. This compulsivity may result from dysfunction in, or imbalance between, two processes: exacerbated motivation for the drug and impaired behavioral control over its use. Although these processes share commonalities, they also implicate distinct brain circuitries and mechanisms. To date, most treatment strategies for drug addiction have focused on reducing exacerbated motivation for drug. Very few pharmacological treatments for drug addiction currently exist, and they often have poor or heterogeneous therapeutic efficacy.

This chapter presents clinical and preclinical evidence concerning the use of surgical intervention as a treatment for drug addiction, including brain lesions and deep brain stimulation (DBS). Ablative surgery of a variety of brain structures was conducted for the treatment of addiction until the 2000s. The dramatic side effects eventually led to a resistance to use of this strategy for psychiatric disorders. More recently, in attempt to reduce some of the negative side effects of lesions, very localized brain ablations via concentrated high-intensity cobalt radiation have been conducted in the treatment of psychiatric disorders [1–4]. To the best of our knowledge, there are no studies which have used this so-called gamma knife surgery approach in the treatment of drug addiction.

In the 1990s, deep brain surgical therapy was revived for the treatment of neurological disorders, such as Parkinson's disease. The procedure typically consists of delivering high-frequency electrical stimulation (130 Hz) to the target brain region. A number of motor symptoms have been successfully alleviated by electrical DBS applied to the subthalamic nucleus (STN) [5]. However, the therapeutic mechanisms of high-frequency DBS are still unclear. As DBS often recapitulates the effects of brain lesions in preclinical models, it is commonly held that the therapeutic effects of DBS are the result of reduced activity of the targeted brain structure. However, DBS also likely impacts the activity of fibers of passage and exerts more widespread effects [6]. More research is needed to better understand how DBS works and how it is best applied. In the case of addiction, questions remain regarding the optimal brain target for DBS [7], and whether or not high- or low-frequency stimulation is preferable, as behavioral effects can differ across frequency [8].

The present chapter reviews a variety of surgical interventions that have been investigated in the structures of the classical reward circuit and proposed as treatments for addiction. It will also highlight a more recent view of this circuitry that includes the STN, a cerebral structure involved in both motivation and inhibitory control. We will discuss results examining manipulations of the STN on motivational processes and addiction-related behaviors that implicate this brain region as a potentially viable treatment target.

After our first publications in rats which suggested that STN inactivation, preferably by DBS, could represent an interesting strategy for the treatment of addiction [9–11], there was some controversy over whether or not it was too early to consider DBS as a treatment for addiction [12]. The use of an invasive procedure for refractory substance abuse disorders raises ethical issues and is difficult because dropout rates tend to be very high before surgery, and the number of referrals is low [13]. However, in extreme cases when no other treatment has proven efficacious, surgery may provide the best option. We will, however, also review certain noninvasive brain stimulation strategies developed in recent years.

TREATING ADDICTION BY TARGETING THE CLASSIC REWARD SYSTEM

Addiction involves an exacerbated motivation for continued drug use, which lead researchers to develop treatment strategies that target the so-called "reward circuit". This circuit comprises a set of interconnected cortical and subcortical

Neural Mechanisms of Addiction. https://doi.org/10.1016/B978-0-12-812202-0.00019-1
Copyright © 2019 Elsevier Inc. All rights reserved.

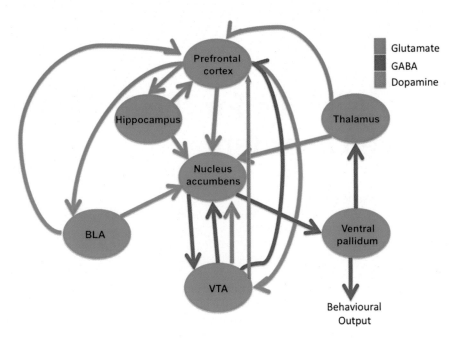

Adapted from Wolf (2002) from Mogenson (1980)

FIG. 1   Schematic representation of the classical reward circuit, positioning the NAc in the center as an interface between motivation to action. *BLA*, basolateral amygdala; *VTA*, ventral tegmental area.

structures that are discussed extensively in other chapters of this book. In brief, a brain structure called the nucleus accumbens (NAc) is posited to be central to the reward circuit, serving as an interface between motivation and action [14]. The NAc receives excitatory glutamatergic inputs from the prefrontal cortex (PFC), the basolateral amygdala (BLA), the hippocampus, and the thalamus, in addition to critical dopaminergic (DA) inputs from the ventral tegmental area (VTA) (Fig. 1). The NAc then primarily sends output information to the ventral pallidum (VP) and back to the VTA. Functionally, the reward circuit has been demonstrated to subserve the ability to acquire and/or retrieve associations between rewarding and predictive cues, such as specific places, stimuli, motor programs, or internal states that initiate and/or energize behavior aimed at obtaining further reward.

Nucleus Accumbens

Preclinical Studies

Given the central role of the NAc in the reward circuit, several surgical strategies to treat addiction have focused on this region. There is considerable preclinical evidence to support a role for the NAc in mediating the motivational effects of conditioned stimuli associated with the drug. Lesions and pharmacological inactivation data of the NAc impair the development [15], the maintenance [16], or the reinstatement of drug seeking by drug associated cues [17]. The NAc is a heterogeneous structure classically divided into two subregions according to morphological, histochemical, functional, and connectivity studies: the NAc core and NAc shell [18,19]. Other reviews have extensively reported the effect of lesions or pharmacological inactivation of these subregions of the NAc on drug motivated behaviors in preclinical models (for review [18]) and therefore will not be addressed here. Instead we will focus on preclinical studies reporting the effects of high-frequency DBS of either or both of these subregions on behaviors relevant to addiction.

In rats, DBS of the core region of the NAc reduces the conditioned place preference induced by morphine. Conditioned place preference is where animals display a preference for the environment previously associated with a drug [20]. DBS of the core of the NAc also attenuates the ability of heroin-associated cues and drug injection to reinstate heroin self-administration [21] and reduces alcohol consumption in a two bottle-choice procedure when alcohol is presented in free choice with water [22]. DBS of the shell of the NAc also reduces alcohol consumption in a two bottle-choice procedure [22] and reduces (though at higher current than usually applied) the preference and consumption of alcohol in alcohol-preferring rats [23]. It also reduces methamphetamine self-administration [24]. Altogether these studies suggest that DBS, at high frequency, targeting the core or the shell of the NAc reduces motivated behavior of drugs of different classes. However, recent data have questioned the relevance of high-frequency stimulation of the NAc for the treatment of addictive behavior [25]. High-frequency DBS of the dorsal-NAc enhances morphine-conditioned place preference, whereas low-frequency DBS reduces it [8]. In addition, only low-frequency (but not high-frequency) DBS of the NAc, refined by selective blockade of dopamine D1 receptors, results in a long-lasting abolishment of behavioral sensitization [26], which refers to the increased locomotor stimulating effect of the drug with repeated injections.

As we have discussed previously [7], the use of DBS in the NAc, at either high or low frequency, to treat addiction should be considered with caution as this approach might lack specificity and lead to unwanted side effects. Accumbens neurons react to cues associated with both natural and drug rewards, and these neurons are not spatially organized in a way to allow for population-specific targeting by current stimulation methods [27]. Questions have been raised as to whether reduction of motivated behavior by DBS can be specific to drugs or will also extend to natural rewards. van der Plasse et al. [28] showed that NAc core DBS does not affect the motivation or the consumption of food. Vassoler et al. [29] also found that DBS applied at the level of NAc medial shell could reduce relapse to cocaine without reducing the desire for food. However, these results are difficult to reconcile with data showing that pharmacological inactivation of the NAc core or medial shell attenuates reinstatement of sugar-seeking behavior [30] or other food-seeking behaviors (for review [31]). The discrepancies between the effect of pharmacological inactivation and DBS of the NAc on seeking behavior for natural reward require further investigation.

Other side effects of DBS of the NAc may include exacerbated impulsivity. While some authors have suggested that the effects of NAc manipulations on impulsivity may be related to the their effect on motivation [32], other authors have proposed that the NAc is important for impulse control [33], which could be disrupted by DBS. In fact, DBS of the NAc in rats has been shown to affect different aspects of impulsivity [34,35].

In addition, the NAc, along with many of the other brain structures of the reward system, contributes not only to excitatory incentive but is also involved in inhibitory incentive [36]. Hence, such general manipulations may act to reduce the incentive for avoiding the negative aspects of drug seeking, which is important for achieving and maintaining abstinence. To the best of our knowledge, no studies have investigated the effect of DBS of the NAc in drug-motivated behavior in the face of adverse consequences.

Clinical Studies

Clinical interventions by lesion of the NAc have been extensively performed in China between 2000 and 2004 in heroin-dependent patients. A follow-up study reported that the abstinence rate dropped from 75% to 20% within 4 years and the procedure was associated with numerous side effects including deficits in concentration, short-term memory, sexual desire, and in various interests [37–39]. These results suggest that, as a treatment for heroin addiction, lesions of the NAc are of questionable effectiveness and safety. The procedure has since been discontinued. On the other hand, DBS of the NAc has been used to successfully treat severe anxiety disorder, depressive disorder, Gilles de la Tourette's syndrome, and obsessive-compulsive disorders (OCD). Notably, OCD and Gilles de la Tourette's syndrome are characterized by intrusive and disruptive thoughts, which lead to the enactment of compulsive behaviors, possibly similar to craving in addiction. The measured success of DBS treatment in OCD and Gilles de la Tourette's suggests that a similar strategy might be effective in drug addiction [40].

Indeed, case studies in patients with OCD or Gilles de la Tourette's syndrome revealed that patients with comorbid alcohol [41,42] or tobacco dependency [40] exhibited a reduction in their drug use. DBS was most effective in helping individuals abstain from tobacco use who were initially less dependent and more motivated to quit [43]. More recently, DBS targeting the NAc has been attempted as a primary indication for the treatment for alcohol addiction in five patients [44,45], for heroin addiction in four patients [46–48], and for cocaine addiction in one patient [49]. Follow-up studies of these patients have reported a general reduction in craving for the drug, but have also highlighted some of the limitations of the procedure. For instance, the patients treated with DBS of the NAc for heroin addiction reported the occasional consumption of other psychotropic substances arising out of boredom or stress [46]. Similarly, in those treated for alcohol dependence, the older patients were observed to relapse in order to relieve perceived stress [44]. Moreover, two out of the five patients may have continued compulsive alcohol use as they died prematurely, most likely from alcohol use. Most of clinical applications of DBS of the NAc for treatment of drug addiction are limited by small samples and the lack of double-blind, sham-controlled designs. To date, results from the few trials that have been conducted suggest that DBS of the NAc may be effective in reducing craving but have more equivocal effectiveness at preventing compulsive drug use. These preliminary clinical outcomes of DBS of the NAc are perhaps not surprising. Considering the role of the NAc in motivation, targeting the NAc is likely to generally affect motivation processes or craving, but not the specific processes affected in drug-addicted individuals, that is, the ability to refrain from using drug compulsively. There is no clear evidence of specific alteration of the NAc in human drug users, who instead tend to show increased gray matter in the dorsal striatum [50]. Moreover, preclinical studies in monkeys have demonstrated alteration of striatal transmission, which is diffuse in the ventral striatum and only associated with recreational drug use; more extensive changes to the more dorsal part of the striatum are observed when compulsive drug use emerges [51].

It is also important to consider that side effects of DBS in the NAc have also been observed including limbic symptoms, such as mania and depression [52]. It also induced impulsivity in two OCD patients [53] and a transient manic-like episode in a patient with Tourette syndrome [41,42]. Forgetfulness or difficulty in finding words have also been reported [54]. Finally, in substance-dependent patients, the induction of a hypomanic state directly after initiation of the DBS [44], and a reduction in sex drive have been reported [49].

Altogether, these results question the viability of targeting the NAc in the treatment of drug addiction. Alternative strategies targeting other structures might offer better efficacy and safety. Some of these are considered below.

Dorsal Striatum

The dorsal striatum was originally considered a motor structure but extended preclinical and clinical studies have demonstrated its involvement in other aspects of motivated behavior. The lateral dorsal striatum shows functional and structural changes in human drug using populations [50]. In rats, the dorsal striatum becomes increasingly involved after a protracted history of drug use. Following extended drug-self administration, drug seeking is no longer sensitive to devaluation of the rewarding properties of the drug but is strongly increased by the presence of drug-associated cues. The sensitivity of instrumental responding to devaluation of the outcome or to Pavlovian influences has been shown to depend on the activity of the dorsal striatum (for review [55]). Pharmacological inactivation of the dorsolateral striatum restores the sensitivity of cocaine or alcohol seeking to the devaluation of drug taking (for review [55]). In addition, pharmacological inactivation of the dorsolateral striatum reduces compulsive cocaine seeking when associated with probabilistic foot-shock punishment, but not when under unpunished conditions [56]. However, DBS of the dorsal striatum, though not specifically targeting its lateral part, failed to alter priming-induced reinstatement of cocaine seeking in rats [29]. It is important to note that these effects were assessed in animals after only a short period of cocaine history, before compulsive cocaine intake had emerged [57,58]. A few other reservations may also limit the application of dorsal striatal DBS. Stimulation of the habit-related regions of the dorsal striatum may propagate to neighboring motor regions and evoke hypokinetic symptoms similar to parkinsonian motor effects. Such undesirable side effects have not yet been reported in the preclinical assessment of the DBS in the dorsal striatum but should be carefully reviewed in further studies.

Amygdala

The amygdala is one of the numerous input regions to the NAc and is thought to provide information about stimulus-outcome associations. Inactivating the BLA in rats reduces cue-induced and context-induced reinstatement of cocaine seeking. This is in accordance with clinical data reporting activation of the amygdala during presentation of drug-associated cues that elicit relapse (for review [59]). On the basis of these findings, Langevin [59] has proposed DBS of the amygdala for the treatment of addiction. However, similar to the NAc, amygdala manipulations are likely to affect both drug and natural reward processes, as well as aversive incentives important for drug cessation. DBS of the amygdala actually affects both natural reward and aversive processes. Applied in the central amygdala, it decreases hedonic and approach responses to natural reward, while increasing aversive responses to quinine [60]. Applied in the BLA, it alleviates fear conditioning [61].

The extended amygdala is responsible for the negative state of withdrawal that might provide a negative motivational incentive to engage in compulsive drug use [62]. Chronic drug use and physical dependence can lead to a state of anxiety and stress during abstinence, which can then lead to a compulsive desire to use drug to alleviate this negative state. In rats, pharmacological inactivation of the extended amygdala can block the anxiogenic effect induced after self-administration of cocaine [63]. In addition, BLA pharmacological inactivation, while reducing relapse provoked by reexposure to the drug self-administration environment after extinction, actually increases relapse after punishment [64]. Similar effects of amygdala inactivation on facilitating compulsive cocaine seeking have been found in related studies [65,66]. Hence, disrupting BLA activity may reduce craving for drug, but may also preclude the negative incentive to abstain from drugs, though this may be specific to negative consequences that induce fear, such as foot shock.

Human alcoholics and cocaine addicts show reduced amygdala volume, which is associated with stronger cravings and a greater chance of relapse [67,68]. However, considering the preclinical evidence, targeting the amygdala for DBS in the treatment of addictive disorder might lack efficacy and safety.

VP and Thalamus

The behavioral effects of surgical manipulation of the VP and the thalamus, despite their strategic location within the reward system, have received little attention. Studies in laboratory animals have proposed the thalamus as an information hub or relay between different subcortical areas and the cerebral cortex, and the VP as the "final common pathway" for drug-seeking behavior [69] and for reward processing in general [70].

Preclinical studies in the field of reward and addiction have generally concentrated their attention on the paraventricular nucleus of the thalamus (PVT). PVT neurons in rats showed increased activity to the reexposure to an alcohol self-administration environment or presentation of discriminative stimuli previously linked with alcohol reward [71,72]. Lesion of the PVT blocks cocaine sensitization [73] and its pharmacological inactivation prevents context-induces reinstatement of alcohol seeking [74] and cocaine-primed reinstatement [75]. In terms of the VP, its activity is sensitive to fluctuating self-administered cocaine levels in rats [76]. Lesion of the VP decreases self-administration of cocaine but not the motivation for cocaine measured in a progressive ratio schedule of reinforcement [77]. Pharmacological inactivation of the VP decreases reinstatement of cocaine or heroin seeking triggered by the drug [78,79]. It also blocks reinstatement of cocaine seeking triggered by foot shock [80] and heroin seeking triggered by drug-associated cues. [79].

However, similar to the NAc and amygdala, the PVT and the VP are implicated in both drug and natural reward processes, as well as in aversive processes important for cessation of drug seeking. Morphological and functional changes in thalamic structures have been reported in drug-dependent individuals [81,82]. However, lesions of the PVT or pharmacological inactivation of the PVT with the GABA agonist muscimol increases food intake in rats [83,84] and attenuates conditioned fear to a tone [85,86].

Several clinical reports point to VP involvement in disorders of motivation [87]. In the context of addiction, cocaine-using subjects treated with baclofen (a GABAB receptor agonist) displayed significantly less activation in a large interconnected bilateral cluster comprising the VP in response to subliminal cocaine (vs neutral) cues, but not sexual or aversive (vs neutral) cues, relative to placebo-treated participants [88]. In preclinical studies, lesion of the VP attenuates conditioned place preference induced by sucrose [89] and its pharmacological inactivation decreases reinstatement of food-seeking triggered by food [78]. Moreover, pharmacological inactivation of the VP in monkeys increases the number of errors during aversive trials in an approach/avoidance task [90] and reduced conditioned aversion in rats [91]. DBS of the VP reduces seizures without affecting gross locomotor activity [92] but no studies have investigated its effect on motivated behaviors.

Given the paucity of preclinical data, it is difficult at this time to predict the effects of DBS of the VP or the PVN in drug-dependent individuals but given their general effect on motivation, targeting the PVT and VP for stimulation-based therapies is likely to lead to unwanted side effects, much like the other subcortical regions that have been discussed.

Prefrontal Cortex

Surgical Interventions

Increasing evidence suggests that poor inhibitory control due to suboptimal functioning of prefrontal cortical structures contributes to addiction [93–95]. Individuals addicted to cocaine or other drugs of abuse show decreased gray matter volume in several subregions of the PFC, including the orbitofrontal cortex (OFC), anterior cingulate cortex (ACC), and insular cortex (IC) [50,93,96,97]. Preclinical animal models have demonstrated a progressive reduction in prefrontal cortical metabolic activity over the course of long cocaine self-administration history [51,98].

From 1962 to 2005, cingulotomies were performed for the treatment of intractable pain and for OCD with comorbid opiate dependence. Cingulotomies were also performed in cases where the primary indication was for drug dependence in an effort to interrupt obsessional thoughts about drug use. Depending on the study, abstinence rates between 62% and 100% have been reported from 3 to 374 patients at follow ups ranging from 2 months to 15 years. Notably, no long-term complications were described. However, side effects were not thoroughly documented at that time and dropout rates from studies were high. Only recently reports have revealed cases of patients having undergone cingulotomy and being impaired in motivation, attention and executive functions [for review [99]]. In addition, metaanalysis of cingulotomy effects over several psychiatric disorders have revealed that this surgical treatment has very little efficacy for treating addictive behaviors. This led the authors to raise doubts about using DBS of the cingulate cortex for the treatment of addiction [100], even though DBS is likely to have fewer side effects than a complete surgical ablation.

Preclinical studies have demonstrated opposing effects of ventral and dorsal medial PFC lesions on relapse following extinction. Pharmacological inactivation of the ventral regions enhances reinstatement, while pharmacological inactivation of the dorsal prevents reinstatement [101–103]. Other results also indicate that both ventral and dorsal mPFC regions can drive and/or inhibit drug seeking (and other types of behaviors) depending on a range of factors including the behavioral context, the drug history of the animal, and the type of drug investigated (for review [104]).

For example, after rats have been punished, prelimbic cortex lesioned animals show a decrease in seeking responses during relapse, whereas those with anterior IC lesions show an increase in cocaine seeking [65]. PFC DBS in rats reduces the effect of foot shock on anxiety-like behavior and promotes extinction of conditioned fear [105], suggesting that stimulation may affect mood and cognitive control. However, very recent data have shown that DBS of the lateral OFC increases compulsive behavior as measured in a reversal learning paradigm [106], indicating that different effects may be observed depending on the cortical region targeted.

There are only a few preclinical studies investigating DBS of the PFC, likely because such invasive procedures may not be as necessary for these structures because they are easily accessible by other noninvasive procedures, such as transcranial magnetic stimulation (TMS) or transcranial direct-current stimulation (tDCS). In fact, electrical stimulations of the medial PFC (prelimbic), not applied using standard DBS parameters but at parameters mimicking TMS, reduce the consumption and motivation for cocaine but not for sucrose. This suggests that TMS of the PFC could represent a strategy for the treatment of addiction [107]. However, these neuromodulatory procedures do not access deep cortical structures, such as the IC or the OFC, making these two intriguing targets for treatment, particularly considering that there is substantial evidence that these regions are involved in the pathology of addiction.

Noninvasive Interventions

Transcranial Direct-Current Stimulation

tDCS consists of delivering pulses of low current via electrodes positioned on the scalp. tDCS is thought to act by changing neural resting membrane potentials and hence their excitability [108,109], even after the stimulation has ended [110,111]. The technique is relatively safe with reported side effects mainly limited to local tingling and skin irritation. Most clinical trials of therapeutic tDCS for addiction to date have targeted the dorsolateral PFC (DLPFC) and used anodal electrodes that generally increase cortical excitability. The majority of these studies report a reduction in craving after tDCS, whether it is for tobacco [112–114], cocaine [115–117], cannabis [118], alcohol [110,111,119,120], heroin [121], or methamphetamine [122]. In contrast, others report no change in craving but rather a decrease in negative affect [123] or the amount of drug consumed (cigarettes) [124,125].

The beneficial effects of stimulation on reducing craving are typically observed at rest, in the presence of drug-associated cues. The notable exception is for methamphetamine, for which the stimulation increases craving induced by the cues [122]. Moreover, there is no relationship between the effect of tDCS on craving and its effect on relapse. For example, da Silva et al. [119] have reported that tDCS reduces craving for alcohol, but with a trend for increased propensity to relapse after stimulation. Conversely, Klauss et al. [126] found that stimulation has no effect on craving but reduces the propensity to relapse.

There is also doubt about whether stimulation above the DLPFC has an effect on the DLPFC itself, or on other prefrontal areas ([126a,126b,126c]). Nevertheless, activation does result in a reduction in risky behaviors [126d] and in an increase of rejection of cigarettes offers at the Ultimatum Game [124], suggesting that the procedure could help addicted individuals regain control over drug use. However, tDCS has also been shown to increase risky behaviors in marijuana users ([118]). Further studies are required to understand the effect of tDCS on decision making in addicted individuals.

The large majority of these studies are preliminary, with small sample sizes, with little patient follow-up, and considerable heterogeneity in the sample populations, study design, tDCS parameters, and outcome measurements. More recently, a large multicentered clinical trial has recently been proposed [127]. Such a large, well-controlled trial will likely help answer the many open questions about the potential efficacy of tDCS of the DLPFC in addiction treatment. In the meantime, while there is good evidence that tDCS can help reduce craving, it is premature to conclude that tDCS over the DLPFC is an effective procedure for reducing relapse.

Very limited preclinical studies have tested the effect of tDCS on addiction-like behaviors, in part due to the challenge of targeting specific regions in the substantially smaller rodent brain. In mice, increases in nicotine-induced place preference and depression-like behavior after chronic nicotine exposure are normalized by repeated tDCS [128]. The same group has also shown that tDCS pretreatment attenuates cocaine-induced locomotor activation and place preference conditioning in mice [129].

Transcranial Magnetic Stimulation

TMS uses a magnetic field generator, or "coil," connected to a pulse generator, or stimulator, that delivers electric current to the coil. The coil placed near the scalp of the patient then produces small electric currents in the targeted region under the coil via electromagnetic induction. In contrast to tDCS, TMS can generate action potentials, and hence directly activate a brain structure. Efficiency of the stimulation can be directly assessed by motor response to stimulation in the motor cortex. Similarly to tDCS, most studies apply TMS just above the DLPFC. The frequency of stimulation can vary between 1 and 20 Hz, with higher frequencies often resulting in adverse effects and lower frequencies generally reducing the activity of the structure [130,131]. TMS is well tolerated with occasional transient headache and scalp discomfort.

Directly after stimulation, clinical studies have demonstrated a reduction in spontaneous craving for cigarettes [37–39,132,133], and in the number of cigarettes smoked [134]. Studies have also found reduced craving for cocaine [135] but increased cue-induced craving for methamphetamine [37–39]. The beneficial effect of repeated TMS sessions on alcohol craving [136] and the number of cigarettes smoked [137,138] were persistent over prolonged periods of time, even reducing the number of relapse episodes [137–140], and general nicotine dependence. Indeed, the fact that TMS decreases different aspects of impulsivity, strongly suggests [141] that the procedure may help patients regain control over drug use.

Similarly to tDCS, the therapeutic mechanism of TMS in addiction is not well established, but some have hypothesized that TMS modulates glutamatergic or dopaminergic activity in the targeted structure, or that TMS leads to orthodromic or antidromic stimulation of the dopaminergic system [142].

REVISITING THE CLASSICAL REWARD CIRCUIT: ROLE OF STN

Given that the STN has substantial connectivity to various brain regions in the reward circuit, it appears to be positioned similarly to the NAc in its potential to regulate motivated behaviors. Although in the classic version of the reward circuit illustrated in Fig. 1 the STN was absent, there is a growing appreciation that the STN can regulate reward systems. An illustration of the anatomical connectivity of the STN is summarized in Fig. 2 and shows that the STN receives direct excitatory glutamatergic inputs from the PFC, a pathway also called the "hyper-direct pathway," from the thalamus (parafascicular-centromedian complex), and from the BLA (although the recent description of the connectivity between these two structures does not provide information regarding its direction). The STN also has reciprocal connections with the ventral and dorsal pallidum, receiving GABA inputs and sending its glutamatergic outputs back. It also sends projections to the DA-enriched areas (VTA and SNc) and receives DA inputs from both. It is important to note that the NAc and STN can influence VTA, and therefore DA neurons, as well as pallidal output.

It is striking to note that the NAc receives extensive inputs from structures involved in the retrieval of stimulus-outcome associations, such as the hippocampus, suggesting that this nucleus integrates expected value from the environment.

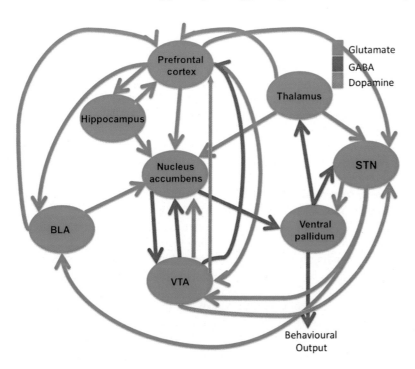

FIG. 2 Revisited schematic representation of the reward circuit showing the interconnection with STN. *BLA*, basolateral amygdala; *VTA*, ventral tegmental area.

In contrast, the STN neither seem to receive direct inputs from the hippocampus nor to be involved in learning and memory processes. This would suggest that the STN is much more involved in integrating information only related to value, as would be required for conflict detection. Two major computational models of STN function posit a crucial role for the STN in response inhibition under conditions in which several responses are possible when conflicting information is presented to the subject [143,144], like when drug users face adverse consequences.

STRATEGIES OF TREATMENTS AIMED AT THE STN

Lesions

Excitotoxic lesions of the STN increase perseverative responses oriented toward a food magazine associated with reward collection in an attentional task [145], suggesting a higher level of motivation for the food reward. This effect was further confirmed by other labs in studies showing that STN lesions increase incentive motivation for food reward [146,147]. Intriguingly, STN lesions reduce motivation to work for cocaine on a PR schedule of reinforcement, while increasing motivation for sweet food reward [9]. Similar conclusions were later obtained from other groups showing that STN pharmacological inactivation reduced economic demand for cocaine and drug seeking during extinction [148] although it was not seen in rats with a poor level of acquisition of the self-administration procedure for which the lesion seemed to have a facilitating effect [149]. The opposite effect for sweet food and cocaine was observed using the conditioned place preference test [9]. These results cannot be attributed to connectivity of the STN within the reward circuit. However, they are particularly interesting if one considers that the aim of a treatment strategy for addiction is to reduce motivation for the drug without reducing all other forms of motivation, a challenge that is difficult to overcome when targeting the NAc and certain other subcortical structures.

The effects of STN lesions are not limited to cocaine-related behaviors. Indeed, lesion of the STN modulates motivation for alcohol though the effect depends on the initial level of individual consumption [150]. In addition, oxytocin can modulate the STN to reduce reinstatement of methamphetamine seeking [151]. Moreover, pharmacological inactivation of the VP output pathways to the STN prevents context-induced reinstatement of alcohol seeking and reacquisition of alcohol self-administration, indicating that disrupting STN activity may be sufficient to reduce drug use [152]. Finally, recent evidence suggests that the STN is not only involved in reinforcement by drugs of abuse but also in the development of the pathological features of drug addiction. The loss of control over drug intake has been operationalized in rats by escalation of intake. STN lesion prevents the exacerbated increases in ethanol intake observed during intermittent ethanol access and after a long period of ethanol deprivation and also reduces the motivation to work for ethanol after escalated intake [153].

Deep Brain Stimulation

Given that lesions of the STN show opposite effects on the motivation for cocaine and sugar [9], the structure may represent an interesting target for the treatment of addiction. DBS of the STN is already routinely used in Parkinsonian patients to treat their motor symptoms with great success [5,154]. The treatment consists of implanting stimulating electrodes and applying chronic high-frequency stimulation (HFS). Reports from some Parkinsonian patients suggest that STN stimulation decreases addiction-related behavior toward DAergic treatment after DBS [155–159].

DBS of the STN also has beneficial effects in psychiatric disorders, such as OCD [160,161], in line with preclinical data demonstrating that the STN is involved in nonmotor processes including response inhibition and impulsivity [32,162,163]. To date, there are no reports of STN DBS effects on any form of addiction in OCD patients, but alleviation of compulsion by DBS of the STN is highly interesting when considering the compulsive aspects of addiction [164].

We have tested the effects of bilateral STN HFS on the rewarding properties of cocaine and sugar in rats using the conditioned place preference test and a PR schedule of reinforcement. We have found that bilateral STN HFS replicates the effects of STN lesions: reduced motivation and responding for cocaine and increased motivation and responding for sweet food [11]. Accordingly, the reduction of the rewarding effect of cocaine induced by STN DBS is associated with the reduction in the cellular response to cocaine in the striatum [165]. However, drug addiction is not solely defined by an increase in motivation for the drug, but also by the loss of control over drug intake and the maintenance of drug use at the expense of other reward and despite adverse consequences (DSMV). More recently, we have used a rat model of escalation of drug intake over extended access to examine the effects of STN DBS on the loss of control over drug intake. We have shown that STN HFS can prevent cocaine escalation and reduce reescalation after protracted abstinence [165a]. This latter effect was also shown for reescalation of heroin intake [166]. Additional studies are currently investigating the effect of STN DBS on other features of addictive-like behavior in rats.

CONCLUSION

Craving as an exacerbated motivation to use drugs is the most common feature analyzed in clinical trials aimed at the treatment of substance use disorders. But this measure is somewhat limited in its ability to predict the efficacy of various treatments in the prevention of relapse (this being the ultimate goal for addiction treatment). Clinical trials of NAc DBS or tDCS treatment for drug-dependent patients similarly show that reports of reduced craving in the laboratory are not good indicators of long-term outcomes for addiction. The fact that some patients relapse or continue to use drugs compulsively after these treatments, suggests that NAc stimulation strategies are of equivocal efficacy. Further studies are needed that include measures of drug addiction beyond exacerbated desire for the drug (or craving), such as maintenance of drug use at the expense of other rewards and DSMV.

The "reward system" is not only involved in orienting responses toward rewards, but also plays a role in regulating avoidance of aversive stimuli. Recent studies report that in the insular/orbital cortex [167], the shell of the NAc [167], the VP [168], and the amygdala [169] neurons are spatially organized such that those regulating reward and aversive processes are located in different parts of the structure, as opposed to being intermixed. Thus, it may be possible to target spatially defined subcomponents of the reward system to induce specific outcomes. However, we think that more specific spatial targeting may still have unwanted side effects because motivation for nondrug rewards is likely to still be affected [174,175]. An alternative to targeting the reinforcing effects of a drug is to identify a strategy that allows the individual to recognize the conflict between the positive and negative aspects of drug use in order to restore control over their drug using behavior.

Toward this goal, considerable evidence has now emerged implicating involvement of the STN in regulating conflict processing [143,144] and behavioral control [32]. Beneficial effects of STN manipulations on choice impulsivity and risk taking in rats [170,171], on compulsive behavior in OCD patients [160,161], and for impulse control disorders induced by LDOPA in PD patients [155–159], further support targeting the STN for addictive disorders. Our recent results in the rat confirm that STN manipulations can regulate compulsive drug seeking [159a]. Although STN DBS is an invasive procedure and difficult to offer to large numbers of patients, it is widely performed in the context of Parkinson's disease, making it feasible Alternatively, recent developments in noninvasive DBS may provide new opportunities to target the STN with HFS without the risks associated with the electrode implantation [172]. Further studies will be necessary to evaluate these possibilities. Meanwhile, other forms of noninvasive strategies, such as TMS, are moving forward in an effort to restore the abnormal cortical activity of drug-dependent individuals [173].

ACKNOWLEDGMENT

The authors would like to thank Adam Mar for helpful comments and revision of the manuscript.

REFERENCES

[1] Batistuzzo MC, Hoexter MQ, Taub A, Gentil AF, Cesar RC, Joaquim MA, D'Alcante CC, McLaughlin NC, Canteras MM, Shavitt RG, Savage CR, Greenberg BD, Noren G, Miguel EC, Lopes AC. Visuospatial memory improvement after gamma ventral capsulotomy in treatment refractory obsessive-compulsive disorder patients. Neuropsychopharmacology 2015;40(8):1837–45.

[2] Leveque M, Carron R, Regis J. Radiosurgery for the treatment of psychiatric disorders: a review. World Neurosurg 2013;80(3–4). S32.e31–39.

[3] Lopes AC, Greenberg BD, Noren G, Canteras MM, Busatto GF, de Mathis ME, Taub A, D'Alcante CC, Hoexter MQ, Gouvea FS, Cecconi JP, Gentil AF, Ferrao YA, Fuentes D, de Castro CC, Leite CC, Salvajoli JV, Duran FL, Rasmussen S, Miguel EC. Treatment of resistant obsessive-compulsive disorder with ventral capsular/ventral striatal gamma capsulotomy: a pilot prospective study. J Neuropsychiatry Clin Neurosci 2009;21(4):381–92.

[4] Spofford CM, McLaughlin NC, Penzel F, Rasmussen SA, Greenberg BD. OCD behavior therapy before and after gamma ventral capsulotomy: case report. Neurocase 2014;20(1):42–5.

[5] Limousin P, Pollak P, Benazzouz A, Hoffmann D, Le Bas JF, Broussolle E, Perret JE, Benabid AL. Effect of parkinsonian signs and symptoms of bilateral subthalamic nucleus stimulation. Lancet 1995;345(8942):91–5.

[6] Gubellini P, Salin P, Kerkerian-Le Goff L, Baunez C. Deep brain stimulation in neurological diseases and experimental models: From molecule to complex behavior. Prog Neurobiol 2009;89(1):79–123.

[7] Pelloux Y, Baunez C. Deep brain stimulation for addiction: why the subthalamic nucleus should be favored. Curr Opin Neurobiol 2013;23 (4):713–20.

[8] Martinez-Rivera FJ, Rodriguez-Romaguera J, Lloret-Torres ME, Do Monte FH, Quirk GJ, Barreto-Estrada JL. Bidirectional modulation of extinction of drug seeking by deep brain stimulation of the ventral striatum. Biol Psychiatry 2016;80(9):682–90.

[9] Baunez C, Dias C, Cador M, Amalric M. The subthalamic nucleus exerts opposite control on cocaine and 'natural' rewards. Nat Neurosci 2005;8 (4):484–9.

[10] Krack P, Hariz MI, Baunez C, Guridi J, Obeso JA. Deep brain stimulation: from neurology to psychiatry? Trends Neurosci 2010;33 (10):474–84.

[11] Rouaud T, Lardeux S, Panayotis N, Paleressompoulle D, Cador M, Baunez C. Reducing the desire for cocaine with subthalamic nucleus deep brain stimulation. Proc Natl Acad Sci U S A 2010;107(3):1196–200.

[12] Carter A, Hall W. Proposals to trial deep brain stimulation to treat addiction are premature. Addiction 2011;106(2):235–7.

[13] Luigjes J, van den Brink W, Schuurman PR, Kuhn J, Denys D. Is deep brain stimulation a treatment option for addiction? Addiction 2015;110 (4):547–8.

[14] Mogenson GJ, Jones DL, Yim CY. From motivation to action: Functional interface between the limbic system and the motor system. Prog Neurobiol 1980;14(2–3):69–97.

[15] Robinson TE, Berridge KC. The neural basis of drug craving: An incentive-sensitization theory of addiction. Brain Res Brain Res Rev 1993;18 (3):247–91.

[16] Everitt BJ, Dickinson A, Robbins TW. The neuropsychological basis of addictive behaviour. Brain Res Brain Res Rev 2001;36(2–3):129–38.

[17] Kalivas PW, McFarland K. Brain circuitry and the reinstatement of cocaine-seeking behavior. Psychopharmacology (Berl) 2003;168 (1–2):44–56.

[18] Meredith GE, Baldo BA, Andrezjewski ME, Kelley AE. The structural basis for mapping behavior onto the ventral striatum and its subdivisions. Brain Struct Funct 2008;213(1–2):17–27.

[19] Voorn P, Gerfen CR, Groenewegen HJ. Compartmental organization of the ventral striatum of the rat: immunohistochemical distribution of enkephalin, substance P, dopamine, and calcium-binding protein. J Comp Neurol 1989;289(2):189–201.

[20] Liu HY, Jin J, Tang JS, Sun WX, Jia H, Yang XP, Cui JM, Wang CG. Chronic deep brain stimulation in the rat nucleus accumbens and its effect on morphine reinforcement. Addict Biol 2008;13(1):40–6.

[21] Guo L, Zhou H, Wang R, Xu J, Zhou W, Zhang F, Tang S, Liu H, Jiang J. DBS of nucleus accumbens on heroin seeking behaviors in self-administering rats. Drug Alcohol Depend 2013;129(1–2):70–81.

[22] Knapp CM, Tozier L, Pak A, Ciraulo DA, Kornetsky C. Deep brain stimulation of the nucleus accumbens reduces ethanol consumption in rats. Pharmacol Biochem Behav 2009;92(3):474–9.

[23] Henderson MB, Green AI, Bradford PS, Chau DT, Roberts DW, Leiter JC. Deep brain stimulation of the nucleus accumbens reduces alcohol intake in alcohol-preferring rats. Neurosurg Focus 2010;29(2):E12.

[24] Batra V, Tran TL, Caputo J, Guerin GF, Goeders NE, Wilden J. Intermittent bilateral deep brain stimulation of the nucleus accumbens shell reduces intravenous methamphetamine intake and seeking in Wistar rats. J Neurosurg 2017;126(4):1339–50.

[25] Creed M, Bonci A, Leggio L. Modulating morphine context-induced drug memory with deep brain stimulation: more research questions by lowering stimulation frequencies? Biol Psychiatry 2016;80(9):647–9.

[26] Creed M, Pascoli VJ, Luscher C. Addiction therapy. Refining deep brain stimulation to emulate optogenetic treatment of synaptic pathology. Science 2015;347(6222):659–64.

[27] Carelli RM, Wondolowski J. Anatomic distribution of reinforcer selective cell firing in the core and shell of the nucleus accumbens. Synapse 2006;59(2):69–73.

[28] van der Plasse G, Schrama R, van Seters SP, Vanderschuren LJ, Westenberg HG. Deep brain stimulation reveals a dissociation of consummatory and motivated behaviour in the medial and lateral nucleus accumbens shell of the rat. PLoS One 2012;7(3):e33455.

[29] Vassoler FM, Schmidt HD, Gerard ME, Famous KR, Ciraulo DA, Kornetsky C, Knapp CM, Pierce RC. Deep brain stimulation of the nucleus accumbens shell attenuates cocaine priming-induced reinstatement of drug seeking in rats. J Neurosci 2008;28(35):8735–9.

[30] Lin P, Pratt WE. Inactivation of the nucleus accumbens core or medial shell attenuates reinstatement of sugar-seeking behavior following sugar priming or exposure to food-associated cues. PLoS One 2014;9:.

[31] Baldo BA, Pratt WE, Will MJ, Hanlon EC, Bakshi VP, et al. Principles of motivation revealed by the diverse functions of neuropharmacological and neuroanatomical substrates underlying feeding behavior. Neurosci Biobehav Rev 2013;37:1985–98.

[32] Eagle DM, Baunez C. Is there an inhibitory-response-control system in the rat? Evidence from anatomical and pharmacological studies of behavioral inhibition. Neurosci Biobehav Rev 2010;34(1):50–72.

[33] Cardinal RN. Neural systems implicated in delayed and probabilistic reinforcement. Neural Netw 2006;19(8):1277–301.

[34] Schippers MC, Bruinsma B, Gaastra M, Mesman TI, Denys D, De Vries TJ, Pattij T. Deep brain stimulation of the nucleus Accumbens Core affects trait impulsivity in a baseline-dependent manner. Front Behav Neurosci 2017;11:52.

[35] Sesia T, Temel Y, Lim LW, Blokland A, Steinbusch HW, Visser-Vandewalle V. Deep brain stimulation of the nucleus accumbens core and shell: opposite effects on impulsive action. Exp Neurol 2008;214(1):135–9.

[36] Johansson AK, Hansen S. Increased alcohol intake and behavioral disinhibition in rats with ventral striatal neuron loss. Physiol Behav 2000;70 (5):453–63.

[37] Li N, Wang J, Wang XL, Chang CW, Ge SN, Gao L, Wu HM, Zhao HK, Geng N, Gao GD. Nucleus accumbens surgery for addiction. World Neurosurg 2013;80(3–4). S28.e29–19.

[38] Li X, Hartwell KJ, Owens M, Lematty T, Borckardt JJ, Hanlon CA, Brady KT, George MS. Repetitive transcranial magnetic stimulation of the dorsolateral prefrontal cortex reduces nicotine cue craving. Biol Psychiatry 2013;73(8):714–20.

[39] Li X, Malcolm RJ, Huebner K, Hanlon CA, Taylor JJ, Brady KT, George MS, See RE. Low frequency repetitive transcranial magnetic stimulation of the left dorsolateral prefrontal cortex transiently increases cue-induced craving for methamphetamine: a preliminary study. Drug Alcohol Depend 2013;133(2):641–6.

[40] Mantione M, van de Brink W, Schuurman PR, Denys D. Smoking cessation and weight loss after chronic deep brain stimulation of the nucleus accumbens: therapeutic and research implications: case report. Neurosurgery 2010;66(1):E218. discussion E218.

[41] Kuhn J, Lenartz D, Huff W, Lee S, Koulousakis A, Klosterkoetter J, Sturm V. Remission of alcohol dependency following deep brain stimulation of the nucleus accumbens: Valuable therapeutic implications? J Neurol Neurosurg Psychiatry 2007;78(10):1152–3.

[42] Kuhn J, Lenartz D, Mai JK, Huff W, Lee SH, Koulousakis A, Klosterkoetter J, Sturm V. Deep brain stimulation of the nucleus accumbens and the internal capsule in therapeutically refractory Tourette-syndrome. J Neurol 2007;254(7):963–5.

[43] Kuhn J, Bauer R, Pohl S, Lenartz D, Huff W, Kim EH, Klosterkoetter J, Sturm V. Observations on unaided smoking cessation after deep brain stimulation of the nucleus accumbens. Eur Addict Res 2009;15(4):196–201.

[44] Muller UJ, Sturm V, Voges J, Heinze HJ, Galazky I, Buntjen L, Heldmann M, Frodl T, Steiner J, Bogerts B. Nucleus Accumbens deep brain stimulation for alcohol addiction—safety and clinical long-term results of a pilot trial. Pharmacopsychiatry 2016;49(4):170–3.

[45] Voges J, Muller U, Bogerts B, Munte T, Heinze HJ. Deep brain stimulation surgery for alcohol addiction. World Neurosurg 2013;80(3–4). S28. e21–31.

[46] Kuhn J, Moller M, Treppmann JF, Bartsch C, Lenartz D, Gruendler TO, Maarouf M, Brosig A, Barnikol UB, Klosterkotter J, Sturm V. Deep brain stimulation of the nucleus accumbens and its usefulness in severe opioid addiction. Mol Psychiatry 2014;19(2):145–6.

[47] Valencia-Alfonso CE, Luigjes J, Smolders R, Cohen MX, Levar N, Mazaheri A, van den Munckhof P, Schuurman PR, van den Brink W, Denys D. Effective deep brain stimulation in heroin addiction: a case report with complementary intracranial electroencephalogram. Biol Psychiatry 2012;71 (8):e35–37.

[48] Zhou H, Xu J, Jiang J. Deep brain stimulation of nucleus accumbens on heroin-seeking behaviors: a case report. Biol Psychiatry 2011;69(11): e41–42.

[49] Goncalves-Ferreira A, do Couto FS, Rainha Campos A, Lucas Neto LP, Goncalves-Ferreira D, Teixeira J. Deep brain stimulation for refractory cocaine dependence. Biol Psychiatry 2016;79(11):e87–89.

[50] Ersche KD, Barnes A, Jones PS, Morein-Zamir S, Robbins TW, Bullmore ET. Abnormal structure of frontostriatal brain systems is associated with aspects of impulsivity and compulsivity in cocaine dependence. Brain 2011;134(Pt 7):2013–24.

[51] Porrino LJ, Smith HR, Nader MA, Beveridge TJ. The effects of cocaine: a shifting target over the course of addiction. Prog Neuropsychopharmacol Biol Psychiatry 2007;31(8):1593–600.

[52] Saleh C, Okun MS. Clinical review of deep brain stimulation and its effects on limbic basal ganglia circuitry. Front Biosci 2008;13:5708–31.

[53] Luigjes J, Mantione M, van den Brink W, Schuurman PR, van den Munckhof P, Denys D. Deep brain stimulation increases impulsivity in two patients with obsessive-compulsive disorder. Int Clin Psychopharmacol 2011;26(6):338–40.

[54] Denys D, Mantione M, Figee M, van den Munckhof P, Koerselman F, Westenberg H, Bosch A, Schuurman R. Deep brain stimulation of the nucleus accumbens for treatment-refractory obsessive-compulsive disorder. Arch Gen Psychiatry 2010;67(10):1061–8.

[55] Hogarth L, Balleine BW, Corbit LH, Killcross S. Associative learning mechanisms underpinning the transition from recreational drug use to addiction. Ann N Y Acad Sci 2013;1282:12–24.

[56] Jonkman S, Pelloux Y, Everitt BJ. Differential roles of the dorsolateral and midlateral striatum in punished cocaine seeking. J Neurosci 2012;32 (13):4645–50.

[57] Pelloux Y, Dilleen R, Economidou D, Theobald D, Everitt BJ. Reduced forebrain serotonin transmission is causally involved in the development of compulsive cocaine seeking in rats. Neuropsychopharmacology 2012;37(11):2505–14.

[58] Pelloux Y, Everitt BJ, Dickinson A. Compulsive drug seeking by rats under punishment: effects of drug taking history. Psychopharmacology (Berl) 2007;194(1):127–37.

[59] Langevin JP. The amygdala as a target for behavior surgery. Surg Neurol Int 2012;3(Suppl 1):S40–46.

[60] Ross SE, Lehmann Levin E, Itoga CA, Schoen CB, Selmane R, Aldridge JW. Deep brain stimulation in the central nucleus of the amygdala decreases 'wanting' and 'liking' of food rewards. Eur J Neurosci 2016;44(7):2431–45.

[61] Sui L, Huang S, Peng B, Ren J, Tian F, Wang Y. Deep brain stimulation of the amygdala alleviates fear conditioning-induced alterations in synaptic plasticity in the cortical-amygdala pathway and fear memory. J Neural Transm (Vienna) 2014;121(7):773–82.

[62] Koob GF, Volkow ND. Neurocircuitry of addiction. Neuropsychopharmacology 2010;35(1):217–38.

[63] Wenzel JM, Waldroup SA, Haber ZM, Su ZI, Ben-Shahar O, Ettenberg A. Effects of lidocaine-induced inactivation of the bed nucleus of the stria terminalis, the central or the basolateral nucleus of the amygdala on the opponent-process actions of self-administered cocaine in rats. Psychopharmacology (Berl) 2011;217(2):221–30.

[64] Pelloux Y, Minier-Tobero A, Li X, Hoots JK, Bossert JM, Shaham Y. Opposite effects of amygdala inactivation on context-induced relapse to cocaine seeking after suppression of drug self-administration by extinction versus punishment. J Neurosci 2018;38(1):51–9.

[65] Pelloux Y, Murray JE, Everitt BJ. Differential roles of the prefrontal cortical subregions and basolateral amygdala in compulsive cocaine seeking and relapse after voluntary abstinence in rats. Eur J Neurosci 2013;38(7):3018–26.

[66] Xue Y, Steketee JD, Sun W. Inactivation of the central nucleus of the amygdala reduces the effect of punishment on cocaine self-administration in rats. Eur J Neurosci 2012;35(5):775–83.

[67] Hill SY, De Bellis MD, Keshavan MS, Lowers L, Shen S, Hall J, Pitts T. Right amygdala volume in adolescent and young adult offspring from families at high risk for developing alcoholism. Biol Psychiatry 2001;49(11):894–905.

[68] Wrase J, Makris N, Braus DF, Mann K, Smolka MN, Kennedy DN, Caviness VS, Hodge SM, Tang L, Albaugh M, Ziegler DA, Davis OC, Kissling C, Schumann G, Breiter HC, Heinz A. Amygdala volume associated with alcohol abuse relapse and craving. Am J Psychiatry 2008;165(9):1179–84.

[69] Kalivas PW, Volkow ND. The neural basis of addiction: a pathology of motivation and choice. Am J Psychiatry 2005;162(8):1403–13.

[70] Smith KS, Tindell AJ, Aldridge JW, Berridge KC. Ventral pallidum roles in reward and motivation. Behav Brain Res 2009;196(2):155–67.

[71] Dayas CV, McGranahan TM, Martin-Fardon R, Weiss F. Stimuli linked to ethanol availability activate hypothalamic CART and orexin neurons in a reinstatement model of relapse. Biol Psychiatry 2008;63(2):152–7.

[72] Wedzony K, Koros E, Czyrak A, Chocyk A, Czepiel K, Fijal K, Mackowiak M, Rogowski A, Kostowski W, Bienkowski P. Different pattern of brain c-Fos expression following re-exposure to ethanol or sucrose self-administration environment. Naunyn Schmiedebergs Arch Pharmacol 2003;368 (5):331–41.

[73] Young CD, Deutch AY. The effects of thalamic paraventricular nucleus lesions on cocaine-induced locomotor activity and sensitization. Pharmacol Biochem Behav 1998;60(3):753–8.

[74] Hamlin AS, Clemens KJ, Choi EA, McNally GP. Paraventricular thalamus mediates context-induced reinstatement (renewal) of extinguished reward seeking. Eur J Neurosci 2009;29(4):802–12.

[75] James MH, Charnley JL, Jones E, Levi EM, Yeoh JW, Flynn JR, Smith DW, Dayas CV. Cocaine- and amphetamine-regulated transcript (CART) signaling within the paraventricular thalamus modulates cocaine-seeking behaviour. PLoS One 2010;5(9).

[76] Root DH, Fabbricatore AT, Ma S, Barker DJ, West MO. Rapid phasic activity of ventral pallidal neurons during cocaine self-administration. Synapse 2010;64(9):704–13.

[77] Robledo P, Koob GF. Two discrete nucleus accumbens projection areas differentially mediate cocaine self-administration in the rat. Behav Brain Res 1993;55(2):159–66.

[78] McFarland K, Kalivas PW. The circuitry mediating cocaine-induced reinstatement of drug-seeking behavior. J Neurosci 2001;21(21):8655–63.

[79] Rogers JL, Ghee S, See RE. The neural circuitry underlying reinstatement of heroin-seeking behavior in an animal model of relapse. Neuroscience 2008;151(2):579–88.

[80] McFarland K, Davidge SB, Lapish CC, Kalivas PW. Limbic and motor circuitry underlying footshock-induced reinstatement of cocaine-seeking behavior. J Neurosci 2004;24(7):1551–60.

[81] Denier N, Schmidt A, Gerber H, Vogel M, Huber CG, Lang UE, Riecher-Rossler A, Wiesbeck GA, Radue EW, Walter M, Borgwardt S. Abnormal functional integration of thalamic low frequency oscillation in the BOLD signal after acute heroin treatment. Hum Brain Mapp 2015;36 (12):5287–300.

[82] Moeller SJ, Tomasi D, Woicik PA, Maloney T, Alia-Klein N, Honorio J, Telang F, Wang GJ, Wang R, Sinha R, Carise D, Astone-Twerell J, Bolger J, Volkow ND, Goldstein RZ. Enhanced midbrain response at 6-month follow-up in cocaine addiction, association with reduced drug-related choice. Addict Biol 2012;17(6):1013–25.

[83] Bhatnagar S, Dallman MF. The paraventricular nucleus of the thalamus alters rhythms in core temperature and energy balance in a state-dependent manner. Brain Res 1999;851(1–2):66–75.

[84] Stratford TR, Wirtshafter D. Injections of muscimol into the paraventricular thalamic nucleus, but not mediodorsal thalamic nuclei, induce feeding in rats. Brain Res 2013;1490:128–33.

[85] Li Y, Dong X, Li S, Kirouac GJ. Lesions of the posterior paraventricular nucleus of the thalamus attenuate fear expression. Front Behav Neurosci 2014;8:94.

[86] Padilla-Coreano N, Do-Monte FH, Quirk GJ. A time-dependent role of midline thalamic nuclei in the retrieval of fear memory. Neuropharmacology 2012;62(1):457–63.

[87] Napier TC, Mickiewicz AL. The role of the ventral pallidum in psychiatric disorders. Neuropsychopharmacology 2010;35(1):337.

[88] Young KA, Franklin TR, Roberts DC, Jagannathan K, Suh JJ, Wetherill RR, Wang Z, Kampman KM, O'Brien CP, Childress AR. Nipping cue reactivity in the bud: Baclofen prevents limbic activation elicited by subliminal drug cues. J Neurosci 2014;34(14):5038–43.

[89] McAlonan GM, Robbins TW, Everitt BJ. Effects of medial dorsal thalamic and ventral pallidal lesions on the acquisition of a conditioned place preference: further evidence for the involvement of the ventral striatopallidal system in reward-related processes. Neuroscience 1993;52 (3):605–20.

[90] Saga Y, Richard A, Sgambato-Faure V, Hoshi E, Tobler PN, Tremblay L. Ventral pallidum encodes contextual information and controls aversive behaviors. Cereb Cortex 2017;27(4):2528–43.

[91] Inui T, Shimura T, Yamamoto T. The role of the ventral pallidum GABAergic system in conditioned taste aversion: effects of microinjections of a GABAA receptor antagonist on taste palatability of a conditioned stimulus. Brain Res 2007;1164:117–24.

[92] Yu W, Walling I, Smith AB, Ramirez-Zamora A, Pilitsis JG, Shin DS. Deep brain stimulation of the ventral pallidum attenuates epileptiform activity and seizing behavior in pilocarpine-treated rats. Brain Stimul 2016;9(2):285–95.

[93] Ersche KD, Jones PS, Williams GB, Turton AJ, Robbins TW, Bullmore ET. Abnormal brain structure implicated in stimulant drug addiction. Science 2012;335(6068):601–4.

[94] Naqvi NH, Bechara A. The hidden island of addiction: the insula. Trends Neurosci 2009;32(1):56–67.

[95] Volkow ND, Fowler JS. Addiction, a disease of compulsion and drive: involvement of the orbitofrontal cortex. Cereb Cortex 2000;10(3):318–25.

[96] Franklin TR, Acton PD, Maldjian JA, Gray JD, Croft JR, Dackis CA, O'Brien CP, Childress AR. Decreased gray matter concentration in the insular, orbitofrontal, cingulate, and temporal cortices of cocaine patients. Biol Psychiatry 2002;51(2):134–42.

[97] Matochik JA, London ED, Eldreth DA, Cadet JL, Bolla KI. Frontal cortical tissue composition in abstinent cocaine abusers: a magnetic resonance imaging study. Neuroimage 2003;19(3):1095–102.

[98] Macey DJ, Rice WN, Freedland CS, Whitlow CT, Porrino LJ. Patterns of functional activity associated with cocaine self-administration in the rat change over time. Psychopharmacology (Berl) 2004;172(4):384–92.

[99] Dougherty DD, Baer L, Cosgrove GR, Cassem EH, Price BH, Nierenberg AA, Jenike MA, Rauch SL. Prospective long-term follow-up of 44 patients who received cingulotomy for treatment-refractory obsessive-compulsive disorder. Am J Psychiatry 2002;159(2):269–75.

[100] Leiphart JW, Valone 3rd FH. Stereotactic lesions for the treatment of psychiatric disorders. J Neurosurg 2010;113(6):1204–11.

[101] LaLumiere RT, Niehoff KE, Kalivas PW. The infralimbic cortex regulates the consolidation of extinction after cocaine self-administration. Learn Mem 2010;17(4):168–75.

[102] Peters J, Kalivas PW, Quirk GJ. Extinction circuits for fear and addiction overlap in prefrontal cortex. Learn Mem 2009;16(5):279–88.

[103] Peters J, LaLumiere RT, Kalivas PW. Infralimbic prefrontal cortex is responsible for inhibiting cocaine seeking in extinguished rats. J Neurosci 2008;28(23):6046–53.

[104] Moorman DE, James MH, McGlinchey EM, Aston-Jones G. Differential roles of medial prefrontal subregions in the regulation of drug seeking. Brain Res 2015;1628(Pt A):130–46.

[105] Reznikov R, Bambico F, Diwan M, Raymond RJ, Nashed MG, Nobrega JN, Hamani C. Prefrontal cortex deep brain stimulation improves fear and anxiety-like behaviour and reduces basolateral amygdala activity in a preclinical model of post-traumatic stress disorder. Neuropsychopharmacology 2018;43(5):1099–106.

[106] Klanker M, Post G, Joosten R, Feenstra M, Denys D. Deep brain stimulation in the lateral orbitofrontal cortex impairs spatial reversal learning. Behav Brain Res 2013;245:7–12.

[107] Levy D, Shabat-Simon M, Shalev U, Barnea-Ygael N, Cooper A, Zangen A. Repeated electrical stimulation of reward-related brain regions affects cocaine but not "natural" reinforcement. J Neurosci 2007;27(51):14179–89.

[108] Creutzfeldt OD, Fromm GH, Kapp H. Influence of transcortical d-c currents on cortical neuronal activity. Exp Neurol 1962;5:436–52.

[109] Purpura DP, McMurtry JG. Intracellular activities and evoked potential changes during polarization of motor cortex. J Neurophysiol 1965;28:166–85.

[110] Boggio PS, Rigonatti SP, Ribeiro RB, Myczkowski ML, Nitsche MA, Pascual-Leone A, Fregni F. A randomized, double-blind clinical trial on the efficacy of cortical direct current stimulation for the treatment of major depression. Int J Neuropsychopharmacol 2008;11(2):249–54.

[111] Boggio PS, Sultani N, Fecteau S, Merabet L, Mecca T, Pascual-Leone A, Basaglia A, Fregni F. Prefrontal cortex modulation using transcranial DC stimulation reduces alcohol craving: a double-blind, sham-controlled study. Drug Alcohol Depend 2008;92(1–3):55–60.

[112] Boggio PS, Liguori P, Sultani N, Rezende L, Fecteau S, Fregni F. Cumulative priming effects of cortical stimulation on smoking cue-induced craving. Neurosci Lett 2009;463(1):82–6.

[113] Fregni F, Liguori P, Fecteau S, Nitsche MA, Pascual-Leone A, Boggio PS. Cortical stimulation of the prefrontal cortex with transcranial direct current stimulation reduces cue-provoked smoking craving: a randomized, sham-controlled study. J Clin Psychiatry 2008;69(1):32–40.

[114] Yang LZ, Shi B, Li H, Zhang W, Liu Y, Wang H, Zhou Y, Wang Y, Lv W, Ji X, Hudak J, Zhou Y, Fallgatter AJ, Zhang X. Electrical stimulation reduces smokers' craving by modulating the coupling between dorsal lateral prefrontal cortex and parahippocampal gyrus. Soc Cogn Affect Neurosci 2017;12(8):1296–302.

[115] Batista EK, Klauss J, Fregni F, Nitsche MA, Nakamura-Palacios EM. A randomized placebo-controlled trial of targeted prefrontal cortex modulation with bilateral tDCS in patients with crack-cocaine dependence. Int J Neuropsychopharmacol 2015;18(12).

[116] de Almeida Ramos R, Taiar I, Trevizol AP, Shiozawa P, Cordeiro Q. Effect of a ten-day prefrontal transcranial direct current stimulation protocol for crack craving: a proof-of-concept trial. J ECT 2016;32(3):e8–9.

[117] Nakamura-Palacios EM, Lopes IB, Souza RA, Klauss J, Batista EK, Conti CL, Moscon JA, de Souza RS. Ventral medial prefrontal cortex (vmPFC) as a target of the dorsolateral prefrontal modulation by transcranial direct current stimulation (tDCS) in drug addiction. J Neural Transm (Vienna) 2016;123(10):1179–94.

[118] Boggio PS, Zaghi S, Villani AB, Fecteau S, Pascual-Leone A, Fregni F. Modulation of risk-taking in marijuana users by transcranial direct current stimulation (tDCS) of the dorsolateral prefrontal cortex (DLPFC). Drug Alcohol Depend 2010;112(3):220–5.

[119] da Silva MC, Conti CL, Klauss J, Alves LG, do Nascimento Cavalcante HM, Fregni F, Nitsche MA, Nakamura-Palacios EM. Behavioral effects of transcranial direct current stimulation (tDCS) induced dorsolateral prefrontal cortex plasticity in alcohol dependence. J Physiol Paris 2013;107 (6):493–502.

[120] den Uyl TE, Gladwin TE, Wiers RW. Transcranial direct current stimulation, implicit alcohol associations and craving. Biol Psychol 2015;105:37–42.

[121] Wang Y, Shen Y, Cao X, Shan C, Pan J, He H, Ma Y, Yuan TF. Transcranial direct current stimulation of the frontal-parietal-temporal area attenuates cue-induced craving for heroin. J Psychiatr Res 2016;79:1–3.

[122] Shahbabaie A, Golesorkhi M, Zamanian B, Ebrahimpoor M, Keshvari F, Nejati V, Fregni F, Ekhtiari H. State dependent effect of transcranial direct current stimulation (tDCS) on methamphetamine craving. Int J Neuropsychopharmacol 2014;17(10):1591–8.

[123] Xu J, Fregni F, Brody AL, Rahman AS. Transcranial direct current stimulation reduces negative affect but not cigarette craving in overnight abstinent smokers. Front Psychiatry 2013;4:112.

[124] Fecteau S, Agosta S, Hone-Blanchet A, Fregni F, Boggio P, Ciraulo D, Pascual-Leone A. Modulation of smoking and decision-making behaviors with transcranial direct current stimulation in tobacco smokers: A preliminary study. Drug Alcohol Depend 2014;140:78–84.

[125] Meng Z, Liu C, Yu C, Ma Y. Transcranial direct current stimulation of the frontal-parietal-temporal area attenuates smoking behavior. J Psychiatr Res 2014;54:19–25.

[126] Klauss J, Penido Pinheiro LC, Silva Merlo BL, de Almeida Correia Santos G, Fregni F, Nitsche MA, Miyuki Nakamura-Palacios E. A randomized controlled trial of targeted prefrontal cortex modulation with tDCS in patients with alcohol dependence. Int J Neuropsychopharmacol 2014;17 (11):1793–803.

[126a] Conti CL, Moscon JA, Fregni F, Nitsche MA, Nakamura-Palacios EM. Cognitive related electrophysiological changes induced by non-invasive cortical electrical stimulation in crack-cocaine addiction. Int J Neuropsychopharmacol 2014;17(9):1465–75.

[126b] Nakamura-Palacios EM, de Almeida Benevides MC, da Penha Zago-Gomes M, de Oliveira RW, de Vasconcellos VF, de Castro LN, da Silva MC, Ramos PA, Fregni F. Auditory event-related potentials (P3) and cognitive changes induced by frontal direct current stimulation in alcoholics according to Lesch alcoholism typology. Int J Neuropsychopharmacol 2012;15(5):601–16.

[126c] Conti CL, Nakamura-Palacios EM. Bilateral transcranial direct current stimulation over dorsolateral prefrontal cortex changes the drug-cued reactivity in the anterior cingulate cortex of crack-cocaine addicts. Brain Stimul 2014;7(1):130–2.

[126d] Gorini A, Lucchiari C, Russell-Edu W, Pravettoni G. Modulation of risky choices in recently abstinent dependent cocaine users: a transcranial direct-current stimulation study. Front Hum Neurosci 2014;8:661.

[127] Trojak B, Soudry-Faure A, Abello N, Carpentier M, Jonval L, Allard C, Sabsevari F, Blaise E, Ponavoy E, Bonin B, Meille V, Chauvet-Gelinier JC. Efficacy of transcranial direct current stimulation (tDCS) in reducing consumption in patients with alcohol use disorders: study protocol for a randomized controlled trial. Trials 2016;17(1):250.

[128] Pedron S, Monnin J, Haffen E, Sechter D, Van Waes V. Repeated transcranial direct current stimulation prevents abnormal behaviors associated with abstinence from chronic nicotine consumption. Neuropsychopharmacology 2014;39(4):981–8.

[129] Pedron S, Beverley J, Haffen E, Andrieu P, Steiner H, Van Waes V. Transcranial direct current stimulation produces long-lasting attenuation of cocaine-induced behavioral responses and gene regulation in corticostriatal circuits. Addict Biol 2017;22(5):1267–78.

[130] Speer AM, Kimbrell TA, Wassermann EM, Repella JD, Willis MW, Herscovitch P, Post RM. Opposite effects of high and low frequency rTMS on regional brain activity in depressed patients. Biol Psychiatry 2000;48(12):1133–41.

[131] Ziemann U, Paulus W, Nitsche MA, Pascual-Leone A, Byblow WD, Berardelli A, Siebner HR, Classen J, Cohen LG, Rothwell JC. Consensus: motor cortex plasticity protocols. Brain Stimul 2008;1(3):164–82.

[132] Hayashi T, Ko JH, Strafella AP, Dagher A. Dorsolateral prefrontal and orbitofrontal cortex interactions during self-control of cigarette craving. Proc Natl Acad Sci U S A 2013;110(11):4422–7.

[133] Pripfl J, Tomova L, Riecansky I, Lamm C. Transcranial magnetic stimulation of the left dorsolateral prefrontal cortex decreases cue-induced nicotine craving and EEG delta power. Brain Stimul 2014;7(2):226–33.

[134] Eichhammer P, Johann M, Kharraz A, Binder H, Pittrow D, Wodarz N, Hajak G. High-frequency repetitive transcranial magnetic stimulation decreases cigarette smoking. J Clin Psychiatry 2003;64(8):951–3.

[135] Camprodon JA, Martinez-Raga J, Alonso-Alonso M, Shih MC, Pascual-Leone A. One session of high frequency repetitive transcranial magnetic stimulation (rTMS) to the right prefrontal cortex transiently reduces cocaine craving. Drug Alcohol Depend 2007;86(1):91–4.

[136] Mishra BR, Nizamie SH, Das B, Praharaj SK. Efficacy of repetitive transcranial magnetic stimulation in alcohol dependence: a sham-controlled study. Addiction 2010;105(1):49–55.

[137] Amiaz R, Levy D, Vainiger D, Grunhaus L, Zangen A. Repeated high-frequency transcranial magnetic stimulation over the dorsolateral prefrontal cortex reduces cigarette craving and consumption. Addiction 2009;104(4):653–60.

[138] Dinur-Klein L, Dannon P, Hadar A, Rosenberg O, Roth Y, Kotler M, Zangen A. Smoking cessation induced by deep repetitive transcranial magnetic stimulation of the prefrontal and insular cortices: a prospective, randomized controlled trial. Biol Psychiatry 2014;76(9):742–9.

[139] Dieler AC, Dresler T, Joachim K, Deckert J, Herrmann MJ, Fallgatter AJ. Can intermittent theta burst stimulation as add-on to psychotherapy improve nicotine abstinence? Results from a pilot study. Eur Addict Res 2014;20(5):248–53.

[140] Trojak B, Meille V, Achab S, Lalanne L, Poquet H, Ponavoy E, Blaise E, Bonin B, Chauvet-Gelinier JC. Transcranial magnetic stimulation combined with nicotine replacement therapy for smoking cessation: a randomized controlled trial. Brain Stimul 2015;8(6):1168–74.

[141] Sheffer CE, Mennemeier M, Landes RD, Bickel WK, Brackman S, Dornhoffer J, Kimbrell T, Brown G. Neuromodulation of delay discounting, the reflection effect, and cigarette consumption. J Subst Abuse Treat 2013;45(2):206–14.

[142] Ceccanti M, Inghilleri M, Attilia ML, Raccah R, Fiore M, Zangen A, Ceccanti M. Deep TMS on alcoholics: Effects on cortisolemia and dopamine pathway modulation. A pilot study. Can J Physiol Pharmacol 2015;93(4):283–90.

[143] Bogacz R, Gurney K. The basal ganglia and cortex implement optimal decision making between alternative actions. Neural Comput 2007;19(2):442–77.

[144] Frank MJ. Hold your horses: A dynamic computational role for the subthalamic nucleus in decision making. Neural Netw 2006;19(8):1120–36.

[145] Baunez C, Robbins TW. Bilateral lesions of the subthalamic nucleus induce multiple deficits in an attentional task in rats. Eur J Neurosci 1997;9(10):2086–99.

[146] Baunez C, Amalric M, Robbins TW. Enhanced food-related motivation after bilateral lesions of the subthalamic nucleus. J Neurosci 2002;22(2):562–8.

[147] Bezzina G, Boon FS, Hampson CL, Cheung TH, Body S, Bradshaw CM, Szabadi E, Anderson IM, Deakin JF. Effect of quinolinic acid-induced lesions of the subthalamic nucleus on performance on a progressive-ratio schedule of reinforcement: a quantitative analysis. Behav Brain Res 2008;195(2):223–30.

[148] Bentzley BS, Aston-Jones G. Inhibiting subthalamic nucleus decreases cocaine demand and relapse: therapeutic potential. Addict Biol 2017;22(4):946–57.

[149] Uslaner JM, Yang P, Robinson TE. Subthalamic nucleus lesions enhance the psychomotor-activating, incentive motivational, and neurobiological effects of cocaine. J Neurosci 2005;25(37):8407–15.

[150] Lardeux S, Baunez C. Alcohol preference influences the subthalamic nucleus control on motivation for alcohol in rats. Neuropsychopharmacology 2008;33(3):634–42.

[151] Baracz SJ, Everett NA, Cornish JL. The involvement of oxytocin in the subthalamic nucleus on relapse to methamphetamine-seeking behaviour. PLoS One 2015;10(8).

[152] Prasad AA, McNally GP. Ventral pallidum output pathways in context-induced reinstatement of alcohol seeking. J Neurosci 2016;36(46):11716–26.

[153] Pelloux Y, Baunez C. Targeting the subthalamic nucleus in a preclinical model of alcohol use disorder. Psychopharmacology (Berl) 2017;234 (14):2127–37.

[154] Moro E, Lozano AM, Pollak P, Agid Y, Rehncrona S, Volkmann J, Kulisevsky J, Obeso JA, Albanese A, Hariz MI, Quinn NP, Speelman JD, Benabid AL, Fraix V, Mendes A, Welter ML, Houeto JL, Cornu P, Dormont D, Tornqvist AL, Ekberg R, Schnitzler A, Timmermann L, Wojtecki L, Gironell A, Rodriguez-Oroz MC, Guridi J, Bentivoglio AR, Contarino MF, Romito L, Scerrati M, Janssens M, Lang AE. Long-term results of a multicenter study on subthalamic and pallidal stimulation in Parkinson's disease. Mov Disord 2010;25(5):578–86.

[155] Eusebio A, Witjas T, Cohen J, Fluchere F, Jouve E, Regis J, Azulay JP. Subthalamic nucleus stimulation and compulsive use of dopaminergic medication in Parkinson's disease. J Neurol Neurosurg Psychiatry 2013;84(8):868–74.

[156] Knobel D, Aybek S, Pollo C, Vingerhoets FJ, Berney A. Rapid resolution of dopamine dysregulation syndrome (DDS) after subthalamic DBS for Parkinson disease (PD): a case report. Cogn Behav Neurol 2008;21(3):187–9.

[157] Lhommee E, Klinger H, Thobois S, Schmitt E, Ardouin C, Bichon A, Kistner A, Fraix V, Xie J, Aya Kombo M, Chabardes S, Seigneuret E, Benabid AL, Mertens P, Polo G, Carnicella S, Quesada JL, Bosson JL, Broussolle E, Pollak P, Krack P. Subthalamic stimulation in Parkinson's disease: restoring the balance of motivated behaviours. Brain 2012;135(Pt 5):1463–77.

[158] Lim SY, O'Sullivan SS, Kotschet K, Gallagher DA, Lacey C, Lawrence AD, Lees AJ, O'Sullivan DJ, Peppard RF, Rodrigues JP, Schrag A, Silberstein P, Tisch S, Evans AH. Dopamine dysregulation syndrome, impulse control disorders and punding after deep brain stimulation surgery for Parkinson's disease. J Clin Neurosci 2009;16(9):1148–52.

[159] Witjas T, Baunez C, Henry JM, Delfini M, Regis J, Cherif AA, Peragut JC, Azulay JP. Addiction in Parkinson's disease: impact of subthalamic nucleus deep brain stimulation. Mov Disord 2005;20(8):1052–5.

[159a] Degoulet M, Tiran-Cappello A, Baunez C*, Pelloux Y*. Reversing compulsive cocaine seeking with low frequency deep brain stimulation of the subthalamic nucleus.

[160] Mallet L, Mesnage V, Houeto JL, Pelissolo A, Yelnik J, Behar C, Gargiulo M, Welter ML, Bonnet AM, Pillon B, Cornu P, Dormont D, Pidoux B, Allilaire JF, Agid Y. Compulsions, Parkinson's disease, and stimulation. Lancet 2002;360(9342):1302–4.

[161] Mallet L, Polosan M, Jaafari N, Baup N, Welter ML, Fontaine D, du Montcel ST, Yelnik J, Chereau I, Arbus C, Raoul S, Aouizerate B, Damier P, Chabardes S, Czernecki V, Ardouin C, Krebs MO, Bardinet E, Chaynes P, Burbaud P, Cornu P, Derost P, Bougerol T, Bataille B, Mattei V, Dormont D, Devaux B, Verin M, Houeto JL, Pollak P, Benabid AL, Agid Y, Krack P, Millet B, Pelissolo A. Subthalamic nucleus stimulation in severe obsessive-compulsive disorder. N Engl J Med 2008;359(20):2121–34.

[162] Pelloux Y, Meffre J, Giorla E, Baunez C. The subthalamic nucleus keeps you high on emotion: behavioral consequences of its inactivation. Front Behav Neurosci 2014;8:414.

[163] Peron J, Fruhholz S, Verin M, Grandjean D. Subthalamic nucleus: a key structure for emotional component synchronization in humans. Neurosci Biobehav Rev 2013;37(3):358–73.

[164] Baunez C, Yelnik J, Mallet L. Six questions on the subthalamic nucleus: Lessons from animal models and from stimulated patients. Neuroscience 2011;198:193–204.

[165] Hachem-Delaunay S, Fournier ML, Cohen C, Bonneau N, Cador M, Baunez C, Le Moine C. Subthalamic nucleus high-frequency stimulation modulates neuronal reactivity to cocaine within the reward circuit. Neurobiol Dis 2015;80:54–62.

[165a] Pelloux Y*, Degoulet M*, Tiran-Cappello A, Cohen C, Lardeux S, George O, Koob GF, Ahmed SH, Baunez C. Subthalamic nucleus inactivation prevents and reverses escalated cocaine use. Mol Psychiatry 2018 (in press).

[166] Wade CL, Kallupi M, Hernandez DO, Breysse E, de Guglielmo G, Crawford E, Koob GF, Schweitzer P, Baunez C, George O. High-frequency stimulation of the subthalamic nucleus blocks compulsive-like re-escalation of heroin taking in rats. Neuropsychopharmacology 2017;42 (9):1850–9.

[167] Berridge KC, Kringelbach ML. Neuroscience of affect: brain mechanisms of pleasure and displeasure. Curr Opin Neurobiol 2013;23(3):294–303.

[168] Root DH, Ma S, Barker DJ, Megehee L, Striano BM, Ralston CM, Fabbricatore AT, West MO. Differential roles of ventral pallidum subregions during cocaine self-administration behaviors. J Comp Neurol 2013;521(3):558–88.

[169] McLaughlin RJ, Floresco SB. The role of different subregions of the basolateral amygdala in cue-induced reinstatement and extinction of food-seeking behavior. Neuroscience 2007;146(4):1484–94.

[170] Adams WK, Vonder Haar C, Tremblay M, Cocker PJ, Silveira MM, Kaur S, Baunez C, Winstanley CA. Deep-brain stimulation of the subthalamic nucleus selectively decreases risky choice in risk-preferring rats. eNeuro 2017;4(4).

[171] Winstanley CA, Baunez C, Theobald DE, Robbins TW. Lesions to the subthalamic nucleus decrease impulsive choice but impair autoshaping in rats: the importance of the basal ganglia in Pavlovian conditioning and impulse control. Eur J Neurosci 2005;21(11):3107–16.

[172] Grossman N, Bono D, Dedic N, Kodandaramaiah SB, Rudenko A, Suk HJ, Cassara AM, Neufeld E, Kuster N, Tsai LH, Pascual-Leone A, Boyden ES. Noninvasive deep brain stimulation via temporally interfering electric fields. Cell 2017;169(6). 1029–1041.e1016.

[173] Diana M, Raij T, Melis M, Nummenmaa A, Leggio L, Bonci A. Rehabilitating the addicted brain with transcranial magnetic stimulation. Nat Rev Neurosci 2017;18(11):685–93.

[174] Prinz P, Kobelt P, Scharner S, Goebel-Stengel M, Harnack D, Faust K, Winter Y, Rose M, Stengel A. Deep brain stimulation alters light phase food intake microstructure in rats. J Physiol Pharmacol 2017;68(3):345–54.

[175] Zhang C, Wei NL, Wang Y, Wang X, Zhang JG, Zhang K. Deep brain stimulation of the nucleus accumbens shell induces anti-obesity effects in obese rats with alteration of dopamine neurotransmission. Neurosci Lett 2015;589:1–6.

Chapter 20

Conclusions and Future Directions

Mary M. Torregrossa

Department of Psychiatry, University of Pittsburgh, Pittsburgh, PA, United States

The purpose of this book is to provide the addiction field with an up to date collection of reviews summarizing the latest neuroscience research on the biological mechanisms underlying the development and persistence of addictive disorders. Neuroscience research has seen broad technological gains in the past 5–10 years that have allowed a much more precise dissection of the neural circuits and molecular mechanisms regulating the development of addiction. In addition to improved technologies, the animal models used to study addiction have become more sophisticated in recent years, allowing for investigation of the different factors that underlie drug use, including positive/rewarding effects, relief of negative affective states, compulsivity, impulsivity, and impaired decision making. With this increase in the specificity of the behavioral processes being studied, we are likely to see a surge in treatment strategies that are tailored to addressing the factors that drive drug seeking in each individual (discussed in Chapters 2, 10–12, and 19).

Another intriguing aspect of ongoing research is the revelation that different classes of drugs of abuse (e.g., opiates vs stimulants) can not only elicit different forms of plasticity in the brain, but can elicit similar types of plasticity via different mechanisms. For example, numerous recent studies have taken advantage of the latest genetic tools available to dissect how drugs of abuse regulate neural plasticity in the nucleus accumbens, a region that is important for motivating behavior associated with rewards. The nucleus accumbens is known to critically regulate drug use, particularly through the actions of dopamine in this region. Emerging studies are finding, however, that drug exposure results in different neuroadaptations depending on the specific nucleus accumbens cell type examined. The nucleus accumbens contains two types of medium spiny neurons, those that primarily express the dopamine D1-type receptor and those that primarily express the D2-type receptor. The receptors couple to excitatory and inhibitory G-proteins, respectively. Thus, dopamine release leads to activation of D1 cells and inhibition of D2 cells. Recent studies discussed in this book (see Chapters 3, 4, and 18) have found that after chronic drug exposure, D1 and D2 cells undergo different changes in gene expression, and that the exact molecular alterations can depend on the drug class studied. For example, cocaine seems to "awaken" previously silent synapses onto D1 neurons, while morphine silences synapses onto D2 neurons. Thus, in both cases there is a magnification of dopamine's effects on the activity of these cells. As such, activating D2 cells or inhibiting D1 cells may be a strategy to reduce drug-induced behaviors.

In addition to an increase in our classical understanding of dopamine's role in addiction, numerous studies have uncovered important roles for previously underappreciated neurobiological and physiological systems in addiction. First, other neurotransmitter systems, such as glutamate (Chapter 5), neuropeptides (Chapter 14), and norepinephrine (Chapter 15) are receiving increasing attention in the addiction field and have been shown to not only impact the development of substance use disorders, but are likely critical for treatment. In addition, to classical neurotransmitter-based signaling at the synapse, newer studies are finding that addiction results in adaptations in neural morphology (Chapter 9), glial cells (Chapter 16), and the extracellular matrix (Chapter 17) that all can regulate synaptic signaling, but that may also offer synapse-independent options for developing novel treatments. For example, manipulation of the actin cytoskeleton that regulates dendritic spine morphology may be an alternative way to reverse some effects of chronic drug use. Moreover, many exciting treatments currently being developed act on glial cell transporters and restore neural homeostasis. Increased investigation of all the neurobiological factors that can lead to compulsive drug use is sure to grow in the coming years and may lead to novel therapeutic strategies.

Novel treatments may also come from targeting physiological systems affected by chronic drug use that are not linked as closely with the direct rewarding effects of the drug. For example, drug use can cause severe disruptions in circadian rhythms and sleep (Chapter 13), and disrupted rhythms can be a risk factor for developing a substance use disorder. Thus, exciting new research suggests that normalizing rhythms and sleep can reverse some drug-induced neuroadaptations and drug-mediated behaviors. Another area of research is aimed at understanding the brain mechanisms regulating

Neural Mechanisms of Addiction. https://doi.org/10.1016/B978-0-12-812202-0.00020-8
Copyright © 2019 Elsevier Inc. All rights reserved.

the interoceptive effects of abused drugs (Chapter 7). Ingestion of drugs not only leads to euphoric effects, but also leads to change in the internal state that the brain can detect. This can include changes in heart rate, blood pressure, arousal, etc., and the perception of this interoceptive state, or the absence of this state can become a cue that drives compulsive drug use. The insular cortex is a part of the brain thought to regulate interoception, and interestingly, damage to the insula, usually due to stroke, has been found to reduce or eliminate smoking in previously heavy smokers. The mechanism for this effect remains unclear, but it may be that eliminating the internal perception of smoking abstinence (i.e., withdrawal, craving) removes the drive to continue smoking. Future research will need to determine if similar results can be obtained by less drastic manipulations and for other drugs of abuse.

An additional factor that drives drug use and relapse and that could be targeted for treatment is the learning and memory system. Encountering stimuli in the environment that are associated with drug use can induce retrieval of memories that can lead to increased thoughts about using, ultimately driving someone to take drugs again. Research over the past 10 years has suggested that specifically targeting the integrity of these memories may be an effective strategy for reducing relapse. This approach usually combines a behavioral therapy where drug using memories are either "reactivated" briefly, followed by administration of an amnestic agent to interfere with memory reconsolidation; or the memories are recalled over many sessions in the absence of drug to induce extinction learning. Theoretically, either extinction learning or reconsolidation blockade could be effective approaches for impairing future retrieval of the memory, and thus reducing craving and relapse. Unfortunately, to date, clinical application of these strategies has met with limited success. However, ongoing preclinical work is being conducted to identify the intracellular and brain wide mechanisms encoding these memories (Chapters 6 and 8), which may provide novel pharmacological targets for more precise and long-lasting disruption of drug memories.

Overall, the future of addiction research is very promising, and the study of the neuroscience of addiction continues to progress at an extremely fast rate. These studies are not only opening the door for novel treatments, but are uncovering fundamental brain mechanisms regulating, reward, aversion, motivation, stress, decision-making, and learning and memory. In addition, studies from other fields, including cell biology, genetics, epigenetics, microscopy, computational biology, and engineering are reshaping the way that neuroscientists study the addicted brain. With these tools and the development of neurostimulation techniques (Chapter 19), there is growing hope that drug addiction may be treated in a targeted, circuit-based, and individualized manner to reduce the burden of this disease on society.

Index

Note: Page numbers followed by *f* indicate figures.

Printed in the United States
By Bookmasters